Lecture Notes in Computer Science 15557

The series Lecture Notes in Computer Science (LNCS), including its subseries Lecture Notes in Artificial Intelligence (LNAI) and Lecture Notes in Bioinformatics (LNBI), has established itself as a medium for the publication of new developments in computer science and information technology research, teaching, and education.

LNCS enjoys close cooperation with the computer science R & D community, the series counts many renowned academics among its volume editors and paper authors, and collaborates with prestigious societies. Its mission is to serve this international community by providing an invaluable service, mainly focused on the publication of conference and workshop proceedings and postproceedings. LNCS commenced publication in 1973.

Dhananjay Singh · Jan-Willem van 't Klooster ·
Uma Shanker Tiwary
Editors

Intelligent Human Computer Interaction

16th International Conference, IHCI 2024
Twente, The Netherlands, November 13–16, 2024
Revised Selected Papers, Part I

Springer

Editors
Dhananjay Singh
Penn State University
Pennsylvania, PA, USA

Jan-Willem van 't Klooster
University of Twente
Enschede, The Netherlands

Uma Shanker Tiwary
Indian Institute of Information Technology
Allahabad, Uttar Pradesh, India

ISSN 0302-9743 ISSN 1611-3349 (electronic)
Lecture Notes in Computer Science
ISBN 978-3-031-88704-8 ISBN 978-3-031-88705-5 (eBook)
https://doi.org/10.1007/978-3-031-88705-5

This Springer imprint is published by the registered company Springer Nature Switzerland AG
The registered company address is: Gewerbestrasse 11, 6330 Cham, Switzerland

Preface

IHCI is an annual international conference in the Human-Computer Interaction field, where we explore research challenges emerging in the complex interaction between machine intelligence and human intelligence.

It is a privilege to present the proceedings of the 16th International Conference on Intelligent Human Computer Interaction (IHCI 2024), organized on site and online by the University of Twente and the BMS Lab during November 13–15, 2024, at Enschede, Netherlands. This sixteenth instance of the conference was on the theme of "Meaningful Human Computer Interaction", having 4 special sessions related to the main theme of the conference. This year out of 107 submitted papers, 58 papers were accepted for oral presentation and publication by the program committee, in each case based on the recommendations of at least 2 expert reviewers in a double-blind review process. The 16th IHCI conference included keynote and invited talks with powerful expert session chairs who have worked in both industry and academia. It attracted more than 200 participants from more than 25 countries.

IHCI has emerged as the foremost worldwide gathering of the field's academic researchers, graduate students, top research think tanks, and industry technology developers. Therefore, we believe that the biggest benefit to the participant is the actualization of their goals in the field of HCI. That will ultimately lead to greater success in business, which is ultimately beneficial to society.

Moreover, our warm gratitude should be given to all the authors who submitted their work to IHCI 2024. We are grateful to the technical program committee (TPC) and organizing committee for their immeasurable efforts to ensure the success of this conference. Finally, we would like to thank our speakers, authors, and participants for their contribution to making IHCI 2024 a stimulating and productive conference. This IHCI conference series cannot achieve yearly milestones without their continued support in future.

November 2024

Dhananjay Singh
Jan-Willem van 't Klooster

Organization

General Chairs

Dhananjay Singh	Penn State University, USA
Jan-Willem van 't Klooster	University of Twente, Netherlands
Teodora Spirova	University of Twente, Netherlands

Steering Committee

Tanya Bondarouk	University of Twente, Netherlands
Jan-Willem van 't Klooster	University of Twente, Netherlands
Jan Treur	Vrije Universiteit Amsterdam, Netherlands
Uma Shanker Tiwary	Indian Institute of Information Technology, Allahabad, India

Advisory Chairs

Ajay Gupta	Western Michigan University, USA
Benno Pals	University of Twente, Netherlands
Wan Young Chung	Pukyong National University, South Korea
Jan Treur	Vrije Universiteit Amsterdam, Netherlands
Venkatasubramanian Ganesan	National Institute of Mental Health & Neurosciences, India

Technical Program Chairs

Uma Shanker Tiwary	Indian Institute of Information Technology, Allahabad, India
Dhananjay Singh	Penn State University, USA
Jan-Willem van 't Klooster	University of Twente, Netherlands

Program Chairs

Teodora Spirova	University of Twente, Netherlands
Rajiv Singh	Banasthali Vidyapith, India
Madhusudan Singh	Long Island University Brooklyn, USA

Best Paper Award Chairs

Jong-Hoon Kim	Kent State University, USA
David (Bong Jun) Choi	Soongsil University, South Korea

Special Session/Workshop Chairs

Sang-Joong Jung	Dongseo University, South Korea
Mohd Helmy Bin Abd Wahab	Tun Hussein Onn Malaysia, Malaysia

Tutorial/Panel Session Chairs

Hanumant Singh Shekhawat	Indian Institute of Technology, Guwahati, India
Maria Weber	Saint Louis University, USA

Invited/Lightning Talk Chairs

Nagamani Molakatala	University of Hyderabad, India
Ikechi Ukaegbu	Nazarbayev University, Kazakhstan

Publicity Chairs

Annamaria Szakonyi	Saint Louis University, USA
Sanjay Singh	Motilal Nehru National Institute of Technology Allahabad, India

Student Paper Session Chairs

Tatiana Cardona	Saint Louis University, USA
Ajit Kumar	Soongsil University, South Korea

Poster Session Chairs

YoungSil Lee	Dongseo University, South Korea
Irish Singh	Penn State University, USA

Student Hack-A-Thon Chairs

Naagmani Molakatala	University of Hyderabad, India
Swati Nigam	Banasthali Vidyapith, India

Industrial Chair

Mario Jose Divan	Intel Corporation, USA

Registration Chairs

Teodora Spirova	University of Twente, Netherlands
Shashi Yadav	Coikism Research Foundation, India

Local Organising Chairs

Teodora Spirova	University of Twente, Netherlands
Benno Pals	University of Twente, Netherlands
Lorette Bosch	University of Twente, Netherlands
Lisa Du Plessis	University of Twente, Netherlands
Gita Ni Made Anandita Dewi	University of Twente, Netherlands
Shaunie Schutten	University of Twente, Netherlands
Iris ten Klooster	University of Twente, Netherlands
Simone Borsci	University of Twente, Netherlands

Keynote Speakers

Bernard P. Veldkamp	University of Twente, Netherlands
Anne-Marie Brouwer	Netherlands Organization for Applied Scientific Research, Netherlands
Sandra Fischer	University of Münster, Germany
Enrico Gianluca Caiani	Politecnico di Milano, Italy

Invited Talks

Simone Borsci	University of Twente, Netherlands
Carolina Herrando	University of Zaragoza, Spain
Krista Hoek	Leiden University Medical Centre, Netherlands
Wolter Pieters	Radboud University, Netherlands
Lisette J. E. W. C. van Gemert-Pijnen	University of Twente, Netherlands
Manon Tolhuizen	Thales Netherlands, Netherlands

Special Sessions

Jan Treur	Vrije Universiteit Amsterdam, Netherlands
Peter Roelofsma	Hague University of Applied Sciences, Netherlands
Guoray Cai	Penn State University, USA
Dhananjay Singh	Penn State University, USA
Graziela Fusaro	Long Island University, USA
Madhusudan Singh	Long Island University, USA
Shreya Nandhivargam	Saint Louis University, USA
Irish Singh	Pennsylvania State University, USA
Maria Weber	Saint Louis University, USA
Annamaria Szakonyi	Saint Louis University, USA
Tatiana Cardona	Saint Louis University, USA

Reviewer Committee

Abhay Kumar Rai	Central University of Rajasthan, India
Aditi Singh	Cleveland State University, USA
Aishvarya Garg	Banasthali Vidyapith, India
Andres Navarro-Newball	Pontificia Universidad Javeriana, Colombia

Bhawana Tyagi	Banasthali Vidyapith, India
David (Bong Jun) Choi	Soongsil University, South Korea
Dhananjay Singh	Penn State University, USA
Hasan Tinmaz	Woosong University, South Korea
Jong Hoon Kim	Kent State University, USA
Madhusudan Singh	Long Island University Brooklyn, USA
Manju Khari	JNU, India
Manoj Kumar Singh	Banaras Hindu University, India
Maria Weber	Saint Louis University, USA
Nagamani Molakatala	University of Hyderabad, India
Nidhi Srivastav	Swami Keshvanand Institute of Technology, Management & Gramothan, India
Pradeep Kumar Gupta	Mohan Babu University, India
Pooja Gupta	Banasthali Vidyapith, India
Rajiv Singh	Banasthali Vidyapith, India
Redwanul Haque Sourave	Kent State University, USA
Ravindra Hegadi	Central University of Karnataka, India
Ritu Chauhan	Amity University Noida, India
Saifuddin Mahmud	Bradley University, USA
Sakshi Indolia	Narsee Monjee Institute of Management Studies, India
Siddharth Singh	University of Lucknow, India
Sonam Seth	Banasthali Vidyapith, India
Sudha Morwal	Banasthali Vidyapith, India
Sudhakar Mishra	Indian Institute of Technology Kanpur, India
Susmita Bhattacharya	Indian Institute of Technology Guwahati, India
Swati Nigam	Banasthali Vidyapith, India
Tatiana Cardona	Saint Louis University, USA

Contents – Part I

Human Factors in Cybersecurity

Contents – Part II

AI Centric HCI

Learning in Healthcare and Wellbeing

A VR Experience Exposing the Hidden Curriculum of the Operating Room: A Case Study

Krista Hoek(✉), Monique van Velzen, Jos van der Hage, and Elise Sarton

Department of Anaesthesiology, LUMC, Albinusdreef 2, 2333ZA Leiden, The Netherlands
k.b.hoek@lumc.nl

1 Introduction

Clinical skills learning (CSL) during the preclinical years serves the purpose of equipping medical students with essential clinical competencies (Shuid et al., 2015). These include skills in clinical examination, mastering basic medical procedures, refining history-taking techniques, and enhancing clinical communication abilities. By integrating CSL into the early stages of medical education, students have the opportunity to familiarize themselves with crucial aspects of patient care before entering the clinical setting.

However, alongside formal instruction, there is often an overlooked yet influential aspect of medical education known as the hidden curriculum (Hafferty & Franks, 1994). First developed by Hafferty in 1994, the hidden curriculum encompasses the influences within the organizational and cultural fabric of medical institutions that shape students' learning experiences (Hafferty & Franks, 1994). While formal education focuses on explicit learning objectives, the hidden curriculum imparts implicit messages through everyday practices, norms, and interactions within the medical community. These implicit learning patterns underscore the importance of ample opportunities for experience in clinical settings. Where the hidden curriculum encompasses a wide range of implicit lessons and values, the *etiquette* of a work environment refers to the expected behaviors and norms within a particular workspace.

Increasing levels of burnout among medical students have raised concerns, emphasizing the need for educators to grasp contemporary junior doctor perspectives during their transition from school banks to the work floor (Thun-Hohenstein et al., 2021). Managing challenges and transitions stimulates learning, but understanding these perspectives is crucial for preparing and supporting medical graduates.

Unfortunately, logistical challenges compounded by a high number of medical students seeking clerkships in popular specialisms such as surgery and anesthesiology may lead to waiting lists for placements in the OR, further hindering medical students' access to essential hands-on experiences.

The COVID-19 pandemic worsened existing challenges. For logistical and hygienic reasons, medical students are often prohibited from participating in hospital rotations and operating room (OR) experiences, leading to the cancellation of clerkships and hands-on clinical training opportunities (Jhajj et al., 2022). This disruption highlighted the importance of innovative approaches to clinical skills acquisition.

D. Singh et al. (Eds.): IHCI 2024, LNCS 15557, pp. 3–17, 2025.
https://doi.org/10.1007/978-3-031-88705-5_1

Medical education has increasingly embraced experiential learning as a means of enhancing clinical skill acquisition. Immersive technologies, particularly virtual reality (VR), have gained prominence in this regard (Bracq et al., 2019; Cannon-Bowers, 2008; Gormley & Murphy; Le Duff et al., 2023; Pottle, 2019). VR aligns with Constructivist Theory, positing that learners actively construct knowledge by building upon existing mental frameworks (Anthony, 1996; Whitman, 1993). In immersive VR environments, students engage in hands-on exploration, facilitating deeper understanding and meaning-making from their experience. Moreover, VR-based immersive learning offers unique advantages by leveraging various sensory modalities to enhance engagement and *embodiment*. Through the phenomenon of embodiment, VR users can visually substitute their real bodies with life-sized virtual representations, experiencing scenarios from a first-person perspective. This concept, known as the *Proteus effect*, enables users to momentarily inhabit virtual personas, fostering a heightened sense of presence and emotional immersion (Bian et al., 2015). The integration of multisensory and motor systems during VR experiences further enhances cognitive processing, leading to realistic emotional and neurocognitive responses (Bem, 1972; Navarro et al., 2022). For instance, individuals who embody avatars with physically attractive features tend to engage in more confident and assertive behavior within the virtual space, mirroring the perceived traits of their avatars (Fox et al., 2013). This explains how a VR-embodied experience may be more influential than simply watching a video.

Considering the challenges posed by traditional clinical training methods, VR represents a promising solution. Unlike conventional approaches, VR transcends spatial and temporal constraints, allowing medical students to access immersive learning experiences remotely. By simulating scenarios akin to those encountered in the OR, VR offers students invaluable exposure that is critical for their future medical careers.

This paper delineates the development of a dual VR experience, offering insights into the unspoken norms and hidden curriculum of the OR from a first-person perspective. It was part of a blended learning project in which an existing e-learning module was enhanced.

2 Methods

This prospective exploratory study was carried out at the Leiden University Medical Center (LUMC, Leiden, the Netherlands), a large tertiary academic teaching hospital.

Informed consent was obtained prior to inclusion, participation was voluntary, and ethics and privacy rights were in alignment with the Declaration of Helsinki and GDPR guidelines, respectively. Data collection and development were conducted between July 2020 and February 2021. Participants received no financial compensation for participation.

2.1 Participants and Procedures

Participants were asked to join in crafting the dual VR experiences after an initial email invitation describing the research project followed by an in-person meeting with eligible participants. Involvement included tasks such as crafting content, ensuring its accuracy, and filming and refining the VR experiences, as outlined in Fig. 1. The participants included a panel of medical students, OR assistants, surgeons, anesthesiologists, and educators.

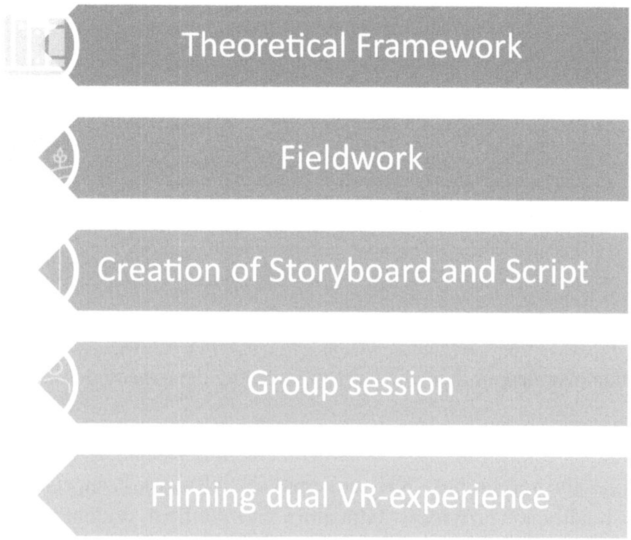

Fig. 1. Structure of the study

2.2 Theoretic Framework and Content Creation

To construct the storyboard and script depicting the typical journey of a new medical student who starts working at the OR center, we began by establishing a theoretical framework grounded in seminal literature within the relevant field. Specifically, we drew upon landmark qualitative and quantitative studies depicting typical challenges of medical students debuting their careers in the OR, as well as articles on the hidden curriculum in this specific context (Alexander et al., 2014; Laraia et al., 2024; Sturman et al., 2017; Surmon et al., 2016; Thun-Hohenstein et al., 2021).

As Surmon et al. (2016) state, possible negative influences on the preparedness of medical students may involve a) uncertainty, b) disconnection, and c) feeling a part of the team (Surmon et al., 2016). Uncertainty and insecurity of medical students in terms of what is expected of them occur frequently. In earlier studies, students cited the existence of unwritten rules or a hidden curriculum lacking familiarity with the clinical environment (Shacklady et al., 2009). Disconnection between the preclinical and clinical curricula contributes to challenges in the preparedness of medical students. As students often feel unsure of hierarchy and how professionals work together, these preoccupations may impede feelings as a part of the team. Throughout content creation, efforts were made to expose medical students to elements of the *etiquette* and hidden curriculum that potentially mitigate these negative influences.

As demonstrated by the work of Laraia et al. (2024), healthcare providers working in the operating room express concerns that students adopt a too passive role, lack situational awareness regarding when or when not to ask questions, and may lack understanding of sterility and the operating room workflow (Laraia et al., 2024). A schematic overview of these themes as described by Laraia et al. (2024) from the OR can be found in Fig. 2 (Laraia et al., 2024).

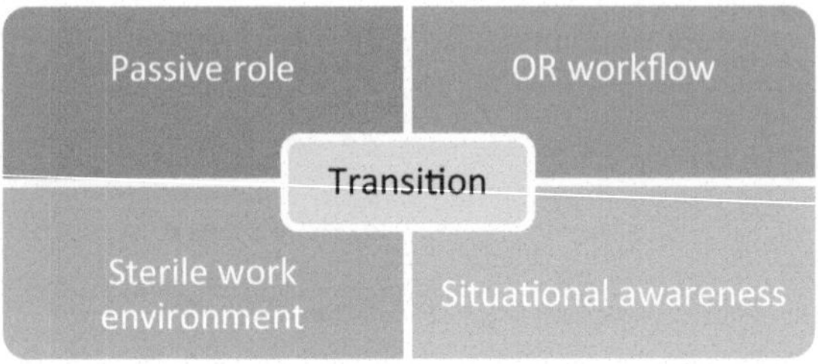

Fig. 2. Schematic overview of thematic barriers impeding the transition to the clinical environment

Furthermore, the researcher (KH) engaged in fieldwork comprising informal discussions with healthcare providers, educators and medical students supplemented by observational data collection. This approach facilitated the development of an initial script outlining medical student journeys.

Two VR experiences were to be developed: one in which a colleague medical student fails to adhere to a rule or follow OR *etiquette*, and the other portraying a colleague medical student who respects the rules and *etiquette* of the OR. The specific communication styles, attitudes and professional conduct of the OR needed to be shown, exposing important elements of the hidden curriculum of the OR implicitly. The two VR experiences differed only slightly, requiring the learner to focus and decipher how these differences would impact a workday. We decided on the epistemological assumption to let the learner *feel* these differences, rather than explaining them, by adhering to constructivist theory (Anthony, 1996).

2.3 Content Validation

Group sessions are an adapted tool for validating storyboards and scripts (Stewart et al., 2007). Group input, such as photographs, focused and selective observation notes, reflective notes and commentaries, was used to write and adapt the medical student journey, which was used for final validation until a consensus was reached; that is, until the final script was obtained, all group members provided consent (Briggs et al., 2005).

The specific learning objectives were verified once more by the educators during a final session.

2.4 Filming and Editing

Requesting that healthcare providers portray their own roles offers several benefits, including cost-effectiveness and easy identification with their professional responsibilities. We tasked the participants with embodying their respective professional roles; for instance, a surgeon would portray the role of a surgeon, a medical student was asked to play a medical student, and so forth. An anaesthetic nurse volunteered to play the patient, as she had already been a patient in our hospital before.

The shooting of the VR-experiences was done with a *GoPro®* camera. Editing and stitching were performed using *PTGUI®*, a panoramic image stitching software, as well as *Pano2VR®*, a virtual tour software that converts 360° into interactive experiences.

3 Results

3.1 Participants and Procedure

Seven healthcare professionals participated in the development of the VR-experiences. This included one OR- assistant, two surgeons, two anaesthetic nurses, and two anaesthesiologist. Two medical students participated, as did two educationalists and one VR cameraman/editor, as shown in Table 1. Participants were invited to participate during the development process of the VR experiences, including content creation, validation through a group session, and filming of the VR experiences.

Table 1. Demographics

Role	Number
Medical student	2
OR assistant	1
Anaesthesiologist	2
Surgeon	2
Anaesthetic nurse	2
Educationalist	2
VR cameraman and editor	1

3.2 Development of the VR-Experiences

Data collection and development were conducted between July 2020 and February 2021. The final durations of the experiences were seven minutes and 32 s and seven minutes and 26 s, respectively. The narrative environment required consistency with reality, employing objects, characters, and actions to enhance the feeling of immersive *presence*.

A blueprint framework was written describing the purpose of the project, the current education, the learning objectives (cognitive, affective, psychometric), the priming of prior knowledge, and when the education was deemed to be successful, as shown in Fig. 3 and detailed in Annex 1.

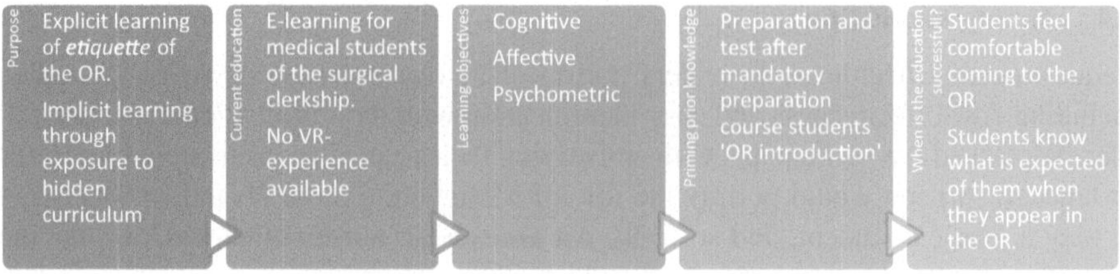

Fig. 3. Blueprint

3.2.1 Script and Storyboard

The fieldwork and group sessions further specified the themes, as shown in Fig. 2, into specific examples to be used for the storyboard, as described below.

Passive Role

As students seem lost and do not seem to show any initiative, in the first experience, the medical student interpreted this passive role by not asking any questions, whereas in the second experience, the student actively asked the surgeon if and how he might participate during the procedure.

OR Workflow

Students may enter or exit the OR at unfortunate moments, such as team briefing, debriefing or anaesthesia induction. In the first experience, the medical student enters during induction and is corrected by the anaesthesiologist afterwards. In the second video, the medical student awaits the end of induction before entering the OR.

Sterile Work Environment

Healthcare professionals express dissatisfaction with respect to sterility among medical students. In the first experience, the medical student enters during a prosthetic surgical procedure, a procedure where door movements need to be as limited as possible. He touched the sterile working space by accident. In the second video, the medical student asks for the right information and changes his schedule to be able to stay for another surgical nonprosthetic surgery. Additionally, in the second video, the medical student assures the correct distance from the sterile working space.

Situational Awareness

Introducing themselves needs to be done at the right moment without interrupting healthcare providers. The same applies to asking questions, which healthcare providers encourage, provided that the clinical situation permits. In the first video, the medical student tries to introduce himself after entering the operating room, while the induction of anaesthesia has already started, interrupting this process. In the second video, the medical student enters the room, checks if the clinical situation is 'under control' before introducing himself, and asks how he might assist.

The storyboard provides an overview of the different scenes, as shown in Fig. 4. The full script of the dual VR experiences is available in Annex 2.

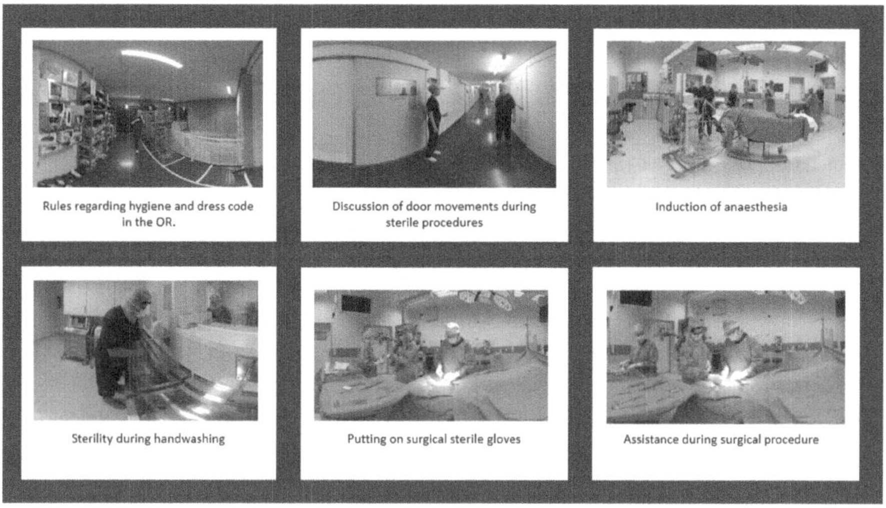

Fig. 4.

An overview of the main differences in interactions and shadow actions between the two experiences can also be found in Table 2. We crafted a script featuring 'shadow actions' that were observable when the learner shifted his or her gaze. Table 2 provides an overview of the key distinctions in setting, direct interactions, and shadow actions.

4 Discussion and Conclusion

4.1 Discussion

This manuscript describes the development of a dual VR experience for medical students in the OR. Through concrete experimental learning, medical students may *feel* what it is like to start working in the OR, preparing themselves for this important transition. As medical students commonly describe this transition as mentally, physically, and emotionally demanding, this dual VR experience may ease this transition, positively affecting their preparedness (Alexander et al., 2014; Sturman et al., 2017).

We deliberately incorporated nuanced shifts in language and interactions, informed by conceptual frameworks and prior scientific investigations, as referenced in our manuscript. Moreover, the participation of healthcare providers, medical students, and educationalists further enhanced the internal validity of the experiences.

Explicit learning objectives of the of *etiquette* included respecting the dress code, presenting themselves appropriately upon entering the operating room, and being mindful of entrance and exit, as well as door movements, in relation to specific moments of anaesthesia and surgery. Additionally, respecting an appropriate distance from coworkers who are working in sterile conditions was emphasized. The implicit learning objectives of the hidden curriculum included the communication style of the OR and the professional conduct, roles and attitudes of different health care providers working in the OR.

Table 2. Overview of the main differences in interactions and shadow actions

	First experience	Second experience
Interactions		
Entrance OR	Colleague medical student 1 forgot to take out earrings	Respect of dress code of the OR
Operation room	Colleague medical student 1 was refused access to an OR where prosthetic surgery was performed Colleague medical student 1verifies OR program beforehand	Colleague medical student 1 enters OR during anaesthesia induction Colleague medical student 1 checks entering the OR after anaesthesia induction
Transitional space OR	Colleague medical student 2 disrespects hand washing procedures	Colleague medical student 2 respects hand washing procedures
Shadow actions		
Entrance OR	Other healthcare professionals pass through the entrance of the OR	
Operation room	Anaesthesiologist and anaesthetic nurse intubate the patient, adapt monitoring and positioning of the patient OR nurse 1 starts prepping the patient OR nurse 2 starts handing over sterile equipment as needed	

The development of this dual VR experience was based on certain epistemological and methodological assumptions. For instance, we chose 360-degree videos with real humans instead of computer-generated images to make the video look as realistic as possible. However, this impeded interactions with the virtual environment. Learners have the ability to observe their surroundings; however, due to the recorded nature of their experiences, learners are not able to freely move around as they would in a virtual environment utilizing computer-generated 3D video imagery.

Additionally, one assumption underlying this project was that the dual VR experiences, created using 360-degree video, could effectively mimic real-life operating room experiences, which are known to ease the transition from academic learning to clinical practice (Shacklady et al., 2009). While 360-degree videos offer a high degree of realism by featuring real people and settings, this approach also introduces limitations. Unlike fully interactive VR environments, 360-degree video restricts the ability of learners to interact with the environment or make choices within the scenario. As a result, students may observe OR dynamics but cannot practice decision-making or physical interactions as they would in a fully immersive, interactive VR setting. This trade-off between realism and interactivity is an important consideration for future iterations of this training method.

4.2 Future Directions

The results of this study have the potential to enhance ongoing research and healthcare education initiatives aiming to integrate experiential learning via VR.

Additional validation could involve assessing the first-hand experiences of medical students who engage in dual VR, along with evaluating the impact of integrating this dual VR experience into the introductory course of medical students' first surgical clerkships. Future studies should also examine the scalability of VR experiences across different medical fields and training levels.

After further validation, this education could be utilized for Bachelor's and Master's students in Clinical Technology, as well as integrated into the Bachelor of Medicine curriculum for students receiving education in the OR during their studies. Subsequently, it could be made accessible to all hospital employees who spend time in the OR as part of their job duties. Moreover, this education could be offered through the corporate training learning management system.

4.3 Conclusion

Utilizing immersive VR technology enables the creation of an interactive and captivating virtual environment, providing learners with the opportunity to engage with simulated realities. In this particular case study, medical students can immerse themselves on a typical day in the OR, where they encounter explicit etiquette and implicitly learn about the hidden curriculum of the OR.

The logistical and hygienic challenges that hinder the transition from school desks to the work floor may be overcome by the use of VR programs, such as the one described in this case study.

The development of this dual VR experience entails meticulous attention to content creation and validation. The findings and insights presented herein contribute to the growing knowledge domain of educational VR development. They underscore the feasibility and potential of harnessing immersive technologies to cultivate engaging and impactful virtual experiences.

Acknowledgements. We thank all the healthcare providers, educators, and medical students who participated in the development of the VR experiences. In particular, we thank the Directorate of Education and Training of Leiden University.

Notes on Contributors. K. HOEK is a clinician specializing in education and communication within anesthesiology and intensive care and conducted this research. She was responsible for **conceptualization, methodology, investigation, data curation, writing – original draft, writing – review & editing, visualization, and project** administration.

M. van Velzen is an educationalist, clinical research coordinator, and assistant professor at the Department of Anaesthesiology of the Leiden University Medical Center. She contributed to **validation, and supervision.**

J. van der HAGE Jos van der Hage is a professor of **surgery**, specializing in intracurricular education in surgery. He cosupervised the research and was responsible for **validation, review & editing, and project administration**.

E. SARTON is a Professor of Anaesthesiology specializing in education in medical training. She was responsible for **supervision, and validation.**

Disclosure of Interests. This research was subsidized by the Leiden Learning and Innovation Center. The authors have no further conflicts of interest to declare.

Annex 1: Blueprint

Purpose

An evaluation with students revealed that existing e-learning, which only contains a few situation photos of an operating room (OR), lacks sufficient realism to give students a sense of how an OR actually functions. This blended learning solution consists of two components:

First, we aim to increase the realism of existing e-learning with a virtual OR tour. Existing static photos of the OR will be replaced by 360° panoramas that are made interactive. New e-learning should be playable without additional requirements for all students' and staff members' laptops. E-learning will be offered through our own educational systems (MedischOnderwijs.nl, Blackboard, and corporate training) within the surgical clerkship period (300 students per year).

Second, the instructions will be expanded with a virtual OR tour using 12 Oculus Quest devices (already available within the LUMC). This education will be implemented in the Monday morning education of the surgical clerkship (12 students per month).

Initial Situation

Initially, blend-it education is for medical students starting their surgical clerkships. E-learning consists of a virtual OR tour without interactive options. It is included in the mandatory preparation for any medical student starting their surgical clerkship. It has explanatory notes on OR instructions detailing the rules of surgical hygienic procedures and sterile work. The e-learning is validated when the corresponding tests have been successfully answered.

What Should a Student Remember from Your Course 10 years Later?

Only a small portion of doctors will be actively involved in the operating department in their eventual careers. However, (especially intramurally working) specialists often interact with colleagues in the operating department. Extramurally working doctors will speak to patients after surgery. For both groups of doctors, the layout, manner, and culture of working in an OR, as well as the *etiquette* that applies in an OR, will remain more or less familiar.

When is Your Education Successful?

Enthusiastic responses from students.

Students feel comfortable coming to the OR.

Students know what is expected of them when they appear in the OR.

Learning Objectives

Cognitive: Achieve summative assessment: e-learning.

Affective: Familiarize with OR environment and culture.

Psychomotor: Knowledge about sterile work.

- Hand washing and sterilization
- Putting on gloves and gown
- Setting up the sterile operating table
- Transferring the patient
- Positioning of the operating room team

Detailed Description of the Current Education

The surgical clerkship period consisted of 2 weeks of preparatory education prior to surgical clerkship. One of the components is the OR instruction. The OR instruction can be used outside the framework of the surgical clerkship as (mandatory) preparation for students/researchers who need to be in the OR for education (half minor) or research purposes. The number of credits will thus be part of the specific activity for which "the virtual OR tour" will be used.

What Should they Still Know in 10 years? (Actually in 1 Week)

How a student should present themselves in the OR.

What the world in the OR is like (for when the student does NOT end up working in the OR in their career)?

Activation of Prior Knowledge

Preknowledge test after mandatory preparation of students' presentations, available on Brightspace about OR attire and sterile work (e-learning within Brightspace and OR instruction).

Annex 2 Script

The first-person perspective of medical students is called 'Camero'.

SCENE 1: GETTING DRESSED

CAMERO stands in the hallway next to the changing room, another medical student returns to the hallway. She has her earrings in her hand.

MEDICAL STUDENT 2

I forgot to take off my earrings, so clumsy, definitely not allowed! I'll quickly put them away! I'll see you at the entrance of OR-1!

SCENE 2: ON THE WAY TO THE OR

CAMERO walks towards the OR. He passes by the recovery room, PACU, and holding area.

SCENE 3: OR 1

CAMERO is in OR-1. The anesthesiologist is starting the induction (administering anesthesia). Then medical student 2 suddenly enters. The anesthesiologist looks annoyed at medical student 2 and continues with the induction.

ANESTHESIOLOGIST

Very good, ma'am, take another deep breath. In the background, you can hear some noises, that's all part of it. We'll take good care of you, we'll see you again after the surgery.

The patient takes another deep breath and closes her eyes.

The anesthesiologist intubates the patient.

Meanwhile, the surgical team enters and the OR assistant rolls a sterile cart next to the patient.

ANESTHESIOLOGIST

It's not very convenient for you to come in while we're in the middle of the induction.

This causes noise and unnecessary extra tension for the patient, could you please pay attention to this next time?

MEDICAL STUDENT 2

My apologies, I didn't realize I need to be more careful when entering.

Medical student 2 walks towards the patient and accidentally touches the sterile cart.

OR ASSISTANT

Hey, hey! You need to be careful not to touch anything sterile, okay? Now I have to get new supplies!

MEDICAL STUDENT 2

Sorry, sorry.

Medical student 2 awkwardly walks back to a corner and watches as the setup continues. The surgical assistant cleans the patient's right arm, and eventually, the surgeon enters.

SURGEON

Alright. Are you all set? I'm eager to get started. Shall we put on some music?

The surgeon stands by the table and then looks at the medical student.

SURGEON

Hello, who are you?

MEDICAL STUDENT 2

Oh, hello, we're the two medical students who were allowed to shadow today. Is it okay if we observe during the surgery?

SURGEON

Oh, how nice that you're here, but we're going to insert a prosthesis, so that means you can't leave the room until 2 p.m. Do you have afternoon classes?

MEDICAL STUDENT 2

Oh, we didn't know that… Our classes start at 2 p.m. so it's not very convenient, right?

She looks questioningly at CAMERO…

SURGEON

No, I thought as much. Then you'd better go and see if you can join on OR 5. And you (he addresses CAMERO now) can maybe go and see on OR-3. That might be more convenient for you today.

MEDICAL STUDENT 2

Thank you, we'll go and check that out.

SCENE 4: MEDICAL STUDENT LEAVES THE OR

The medical students are in the intermediate area.

Medical student 1 takes off her nose mask and throws it away, then disinfects her hands.

SCENE 5: MEDICAL STUDENT IN THE HALLWAY

They walk towards the corridor.

They meet another medical student who looks at them questioningly.

MEDICAL STUDENT 3

Are you leaving the OR already?

MEDICAL STUDENT 2

Yes, we didn't check beforehand if this was an OR where we could shadow, but they're placing a prosthesis so no one is allowed in or out during the surgery.

MEDICAL STUDENT 3

Fortunately, you found out in time. Imagine if you had missed this afternoon's class!

MEDICAL STUDENT 2

Yes, they were a bit grumpy because I didn't introduce myself, but I didn't want to disturb them.

MEDICAL STUDENT 3

Yeah, that can be difficult, huh? But always do it, otherwise you'll always get complaints.

MEDICAL STUDENT 2

Yeah exactly, and I actually understand it, but oh well, I'm going to OR-5 now, see you later.

SCENE 6: MEDICAL STUDENT GOES TO SECOND OR

Medical student 3 walks towards OR-5.

Medical student 2 is already in front of OR-3. An OR assistant is just walking out.

OR ASSISTANT

Are you a medical student? You can come in now. The patient has just been intubated and we're still preparing.

SCENE 7: MEDICAL STUDENT IN OR-3

Background music is playing already, the surgeon is about to commence the surgery and looks up at the medical student.

The OR assistant walks back in and walks around the medical student.

OR ASSISTANT:

This is CAMERO, a medical student from surgery, he's here to observe our operation this morning. That's okay for this operation, right?

CAMERO:

Hello everyone, thanks for letting me observe.

SURGEON:

Yes, that's fine, in fact, can you already scrub in? You can also gown up. Stef, could you assist Guido? Sebastian, the new trainee OR assistant, has already started scrubbing.

SCENE 8 SCRUBBING

The surgical assistant observes as the medical student and the trainee OR assistant scrub their hands sterile.

TRAINEE OR ASSISTANT:

Hi, I've had the training for scrubbing, but I haven't done it many times yet. If you see any areas for improvement, please let me know!

STEF:

That's alright, but you're doing fine!

CAMERO and the trainee OR assistant scrub their hands.

SCENE 9 GOWNING UP IN THE OR

Camero stands by the table.

He puts on an OR gown.

Sterile gloves are handed to him.

SURGEON:

Glad you're here, could you hold this for a moment?

CAMERA ZOOMS OUT, IF POSSIBLE SHOW LAPAROSCOPY OR SOMETHING SIMILAR SO WE CAN REALLY SEE THE OPERATION HAS BEGUN… END.

References

Alexander, C., Millar, J., Szmidt, N., Hanlon, K., Cleland, J.: Can new doctors be prepared for practice? A review. Clin. Teach. **11**(3), 188–192 (2014). https://doi.org/10.1111/tct.12127

Anthony, G.: Active learning in a constructivist framework. Educ. Stud. Math. **31**(4), 349–369 (1996). http://www.jstor.org/stable/3482969

Bem, D.: Self-perception Theory, vol. 6. Academic Press (1972)

Bian, Y., Zhou, C., Tian, Y., Wang, P., Gao, F.: The proteus effect: influence of avatar appearance on social interaction in virtual environments. In: Stephanidis, C. (ed.) HCI 2015. CCIS, vol. 529, pp. 78–83. Springer, Cham (2015). https://doi.org/10.1007/978-3-319-21383-5_13

Bracq, M.S., et al.: Learning procedural skills with a virtual reality simulator: an acceptability study. Nurse Educ. Today **79**, 153–160 (2019). https://doi.org/10.1016/j.nedt.2019.05.026

Briggs, R., Kolfschoten, G., de Vreede, G.-J.: Toward a theoretical model of consensus building (2005)

Cannon-Bowers, J.A.: Recent advances in scenario-based training for medical education. Curr. Opin. Anaesthesiol. **21**(6), 784–789 (2008). https://doi.org/10.1097/ACO.0b013e3283184435

Fox, J., Bailenson, J., Tricase, L.: The embodiment of sexualized virtual selves: the Proteus effect and experiences of self-objectification via avatars. Comput. Hum. Behav. **29**, 930–938 (2013). https://doi.org/10.1016/j.chb.2012.12.027

Gormley, G.J., Murphy, P.: When I say … simulation. Med. Educ. (n/a). https://doi.org/10.1111/medu.15165

Hafferty, F.W., Franks, R.: The hidden curriculum, ethics teaching, and the structure of medical education. Acad. Med. **69**(11), 861–871 (1994). https://doi.org/10.1097/00001888-199411000-00001

Jhajj, S., et al.: Impact of Covid-19 on medical students around the globe. J. Community Hosp. Intern. Med. Perspect **12**(4), 1–6 (2022). https://doi.org/10.55729/2000-9666.1082

Laraia, K.N., et al.: Analysis of operating room personnel perspectives to enhance medical student readiness for surgery. J. Surg. Educ. **81**(1), 70–75 (2024). https://doi.org/10.1016/j.jsurg.2023.09.011

Le Duff, M., et al.: Virtual reality environments to train soft skills in medical and nursing education: a technical feasibility study between France and Japan. Int. J. Comput. Assist. Radiol. Surg. (2023). https://doi.org/10.1007/s11548-023-02834-0

Navarro, J., Peña, J., Cebolla, A., Baños, R.: Can avatar appearance influence physical activity? User-avatar similarity and Proteus effects on cardiac frequency and step counts. Health Commun. **37**(2), 222–229 (2022). https://doi.org/10.1080/10410236.2020.1834194

Pottle, J.: Virtual reality and the transformation of medical education. Future Healthc. J. **6**(3), 181–185 (2019). https://doi.org/10.7861/fhj.2019-0036

Shacklady, J., Holmes, E., Mason, G., Davies, I., Dornan, T.: Maturity and medical students' ease of transition into the clinical environment. Med. Teach. **31**(7), 621–626 (2009). https://doi.org/10.1080/01421590802203496

Shuid, A.N., et al.: Effect of early clinical skills teaching on 3rd year medical students' learning: the student perspective. J. Taibah Univ. Med. Sci. **10**(1), 26–32 (2015). https://doi.org/10.1016/j.jtumed.2014.12.004

Stewart, D., Shamdasani, P., Rook, D.: Focus Groups. 2nd edn. (2007). https://doi.org/10.4135/9781412991841

Sturman, N., Tan, Z., Turner, J.: "A steep learning curve": junior doctor perspectives on the transition from medical student to the health-care workplace. BMC Med. Educ. **17**(1), 92 (2017). https://doi.org/10.1186/s12909-017-0931-2

Surmon, L., Bialocerkowski, A., Hu, W.: Perceptions of preparedness for the first medical clerkship: a systematic review and synthesis. BMC Med. Educ. **16**(1), 89 (2016). https://doi.org/10.1186/s12909-016-0615-3

Thun-Hohenstein, L., Höbinger-Ablasser, C., Geyerhofer, S., Lampert, K., Schreuer, M., Fritz, C.: Burnout in medical students. Neuropsychiatry **35**(1), 17–27 (2021). https://doi.org/10.1007/s40211-020-00359-5. (Burnout bei Medizinstudent*innen)

Whitman, N.: A review of constructivism: understanding and using a relatively new theory. Fam. Med. **25**(8), 517–521 (1993)

LifeCharger SynCare Mobile App: A Preliminary Usability Study

Claudio Pighini[1,2](✉), Ambra Cesareo[2], Andrea Migliavacca[2], and Enrico Gianluca Caiani[1,3]

[1] Electronics, Information and Bioengineering Department, Politecnico di Milano, 20133 Milan, Italy
claudio.pighini@polimi.it
[2] LifeCharger srl, Milan, Italy
[3] Department of Cardiology, IRCCS Istituto Auxologico Italiano, Piazzale Brescia 20, 20149 Milan, Italy

1 Introduction

1.1 Overview

Over the last decade, the world has witnessed the growth of digital health through the emergence of social media platforms, smartphones, mobile applications, wearable devices, cloud-based data platforms, real-world evidence (RWE) studies, and similar technologies. Consequently, general well-being and health monitoring are gradually transitioning from being confined to hospitals and clinics to becoming part of the widespread digital world via various smartphone applications [1].

Digital health technologies that engage patients in their health and well-being through smartphones (or tablets) mobile applications are known as mobile health (mHealth) apps. These apps not only improve users' eHealth literacy by providing health and wellness information and suggestions, but also can deliver digital health services, allowing the collection of real-world data (RWD), thereby enriching and enhancing end-users RWE [2].

To exploit such new tools, innovative models are studied in the digital health domain, creating new user-centric platforms to enhance the value of RWD and optimize the potential of professionals from traditional healthcare settings.

1.2 LifeCharger Syncare Ecosystem

The LifeCharger (LC) SynCare [3] is a complex ecosystem, that aims at increasing the therapeutic alliance between professionals and patients through the delivery of digital health & wellness services.

The LC SynCare ecosystem purposes are dependent on the specific application field in which it is deployed. For example, a psychologist could provide a service for remote psychological consultations and provide personalized mental wellbeing suggestions through the LC mobile App, that represents the patient's front-end, while a speech

D. Singh et al. (Eds.): IHCI 2024, LNCS 15557, pp. 18–28, 2025.
https://doi.org/10.1007/978-3-031-88705-5_2

therapist could remotely provide speech training and prescribe different personalized exercises shown through the same app.

Indeed, the LC SynCare Ecosystem is a general-purpose platform acting as service delivery that can be specialized through the activation of different services, according to their different purposes. By performing the different prescribed activities (e.g., measure physiological parameter, fill-in questionnaires and evaluation scales, report symptoms, etc.), the patients allow the professionals linked to the services to monitor their performance and health status.

In this way, the LC SynCare ecosystem purpose relies on the activated services, that are the real products offered to the end-users, by monitoring how the user engaged with such services.

To allow the fruition of these services, the LC SynCare Ecosystem architecture relies basically on 3 levels, declined into 3 different software application: 1) the mobile App, that represents the end-user front-end and the patient's gate to the LC SynCare Ecosystem; 2) the Clinical Dashboard, that represents the professional front-end and the software to prescribe and monitor user activities; and 3) the back-end, including the Data Analytics module, where the user data collected through the App are processed whose results will be shown on the Clinical Dashboard.

The dataflow, described by the diagram of Fig. 1, starts from the patient that collects data by using the LC App. If the patient signed the digital consents approving the data sharing with third parties (LifeCharger included), data can be shared by the LC App with the back-end for analysis. Every time a consent is signed, the relevant information is saved into the Ethereum Blockchain as a Smart Contract, thus allowing a transparent tracking of signed consents [3].

The Data Analytics module processes the data to be visualized by professionals through the Clinical Dashboard. Professionals can thus use the Clinical Dashboard to visualize the data of their patients, to monitor their situation and potentially send them feedback and action plan, thus closing the loop.

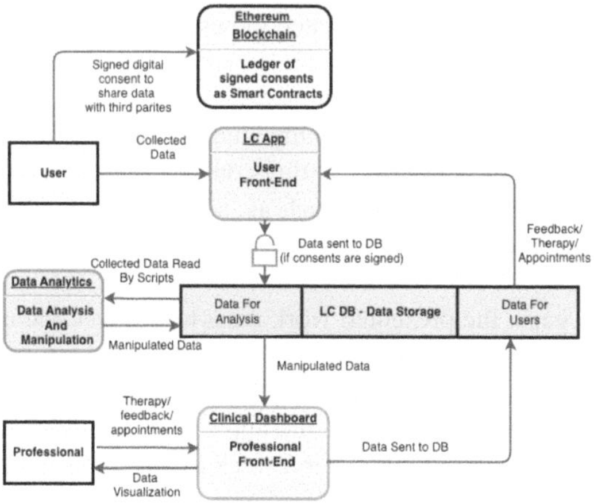

Fig. 1. Block diagram representing the data flow passing through the LifeCharger SynCare Ecosystem modules.

1.3 Objective and Research Gap

For mHealth-based digital health services to be utilized over time, as this technology is patient-oriented and smartphone-based, and frequently in the hands of patients, is important to understand the level of perceived quality they offer to the end-users, with a specific focus to evaluate its quality in terms of usability and acceptability [4].

In recent years, besides the typical methods available to evaluate usability and acceptability of a mobile app, such as the System Usability Scale (SUS) and Post-Study System Usability Questionnaire (PSSUQ) [5], a more specific questionnaire has gained relevance: the user version of the Mobile Application Rating Scale (uMARS) [6]. This questionnaire assesses dimensions of engagement, functionality, aesthetics, information, and subjective quality by a 5-point scale, and it has been largely used to evaluate different mHealth apps, making possible comparisons in literature between different usability studies and results. In [5], a group of 118 anesthesiologists evaluated an applet-based National Remote Emergency System for Malignant Hyperthermia (MH), using a modified version of uMARS, to deal with MH crises. The evaluation was done after completing some tasks in a simulative utilization. The uMARS was then compared with the results obtained also by the PSSUQ and SUS questionnaires, demonstrating high reliability and validity. In [7], a pilot study was conducted to evaluate usability and acceptability of the FeverDx mobile app with a modified version of uMARS, that was administered to 20 general practitioners to evaluate a simulated emergency consultation of cases of acute febrile syndromes such as Zika, dengue, chikungunya, and other common infections in tropical areas (leptospirosis). In [8], the PREVENTION e-platform, that provides accessible and evidence-based health information tailored to different Breast Cancer (BC) risk levels, was evaluated via uMARS by 30 women. This was done by assigning randomly the 30 women to a BC risk level, and they could access only the info of the assigned section. In [9], 126 rheumatic patients evaluated Rheuma Auszeit self-management mobile app with a simplified version of uMARS after at leat 10 min using a tablet. Finally, in [10] 26 patients assessed Convoy-Pal, that is a mobile intervention to deliver self-management tools and palliative care to older adults with advanced heart failure. The patients evaluated the usability of the Convoy-Pal afeter an interview between 40 and 60 min, where tools and characteristics of the system were shown.

As briefly described, the mHealth usability studies often do not consider initial user interaction, or the impact of real-world settings on short-term engagement. Moreover, the basic functions of the applications (such as, self-reporting, filling questionnaires, etc…) are not evaluated separately from that used within the provided digital health or wellness service.

To address these gaps, the presented work aims to evaluate the usability of the LC App specifically focusing on its basic functions, without any additional service activated. By using the validated Italian version of the uMARS questionnaire [11], this study aims at providing a preliminary assessment of the app's usability, objective and subjective quality, and perceived impact on users after a week of autonomous use.

More specifically, this work is a part of a larger study that aims to evaluate the LC App in 3 different use cases: 1) The LC App used with its basic functions, i.e. without additional services activated; 2) the LC App used with one service activated, but without receiving personalized contents and feedback (standalone); and 3) the LC App used with one service activated, also receiving personalized contents and feedback for the patient (interactive).

In terms of human-computer interaction, this study can represent a proposal for a model of evaluation that investigates the interaction between end-users and mHealth technologies in real-world environments, leaving them free to autonomously use the app, respecting the smartphone ubiquity intrinsic property. Moreover, the comparison of different user-cases let to test each of the developed features offered to the end-users, separately and in the context of health and wellness delivered services.

In this study, the preliminary results extracted from the first use case, assessing the usability of the LC App in its basic functionality as a container, evaluating the interaction and perceived impact of the end-users without any external intervention, or connection, additional information, or contents, will be presented. Such preliminary insights will serve as a baseline for future comparisons with the additional two use cases, designed to further evaluate the app's performance when it is enriched with health and wellness services.

2 Material and Methods

2.1 LifeCharger Mobile App

The LC mobile App in its basic function (free on Google Play and Apple Stores) allows the users to self-manage their therapies, with the possibility to track drug intakes, symptoms, physiological parameters (such as blood pressure, weight, etc…) and physical activities, to fill in, directly In-App, questionnaires in the form of Patient Reported Outcome Measures (PROMs) or surveys, and to receive reminders via push notifications when the activities were programmed (Fig. 2).

The LC App also allows easy reporting and monitoring of symptoms through an intuitive diary feature, helping to maintain a detailed health record. Through the "Summary" functionality, the LC App allows the users to keep under control the trends of their physiological parameters and to monitor therapeutic adherence.

Finally, the "Document archive" allows uploading documents and organizing them in categories in order to create a proper organized archive of medical reports.

From the user interface (UI) point of view, the LC App presents to the user a simple Home page divided into different cards, that shows at first impact the different functions, with a last window containing suggestions or contents to guide the users through the app or to give them some useful hints about their health. Every card is responsive to clicks and create a shortcut to the main App sections (Fig. 2).

Finally, the App presents on the left-upper corner a hamburger button to open the navigation menu to allows the user the exploration of the entire App (Fig. 2).

2.2 Protocol

The study protocol was approved by the Politecnico di Milano ethics committee board (act n. 02/2023). All participants filled-in the informed consent, maintaining the

Fig. 2. Series of LC App (Italian version) screenshots: 1) home page containing different Cards, such as Services, Profile, etc…; 2) the view that shows what type of activity you can program, such as drugs intake, physical activity, appointments (not active in this study), physiological parameters, and questionnaires or PROMS; 3) a view of "La tua giornata", representing your daily with to-do programmed activities and where the user can check if the activity was effectively done; 4) the menu to navigate over the different LC App sections, that was opened using the "hamburger" button on the left-upper corner

anonymity to the researcher responsible for data analysis. All collected data were pseudo-anonymized by associating a randomized unique id, which is saved in an encrypted way by the user LC App.

The volunteer recruitment was done by different digital means, such as social media (LinkedIn, Instagram, Facebook) via post, or email. After the recruitment process, the researcher provided by email to the volunteers the instructions for downloading, installing and using the App, on the iOS or Android smartphone owned by the participant, illustrating the App functionalities and the study procedure.

The volunteer received from the researcher a 6-digit code, that must be inserted in-App to begin the study. At the first run, the LC App shows the login screen, through which the volunteer could register by selecting a username and password, and then entering such credentials to login into the App. An on-board (activation) phase followed the first registration and login, where the user had to sign the digital informed consent that explains what data types and how data are managed by the Syncare ecosystem [3].

Once the users finished the activation process, the LC App lands on the Home Page, divided into different cards representing the main functions of the app. The first card presented inside the Home is the "Services" card, which can be used to buy and activate services that could be delivered through the app. Each service is activated through a 6-digit code, that the user may receive through different channels (e-shop, participation to experimentations, professionals linked to services, etc.). Participants were asked to insert a 6-digit code, provided during the recruitment phase, just to track the effective participation and the beginning of the usability study (Fig. 3).

Participants were then asked to follow the following procedure: 1) fill in the profile modules, including information such as gender age, education level & occupation, height & weight, habits (smoke and alcohol consumption), fill-in an eHealth literacy

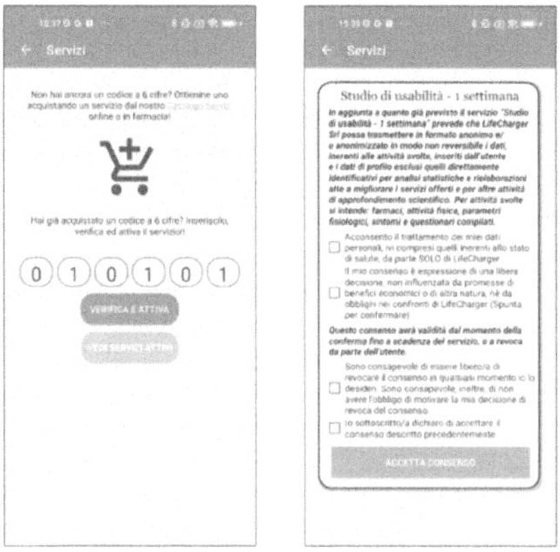

Fig. 3. 1) This view, opened using the Services Card on the home page (Fig. 2, 1)), shows the 6-digits code insertion, that established the beginning of the study; 2) after having inserted the 6-digits code, the LC App showed the In-App digital informed consent (Italian version), that the user has to accept to participate by checking the different mandatory clauses

(eHEALS) questionnaire [12] and a questionnaire to evaluate the grade of adherence to the therapy at the baseline ([13] or similar); 2) insert and schedule their therapies and activities, according to their care plans (e.g. drugs to be taken, physiological parameters to be monitored and measured, etc.), using the dedicated app functionality; 3) Use the app to receive notifications reminding the inserted activities to be performed, and checking the executed activities on the daily diary.

After 7 days of utilization, the patient was asked to fill-in the uMARS questionnaire, used to evaluate the app usability, as showed in the list of to-do activities (Fig. 2, 3)).

In Fig. 4 a block diagram is divided into the different phases of the protocol, illustrating the pathway that the volunteer needs to go through, from the recruitment to the final app usability evaluation, is shown.

2.3 Usability Evaluation

The App usability was evaluated administering to the volunteers the uMARS questionnaire [6], in its Italian uMARS validated version [11]. It included four objective subscales (or Sections), one subjective quality subscale, and a perceived impact section.

The uMARS included four subscales (or "Sections") to evaluate objectively the app quality, covering the following areas: Section A covers the engagement, including five items to evaluate if the platform is entertaining, interesting, customizable, interactive, and appropriate for the target group; Section B the functionality, including four items to assess performance, ease-of-use, navigation, and gestural design; Section C assesses the aesthetics, including three items to evaluate layout, graphics, and visual appeal; while Section D the information, including four items to evaluate the quality and quantity of the information, the correctness of the visuals, and their credibility.

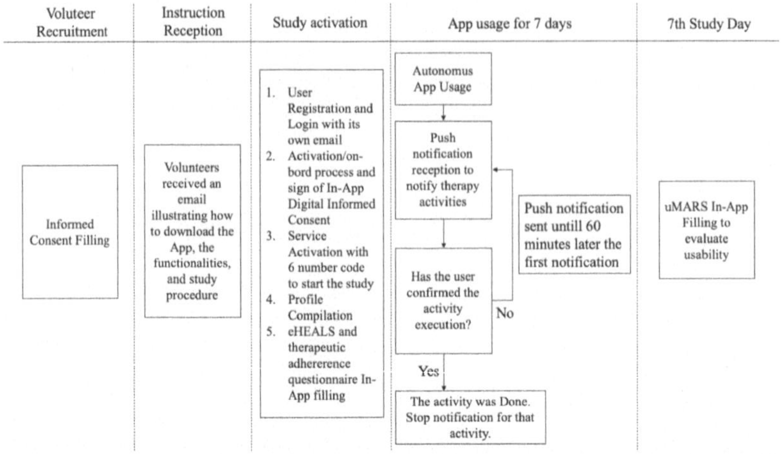

Fig. 4. Study Protocol, from recruitment phase to usability evaluation at the end of the study

The uMARS subjective quality subscale (section E) explores through four items the willingness to recommend the e-platform, willingness to pay for it, anticipated frequency of use, and overall perceived quality.

Lastly, the perceived impact subscale (section F), composed of 6 items, assesses changes in participant's related to awareness, knowledge, attitudes, intentions, help-seeking, and intention to change health behaviors.

Each section included questions with a possible value ranging between 1 (inadequate) and 5 (excellent), with at the end of the questionnaire a free answer box to allow leaving a free-text comment about the app.

2.4 Data Analysis

All anonymized data were extracted from the LC Cloud DB and analyzed using a python script on PyCharm IDE.

To obtain the overall usability uMARS score, the mean of each objective subscale was derived, and their average was computed. This score represents the objective quality of the app.

3 Results

3.1 Participant Recruitment

The study was conducted from February to December 2023. Ten volunteers were recruited, with the reception of the 6-digit code to activate the App after filling the informed consent. Three volunteers never activated the study, so the following preliminary results are relevant to the remaining 7 subjects.

3.2 The LC App Usability Evaluation

The uMARS scores for each objective quality sections and the overall uMARS score are reported in the Table 1, while all sections scores are reported in Fig. 5.

The average of the four-objective sub-scales (Section A to D) represents the overall uMARS score, with a resulting mean value of 3.79, SD = 0.45 (median = 4).

Considering singularly the objective subscales, sections A (engagement), B (Functionality), and D (Information) showed high and similar values (mean 3.86 ± 0.87, 3.86 ± 0.91, and 3.89 ± 1.11, respectively) while section B (Aesthetics) scored lower (3.57 ± 0.90). For the subjective quality subscale (Section E) it was 3.68 ± 1.10, while for the perceived impact subscale (Section F) it was the lowest (3.33 ± 1.44).

Table 1. Table of overall uMARS score, objective and subjective quality, and perceived impact section scores expressed as Mean and Standard Deviation (SD) as well as Median with 25th and 75th percentiles (25th; 75th)

Measure	Mean (SD)	Median (25th; 75th)
Overall uMARS Score = Objective Quality	3.79 (0.45)	4 (3; 4)
Section A Score (Engagement)	3.86 (0.87)	4 (3; 5)
Section B Score (Functionality)	3.86 (0.91)	4 (3,75; 4)
Section C Score Mean (Aesthetics)	3.57 (0.90)	4 (3; 4)
Section D Score Mean (Information)	3.89 (1.11)	4 (4; 5)
Section E Score Mean (Subjective Quality)	3.68 (1.10)	4 (3; 5)
Section F Score Mean (Perceived Impact)	3.33 (1.44)	4 (2; 4)

4 Discussion

The aim of this preliminary study was to evaluate the LC App usability when the user utilized its basic functions, without activating any additional services.

The LC App was assessed using the administered in-App digital uMARS questionnaire (validated Italian version) after 7 days of app usage. This duration allowed the users to become familiar with the app and its functionalities, to evaluate the objective quality of the end-user front-end, subjective quality, and perceived impact of the technology.

The uMARS has demonstrated to have a good psychometric properties and internal consistency (Cronbach's alpha; $\alpha = 0.70$–0.80) on subscales with high overall consistency ($\alpha = 0.90$) [6].The overall obtained uMARS score of 3.79 can be considered a good result, when compared to other similar usability study of mHealth apps [5, 7–10]. However, the scores for each subscale deserve further consideration.

The engagement, functionality, and aesthetics scores met the expected outcomes of this study. Users positively evaluated the engagement through which the LC App sent push notifications to remind them of different activities, such as taking medication, measuring physiological parameters, or engaging in physical activity. The app sent reminders

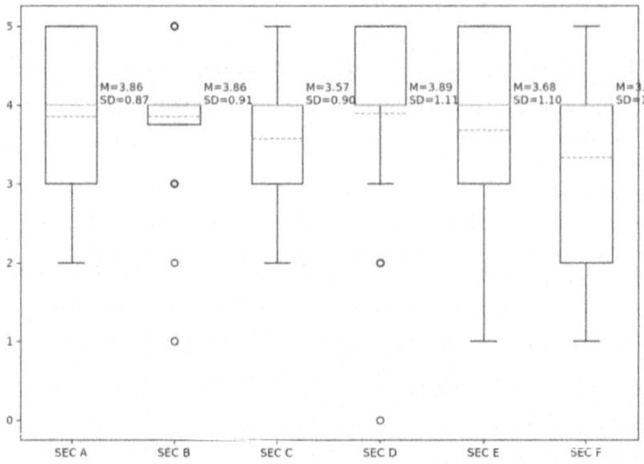

Fig. 5. Boxplot of different section scores (SEC A = Engagement, SEC B = Functionality, SEC C = Aesthetics, SEC D = Information, SEC E = Subjective Quality, SEC F = Perceived Impact). The green dotted lines represent the means of section scores, while the red ones the median values. (Color figure online)

at scheduled times and additional notifications after 5, 15, 30, and 60 min until the user confirmed completion of the activity. This feature contributed to user engagement by encouraging app usage. Similar considerations apply to the functionality section of the app's basic version, that allowed users to access all functionalities without limitations, and the positive score obtained for this subscale reflects the perceived utility of the core functions of the app, that formed the basis of the LC App design [14]. The aesthetics section, while receiving the lowest score among the subscales, still performed well compared to other studies [5, 7–10]. Nevertheless, this result must not be overlooked, as app aesthetics, or visual design, are a key strength for maintaining engagement [4].

Conversely, the information subscale received a higher-than-expected score. In the tested use case, the LC App did not contain additional content or specific information for the users, but only provided advice and suggestions relevant to its functionalities, without considering the user's health status and collected data. Nevertheless, this was considered positively.

As regards the subjective quality subscale results, the first and second questions, assessing the willingness to recommend the e-platform and anticipating the frequency of use, received a very good outcome (i.e., 4), together with a high value (i.e., 3.86) for the last question, reflecting high user satisfaction. The third question, measuring the willingness to pay for the app, received the lowest score (i.e., 2.86), in alignment with the study expectation, as the evaluated LC App without any activated services or additional personalized information is free for users.

Section F assessed changes in participants' perceived health status-related awareness, knowledge, attitudes, intentions (e.g., speak to a physician), as well as intention to change health behaviors (e.g., following the prevention or health-related risk recommendations). This section scored the lowest compared to other sections, likely due to the lack of contents and information about health behaviors or health-related recommendations in the evaluated use case. Comparisons with other studies are challenging, as different studies used modified questions for this section to suit their mobile apps [5, 7], or used different descriptive analysis to present results [8], or they did not include this section at all [9]. Nevertheless, the perceived impact scored a mean value of 4, which is considered a very good result, as it may be related to an increased awareness about health status, possibly induced by the self-management of the therapy or health activities, thus encouraging the users to contact or speak to physicians to better understand their health conditions.

This study had some limitations: in particular, the small sample size (n = 7) limited the validity of the conclusions. However, the study's design, with the user evaluating the app after seven days of its utilization without any real training, strengthen the findings. This approach reinforces the idea that the good scores achieved for LC App functionality reflect a reliable evaluation in real-world scenarios. This finding is noteworthy when compared to most studies that evaluate app usability after a single usage, or without physical interaction via remote interview [5, 7–10].

5 Conclusions and Future Implementations

We presented the preliminary results from the first use case, focusing on the LC App's usability in its free version. We evaluated its user interaction and perceived impact without any external intervention or connection (e.g. involvement of healthcare professional) and without additional information or contents, thus achieving a very good and promising outcome.

These results will constitute the comparison for assessing the LC App in the next two planned service-based use cases, the standalone and the interactive, where users will receive general health information, or additional contents and feedback based on their user history and collected data, respectively.

The standalone use case study is currently on-going, serving as a pilot for testing the LC service called "GrowApp", aiming at assessing child growth across various developmental domains, in collaboration with IRCSS Fondazione Instituto Neurologico Nazionale C. Mondino, involving 120 couples for a total of 240 users.

Upon completing all three use cases, a comprehensive insight into the LC App's performance and user satisfaction across various operational contexts will be obtained, to serve as basis for further improvements. Indeed, mHealth technologies, due to their intrinsic nature, need continuous and iterative developments and upgrades. This study represents a huge possibility to understand and improve LC App functionalities and delivered health & wellness services. Moreover, this is an opportunity to create a positive tendence in using methods, such as uMARS, that can be comparable with other user-centric studies, as well as show a proposal of a comparative study that assesses mHealth technologies.

References

1. Dang, A., Arora, D., Rane, P.: Role of digital therapeutics and the changing future of healthcare. J. Fam. Med. Prim. Care **9**(5), 2207–2213 (2020)

2. Khosla, S., et al.: The alignment of real-world evidence and digital health: realising the opportunity. Ther. Innov. Regul. Sci. **55**, 889–898 (2021)
3. Pighini, C., et al.: SynCare: an innovative remote patient monitoring system secured by cryptography and blockchain. In: Privacy and Identity Management. Between Data Protection and Security (2022)
4. Mathews, S.C., McShea, M.J., Hanley, C.L., Ravitz, A., Labrique, A.B., Cohen, A.B.: Digital health: a path to validation. Nat. Partner J. Digit. Med. **38** (2019)
5. Yu, H., Tan, L., Zhu, T., Deng, X.: A WeChat applet-based national remote emergency system for malignant hyperthermia in China: a usability study (2023)
6. Stoyanov, S.R., Hides, L., Kavanagh, D.J., Wilson, H.: Development and validation of the user version of the mobile application rating scale (uMARS). JMIR mHealth uHealth **4**(2), 72 (2016)
7. Sarita, R., et al.: Acceptability and usability of a mobile application for management and surveillance of vector-borne diseases in Colombia: an implementation study, Egypt. Plos One (2020)
8. Attieh, S., et al.: Perceptions and usability of PREVENTION: a breast cancer risk assessment e-platform. Personalized Med. **13** (2023)
9. Lambrecht, A., et al.: Quality of a supporting mobile app for rheumatic patients: patient-based assessment using the user version of the mobile application scale (uMARS). Front. Med. **8** (2021)
10. Villalobos, J.P., Bull, S.S., Portz, J.D.: Usability and acceptability of a palliative care mobile intervention for older adults with heart failure and caregivers: observational study. JMIR Aging **5**(4) (2022)
11. Morselli, S., Sebastianelli, A., Domnich, A., Gacci, M.: Translation and validation of the Italian version of the user version of the mobile application rating scale (uMARS). J. Prev. Med. Hyg. **62**, 243–248 (2021)
12. Bravo, G., et al.: Validazione della versione italiana del questionario di alfabetizzazione sanitaria digitale (IT-eHEALS). Not. Ist. Super Sanità **31**(7–8), 6–10 (2018)
13. Fabbrini, G., et al.: Adherence to anti-Parkinson drug therapy in the 'rEASON' sample of Italian patients with Parkinson's disease: the linguistic validation of the Italian version of the 'Morisky Medical Adherence Scale-8 Items.' Neurol. Sci. **34**, 2015–2022 (2013)
14. Pighini, C., Cesareo, A., Migliavacca, A., Accardo, M., Guarneri, R.M.: Re-hub-ILITY: a personalized home system and virtual coach to support and empower elderly people with chronic conditions. In: Proceedings of Sixth International Congress on Information and Communication Technology (2021)

From Theory to Practice: Simulation in Anesthesiology Through Active Learning Experiences

Krista Hoek(✉), Chris Martini, Monique van Velzen, and Elise Sarton

Department of Anesthesiology, LUMC, Albinusdreef 2, 2333ZA Leiden, The Netherlands
k.b.hoek@lumc.nl

1 Introduction

Simulation in medical education serves as a cornerstone in preparing anesthesiologists for training for their future roles by equipping them with essential knowledge and skills. Simulation is a methodology, rather than a technology, aimed at substituting or enhancing real experiences with guided experiences. Central to this preparation is the utilization of clinical scenarios to foster effective learning [1–3]. Simulation-based learning experiences (SBLE) have emerged as a pivotal tool, offering a secure environment for trainees to gain practical experience in managing medical emergencies or rare complications within a controlled setting, thus minimizing risks to real patients [4–6].

For anesthesiologists, managing acutely deteriorating patients demands a combination of technical and nontechnical skills within high-stress and unpredictable settings. Technical competencies include skills such as tracheal intubation, difficult airway management, vascular catheter placement, and regional anesthesia. Extensive medical knowledge covers applied physiology, differential diagnoses and drug dosages. Nontechnical skills may include action triage, effective teamwork and crisis resource management (CRM) [7]. Crisis resource management is part of human factor competencies and refers to interpersonal and cognitive behaviors that contribute to optimal team performance. CRM originated in the aviation industry, where human error was established to be a major cause of aviation accidents in the 1970s [8]. Crucially, proficiency in situational awareness, decision-making, teamwork, communication, and leadership, as underscored by the Institute of Medicine Committee on Quality of Health Care, remains indispensable in clinical practice. Improved patient outcomes often depend on the implementation of multitiered rapid response systems, underlining the importance of effective CRM training [9]. Weller established the ecological validity of simulated CRM training in anesthesiology training [10].

The process of learning encompasses internal changes in learners' attitudes, beliefs, behaviors, and knowledge acquisition [2]. Effective medical education demands adaptability from educational developers to meet the evolving demands of the healthcare landscape. Traditional pedagogical approaches, such as didactic lectures and bedside demonstrations, have long been relied upon; however, they present limitations, including minimal learner engagement and insufficient development of practical competencies tailored to diverse learning styles. Recognizing that knowledge is constructed within

D. Singh et al. (Eds.): IHCI 2024, LNCS 15557, pp. 29–40, 2025.
https://doi.org/10.1007/978-3-031-88705-5_3

social contexts, fostering active engagement among learners becomes paramount [2, 11]. Kolb's experiential learning cycle provides insights into the cognitive processes essential for skill development, offering a framework for effective teaching and learning [12].

As technology advances, new modalities like virtual simulations, and game-based learning have emerged. While these methods offer advantages, their add-on effectiveness against traditional teaching modalities such as clinical courses or bedside teaching, and their impact on clinical skill development remains uncertain. Thus, rigorous evaluation of these newer methods becomes imperative as they gain widespread adoption.

This research article seeks to elucidate the rationale for evaluating new simulation modalities amidst limited existing evidence. By exploring educational theoretic frameworks and elucidating the effectiveness of different teaching modalities, educational developers, educational institutions, and policymakers can design curricula that effectively cultivate the future generation of competent anesthesiologists.

2 Design

The current study evaluates two simulation modalities—manikin-based and VR-based—through a comparative analysis focused on their respective advantages in terms of fidelity, educational impact, and practical feasibility. This framework is intended to provide insights for educators and curriculum developers to inform the design and implementation of simulation-based education programs that effectively address the diverse learning needs in anesthesiology training. Additionally, the study examines the current state of simulation education, challenges associated with each modality, and potential solutions for optimizing future applications of these simulation methods.

3 Current State of Simulation Education

3.1 Manikin-Based Simulation

The gold standard of current simulation education in anesthesiology is manikin simulation. It generally involves an initial instructional briefing, explaining the clinical case and roles of participants, a lived simulation experience where participants work on the case study with the guidance of the trainer, followed by a debriefing phase allowing participants and trainers to evaluate the simulation, as shown in Fig. 1.

The initial briefing serves multifaceted pedagogical purposes. First, they are instrumental in elucidating the overarching goals of exercise. By outlining specific learning objectives, participants gain clarity on what skills or knowledge they are expected to acquire or improve upon during the simulation. This enhances participant engagement and facilitates targeted learning outcomes [13]. Moreover, they allocate distinct roles to participants, fostering a collaborative learning environment. Each participant is encouraged to actively contribute their insights, observations, and reflections during the simulation experience. Through this participatory approach, participants may benefit from diverse perspectives offered by their peers.

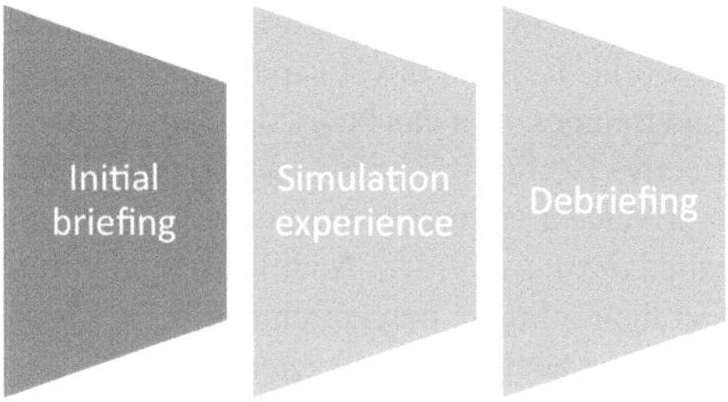

Fig. 1. Manikin simulation

The simulation experience itself, guided by a trained facilitator or trainer, serves as a pivotal pedagogic tool in manikin simulation training. Unlike more traditional didactic methods, simulations immerse participants in realistic scenarios, providing them with hands-on experience in a controlled but authentic environment. Under the guidance of the trainer, participants navigate through various clinical challenges and decision points, applying theoretical knowledge to practical situations. This experiential learning approach enables participants to bridge the gap between theory and practice by practicing their critical thinking, problem-solving, and decision-making skills. By actively engaging in simulated scenarios, participants develop the confidence and competence necessary to handle similar situations in real-world settings. According to the principles of cognitive load theory (CLT) [14], there is a limit on how much information one can process simultaneously, impacting information storage and retrieval. A trainer may not know beforehand how the trainees will perform during the scenario; hence, he/she may not know beforehand how they want the scenario to evolve. This highlights the importance of the adaptability of a scenario that is possible with manikin simulation, enabling a more effective and valuable learning experience.

Following the simulation experience, debriefing sessions offer participants an opportunity to reflect on their lived experience and evaluate their performance. Unlike traditional evaluations, which may focus solely on outcomes, debriefing encourages participants to delve deeper into their thought processes, emotions, and actions during the simulation [6]. During debriefing, both participants and the trainer engage in a comprehensive evaluation of the simulation experience. Participants articulated their thoughts, feelings, and reactions to the scenario, identifying strengths, areas for improvement, and lessons learned. This reflective process may enhance self-awareness self-directed learning. By leveraging the expertise of the trainer, participants gain valuable insights into best practices, evidence-based strategies, and alternative approaches to problem solving. Through this collaborative endeavor, manikin simulation training achieves its pedagogic objectives of enhancing clinical skills, promoting teamwork, and ultimately improving patient outcomes [15, 16].

3.2 VR-Based Simulation and Game-Based Learning

VR is a computer-generated simulation of an environment that immerses users in a simulated three-dimensional, interactive experience. Immersive VR uses a head-mounted display projecting in front of the eyes, allowing users to focus on display without interaction with the outside world. It offers novel capabilities that provide new qualitative support for educators and trainees, as it can be used independent of geography, time and space, and it is cost-effective [9, 17].

Immersive VR can produce a visceral feeling of being in a simulated world, a form of spatial immersion called *presence* [9, 18]. Additionally, in virtual interactions, participants' avatars can affect their attitude, perception and behavior in conscious or unconscious matter, which is known as the *protheus effect* [19, 20]. Users tend to align their behavior with the traits of their avatars, adapting to virtual identities that may differ from their real-life selves. For instance, individuals who embody avatars with physically attractive features tend to engage in more confident and assertive behavior, mirroring the perceived traits of their avatars [21].

Today's VR simulation programs often use preset scenarios [22–25]. Standardized patients (SPs) or standardized scenarios have been utilized for procedural skills assessment and nontechnical skills development [17]. Prewritten scenarios offer a structured approach to medical training and assessment. These scenarios are designed to replicate diverse clinical encounters and challenges, providing learners with opportunities to practice decision-making, critical thinking, and teamwork in a controlled environment. Especially in the case of learning strict protocols, such as cardiopulmonary resuscitation and basic life support training, VR simulation programs are apt to create easy standardized assessment options [26, 27].

Current VR simulation programs generally encompass an initial possibility to learn how to train in VR simulation, which is also called onboarding [28]. After this initialization, participants often receive an online briefing describing the clinical scenario and roles, followed by the VR simulation and evaluative debriefing, as depicted in Fig. 2. The briefing and debriefing are generally digitalized, with little interaction with the trainer [26, 27, 29].

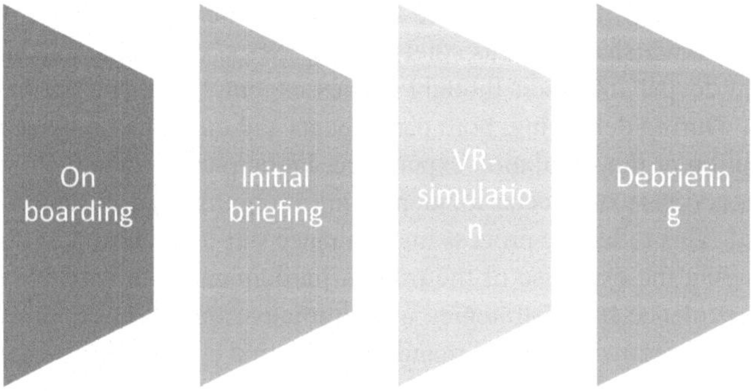

Fig. 2. VR simulation

4 Challenges in Simulation Education

To address the challenges encountered during simulation-based learning experiences, we explore key educational aspects that can impede effective learning experiences and their specific implications for both manikin and VR simulations, as shown in Fig. 3.

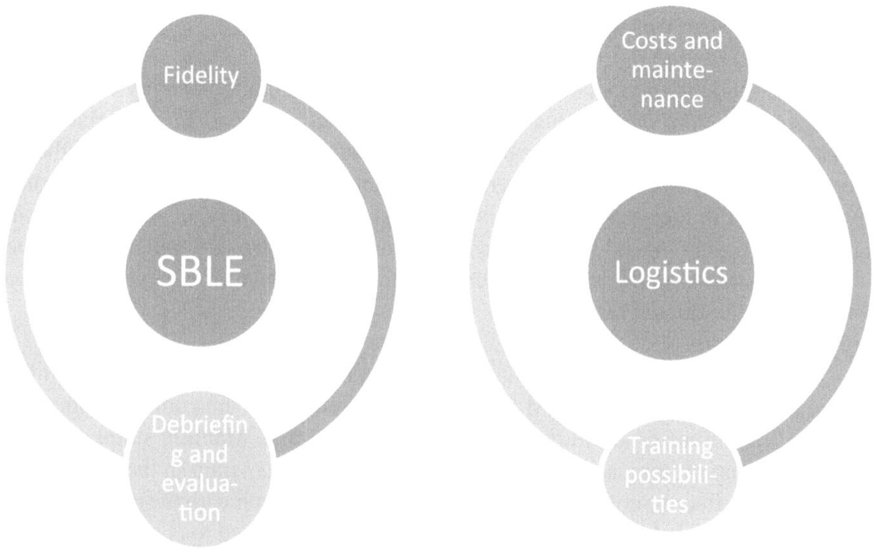

Fig. 3. Schematic view of key elements of simulation-based learning experiences (left) and their logistical challenges (right). SBLE: simulation based learning experience.

4.1 Simulation-Based Learning Experience

By adopting a constructivist framework, realism enhances the active learning process. Both manikin simulation and VR simulation are designed to be as real as possible, enabling the feeling of *presence* and immersion. *Fidelity* refers to how accurately the simulated world reflects the real world. In VR simulations, a higher degree of freedom (DOF) refers to the range of motion or movement. Modern VR allows movement along three axes (x, y, z) and rotation around those axes (6-DOF), providing a high-fidelity, generally well-tolerated system that limits simulation nausea or sickness. Another adaptation to VR can be provided by a sufficient amount of time to adapt to the simulator conditions [30].

Fidelity and Immersion

Higher fidelity simulators are able to provide physical, psychological, and conceptual fidelity [31].

Physical Fidelity

Physical fidelity is facilitated by audio, kinesthetic, and visual elements. Kinesthetic and tactile feedback remains a challenge in both manikin simulation and VR simulation. Technical procedures in simulation often lack the true skin and tissue feedback experienced by real patients. To address this issue, manikin simulations are incorporating more sophisticated materials, such as soft plastic and rubber, to better mimic human skin and tissue. Similarly, VR simulations are advancing with controllers featuring various tactile feedback options and the emergence of hypersensitive haptic gloves to enhance the sense of touch and realism within virtual environments.

Psychological Fidelity

Psychological fidelity pertains to the extent to which a simulation activity elicits emotional responses from participants as if the experience were real. In mannekin-based simulations, achieving psychological fidelity can be challenging due to the inherent artificiality of the physical environment and limitations of simulator manikins, which may hinder full immersion and emotional engagement. Similarly, in VR simulations, challenges in psychological fidelity may arise from issues such as simulation sickness, technical constraints, and limitations in creating truly immersive virtual experiences that evoke genuine emotional responses.

Conceptual Fidelity

Conceptual fidelity reflects the essential principles, behaviors, and relationships inherent in real-world **systems**. As CRM revolves around effective team communication, the conceptual fidelity of group interactions plays a crucial role. Manikin simulation easily enables group training sessions. Up to six participants may train in the same space. This facilitates effective communication, that is, communication with the message being understood as intended. Challenges in manikin simulation related to conceptual fidelity may include limitations in scenario design and the veracity of patient simulators, which may impact the accuracy and authenticity of the simulation. In VR simulations, challenges in conceptual fidelity may stem from the complexity of accurately representing authentic scenarios in virtual environments, requiring careful scenario design and content expert input to ensure fidelity to real-world principles and behaviors. Additionally, CRM training requires high-fidelity avatars that are able to reflect realistic facial expressions and nonverbal and para-verbal elements associated with effective information exchange [32, 33].

Evaluation and Debriefing

Standardized Evaluation

Assessment of clinical competency has evolved, favoring the "performance as competence paradigm" over traditional assessments based solely on theoretical knowledge. Entrusted professional activities (EPAs) translate competence-based training into manageable entities describing the knowledge, skills, and attitudes needed to perform an activity. EPAs have been used for standardized specialist training. Throughout the curriculum, EPAs are being incorporated, and according to explorative studies, they can be effectively evaluated in simulation-based education [34, 35].

However, insufficient research has been conducted on the implementation of EPAs within the curriculum of anesthesia residents in simulation education. Whether EPAs can be successfully used in manikin-based or VR simulations has yet to be validated.

Debriefing

Simulation debriefing is a crucial element of effective learning experiences [9]. Debriefing is a guided reflection on emotional and simulation experiences, learning outcomes, and possible plans for action. It is strongly underpinned by the experiential learning model theory of Kolb, which facilitates reflective observation, abstract conceptualization, and active experimentation [12] (Fig. 4).

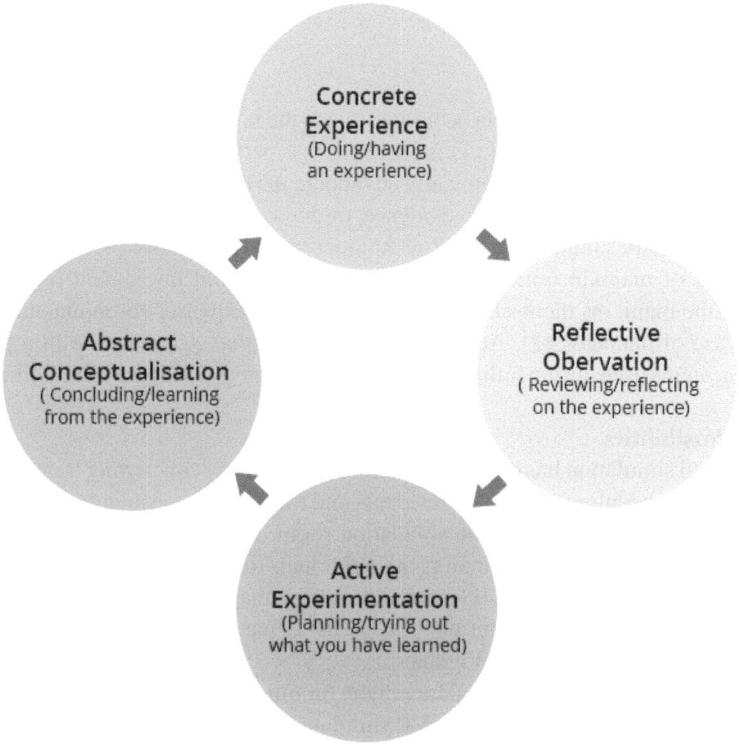

Fig. 4. Kolb's experiential learning model

However, a lack of psychological safety or ineffective debriefing may impede its effectiveness [36]. Debriefing can be facilitator-guided or self-guided, and it can occur during or after a simulation [37]. Debriefing heavily relies on the competence of the debriefer in the case of facilitator-guided debriefing. An effective debriefer acts as a facilitator, guides reflective learning, creates a safe learning environment, and encourages self-reflection. Although several conversational structures may be used, they all highlight the importance of guiding from the beginning, examining events, and summarizing possible learning points [37].

Cognitive aids such as cue cards may facilitate self-assessment practices and are commonly used for anesthesiology training programs.

Traditional manikin-based simulation settings typically involve facilitator-guided in-person debriefing sessions, as participants are physically present. However, VR simulations may involve participants who are not in the same room, who necessitate a virtual debriefing room, or who use self-guided debriefing methods. Research does not prioritize one method over the other, and effective teaching of nontechnical skills can be achieved through formative self-assessment [38].

4.2 Organization

Costs and Maintenance

The adoption of VR technology can be limited due to the high upfront costs of purchasing the hardware and software needed. A cost analysis was performed by Mills et al., indicating that VR training is more economically advantageous than manikin-based simulation [39]. Although there is insufficient evidence to demonstrate clear cost-effectiveness, many research studies indicate that VR techniques are cost-effective [9, 40, 41].

The costs of manikin training are high due to the initial investment required for purchasing the manikins themselves, as well as ongoing expenses for maintenance and repair in case of damage [42]. Additionally, there are costs associated with replacing consumable items used during training sessions, such as medical supplies and equipment.

Training Possibilities

Manikin-based simulation has certain limitations, particularly concerning training logistics. It requires scheduling anesthesia residents and trainers to be available during working hours and securing access to the simulation room, which must remain unoccupied during the training session. Coordinating all individuals to be present in the same place at the same time can be challenging. Hygienic restrictions may further complicate manikin-based simulation, as was evident during the COVID-19 pandemic, when individuals were not allowed to be in a small room together.

In contrast, VR offers the potential for remote training, allowing participants from different centers to participate. Moreover, VR simulations can be conducted independently of real-time participation by trainers because of the availability of asynchronous preset VR simulations. These features alleviate some of the logistical challenges associated with manikin-based training and offer more flexibility in training opportunities.

5 Discussion and Conclusion

5.1 Discussion

In traditional clinical environments, the focus on patient safety often limits student-centered learning opportunities. Simulation activities offer a unique opportunity to tailor learning experiences to students' needs while maintaining a safe environment for skill development. The transferability of learning in simulations is of prime interest to educators. Simulation activities that closely resemble real life add a degree of authenticity that allows learners to use the knowledge and skills they develop in the simulated setting and apply them in the clinical arena.

In discussing the current state of simulation education, it is evident that both manikin-based simulation and VR simulation offer unique advantages and face distinct challenges. Manikin simulation provides a realistic physical representation of team members, fostering hands-on practice of clinical skills and teamwork. However, its fidelity is limited, particularly in terms of psychological and conceptual aspects. Conversely, VR simulation offers an immersive learning environment with significant potential for scalability. The ability to implement standardized scenarios can be advantageous in strictly protocolled procedures such as advanced life support training. However, this standardization may be disadvantageous for more sophisticated training, where the adaptability of the simulation is crucial for optimizing cognitive load and nontechnical skills training, such as CRM training. Additionally, VR simulation involves initial cost barriers and challenges related to simulation sickness and the adaptability of scenarios.

However, educational developers must carefully balance fidelity levels because high fidelity does not always equal better performance [31, 43]. Therefore, thoughtful planning is crucial to avoid the misuse of resources and maintain a balanced cost–benefit ratio in simulation-based education.

The ultimate goal of simulation education should be to improve learning outcomes. Although extensive research indicates that simulation education can enhance residents' skills, knowledge, and behavior, there is limited evidence to suggest that it directly leads to better learning outcomes. In this light, the Kirkpatrick model classifies different levels of educational intervention outcomes and is commonly used to evaluate training [44, 45].

5.2 Conclusion

Simulation-based education, encompassing manikin or VR modalities, is pivotal in medical training, offering a controlled, safe environment for skill acquisition and refinement. While manikin simulations excel in providing realistic, hands-on practice, their limitations in fidelity and high operational costs cannot be overlooked. VR simulations, on the other hand, offer immersive and scalable training solutions, albeit with challenges related to initial costs and achieving high psychological fidelity in team training.

To maximize the benefits of simulation-based education, educational developers must consider the balance between the fidelity of simulations and their practical implementation. High-fidelity simulators should be critically assessed to determine whether they offer significant educational advantages over low-fidelity models. Both mannekin-based simulation and VR simulation can complement each other.

There are several ways in which VR can complement manikin-based simulation education. Collaboration among institutes can expand access to VR training, while standardized scenarios and assessment tools ensure consistency. Integrating simulation training into competency-based frameworks such as EPAs provides a structured approach to skill development.

Future research should focus on integrating simulation training into competency-based frameworks such as EPAs to provide a structured approach to skill development and evaluate improvements in patient outcomes. By leveraging the strengths of both manikin and VR simulations, medical education can be enhanced, which may ultimately lead to improved patient outcomes and more competent anesthesiologists.

Acknowledgments and Disclosure of Interests. The authors have no competing interests to declare that are relevant to the content of this article.

References

1. Yunoki, K., Sakai, T.: The role of simulation training in anesthesiology resident education. J. Anesth. **32**(3), 425–33 (2018)
2. Anthony, G.: Active learning in a constructivist framework. Educ. Stud. Math. **31**(4), 349–69 (1996)
3. Cook, D.A., et al.: Technology-enhanced simulation for health professions education: a systematic review and meta-analysis. JAMA **306**(9), 978–88 (2011)
4. Cannon-Bowers, J.A.: Recent advances in scenario-based training for medical education. Curr. Opin. Anaesthesiol. **21**(6), 784–9 (2008)
5. Gormley, G.J., Murphy, P.: When I say … simulation. Med. Educ. (n/a)
6. Pacheco Granda, F.A., Salik, I.: Simulation training and skill assessment in critical care. StatPearls. Treasure Island (FL) with ineligible companies. Disclosure: Irim Salik declares no relevant financial relationships with ineligible companies: StatPearls Publishing. Copyright © 2023, StatPearls Publishing LLC (2023)
7. Lei, C., Palm, K.: Crisis Resource Management Training in Medical Simulation. StatPearls. Treasure Island (FL): StatPearls Publishing. Copyright © 2024, StatPearls Publishing LLC (2024)
8. Helmreich, R.L., Merritt, A.C., Wilhelm, J.A.: The evolution of crew resource management training in commercial aviation. Int. J. Aviat. Psychol. **9**(1), 19–32 (1999)
9. Institute of Medicine Committee on Quality of Health Care in A. In: Kohn, L.T., Corrigan, J.M., Donaldson, M.S. (eds.) To Err is Human: Building a Safer Health System. National Academies Press (US), Washington (DC). Copyright 2000 by the National Academy of Sciences. All rights reserved (2000)
10. Weller, J., et al.: Building the evidence on simulation validity: comparison of anesthesiologists' communication patterns in real and simulated cases. Anesthesiology **120**(1), 142–8 (2014)
11. Modell, H.I.: Preparing students to participate in an active learning environment. Am. J. Physiol. **270**(6 Pt 3), S69-77 (1996)
12. Kolb, D.: Experiential Learning: Experience as the Source of Learning and Development: Eaglewood Cliffs (1984)
13. Hughes, P.G., Hughes, K.E.: Briefing Prior to Simulation Activity. StatPearls. Treasure Island (FL): StatPearls Publishing. Copyright © 2024, StatPearls Publishing LLC (2024)
14. Reedy, G.B.: Using cognitive load theory to inform simulation design and practice. Clin. Simul. Nurs. **11**(8), 355–60 (2015)
15. Amitai, Z., Small, S.D., Wolpe, P.R.: Patient safety and simulation-based medical education. Med Teach. **22**(5), 489–495 (2000)
16. Henriksen, K.: Advances in patient safety. In: Henriksen, K., Battles, J.B., Keyes, M.A., Grady, M.L. (eds.) Advances in Patient Safety: New Directions and Alternative Approaches (Vol 3: Performance and Tools). Agency for Healthcare Research and Quality (US), Rockville, MD (2008)
17. Pottle, J.: Virtual reality and the transformation of medical education. Future Healthc J. **6**(3), 181–5 (2019)
18. Liaw, S.Y., Chan, S.W.-C., Chen, F.-G., Hooi, S.C., Siau, C.: Comparison of virtual patient simulation with mannequin-based simulation for improving clinical performances in assessing and managing clinical deterioration: randomized controlled trial. J. Med. Internet Res. **16**(9), e214 (2014)

19. Fuhrt, B.: Immersive virtual reality. In: Furht, B. (ed.) Encyclopedia of Multimedia, pp. 345–346. Springer, Boston, MA, US (2008)
20. Bian, Y., Zhou, C., Tian, Y., Wang, P., Gao, F. (eds.): The Proteus effect: influence of avatar appearance on social interaction in virtual environments. In: HCI (2015)
21. Navarro, J., Peña, J., Cebolla, A., Baños, R.: Can avatar appearance influence physical activity? User-avatar similarity and Proteus effects on cardiac frequency and step counts. Health Commun. **37**(2), 222–9 (2022)
22. Fox, J., Bailenson, J., Tricase, L.: The embodiment of sexualized virtual selves: the Proteus effect and experiences of self-objectification via avatars. Comput. Hum. Behav. **29**, 930–8 (2013)
23. Bracq, M.S., et al.: Learning procedural skills with a virtual reality simulator: an acceptability study. Nurse Educ. Today **79**, 153–60 (2019)
24. Brammer, S.V., Regan, S.L., Collins, C.M., Gillespie, G.L.: Developing innovative virtual reality simulations to increase health care providers' understanding of social determinants of health. J. Continuing Educ. Health Prof. **42**(1) (2022)
25. Macnamara, A.F., Bird, K., Rigby, A., Sathyapalan, T., Hepburn, D.: High-fidelity simulation and virtual reality: an evaluation of medical students' experiences. BMJ Simul. Technol. Enhanc. Learn. **7**(6), 528–35 (2021)
26. Frederiksen, J.G., et al.: Cognitive load and performance in immersive virtual reality versus conventional virtual reality simulation training of laparoscopic surgery: a randomized trial. Surg. Endosc. **34**(3), 1244–52 (2020)
27. Sadeghi, A.H., et al.: Virtual reality simulation training for cardiopulmonary resuscitation after cardiac surgery: face and content validity study. JMIR Serious Games **10**(1), e30456 (2022)
28. Peek, J.J., et al.: Virtual reality simulator versus conventional advanced life support training for cardiopulmonary resuscitation post-cardiac surgery: a randomized controlled trial. J. Cardiovasc. Dev. Dis. **10**(2) (2023)
29. Gupta, S., Wilcocks, K., Matava, C., Wiegelmann, J., Kaustov, L., Alam, F.: Creating a successful virtual reality-based medical simulation environment: tutorial. JMIR Med. Educ. **9**, e41090 (2023)
30. Abelson, J.S., Silverman, E., Banfelder, J., Naides, A., Costa, R., Dakin, G.: Virtual operating room for team training in surgery. Am. J. Surg. **210**(3), 585–90 (2015)
31. Dużmańska, N., Strojny, P., Strojny, A.: Can simulator sickness be avoided? A review on temporal aspects of simulator sickness. Front. Psychol. **9**, 2132 (2018)
32. Carey, J.M., Rossler, K.: The How When Why of High Fidelity Simulation. StatPearls. Treasure Island (FL): StatPearls Publishing. Copyright © 2024, StatPearls Publishing LLC (2024)
33. Dunbar, R.I.M., Robledo, J.P., Tamarit, I., Cross, I., Smith, E.: Nonverbal auditory cues allow relationship quality to be inferred during conversations. J. Nonverbal Behav. **46**(1), 1–18 (2022)
34. McDonnell, R., Breidt, M., Bülthoff, H.H.: Render Me real? Investigating the effect of render style on the perception of animated virtual humans. ACM Trans. Graph. **31**(4) (2012). Article 91
35. Herrigel, D.J., Donovan, C., Goodman, E., Pradhan, A., Bridgeman, M.M., Hogshire, L., et al.: Simulation as a platform for development of entrustable professional activities: a modular, longitudinal approach. Cureus **12**(10), e11098 (2020)
36. Dwyer, T., Wadey, V., Archibald, D., Kraemer, W., Shantz, J.S., Townley, J., et al.: Cognitive and psychomotor entrustable professional activities: can simulators help assess competency in trainees? Clin. Orthop. Relat. Res. **474**(4), 926–34 (2016)
37. Nyoni, C.N., et al.: Health sciences educator's simulation debriefing practice needs: a mixed methods study. J. Educ. Health Promot. **12**, 55 (2023)

38. Kolbe, M., Grande, B., Spahn, D.R.: Briefing and debriefing during simulation-based training and beyond: content, structure, attitude and setting. Best Pract. Res. Clin. Anaesthesiol. **29**(1), 87–96 (2015)
39. Sawyer, T., Eppich, W., Brett-Fleegler, M., Grant, V., Cheng, A.: More than one way to debrief: a critical review of healthcare simulation debriefing methods. Simul. Healthc. **11**(3) (2016)
40. Boet, S., Bould, M.D., Bruppacher, H.R., Desjardins, F., Chandra, D.B., Naik, V.N.: Looking in the mirror: self-debriefing versus instructor debriefing for simulated crises. Crit. Care Med. **39**(6), 1377–81 (2011)
41. Mills, B., et al.: Virtual reality triage training can provide comparable simulation efficacy for paramedicine students compared to live simulation-based scenarios. Prehosp. Emerg. Care **24**(4), 525–36 (2020)
42. Abbas, J.R., et al.: Virtual reality in simulation-based emergency skills training: a systematic review with a narrative synthesis. Resusc. Plus. **16**, 100484 (2023)
43. Liaw, S.Y., et al.: Artificial intelligence in virtual reality simulation for interprofessional communication training: Mixed method study. Nurse Educ. Today **122**, 105718 (2023)
44. Savoldelli, G.L., Naik, V.N., Hamstra, S.J., Morgan, P.J.: Les barriéres à l'utilisation de la formation basée sur simulateur. Can. J. Anesth. **52**(9), 944–50 (2005)
45. Chandra, D.B., Savoldelli, G.L., Joo, H.S., Weiss, I.D., Naik, V.N.: Fiberoptic oral intubation: the effect of model fidelity on training for transfer to patient care. Anesthesiology **109**(6), 1007–13 (2008)
46. Falletta, S.V.: Evaluating training programs: the four levels. In: Kirkpatrick, D.L. (ed.) Berrett-Koehler Publishers, San Francisco, CA (1996). 229 p. Am. J. Eval. **19**(2), 259–261 (1998)
47. Smidt, A., Balandin, S., Sigafoos, J., Reed, V.A.: The Kirkpatrick model: a useful tool for evaluating training outcomes. J. Intellect. Dev. Disabil. **34**(3), 266–74 (2009)

Methods for Optimizing Diagnostic Data for Patients with a Given Condition

Hakimjon Zaynidinov[1], Nandan Hirpara[2], Dhananjay Singh[2](✉), Sarvar Makhmudjanov[1], and Akmal Aliqulov[1]

[1] Tashkent University of Information Technologies named after Muhammad al Khwarizmi, Tashkent, Uzbekistan

[2] Penn State University, University Park, PA 16802, USA

dsingh@psu.edu

Abstract. Advances in medical diagnostics necessitate innovative approaches to address the complexity of modern healthcare. This study introduces an automated diagnostic information system designed to enhance the accuracy and efficiency of gastroenterological disease diagnosis. Leveraging cosine similarity, the system analyzes high-dimensional signal data to identify relevant patient cases, achieving a diagnostic accuracy of 92% outperforming traditional metrics like Euclidean and Manhattan distances. The system processes diagnostic queries in real-time with an average runtime of 1.2 s, highlighting its clinical utility. By addressing limitations in existing systems, such as the lack of comparative diagnostics and automation, the proposed solution reduces physician workload and streamline decision-making. Its modular architecture ensures adaptability to new data, paving the way for continuous improvements. This research demonstrates the potential of cosine similarity in healthcare diagnostics, offering a scalable and efficient solution to improve patient outcomes and revolutionize medical decision-making.

Keywords: Automated Information Systems · Cosine Similarity · Gastroenterological Signals · Diagnostic Efficiency · Medical Decision-Making

1 Introduction

Automated information and decision-support systems are becoming indispensable in modern healthcare, offering innovative solutions to manage the increasing complexity of medical diagnostics. These systems play a critical role in assisting healthcare professionals by integrating specialized databases, libraries, search engines, telemedicine capabilities, and remote monitoring tools. As diagnostic and treatment processes grow more complex, the demand for systems that can provide accurate, real-time, and evidence-based support continues to rise.

Advancements in computer technology have enabled the development of next-generation diagnostic systems, which move beyond traditional static recommendations to dynamic, data-driven decision-making [1, 2]. These systems can analyze vast amounts of patient data to provide tailored medical recommendations, addressing complex clinical scenarios that may not be captured by conventional diagnostic protocols [3]. Their

D. Singh et al. (Eds.): IHCI 2024, LNCS 15557, pp. 41–53, 2025.
https://doi.org/10.1007/978-3-031-88705-5_4

application spans various medical fields, with significant potential to reduce diagnostic errors, improve treatment strategies, and enhance patient outcomes.

Gastroenterological diseases pose unique diagnostic challenges due to the subtle and often overlapping nature of their symptoms. Traditional diagnostic workflows heavily rely on manual signal analysis and expert interpretation, which are time-intensive and prone to human error [6, 7]. Automated diagnostic systems tailored for gastroenterological signal analysis provide a promising solution by efficiently processing and analyzing patient data to assist physicians in diagnosing conditions such as gastritis, Crohn's disease, and ulcerative colitis [8, 11].

This study addresses the limitations of existing diagnostic systems, including the absence of comparative diagnostic capabilities, limited automation, and underutilization of advanced similarity metrics for signal analysis [9, 12]. By introducing a diagnostic module that leverages cosine similarity, the proposed system automates the identification of relevant patient cases, enabling more accurate initial diagnoses. Furthermore, the modular design of the system ensures adaptability, allowing continuous updates as new patient data becomes available [5].

The aim of this paper is to present the development and evaluation of this automated diagnostic system, highlighting its accuracy, efficiency, and potential for real-world clinical applications. By bridging the gap between manual and automated diagnostics, this study contributes to the advancement of data-driven medical decision-making and paves the way for future innovations in healthcare technology.

2 System Design and Methodology

2.1 Overview of Previous Studies

Previous research has delved deeply into the recording and analysis of gastroenterological signals, utilizing specialized devices and algorithms to enhance diagnostic processes [4]. These studies typically emphasize a systematic workflow that begins with the use of advanced devices to capture patient signals accurately. Following this, custom algorithms are applied to process raw data, filtering out noise and improving overall data quality [13]. Once processed, the filtered data is stored in structured databases, ensuring it is readily accessible for future analysis [9]. This organized repository of information serves as a critical resource for supporting expert doctors in diagnosing gastroenterological conditions. The general structure of these systems, as explored in earlier studies, encompasses this entire process—from data recording and filtering to storage and diagnostics—and is visually represented in Fig. 1, which highlights the interconnected steps involved in managing gastroenterological signal data.

2.2 Gaps Identified in Previous Research

While previous systems have achieved notable successes, they still exhibit significant limitations that hinder their effectiveness in modern medical diagnostics. One key shortcoming is the absence of mechanisms for comparative diagnostics, which prevents these systems from efficiently matching new patient data with existing diagnostic records

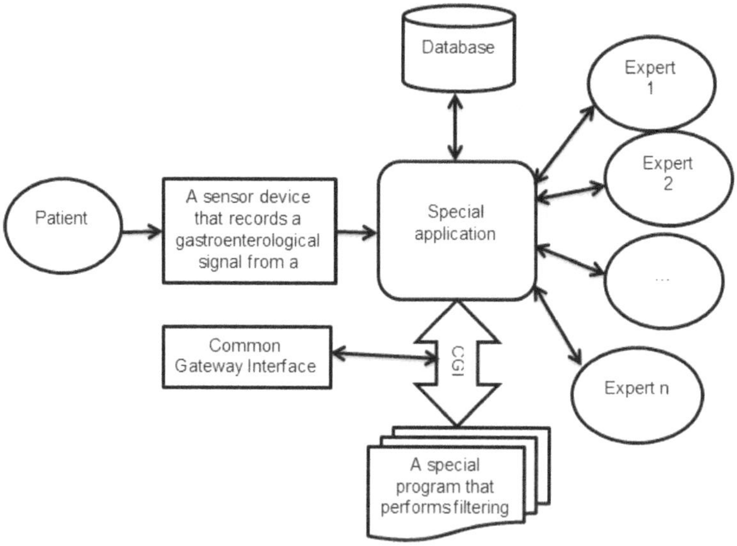

Fig. 1. Structure of the information system for recording, processing and storing gastroenterological signals.

stored in the database. Furthermore, the initial diagnostic process remains largely manual, heavily reliant on expert intervention and thus limiting the scalability and speed of decision-making. Another critical limitation lies in the restricted use of advanced algorithms. Most systems fail to incorporate sophisticated similarity metrics, such as cosine similarity, which are essential for optimized data analysis and comparison in high-dimensional datasets. These gaps collectively underscore the pressing need for innovative systems that can automate diagnostic workflows, enhance decision-making processes, and leverage advanced computational methods to improve accuracy and efficiency.

2.3 Proposed Enhancements to System Design

To address the gaps identified in previous systems, this study introduces an enhanced system architecture that incorporates a new diagnostic module. This module is designed to improve diagnostic accuracy and efficiency through three core features. First, the system employs similarity matching using advanced algorithms, such as cosine similarity, to identify cases in the database with signals resembling those of newly introduced patients. Second, it automates the initial diagnostic process by suggesting potential diagnoses based on the closest-matching cases, which are subsequently validated by a team of expert physicians. Finally, the system's dynamic architecture ensures continuous database updates with new patient cases and diagnostic outcomes, allowing for sustained accuracy and scalability over time.

The updated system architecture is visually represented in Fig. 2, which illustrates the integration of various components. At its core is a special application that serves as the processing hub. This application interfaces with several key modules:

- **Sensor Devices:** These capture gastroenterological signals from patients and relay them to the system for analysis.
- **Filtering Program:** A dedicated program processes raw signal data, removing noise and enhancing its quality before storage.
- **Database Module:** Filtered data is stored in a structured database that can be accessed for comparisons and diagnostics.
- **Proximity Check Module:** This module calculates the similarity between new and existing patient data using cosine similarity, enabling the system to identify the closest-matching cases efficiently.

The special application connects with a **Common Gateway Interface (CGI)** that facilitates seamless data flow between modules and ensures real-time interaction with expert physicians. Experts, represented in the diagram, validate the system-generated diagnostic suggestions, reinforcing the collaborative nature of the workflow.

This interconnected system architecture exemplifies a streamlined approach to diagnostic decision-making. By leveraging automation, advanced algorithms, and expert validation, it effectively bridges the gap between manual and automated processes, thereby enhancing diagnostic reliability and efficiency.

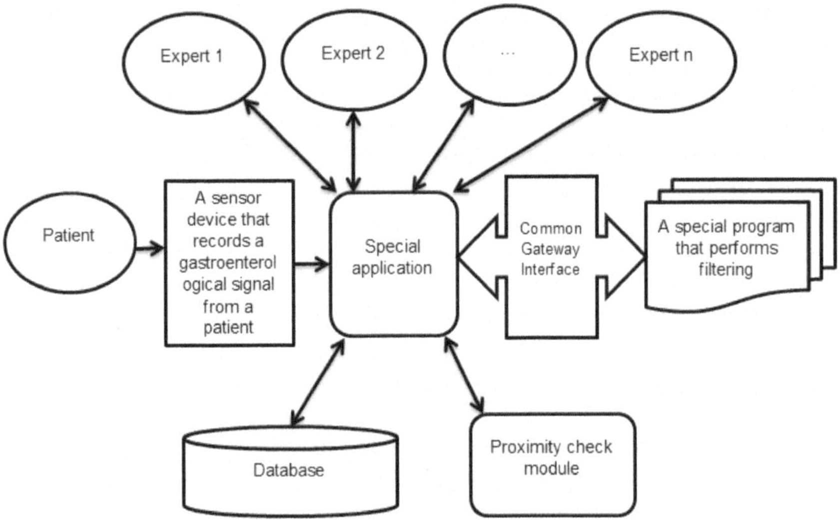

Fig. 2. Structure of the information system that helps patients diagnose.

Additionally, Table 1 provides an example of data for a newly introduced patient, illustrating how the system processes and integrates new patient information.

Table 1. New patient information being loaded.

№	Name	Age	Gender	Unfiltered	Colon	Stomach	Small intestine	Duodenum	Diagnosis
1	Botirov O.	32	male	S1[…]	S2[…]	S3[…]	S4[…]	S5[…]	-

2.4 Updated System Architecture

The enhanced system introduces a systematic process for analyzing new patient data. Filtered signal data from new patients is compared against data from existing cases in the database. This comparison uses advanced similarity metrics, such as cosine similarity, to identify the closest matches.

Table 2 presents an example of how patient information is stored in the system database, displaying how filtered signal data and diagnoses are systematically organized.

Table 2. Table where patient information is stored.

№	Name	Age	Gender	Unfiltered	Colon	Stomach	Small intestine	Duodenum	Diagnosis
1	Azimov O. U.S.	54	male	S[…]	S2[...]	S3[...]	S4[...]	S5[...]	Gastritis
2	Jdanov, U.S.A.	42	male	S[…]	S2[...]	S3[...]	S4[...]	S5[...]	Ulcerative colitis
3	Toirov Y.J.	72	male	S[…]	S2[...]	S3[...]	S4[...]	S5[...]	Crohn's disease
4	Ahalaya A.F.	50	male	S[…]	S2[...]	S3[...]	S4[...]	S5[...]	Enterit
5	Amelina K.R.	42	femalc	S[…]	S2[...]	S3[...]	S4[...]	S5[...]	Diarrhea
6	Amonova O.P.	26	female	S[…]	S2[...]	S3[...]	S4[...]	S5[...]	Piloric Stenosis
7	Andreano V.K.L.	38	male	S[…]	S2[...]	S3[...]	S4[...]	S5[...]	Atropic Gastritis
8	Azamova E.D.	61	male	S[…]	S2[...]	S3[...]	S4[...]	S5[...]	Cylinder
9	Azizova N.Sh.	39	female	S[…]	S2[...]	S3[...]	S4[...]	S5[...]	Helicobacter pylori
10	Baqoev A.S.	27	male	S[…]	S2[...]	S3[...]	S4[...]	S5[...]	Epstein-Barr virus

Figure 3 illustrates how distance measurements, such as cosine similarity, are used to evaluate the similarity between filtered signal data from new patients and existing cases.

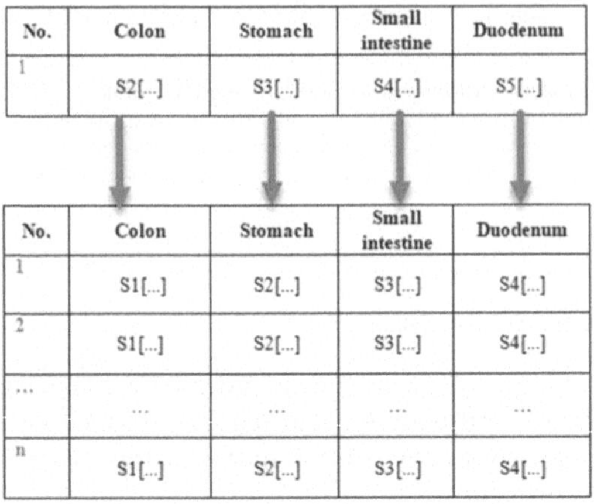

Fig. 3. Distance measurements between filtered signal data from a newly entered patient and filtered signal data from patients in a patient's schedule.

Each data in the corresponding signal data is viewed as one point, and their relationship is compared.

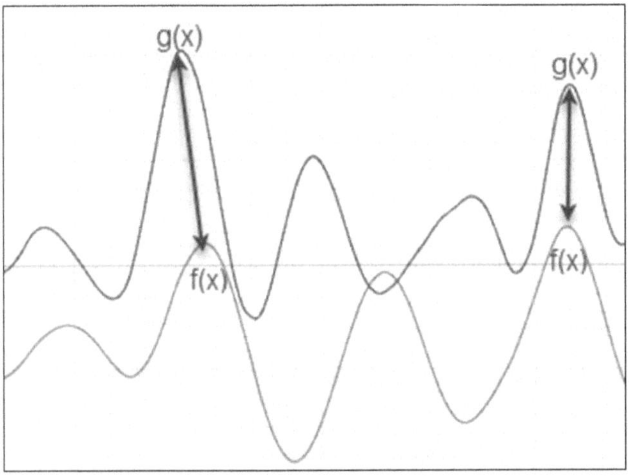

Fig. 4. Distance between two points.

The new module, which will be added to the information system that helps patients diagnose, performs the above task, namely, to verify the closeness of each patient stored in the patient's schedule with the filtered signal data of the newly entered patient.

Further elaborating on the comparison process, Fig. 4 provides a graphical representation of the relationship between the data of new and existing patients. In this figure, **g(x)** represents the signal data of existing patients stored in the database, while **f(x)** corresponds to the signal data of the newly entered patient. The figure shows how cosine similarity is used to measure the distance between these two sets of data points. The smaller the distance, the greater the similarity, enabling the system to determine the most likely diagnosis for the new patient.

This graphical representation emphasizes the importance of directional similarity rather than magnitude, which is particularly effective for high-dimensional datasets like medical signal data. By comparing g(x) and f(x), the system ensures that the diagnostic suggestions are precise and closely aligned with real-world patient patterns.

In summary, the updated system architecture combines structured data storage, advanced similarity metrics, and visualized data comparisons to deliver a robust diagnostic tool. This architecture not only improves the accuracy and efficiency of the diagnostic process but also sets a foundation for future enhancements and scalability.

3 Algorithms and Experimental Results

3.1 Numerical Analysis

The following distance metrics were considered:

d**:** The distance between two points.

x_i**:** The i -th component of the first data point/vector.

y_i**:** The i -th component of the second data point/vector.

n**:** The total number of dimensions or features.

p**:** The order of the distance metric (e.g., p = 1 gives Manhattan distance, p = 2 gives Euclidean distance).

- **Euclidean Distance:** Measures the straight-line distance between two points in multidimensional space. While simple and widely used, it is sensitive to outliers and scale differences.

$$d = \sqrt{\sum_{i=1}^{n}(x_i - y_i)^2} \tag{1}$$

- **Manhattan Distance:** Calculates distance as the sum of absolute differences, often suitable for discrete data or grid-based systems.

$$d = \sum_{i=1}^{n}(x_i - y_i)^2 \tag{2}$$

- **Cosine Similarity:** Evaluates the angle between two vectors, emphasizing directional similarity rather than magnitude. It is effective for high-dimensional and normalized data, making it particularly suitable for signal data.

$$\mathrm{d} = \frac{\sum_{i=1}^{n} X_i * Y_i}{\sqrt{\sum_{i=1}^{n} X_i^2} * \sqrt{\sum_{i=1}^{n} Y_i^2}} \tag{3}$$

- **Minkowski Distance:** A generalization of both Euclidean and Manhattan distances, offering flexibility through its parameterization.

$$d = \sqrt[p]{\sum_{i=1}^{n} (x_i - y_i)^p} \tag{4}$$

The results of the following four methods were compared and the results were obtained from the following view (Fig. 5):

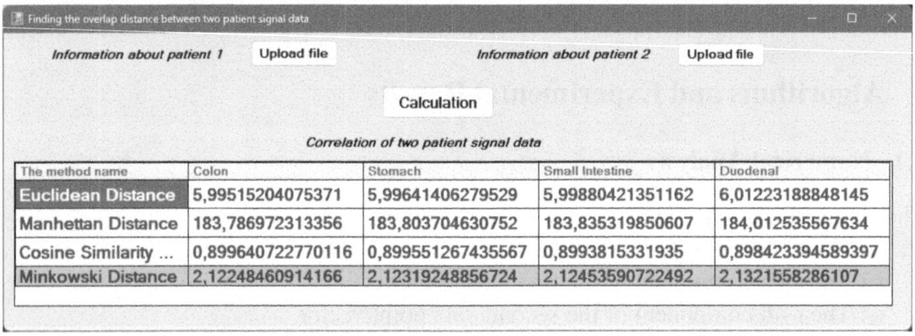

Fig. 5. A comparison of the methods of finding the distance between the two sets.

The program developed for this study evaluates various distance metrics to enhance the functionality of the proposed module within the diagnostic information system. Among these metrics, cosine similarity was identified as the most effective for comparing filtered signal data. Using this metric, the system filters and compares signal data from patients stored in the database, producing a two-dimensional matrix as shown in Fig. 2

Table 3. Results with filtered signal data from the newly entered patient using the distance of corresponding signal data filtered by each patient in the patient schedule to be similar to cosines.

№	Colon Distance 1	Stomach Distance 2	Small intestine Distance 3	Duodenum Distance 4
1	d11	d12	d13	d14
2	d21	d22	d23	d24
...	...	...	...	...
4	dn1	dn2	dn3	dn4

Each d value in Table 3 is found using two sets of elements using the cosine-like distance formula (3). For example, d11 is determined by the intestinal signal data recorded from a new patient being entered into the system to determine the value, using the intestinal signal data of a patient who is included in the system and is 1st in the list and has diagnostic data. Signal data recorded from patients in the system is saved in separate files. To read the signal data in this file from the file and find the d value using them, you first need to absorb this data into a mass.

In this order, the following X mass is produced in Fig. 6:

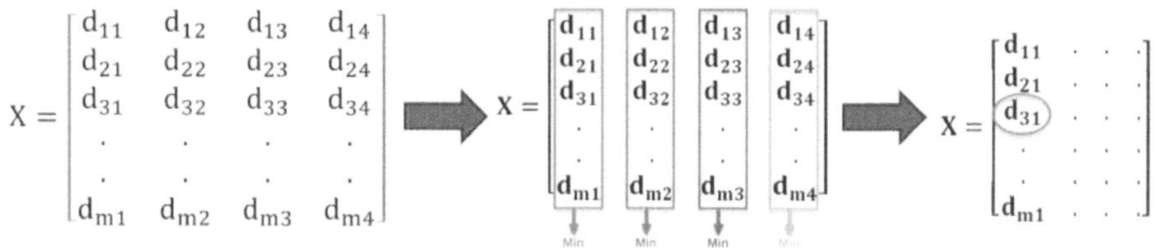

Fig. 6. The resulting two-dimensional mass found with minimums approach.

These minimum values found determine which patient belongs to, and the diagnosis placed on that patient is considered the closest diagnosis to the newly introduced patient.

Selection of Cosine Similarity

Cosine similarity was selected as the optimal metric for this study due to its ability to minimize the impact of variations in signal magnitude and accurately capture the relationship between signal patterns, particularly in high-dimensional datasets [10].

Diagnostic Accuracy

The diagnostic system achieved an overall accuracy of **92%**, with the cosine similarity metric contributing significantly to the system's success in identifying the correct initial diagnosis. Table 4 presents the accuracy metrics for the system using different distance measurement techniques.

Table 4. Accuracy metrics for different distance measurement techniques

Distance Metric	Accuracy (%)	Precision (%)	Recall (%)	F1-Score (%)
Euclidean Distance	80	78	76	77
Manhattan Distance	84	82	80	81
Cosine Similarity	**92**	**90**	**91**	**90.5**
Minkowski Distance	86	85	84	84.5

Runtime Performance

Cosine similarity demonstrated superior computational efficiency compared to other metrics, as shown in Table 5. The system successfully processed diagnostic queries in real-time, with an average runtime of **1.2 s per query** for a database of 100 patients.

Table 5. Runtime performance comparison for distance metrics

Distance Metric	Average Runtime (seconds)
Euclidean Distance	1.8
Manhattan Distance	1.6
Cosine Similarity	1.2
Minkowski Distance	1.5

3.2 Algorithm for Similarity Calculation

The process of identifying the closest-matching patient cases in the database is powered by a custom algorithm, as shown in Fig. 7. This algorithm performs the following steps:

- **Data Preprocessing:** Filters raw signal data to remove noise and normalize values.
- **Similarity Calculation:** For each new patient, the algorithm calculates cosine similarity between their filtered signal data and stored data in the patient database.
- **Ranking Matches:** The similarity scores are ranked in ascending order, with the smallest cosine distance indicating the most similar case.
- **Candidate Selection:** Based on the ranked results, the algorithm identifies up to four closest matches, which are reviewed by expert physicians.

This algorithm ensures efficient and reliable comparison of patient data, leveraging the strengths of cosine similarity to identify diagnostic patterns. By automating the similarity calculation, the diagnostic module reduces the burden on physicians, enabling faster and more accurate initial diagnoses.

3.3 Justification and Practical Application

With the use of Cosine Similarly and the generated algorithm, it is determined which patient has the same diagnosis as the new patient in Table 2. In columns 2, 3, and 4, respectively, the minimum values are found, and the results of the diagnosis of the number of patients they are presenting are also recommended, and in general, the system presents the results of the diagnosis of a minimum of one patient to the newly introduced patient, and the results of the diagnosis of a maximum of 4 patients. The final decision will be made by a special group of experts involved in the system.

Furthermore, Fig. 4 illustrates the process of calculating cosine similarity, while Fig. 3 compares sample patient signals to demonstrate the effectiveness of the metric. The integration of cosine similarity enhances diagnostic precision and streamlines the identification of closely matching cases.

The digital signal data recorded from the patient, which is newly introduced into the information system that helps patients diagnose, is filtered using a module that performs a special filtering in the system and compares with the information that each patient in the patients' schedule in the system has a diagnostic outcome. The comparison is conducted by a newly added proximity check module to the system. The patient closest to the newly introduced patient's signal data will be identified and diagnosed with the

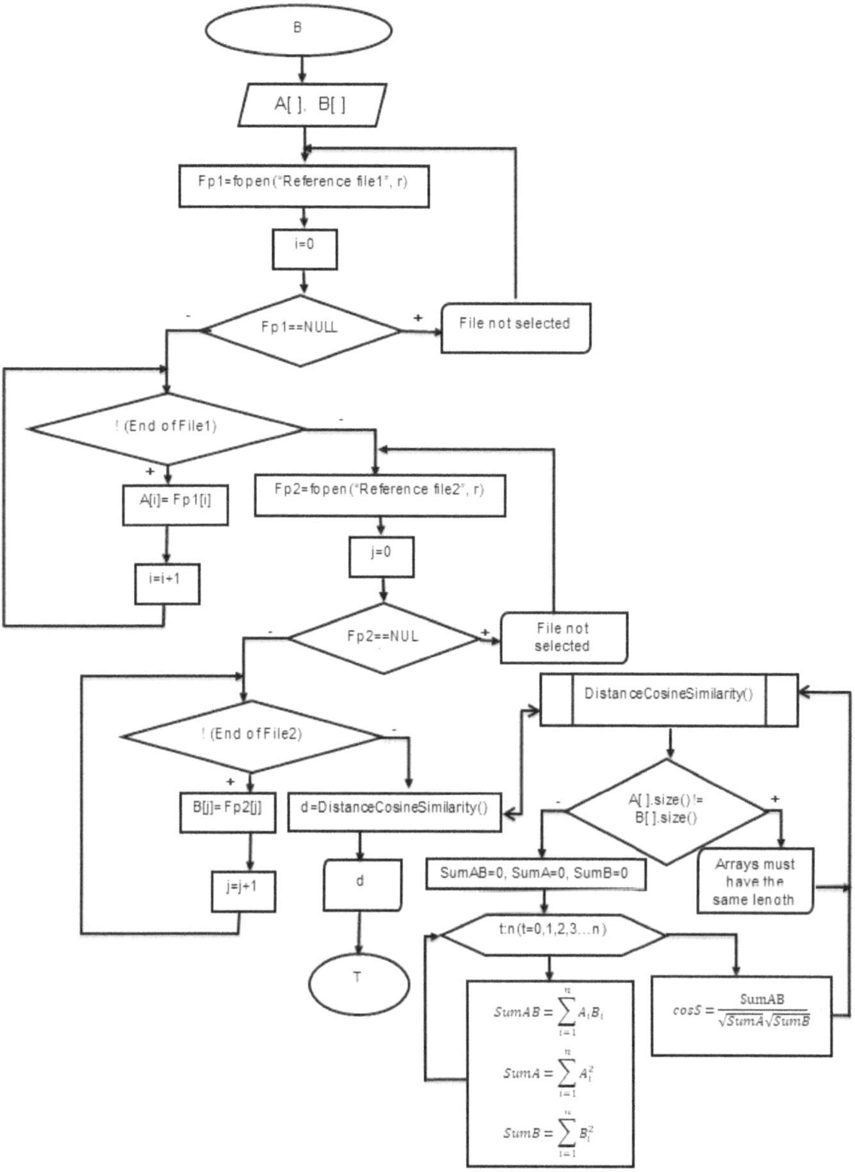

Fig. 7. An algorithm for finding the distance between two sets of Cosines similarity.

patient based on this information, but the final decision will be made by a team of expert doctors involved in the system (Fig. 8).

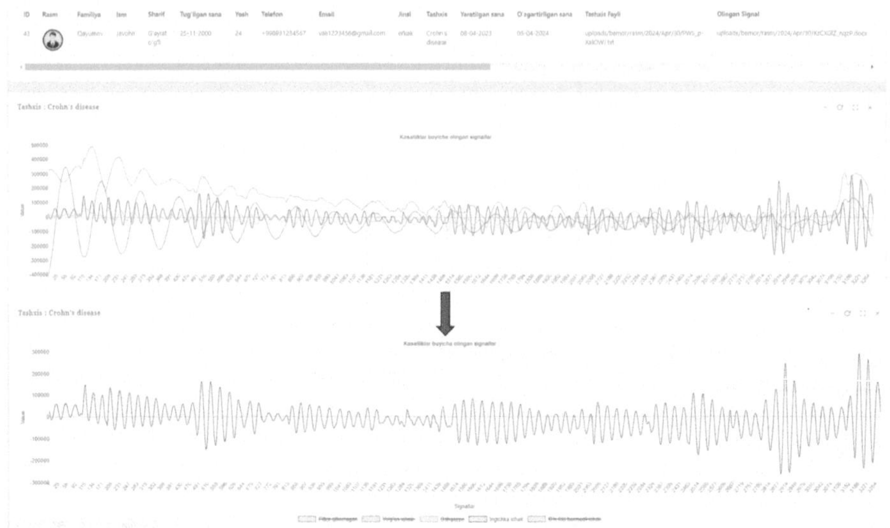

Fig. 8. Preliminary diagnosis of a new patient, which is included in the information system that helps patients diagnoses.

4 Final Remarks

This study introduces a comprehensive diagnostic information system that leverages digital technologies and cosine similarity to enhance gastroenterological diagnostics. By integrating hardware, software, memory, and medical technologies into a single complex model, the system addresses key gaps in automation and comparative diagnostics, achieving a diagnostic accuracy of 92% while processing queries in real-time (1.2 s). This system elevates the diagnostic process by using signals and stored data to identify cases closest to the patient under consideration, enabling efficient initial diagnoses. The hybrid data analysis approach significantly improves decision-making efficiency in distributed systems. Modules developed to determine similarity between datasets and the database facilitate this hybrid approach, with cosine similarity proving to be the most effective metric for comparing high-dimensional signal data. This innovation reduces the burden on physicians by automating routine comparisons while ensuring expert validation for final diagnoses. Additionally, the study demonstrates the transformative potential of digital and mathematical technologies in healthcare, enabling more accurate, scalable, and efficient diagnostic tools. Practical results include the development of software applications to enhance diagnostic effectiveness in distributed systems, enriching patient knowledge bases and addressing global healthcare challenges. Challenges such as data storage for analyzed patients were also resolved, further supporting the system's scalability and adaptability. While promising, future efforts should expand the dataset and integrate advanced machine learning techniques for enhanced signal preprocessing and system robustness. Overall, this research highlights how modern computational methods can revolutionize diagnostics, providing a scalable framework to improve patient outcomes and streamline healthcare decision-making.

References

1. Zaynidinov, H., Makhmudjanov, S., Bismillov, F., Singh, D.: IoT-enabled mobile device for electrogastrography signal processing. In: 12th International Conference on Intelligent Human Computer Interaction (IHCI), Daegu, South Korea, pp. 346–356 (2020). https://doi.org/10.1007/978-3-030-68452-5_36
2. Harerimana, G., Jang, B., Kim, J.W., Park, H.K.: Health big data analytics: a technology survey. IEEE Access **6**, 65661–65678 (2018). https://doi.org/10.1109/ACCESS.2018.2878254
3. Palanisamy, V., Thirunavukarasu, R.: Implications of big data analytics in developing healthcare frameworks – a review. J. King Saud Univ. Comput. Inf. Sci. **31**(4), 415–425 (2019). https://doi.org/10.1016/j.jksuci.2017.12.007
4. Jindal, A., Dua, A., Kumar, N., Das, A.K., Vasilakos, A.V., Rodrigues, J.J.P.C.: Providing healthcare-as-a-service using fuzzy rule-based big data analytics in cloud computing. IEEE J. Biomed. Health Inform. **22**(5), 1367–1378 (2018). https://doi.org/10.1109/JBHI.2018.2799198
5. Nixon, M.S., Aguado, A.S.: Distance, classification, and learning. In: Feature Extraction and Image Processing for Computer Vision, 4th edn. Academic Press (2020). https://doi.org/10.1016/B978-0-12-814976-8.00012-9
6. Lin, M.C., Dung, L.R.: A subsample-based low-power image compressor for capsule gastrointestinal endoscopy. EURASIP J. Adv. Sig. Process. (2011). https://doi.org/10.1155/2011/257095
7. Mohapatra, S., Pati, G.K., Mishra, M., Swarnkar, T.: Gastrointestinal abnormality detection and classification using empirical wavelet transform and deep convolutional neural network from endoscopic images. Ain Shams Eng. J. (2023). https://doi.org/10.1016/j.asej.2022.08.011
8. Hause, R.J., Pritchard, C.C., Shendure, J., Salipante, S.J.: Classification and characterization of microsatellite instability across 18 cancer types. Nat. Med. (2016). https://doi.org/10.1038/nm.4191
9. Butt, M.H.F., Li, J.P., Ji, J., Riaz, W., Anwar, N., Butt, F.F.: Intelligent tumor tissue classification for hybrid healthcare units. Front. Med. (2024). https://doi.org/10.3389/fmed.2024.1385524
10. Barbosa, D.J.C., Ramos, J., Correia, J.H.: Automatic detection of small bowel tumors in capsule endoscopy based on color curvelet covariance statistical texture descriptors. In: Proceedings of the Annual International Conference of the IEEE Engineering in Medicine and Biology Society (EMBC) (2009). http://dei-s2.dei.uminho.pt/pessoas/higino/pampus/EMBC_09_DBarbosa.pdf
11. Khan, M.A., Sahar, N., Khan, W.Z., Alhaisoni, M., Tariq, U.: GestroNet: a framework of saliency estimation and optimal deep learning features based gastrointestinal diseases detection and classification. Diagnostics **12**(11), 2718 (2022). https://doi.org/10.3390/diagnostics12112718
12. Guo, C., Jiang, Z., He, H., Liao, Y., Zhang, D.: Wrist pulse signal acquisition and analysis for disease diagnosis: a review. Comput. Biol. Med. **145**, 105447 (2022). https://doi.org/10.1016/j.compbiomed.2022.105447
13. Meyer-Baese, A., Schmid, V.J.: Pattern Recognition and Signal Analysis in Medical Imaging. CRC Press (2014). https://books.google.com/books?hl=en&id=dPSAAgAAQBAJ

An Exploration of Phantom Touch as a Design Aspect to Enhance Wellbeing and Safety in Social VR

Ágnes Karolina Bakk(✉)

Moholy-Nagy University of Art and Design, Zugligeti út 9-21, Budapest, Hungary
bakk@mome.hu

1 Introduction

Humans have a need for social physical contact and the act of touching each other is an important aspect of it. Touch can "generate positive emotions and modulat[e] negative ones, [and] it also is capable of generating negative emotions" [1]. The question of the sense of touch is an issue that also comes up in the context of commercial social virtual reality (social VR or SVR) platforms, which have been proliferating at an increasing rate and attracting more and more users, especially since 2020, the beginning of the COVID-19 pandemic. 'Social VR' refers to 3D virtual spaces where users can interact with one another using head-mounted VR displays or other VR-capable devices [2, 3]. In these spaces, users can socialize and conduct various types of activities (e.g., gaming, working, and learning). In terms of software and user experience design, social VR spaces are similar to earlier collaborative virtual spaces like Second Life [4]. While other social/online spaces facilitate interactions primarily between acquaintances, SVR provides novel pathways of socializing. While phantom touch isn't widely recognized as a design phenomenon in virtual reality (VR), it can be harnessed to enhance user well-being. Therefore it is important to understand its phenomenon in-depth in order to integrate it into personalized interaction mechanisms and avatar designs. This way social VR experiences can foster a sense of safety and vulnerability, ultimately leading to a more positive social experience.

2 Background

2.1 SocialVR

Social VR offers a high level of the sense of embodiment due to the three-dimensional representation of users as avatars and the full-body control which provides a stronger sense of presence [5]. This is hugely impacting the users' virtual behaviour. Freeman and Acena [6] look at the motives of people to build personal relationships (e.g. friendships), how they foster these and how these relationships affect their online and offline social lives. The authors highlight some key activities for building and maintaining relationships, such as sleeping together (there are sleep worlds in some of these platforms), going

D. Singh et al. (Eds.): IHCI 2024, LNCS 15557, pp. 54–59, 2025.
https://doi.org/10.1007/978-3-031-88705-5_5

to virtual bars, watching movies, dancing, and doing homework. Besides engaging with each other in social VR, users also use third-party applications to maintain relationships. The participants interviewed for this study noted positive effects of social interactions in virtual spaces, including "learning about their gender identity, developing a better mental state, and practicing social skills" in a natural way. In their design suggestions, they list proposals for social VR platform developers such as creating more everyday activities as well as developing more family-friendly spaces where families could spend time together. They also suggest that methods to stay connected when one user is not online would also nurture the relationships further. These relationships become multi-layered also because social VR mainly uses the visual and auditive modalities, and although touch and other haptic feedback modalities are missing from these technologies, users (either voluntarily or by relying on mental imagery) can have an illusory sense of phantom touch which can further enhance the sense of presence.

2.2 The Multisensorial Phenomena of Phantom Touch

Phantom touch (also known as illusory touch perception [7] or phantom touch illusion (PTI) [8] or phantom tactile sensation (PTS) [9] in VR can be described as a touch-like sensation based on the visual stimuli experienced in VR. This phenomenon is under-studied, but it is certainly related to the extent of time that the user spends in VR as well as mental imagery and the influence of chemical substances [10]. While some users report that they can also experience ambiance and environmental stimuli in the form of phantom touch, the most common trigger for the illusory touch is being touched by other people/avatars in social VR platforms [10]. While in VR there are no actual stimuli (such as tactile, haptic) other than auditory and visual stimuli, the users report that they "feel" touched on their body when they saw that their avatar was touched in VR. Users report usually sense of heat or pressure [11, 12] when experiencing phantom touch. The most recurrent body location where users feel the phantom touch are the head and face, while the next commonly mentioned category is described as "full body" [10], and, according to Alexdottir and Yang, people feel the sensation of touch at those parts of their body where nerve endings densely innervate [11], who also mention that there is a relationship between the avatar type used by the user and their ability to sense the virtual touch.

While the phenomena might be easier to study in a laboratory environment, understanding its role in a social VR environment and what effect it has on the users is more challenging. Chen et al. brought their method of inquiry outside of the laboratory context and they analysed Reddit forums on RecRoom and VRChat and they identified three ways how this phenomenon might occur: (1) Feeling it naturally; (2) Intentionally developing the ability to experience it; (3) Feeling it under the effect of a substance. Users evaluate these phantom touches in interpersonal situations on various levels, such as technical, social and the level of feelings. They analysed regular users of VR Chat and RecRoom, that is, the subjects had significant experience with VR and they were not users recruited for a laboratory for a short-term experience. According to the researchers "sensing virtual touches, by inference, can both enhance the "blurriness" and indicate people having a hard time differentiating reality and the virtual world" [9] and this adds to the multisensorial aspect of social VR spaces. This observation can offer novel ways for designing these commercial social spaces and it also enables us to think in advance how to create ethical and safety boundaries in order not to nudge the users in misperceiving the environment.

3 Methodology

We have studied testimonies about phantom touch in regard to romantic relationships, with the aim of reaching a better understanding of how it might affect the users' mental health and what this could mean from a design perspective. In order to understand users' perception of how phantom touch can be beneficial for mental health, we will evaluate how phantom touch can be a tool for manifesting empathy or love. The testimonies are part of a qualitative study that was conducted in 2023, the results of which will be published later in 2024. All aspects of this study including recruitment and interview procedures have been approved by the United Ethical Review Committee for Research in Psychology of Eötvös Lóránd University (ELTE). We conducted semi-structured interviews with 26 participants who responded to questions about their experience of using social VR platforms for social and romantic relationships. Participants were recruited from relevant forums and community groups on websites such as Reddit and Facebook as well as through advertisements on social VR dating applications such as Flirtual and Nevermet. After expressing interest, all participants received full information and time to ask the researcher any questions before self-consenting. Interviews lasted roughly 40–50 min and were conducted between January and March 2023 through text and audio chat on Google Hangouts.

The interview protocol was created by an interdisciplinary team with expertise in psychology, sociology, and media theory relying on the reviewed literature. Questions were designed to explore key concepts related to users' engagement with SVR platforms generally and as a tool for finding or maintaining romantic relationships. Specifically, questions covered the use of avatars, activities done with friends and romantic partners, and perceived differences between VR and in-person relationships. Participants were also asked about issues they face when using SVR platforms for dating or maintaining romantic relationships and their recommendations for improvements of these platforms. A full list of interview questions and protocol is available upon request.

4 Results

In the above-mentioned list of questions, participants were also asked about their perception of phantom touch and how they evaluate it. Out of the 26 respondents, 14 reflected on their experience with phantom touch and 11 stated that they felt the sensation.

Phantom touch can serve as a reflection of digital intimacy and this way it can reveal a complex relationship between perceived and actual tactile sensations. For users, this sense can often mean that the virtual touch transgresses the boundaries of virtuality and can present a convergence between the tangible and the simulated. For instance, P15 stated that "Now whenever like I'm being embraced, or someone holds my hand in VR it feels I get the same feeling as if I were hugging someone in real like", and P26 describes the experience as "being able to feel contact even though there's no-one actually touching you." Other users, such as P19, also emphasized the importance of the phantom touch for making connections, stating about their partner that "it's like [her] brain makes those connection and [she] is actually touching or being patted on the head." This highlights the capacity of VR environments for fostering meaningful relationships that extend beyond verbal communication and joint playful activities. Intimacy within social VR platforms often manifests through roleplay and simulated physical interaction, including activities like kissing and hand-holding. This effectively replicates real-world intimacy, as exemplified by a user's description of their avatar engaging in virtual intercourse (P1, P26). Such immersive experiences blur the lines between the physical and digital realms.

There are also aspects of experience in SVR pointed at by some users which, even though not mentioning phantom touch, are the vicinity of the phenomenon and therefore concern phantom touch as a design aspect. P21 stated that "Sometimes I felt like I was being like, someone was there with me, it was physical", while another participant stated that "We still hold hands, even though he can't feel it," (P1) illustrating the profound emotional connection that can be established despite the absence of physical touch.

P20 also pointed out how the avatar embodiment and the fragility of it can bring unpleasant situations: "the feeling is disgusting [...] if you are a woman, [and] there are a lot of man around you"; P22 similarly expressed a feeling of nervousness "when boys approach for hug", meaning that the avatar embodiment can also have negative effects on the wellbeing of a participant. P19's statements offer evidence for how some users can have an enhanced sense of phantom touch under the influence of various types of chemical substances. These testimonies underline the importance of the implementation of possible moderation rules in relation to phantom touch, as well as the introduction of protocols that can allow users to acknowledge the existence of this phenomena in social VR. It can help in sensitizing some of the users, who either might not have this sense of phantom touch, or those who, deliberately or not, may act in a way to cause this sensation in other users who may not desire it.

5 Discussion

These testimonies indicate that designers of social VR spaces should take into consideration how these spaces affect the mental health of the users. It is important to take phantom touch into consideration as social VR spaces are becoming more prevalent and technological developments may enhance this sensation. The author would invite the designers of these spaces to look into the phenomena of phantom touch and to adapt a more ethical attitude that takes into consideration the effect of multisensorial design on mental health from the following aspects:

Consent: While it is understudied how strongly unwanted touches can create the sense of phantom touch, nevertheless the phenomenon raises the question of consent and agency [13] as the users might not have the capacity to protect their 3D avatars from unwanted touch.

Safety (especially for vulnerable groups): The ability to feel phantom touch can create a heightened sense of vulnerability, especially for women who may be disproportionately targeted by unwanted advances in virtual spaces. Ethical considerations should focus on promoting gender equality and addressing power imbalances in VR. The feeling of being touched without actual physical contact can be disorienting and potentially lead to the feeling of being exploited or discomfort. Ethical considerations should focus on creating safe spaces in VR and establishing guidelines to protect users from unwanted advances or harm [14]. As the phantom touch is also a phenomenon that is enhanced as users are under drug influences, it is important to acknowledge this possibility during design and create safer spaces.

Identity and Embodiment: the sense of being touched in different body parts raises questions about identity and embodiment in VR, and as Alexdottir [11] mentions, the type of the avatar can also have an effect on the user's sense of phantom touch. The testimonies suggest that virtual experiences can have real emotional and sensory impacts, blurring the lines between physical and virtual realities. Ethical considerations should explore the implications of social VR avatar design and how this can change self-perception (besides self-body image), and its consequences for social interactions.

The question of how to approach the phenomenon of phantom touch is an important challenge for the virtual world and for avatar designers, especially given that it interacts with many parameters such as the users' level of VR expertise, their imagination, and their usage of chemical substances. While consent and safety is a well-explored subject in the field of virtual worlds (for instance see [15]), how to incorporate this specific type of speculative multisensoriality of these platforms is a difficult question. Phantom touch is a highly individual and multifaceted phenomenon, and it is important to acknowledge its effects on mental health and on how this raises ethical questions that should be addressed by designers and by co-users. One possible alley for multisensorial design is Virtual-Physical Perceptual Manipulations (VPPMs); according to Tseng et al. [16] these designed manipulations alter a user's multi-sensory perception of their own physical actions and reactions in order to nudge their physical movements. Phantom touch, even though so far not much considered as a design phenomenon, is an aspect of the virtual reality experience that can be manipulated for a similar purpose, with the intent of increasing the wellbeing of the users, also by taking into consideration the vulnerability and the feeling of safety of the users that can be enhanced by integrating interaction mechanisms and avatar designs that offer a more personalized virtual social experience.

Acknowledgments. Supported by the ÚNKP-23-4-II. New National Excellence Program of the Ministry for Culture and Innovation from the source of the National Research, Development and Innovation Fund.

References

1. Hertenstein, M.J., Verkamp, J.M., Kerestes, A.M., Holmes, R.M.: The communicative functions of touch in humans, nonhuman primates, and rats: a review and synthesis of the empirical research. Genet. Soc. Gen. Psychol. Monogr. **132**(1), 5–94 (2006). https://doi.org/10.3200/MONO.132.1.5-94
2. McVeigh-Schultz, J., Márquez Segura, E., Merrill, N., Isbister, K.: What's it mean to 'be social' in VR?: Mapping the social VR design ecology. In: Proceedings of the 2018 ACM Conference Companion Publication on Designing Interactive Systems, Hong Kong, China, pp. 289–294. ACM (2018). https://doi.org/10.1145/3197391.3205451

3. McVeigh-Schultz, J., Kolesnichenko, A., Isbister, K.: Shaping pro-social interaction in VR: an emerging design framework. In: Proceedings of the 2019 CHI Conference on Human Factors in Computing Systems, Glasgow Scotland, UK, pp. 1–12. ACM (2019). https://doi.org/10.1145/3290605.3300794
4. Edd, G.: What can the metaverse learn from second life. IEEE Spectr. (2021). https://spectrum.ieee.org/metaverse-second-life
5. Kilteni, K., Bergstrom, I., Slater, M.: Drumming in immersive virtual reality: the body shapes the way we play. IEEE Trans. Visual Comput. Graphics **19**(4), 597–605 (2013). https://doi.org/10.1109/TVCG.2013.29
6. Freeman, G., Acena, D.: Hugging from a distance: building interpersonal relationships in social virtual reality. In: ACM International Conference on Interactive Media Experiences, Virtual Event USA, pp. 84–95. ACM (2021). https://doi.org/10.1145/3452918.3458805
7. Škola, F., Liarokapis, F.: Examining and enhancing the illusory touch perception in virtual reality using non-invasive brain stimulation. In: Proceedings of the 2019 CHI Conference on Human Factors in Computing Systems, Glasgow Scotland UK, pp. 1–12. ACM (2019). https://doi.org/10.1145/3290605.3300477
8. Pilacinski, A., Metzler, M., Klaes, C.: Phantom touch illusion, an unexpected phenomenological effect of tactile gating in the absence of tactile stimulation. Sci. Rep. **13**(1), 15453 (2023). https://doi.org/10.1038/s41598-023-42683-0
9. Chen, Q., Spapé, M.M., Jacucci, G.: Understanding phantom tactile sensation on commercially available social virtual reality platforms. Proc. ACM Hum.-Comput. Interact. **8**(CSCW1), 1–22 (2024). https://doi.org/10.1145/3637418
10. Huang, B.: Examining illusory touch perception in virtual reality among virtual reality users. EHSS **26**, 1075–1080 (2024). https://doi.org/10.54097/a9tyq152
11. Alexdottir, S., Yang, X.: Phantom Touch phenomenon as a manifestation of the Visual-Auditory-Tactile Synaesthesia and its impact on the users in virtual reality. In: 2022 IEEE International Symposium on Mixed and Augmented Reality Adjunct (ISMAR-Adjunct), Singapore, Singapore, pp. 727–732. IEEE (2022). https://doi.org/10.1109/ISMAR-Adjunct57072.2022.00218
12. emi-folk-music: "Phantom Sense / Touch explained," r/VRChat. https://www.reddit.com/r/VRchat/comments/18hiu27/phantom_sense_touch_explained/
13. Gray, J.E., Carter, M., Egliston, B.: Trust and safety in social VR: current industry practices. In: Governing Social Virtual Reality, pp. 61–75. Springer, Cham (2024). https://doi.org/10.1007/978-3-031-61831-4_6
14. Gray, J.E., Carter, M., Egliston, B.: Conduct harms in social VR: embodied harassment, gender-based harm and toxic cultures. In: Governing Social Virtual Reality, pp. 23–34. Springer, Cham (2024). https://doi.org/10.1007/978-3-031-61831-4_3
15. Zytko, D., Chan, J.: The dating metaverse: why we need to design for consent in social VR. IEEE Trans. Visual Comput. Graphics **29**(5), 2489–2498 (2023). https://doi.org/10.1109/TVCG.2023.3247065
16. Tseng, W.-J., et al.: The dark side of perceptual manipulations in virtual reality. In: CHI Conference on Human Factors in Computing Systems, New Orleans LA USA, pp. 1–15. ACM (2022). https://doi.org/10.1145/3491102.3517728

Predictive Analytics for Stress Management in Nursing: A Machine Learning Approach Using Wearable IoT Devices

Ritu Chauhan[1] and Dhananjay Singh[2(✉)]

[1] Centre for Computational Biology and Bioinformatics, Amity University, Noida, India
[2] College of Information Sciences and Technology, Penn State University, University Park, USA
dsingh@psu.edu

1 Introduction

The rapid advancement of technology has ushered in an era where the human race is surrounded by an unprecedented amount of data. This data is derived from various sources, including wearable devices, smart environments, and IoT (Internet of Things) networks. These technological interventions have revolutionized data collection and utilization, offering significant benefits to society by enabling informed decision-making and personalized recommendations. Particularly in the healthcare sector, IoT has transformed traditional data collection methods, fostering a more integrated approach to health management [1]. Wearables have further amplified this transformation, allowing individuals to monitor key health indicators such as heart rate, sleep patterns, physical activity, and stress levels in real time [2]. Consequently, healthcare practitioners and researchers are increasingly exploring innovative methods to leverage this data for creating supportive and efficient environments [3].

As the volume of data generated continues to grow exponentially, it has surpassed the capacity of human monitoring and intervention. Technologies have evolved to collect, store, and process vast amounts of information from multiple sources, including sensors, wearable devices, and IoT networks. However, while these systems excel at handling and organizing data, there remains a significant gap in transforming raw data into actionable insights [19]. Addressing this gap requires the implementation of intelligent systems capable of identifying patterns, uncovering correlations, and making informed predictions from complex datasets [4]. By applying machine learning (ML) algorithms, it is possible to analyze large datasets, uncover hidden patterns, and create meaningful insights that can predict future trends. This paradigm shift moves beyond mere data storage to enable immediate, actionable insights that inform better decisions, improve predictability, and yield effective, anticipatory approaches across various domains.

In healthcare, the integration of IoT and machine learning technologies lays the foundation for a future where health monitoring is seamlessly embedded into daily life. These advancements enable continuous tracking, early identification of health risks, and timely interventions, promoting healthier lifestyles. However, stress detection through IoT and machine learning comes with its own set of challenges. Identifying relevant patterns from physiological signals often involves filtering out environmental noise and

D. Singh et al. (Eds.): IHCI 2024, LNCS 15557, pp. 60–75, 2025.
https://doi.org/10.1007/978-3-031-88705-5_6

mitigating the impact of faulty sensors. Unlike traditional statistical methods that rely on fixed, linear assumptions, machine learning offers the flexibility to capture intricate, dynamic relationships in high-dimensional data, enhancing the accuracy and reliability of stress detection.

This work addresses the pressing need for intelligent systems to manage stress, particularly within the nursing profession, where high-pressure environments demand innovative solutions. Wearable devices, combined with machine learning, offer a transformative approach to stress detection and management. Section 1 highlights the demand for these intelligent systems in stress management for healthcare professionals. Section 2 reviews previous research on IoT-based stress detection systems, identifying their limitations in scope, applicability, and utility. Section 3 details the dataset, preprocessing techniques, and machine learning models employed, emphasizing their importance in analyzing intricate physiological data. Section 4 evaluates various algorithms from the perspective of stress predictive efficiency, comparing their strengths and weaknesses. Finally, Sect. 5 discusses the outcomes, their effectiveness in real-time applications, and future research directions.

This study contributes to the field by proposing a robust stress detection framework that utilizes machine learning to extract insights from physiological data. It establishes a compelling case for population-specific interventions, paving the way for technology-driven, personalized stress management solutions.

2 Related Work

The rapid spread of IoT (Internet of Things), specifically through wearable sensors, has brought about a revolutionary transformation in continuous health monitoring. Devices such as smartwatches, fitness bands, and specialized biosensors can now measure electrodermal activity (EDA), skin temperature, and heart rate in real-time. These devices serve an important purpose for health monitoring because they are non-invasive, portable and can collect data for prolonged periods. There is an increase in research on the use of these wearable technologies in targeting daily exercise, sleep patterns, levels of stress and other health indicators. Table 1 presents a comprehensive comparative analysis of previous research studies, highlighting gaps in the existing body of literature.

In a rapidly evolving healthcare landscape, a groundbreaking solution has emerged to address the critical challenge of stress detection and management. By integrating IoT technology with wearable devices, this innovation offers real-time stress monitoring using physiological indicators such as heart rate variability (HRV), EDA, and skin temperature. These devices go beyond traditional methods, leveraging advanced machine learning (ML) models to analyze diverse biomarkers and deliver precise, actionable insights.

Tailored specifically for high-pressure environments like healthcare, the solution focuses on job-specific stress detection, particularly for nursing professionals. Real-time data from hospital shifts, including workload and intervention scenarios, is analyzed to provide early predictions and detailed stress classifications. This enables proactive interventions, ensuring continuous stress management and fostering a healthier work environment.

The integration of big data analytics enhances the system's scalability and accuracy, allowing it to adapt to dynamic healthcare settings. Advanced ML techniques, such as tree-based models and ensemble algorithms, are employed to classify stress levels with unprecedented precision. These innovations not only address stress but also pave the way for scalable implementations across diverse occupational settings.

Through this IoT-powered framework, healthcare providers can now shift from reactive to proactive stress management, ensuring well-being and efficiency in demanding environments. This transformative approach exemplifies the potential of technology to blend real-time monitoring, predictive analytics, and human-centric design, ultimately creating a healthier, more resilient workforce.

Table 1. A comparative study of previous literature

Reference	Work	Research Gaps	Our Study
Kelly et al. (2020) [1]	Discussed the impact and implications of IoT for healthcare delivery, focusing on connectivity and patient monitoring.	Limited focus on stress-specific biomarkers and real-time stress detection.	Introduced IoT-integrated wearable solutions targeting stress monitoring with physiological indicators in real-time scenarios.
Hickey et al. (2021) [2]	Reviewed smart devices and wearable technologies for detecting depression, anxiety, and stress using HRV, EEG, and electrodermal activity.	Questionable reliability of electrodermal activity; lack of devices for depression detection using EEG; limited datasets.	Utilized multiple physiological indicators (heart rate, EDA, skin temperature) and employed diverse ML models for enhanced accuracy.
Wang et al. (2018) [3]	Analyzed the capabilities and benefits of big data analytics for healthcare organizations.	Lacked focus on stress detection in specific professions and scenarios.	Leveraged big data analytics for tailored stress detection solutions in healthcare environments.
Bustos-Lopez et al. (2022) [5]	Reviewed physiological signals for engagement detection in educational settings using wearables.	Focused on student engagement, not stress detection; lacked comprehensive analysis of stress-specific biomarkers.	Targeted stress detection in high-pressure healthcare settings with real-time monitoring and diverse biomarkers.
Mishra et al. (2023) [6]	Explored biosensors for continuous, non-invasive stress monitoring using biomarkers and ML algorithms.	General focus on stress indicators; lacked tailored applications for specific professions like nursing.	Tailored for nursing professionals; integrated IoT and ML for job-specific stress management solutions.
Can et al. (2019) [7]	Surveyed stress detection in daily life via smartphones and wearable devices.	Limited studies in real-life, uncontrolled settings; lacked integration of hospital-specific stress factors.	Analyzed real-time data during hospital shifts, including workload and intervention implications.

(*continued*)

Table 1. (*continued*)

Reference	Work	Research Gaps	Our Study
Delmastro et al. (2020) [8]	Explored wearable tech for stress monitoring during motor and cognitive training for older adults; proposed a DSS for therapy personalization.	Stress detection limited to binary classification; lacked scalability for larger, dynamic datasets.	Provided early predictions with detailed classifications; employed multiple ML algorithms for scalable implementations.
Kim et al. (2018) [9]	Surveyed HRV as an objective psychological stress measure and identified changes in HRV variables under stress.	Limited to HRV; lacked real-time analysis and integration with wearable devices.	Offered real-time monitoring of stress using multiple physiological indicators (HRV, electrodermal activity, and skin temperature) integrated with wearable IoT devices.
Thayer et al. (2012) [10]	Explored the brain-heart connection and proposed HRV as a proxy for stress regulation.	Theoretical emphasis; lacked practical, real-time applications for occupational stress detection.	Demonstrated practical applications by integrating IoT sensors with ML models for real-time stress classification in healthcare.
Edwards & Burnard (2003) [11]	Systematically reviewed stress management interventions for mental health nurses, focusing on post-event management techniques like relaxation training.	Lacked integration of wearable technology and ML for real-time stress detection and intervention.	Used wearable IoT devices and ML for proactive stress monitoring and interventions, promoting continuous stress management.
Ali et al. (2012) [12]	Reviewed random forests and decision tree algorithms for various classification and prediction tasks.	Lack of application in stress detection and integration with physiological indicators.	Employed tree-based ML models for detailed stress classification, improving prediction accuracy.

3 Methodology

3.1 Dataset Description

The information used for this research comes from Dryad, a public repository for research data. The data gathered at regular intervals was collected on nurses during their working shifts in a hospital. The data was captured by means of wearing gadgets designed to collect different forms of physiological signals aimed at tracking stress levels. Its main features include skin temperature (TEMP), heart rate (HR), and electrodermal activity (EDA). Additionally, the collection includes orientation (location of the individual in question) information (X, Y, Z axes), date-time stamp (DateTime) as well as a categorical stress label (label) indicating the nurse's reported level of stress aside from these physiological parameters. Each nurse is given a unique identifier (id), thus ensuring anonymity and data security while allowing physiological signals to individual participants.

This dataset contains timestamped data that captures the stress changes across shifts and tasks for nurses. It adds a time dimension allowing an exploration of how activity changes over time. If the analysis is focused on the datetime column, it is possible to observe the stress levels combining periods of maximum workload, phase changes of the work (e.g., working with a patient), or certain events like a crisis. Indeed, it is likely that such changes in stress may have 'high peaks' during a day or at the handover of two shifts or during activities in the intensive care unit. When these temporal trends are combined with physiological data such as EDA, HR, and TEMP, it contributes to understanding stress behaviour across different situations and therefore supporting the design of contextual interventions. As per machine learning practice, the data was divided into 80% which was used for training and 20% which was used for testing the model after the full machine learning cycle. This technique ensures that baskets trained on valid data do not exceed the limits of overfitting since there is data reserved not used for training. In addition, for preliminary testing and adjustments, which were necessary to avoid the issues of the too large dataset with over 11.5 million records, only 1% of the original dataset was sampled. The method supported rapid trial and error behavior with modeling methods but also preserved the variance in the input data. The smaller portion allowed for realizable testing and fitting while the later possibility of using the whole data set made sure model building was exhaustive and precise.

3.2 Exploratory Data Analysis (EDA)

Exploratory data analysis also known as EDA, is crucial for interpreting the structure of a dataset, identifying patterns, and even relating variables. In order to achieve an initial understanding of physiological signal distributions (EDA, HR, TEMP) and their correlations to stress labels, this study utilized EDA. During the EDA process data distributions were displayed; missing values verification was conducted; outliers were pointed out; and correlations among various physiological parameters were studied.

Table 2. Summary statistics of key features measured by IoT device.

Variable	Mean	Median	Variance	Std_Dev
EDA	3.502127	1.157407	31.99645	5.656541
HR	85.7609	83.9	201.5382	14.19642
TEMP	32.23435	32.57	5.109934	2.260516

A correlation matrix was created to analyze relationships between different aspects while considering the complexity of physiological signals and their possible dependency. The correlation matrix was displayed as a heatmap that provided visual clarity on relationships between various orientation data (X, Y, Z) as well as physiological signals such as EDA, HR, and TEMP. Additionally, there was also a feature importance plot that ranked the predictive ability of each feature with respect to predicting stress levels. Table 2 provides a detailed summary of statistical analyses is provided for the key features measured by Internet of Things (IoT) devices, offering valuable insights into their variability, and relevance across various applications.

Some orientation features, although their overall correlations with stress levels were low, were still retained as they may express contextual or behavioral aspects that other variables do not. For example, orientation features may be used to analyse the way an individual holds, moves, or interacts with their surroundings, all of which may play a role in stress but do not have a direct correlation. Such features are useful in building a multifaceted view of the model as they enable machine learning algorithms to seek out intricate and non-linear relationships that do not seem obvious after performing a simple correlation. Furthermore, orientation data can be synergistic with some other features in its role in stress prediction, only when those other features are physiological or contextual do the contributions become apparent.

3.3 Data Pre-processing

Preprocessing is an essential step before feeding data into machine learning tools to ensure that the dataset is free from errors, has been standardized and is ready for analysis. The preprocessing in this study comprised several activities.

Firstly, removing the missing values through either mean or median imputation methods or removal of rows with little amount of data when the missing data was little is done. In the next step, physiological signals (EDA, HR, TEMP) were normalized via min-max scaling so as all features stayed within range thus preventing the models from being biased towards features which have higher values.

Outliers in physiological signals, for example heart rate (HR) sudden abnormal peaks, were handled using statistical methods, such as z-score analysis to identify those data points which are well outside the mean and delete them. This ensured that the data set represented realistic values on physiological range thus improving the reliability of the model built. A rolling window technique was included for temporal feature extraction where a window size of 5 min was selected in a bid to capture short lived changes while dealing with time complexity. Summary statistics such as mean, standard deviation, and variance were taken within each window to make a feature out of a stream of continuous data. Categorical stress level labels were label encoded for tree-based algorithms such as Random Forest and Decision tree as they can manage ordinal relationship quite well. However, one hot encoding was adopted for models that included Logistic regression and K-Nearest Neighbors (KNN), which are more efficient with exclusive binary categorical depictions and avoided introducing unwarranted ordinal assumptions. This preprocessing pipeline configured the data for different machine-learning algorithms, increasing the accuracy and the interpretability of the models developed.

3.4 Machine Learning Algorithms Applied

Models that are tree-based, for instance, Decision Trees and Random Forests invariably produce plots depicting the importance of distinctive features, as the data at each node is always split according to the features, thereby calculating how much each feature contributes to the prediction made by the model. These algorithms were selected in this case because they can deal well with the mixture of numerical, such as EDA, HR, or TEMP, and categorical stress levels data, while at the same time exploring the nonlinear relationship between the features. In particular, the Random Forest technique improves the prediction accuracy by taking an average predicted by multiple decision trees, which helps alleviate the problem of overfitting. Gradient Boosting was also added to this model for the purpose of 'improving' the model by reducing bias and modeling more complex physiological interactions which was built by Random Forest. Stochastic Gradient Descent was used due to its efficiency and speed in processing large volumes of data comparable to what we had. Even though Support Vector Machines and neural networks were potential approaches, tree-based methods were preferred because they were less computationally expensive and more interpretable than SVMs and neural networks, which were sensitive to tuning hyper-parameters. Table 3 highlights the machine learning algorithms applied in predicting stress levels, particularly among nurses, using physiological data collected via IoT devices. Each algorithm brings unique strengths to the analysis.

Table 3 shows different machine learning algorithms applied to predict stress among nurses, through data collected using IoT devices.

Table 3.

ML Algorithm	Strengths	Limitations	Contribution to Predict Stress	Suitability for Stress Detection
Random Forest [12]	Robust to overfitting; handles large datasets; provides feature importance.	Computationally intensive for large-scale applications.	Ensemble learning methods using multiple decision trees to enhance prediction accuracy and reduce overfitting. Used for stress level prediction, with feature importance plotting the relevance of each signal.	Ideal for high-dimensional physiological data with clear insights into feature relevance.
Decision Tree [13], [24]	Easy to interpret; handles both categorical and numerical data; non-linear.	Prone to overfitting, especially with small datasets; lacks generalization.	Non-parametric algorithm forming branches based on feature values, effective for both quantitative and qualitative data but prone to overfitting. Applied for stress level categorization.	Useful for initial stress classification, but needs ensemble methods for improved accuracy.
Gradient Boosting [14]	Excellent for reducing bias; highly accurate; handles complex relationships.	Slow training process; risk of overfitting if not tuned properly.	Sequential ensemble approach where each tree corrects previous errors, reducing bias and enhancing accuracy. Applied for classifying stress by considering interactions between physiological features.	Best suited for refined stress prediction models with multiple features.
Stochastic Gradient Descent (SGD) [15]	Efficient for large-scale datasets; well-suited for linear models.	Requires careful parameter tuning; struggles with non-linear relationships.	Iterative optimization method used in linear models like SVMs, ideal for large datasets as it updates parameters per data point. Applied to enhance hyperplanes for distinguishing stress levels.	Effective for stress detection with linear physiological features; less flexible for complex data.
Naive Bayes [16]	Simple and fast; works well with small datasets and probabilistic features.	Assumes feature independence; less effective for correlated features.	A probabilistic classifier based on Bayes' theorem assuming conditional independence. Used to classify stress levels, benefiting from simplicity and rapid experimentation.	Useful for stress detection when features are mostly independent.

(*continued*)

(*continued*)

ML Algorithm	Strengths	Limitations	Contribution to Predict Stress	Suitability for Stress Detection
K-Nearest Neighbors (KNN) [17]	Simple and intuitive; no training phase required; highly effective with small datasets.	Sensitive to noise; computationally expensive on large datasets; requires scaling.	Instance-based learning method classifying based on distances in feature space. Used to calculate stress levels by evaluating nearest neighbors, though less effective on large datasets.	Excellent for initial experimentation but less practical for real-time stress monitoring.
Logistic Regression [18]	Easy to implement; interpretable; good for binary classification problems.	Struggles with non-linear relationships; limited performance on multi-class tasks.	Analytical model handling multi-class problems, determining stress level probabilities based on physiological parameters. Known for interpretability and quick training.	Suitable for baseline stress level classification.

4 Results

4.1 Exploratory Data Analysis

The connection among these several elements in the dataset e.g. orientation data (X, Y, Z), HR, TEMP and EDA from the correlation heatmap, in Fig. 1, can be analyzed. It shows that correlation values vary from -1 to 1 where a perfect negative correlation is given by -1 while an absence of any correlation amounts to 1 and a perfect positive correlation amounts to 1.

It can be observed from the correlation heatmap that there are strong correlations between the physiological attributes, namely Electrodermal Activity (EDA), Heart Rate (HR), Skin Temperature (TEMP), and the label for stress. EDA and HR are the best predictors of stress in humans again consistent with previous research on detecting it. EDA, which is the most affected by the changes in autonomic nervous system with restrain to voluntary control, and HR represents the level of stress experienced by the body, which also correlates highly with the stress level making them the best predictors. Their rank in the importance graph further shows their importance in developing effective stress prediction systems. This also supports EDA and HR as the most valuable features for any machine learning model seeking to predict stress.

The graph of feature importance, in Fig. 2, demonstrates the factors that are most relevant in predicting stress levels. It is primarily derived from models based on trees like Random Forest and Gradient Boosting. From this assessment, HR and EDA were found to be the most notable parameters, which is in agreement with other research that correlates these kinds of physiological markers with stress responses. This graphs Shows the importance of each factor, this graph shows which physiological signals are most significant for detecting stress. Furthermore, it gives suggestions for improving models based on these key characteristic factors thus aiding accuracy.

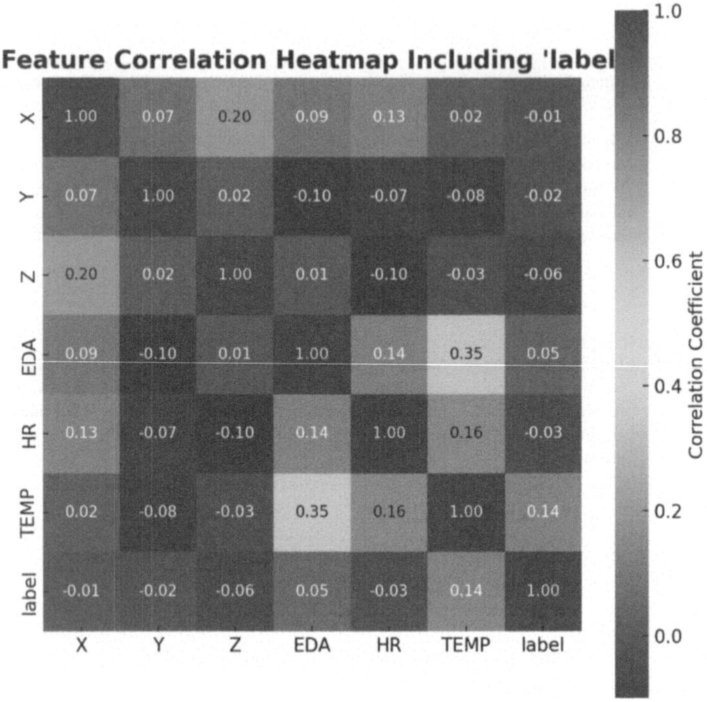

Fig. 1. The heatmap above represents the correlation between all the attributes in our dataset.

4.2 Algorithms Results

Tree-based models, such as Random Forests and Decision Trees, include a feature importance plot as these models assess how much each feature contributes to the nodes in the tree and therefore, they enable estimation of the importance of each feature in making predictions. There are however some algorithms such as Logistic Regression or KNN which do not provide such visual tools for the users because the algorithms are either linear or distance-based and cannot be formatted with the same ease. SVMs are powerful algorithms but are more computationally intensive especially for large datasets, and require careful tuning of hyperparameters, making tree-based methods more suitable for this study.

The models evaluated differently on estimating stress levels of nurses using some machine learning algorithms based on physiological inputs including skin temperature, heart rate and electrodermal activity. By interpreting a correlation heatmap and considering feature importance graph, which provided more insight into how connected each of the features were to stress prediction, this assessment was improved. The model performance comparison, in Table 4, exhibited different accuracies hence underscoring several merits and demerits associated with each approach.

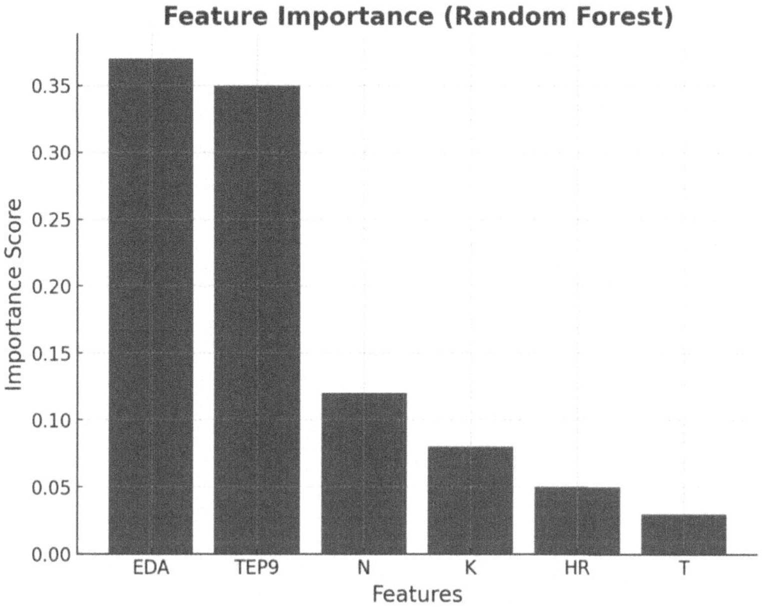

Fig. 2. The graph above represents the Feature importance of all the attributes in our dataset

However, it is important to note that tree-based models, such as Random Forest and Decision Tree, scored highly among the different machine learning models used for predicting levels of stress. There is great confidence in Random Forest with 83% accuracy while Decision Tree had a maximum accuracy of 85%, although this was at the risk of overfitting. The performance of Gradient Boosting was therefore not as high at 77%, indicating that hyperparameter tuning would be necessary for optimum performance. On the other hand, simpler models like Stochastic Gradient Descent (SGD) and Naive Bayes achieved moderate accuracy (74%) but had difficulties due to their independence requirements related to features.

KNN emerged as the most effective with an accuracy of 90%, maybe owing to libraries distinct in the dataset, but its computation expense remains a disadvantage, especially for huge datasets. In contrast, Logistic Regression performed significantly lower at 49% accuracy because it failed to grasp the complex relationships exhibited by the physiological signals. This comparison indicates that Random Forest and KNN which are more complex non-linear models are better suited for stress prediction than simpler ones. The results can be referred to in Table 5.

Table. 4. The classification report for the dataset is presented in the table below, which predicts the accuracy after using the logistic regression classifier

KNN	**Precision**	**Recall**	**F1 Score**	**Support**
Prediction (0)	0.83%	0.76%	0.79%	648149
Prediction (1)	0.81%	0.69%	0.74%	241788
Prediction (2)	0.92%	0.95%	0.94%	2562779
SGD	**Precision**	**Recall**	**F1 Score**	**Support**
Prediction (0)	0.00%	0.00%	0.00%	648149
Prediction (1)	0.00%	0.00%	0.00%	241788
Prediction (2)	0.74%	1.00%	0.85%	2562779
Decision Tree	**Precision**	**Recall**	**F1 Score**	**Support**
Prediction (0)	0.78%	0.56%	0.65%	648149
Prediction (1)	0.84%	0.41%	0.55%	241788
Prediction (2)	0.86%	0.96%	0.91%	2562779
Random Forest	**Precision**	**Recall**	**F1 Score**	**Support**
Prediction (0)	0.89%	0.44%	0.59%	432097
Prediction (1)	0.99%	0.23%	0.37%	160853
Prediction (2)	0.83%	0.99%	0.90%	1708861
Naïve Bayes	**Precision**	**Recall**	**F1 Score**	**Support**
Prediction (0)	0.32%	0.03%	0.05%	648149
Prediction (1)	0.64%	0.03%	0.05%	241788
Prediction (2)	0.74%	0.98%	0.85%	2562779
Gradient Boosting	**Precision**	**Recall**	**F1 Score**	**Support**
Prediction (0)	0.86%	0.14%	0.24%	648149
Prediction (1)	1.00%	0.06%	0.12%	241788
Prediction (2)	0.77%	1.00%	0.87%	2562779
Logistic Regression	**Precision**	**Recall**	**F1 Score**	**Support**
Prediction (0)	0.26%	0.51%	0.34%	432097
Prediction (1)	0.08%	0.16%	0.11%	160853
Prediction (2)	0.79%	0.52%	0.63%	1708861

Table. 5. Comparison between the four machine learning classifiers' accuracies

ML Model	Accuracy
KNN	90%
SGD	74%
Decision Tree	85%
Random Forest	83%
Naïve Bayes	74%
Gradient Boosting	77%
Logistic Regression	49%

Various models were tested using AUC (Area Under the Curve), as illustrated in Fig. 3, and KNN was the best-performing model with an AUC score of 0.98 indicating a strong ability to classify accurately. Random Forest and Decision Tree presented good outcomes with AUCs of 0.90 each. Gradient Boosting performed fairly well at an AUC value of 0.73. On the other hand, SGD and Naive Bayes had really low values of AUC i.e., 0.58 and 0.63 respectively, just like Logistic Regression whose AUC score stands at 0.58. KNN remained undoubtedly the most successful model while Logistic Regression and SGD followed closely behind it.

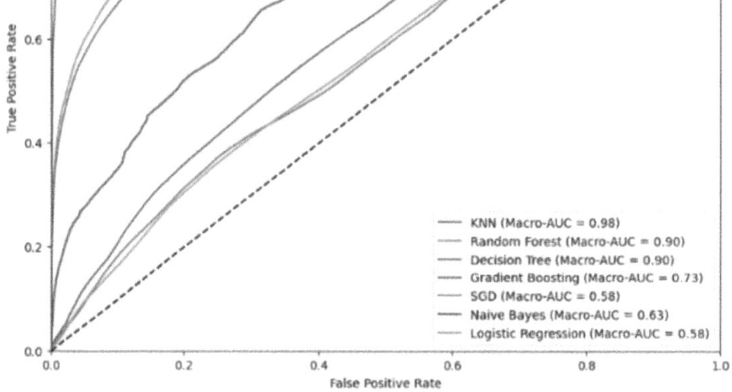

Fig. 3. The graph above represents the ROC curves of all the classifiers we used in our dataset to predict stress level of nurses.

Our study assesses models of machine learning by means of ROC curves and with the aid of metrics such as precision, recall, and F1 score. In addition, accuracy alone cannot suffice for imbalanced datasets considering that precision, recall and F1 scoring provide a fairer view of performance. Such metrics are important in stress prediction as high stress class for instance may not be well represented. Precision calculates true positives within all positive predictions, recall assesses correctly counted actual positives, and f1-score is an average of the two. KNN performed better on the MLR model and achieved an impressive AUC score of 0.98 since it could fit nonlinearities while Logistic Regression could only work with linear assumptions AUC = 0.58. The area under the receiver operating characteristic curves AUC indicates how well models can separate the classes, with higher values indicating better performance. The AUC of KNN indicates that the model can predict stress levels well. The AUC of Logistic Regression, however, illustrates the low memories that will ever be encountered with the algorithm. Therefore, Nonlinear models of KNN and Random Forest are more appropriate in predicting stress in the health care unit.

5 Conclusion

The findings from our research are consistent with previous studies on stress prediction with border and physiological data. However, it extends the use of more IoT sensors for better real-time monitoring of stress in hospital nurses. The best performance was recorded with KNN classifiers achieving 90% accuracy compared to Random Forest at 83%, and Decision Tree at 85% hence proving tree-based classifiers to be effective. More so, there were limitations experienced by simpler models such as Logistic Regression 49% and Naïve Bayes 74% due to their inability to complex non-linear relationships, which can be observed in the data and implies advanced techniques should be employed. The use of more advanced artificial neural networks such as recurrent neural networks (RNNs) or long-short term memory (LSTM) RNNs can help on the prediction performance by learning more complex behaviors of conveniently few time series data. In addition, longitudinal studies that involve measuring stress across long timescales would also enhance the accuracy of prediction and help understand its chronic implications. The implementation of real-time measurement and long-duration lenses approaches will advance the relief of stress among these healthcare workers and the nurses' patients.

References

1. Kelly, J.T., Campbell, K.L., Gong, E., Scuffham, P.: The Internet of Things: impact and implications for health care delivery. J. Med. Internet Res. **22**(11), e20135 (2020)
2. Hickey, B.A., et al.: Smart devices and wearable technologies to detect and monitor mental health conditions and stress: a systematic review. Sensors **21**(10), 3461 (2021)
3. Wang, Y., Kung, L., Byrd, T.A.: Big data analytics: understanding its capabilities and potential benefits for healthcare organizations. Technol. Forecast. Soc. Chang. **126**, 3–13 (2018)
4. Elragal, A., Klischewski, R.: Theory-driven or process-driven prediction? Epistemological challenges of big data analytics. J. Big Data **4**, 1–20 (2017)

5. Bustos-Lopez, M., Cruz-Ramirez, N., Guerra-Hernandez, A., Sánchez-Morales, L.N., Cruz-Ramos, N.A., Alor-Hernandez, G.: Wearables for engagement detection in learning environments: a review. Biosensors **12**(7), 509 (2022)
6. Mishra, A., Agrawal, M., Ali, A., Garg, P.: Uninterrupted real-time cerebral stress level monitoring using wearable biosensors: a review. Biotechnol. Appl. Biochem. **70**(6), 1895–1914 (2023)
7. Can, Y.S., Arnrich, B., Ersoy, C.: Stress detection in daily life scenarios using smart phones and wearable sensors: a survey. J. Biomed. Inform. **92**, 103139 (2019)
8. Delmastro, F., Di Martino, F., Dolciotti, C.: Cognitive training and stress detection in mci frail older people through wearable sensors and machine learning. IEEE Access **8**, 65573–65590 (2020)
9. Kim, H.G., Cheon, E.J., Bai, D.S., Lee, Y.H., Koo, B.H.: Stress and heart rate variability: a meta-analysis and review of the literature. Psychiatry Investig. **15**(3), 235 (2018)
10. Thayer, J.F., Åhs, F., Fredrikson, M., Sollers, J.J., III., Wager, T.D.: A meta-analysis of heart rate variability and neuroimaging studies: implications for heart rate variability as a marker of stress and health. Neurosci. Biobehav. Rev. **36**(2), 747–756 (2012)
11. Edwards, D., Burnard, P.: A systematic review of stress and stress management interventions for mental health nurses. J. Adv. Nurs. **42**(2), 169–200 (2003)
12. Ali, J., Khan, R., Ahmad, N., Maqsood, I.: Random forests and decision trees. Int. J. Comput. Sci. Issues (IJCSI) **9**(5), 272 (2012)
13. Song, Y.Y., Ying, L.U.: Decision tree methods: applications for classification and prediction. Shanghai Arch. Psychiatry **27**(2), 130 (2015)
14. Demir, S., Sahin, E.K.: An investigation of feature selection methods for soil liquefaction prediction based on tree-based ensemble algorithms using AdaBoost, gradient boosting, and XGBoost. Neural Comput. Appl. **35**(4), 3173–3190 (2023)
15. Ren, L.H., Ye, Z.F., Zhao, Y.P.: A modeling method for aero-engine by combining stochastic gradient descent with support vector regression. Aerosp. Sci. Technol. **99**, 105775 (2020)
16. Flach, P.A., Lachiche, N.: Naive Bayesian classification of structured data. Mach. Learn. **57**, 233–269 (2004)
17. Steinbach, M., Tan, P.N.: kNN: k-nearest neighbors. In: The Top Ten Algorithms in Data Mining, pp. 165–176. Chapman and Hall/CRC (2009)
18. Friedman, J., Hastie, T., Tibshirani, R.: Additive logistic regression: a statistical view of boosting (with discussion and a rejoinder by the authors). Ann. Stat. **28**(2), 337–407 (2000)
19. Saggi, M.K., Jain, S.: A survey towards an integration of big data analytics to big insights for value-creation. Inf. Process. Manage. **54**(5), 758–790 (2018)
20. Chauhan, R., Singh, D.: Maximizing accuracy in AI-driven pattern detection in cardiac care. In: Choi, B.J., Singh, D., Tiwary, U.S., Chung, W.Y. (eds.) IHCI 2023. LNCS, vol. 14531, pp. 176–187. Springer, Cham (2024). https://doi.org/10.1007/978-3-031-53827-8_17

Robots and Conversation Agents

Robot-to-Robot Collaborative Knowledge Sharing with Human Operators Under Constraint Resources Through Connotative 2.5D Mapping

Nafiul Alam, M. I. R. Shuvo, and Jong-Hoon Kim(✉)

Advanced Telerobotic Research Lab, Computer Science, Kent State University, Kent, OH, USA
jkim72@kent.edu
https://www.atr.cs.kent.edu/

Abstract. Robotic systems may navigate unexpected locations without prior knowledge. Robots that need position data need SLAM. This work introduces 2.5D mapping, which combines 2D mapping with more information, to overcome 3D mapping's information restrictions. 2.5D mapping requires Intel RealSense D435 RGBD frames. Hitmaps are made from frame items of interest. Hitmap is updated using robot self-position and individuals' mainly error-free center positions. Our robot estimates its position using visual odometry and an inertial measurement unit. Hitmaps are 3D dot maps of self-placements and interest. A 3D map turns 2.5D in two steps. Make a 2D map. A feature vector array with rich information is then created for each map point. Rasterize the 2D map to make a 3D model. 2.5D mapping uses feature vectors and two-dimensional maps. A 2.5D map guides robot mergers. Through the map, people may observe their surroundings. Reducing 3D pictures to 2.5D reduces visual information. Humans get the integrated map. Human operators create 3D maps from 2.5D maps. We created valuable components and significant results for design viability.

Keywords: Robots · Human · SLAM · 2.5D mapping · RGBD frames · Hitmap · Visual Odometry · Intel RealSense D435 · Integrated map

1 Introduction

A robotic system travels across unfamiliar areas without prior knowledge of its surroundings [1,2]. Robots that need position data must use SLAM. Robots use sensor data to create a comprehensive map of their surroundings and estimate their position. Unmanned Air Vehicles (UAVs), Unmanned Ground Vehicles (UGVs), and Unmanned Underwater Vehicles (UUVs) use SLAM to get their real-time spatial coordinates. Most Simultaneous Localization and Mapping (SLAM) methods today use three-dimensional cloud points acquired by Intel RealSense or Lidar cameras.

D. Singh et al. (Eds.): IHCI 2024, LNCS 15557, pp. 79–91, 2025.
https://doi.org/10.1007/978-3-031-88705-5_7

However, 3D cloud point gathering generates lots of data. In a perfect communication medium, sending a lot of data to other robots or operators is easy. However, in high-risk locations including badly damaged structures, radioactive zones, and undiscovered caves, the catastrophic event may cause structural alterations in components. This scenario makes it impossible to accurately describe the places' geographical configuration. The robots need new maps of these sites to navigate and complete jobs. Robots must transmit and compute map data in real time for cartography. Due to limited communication routes after a disaster, transmitting large amounts of data may be difficult. Calculating and transporting with mobile trucks may be difficult.

2 State of Arts

Self-localization and mapping for robots are widely employed in various areas, including agriculture, forestry, construction, and disaster rescue [3,4]. Localization and mapping are employed by mobile robots in practical operations. Search and rescue in disaster environments [5,6]. Deep learning was used by Xiao et al. [7] to improve Dynamic-SLAM, a semantic monocular visual mapping system, in dynamic scenarios For low recall rates, they devised a missed detection compensation technique, selective tracking increased system robustness and accuracy, and a visual Dynamic-SLAM system with features. An SSD convolutional neural network-based object detection thread for robot localization and mapping was constructed via deep learning. A study [8] suggests separating ground points and point clouds to enhance 3D posture estimation. Point clouds are clustered from 2D images.

Vinicio and Lee [9] suggest a 3D mobile robot localization approach using point cloud and Monte Carlo methods. This method recreates an occupancy map utilizing 3D point cloud and 2D features for real-time localization. This work [10] suggests using natural landmarks to enhance mobile robot SLAM performance in broad spaces, lowering the challenge of merging identical landmarks. Search and rescue (SAR) robots need SLAM. Wang et al. [11] developed a 2D map-based 3D OOI reconstruction approach for SAR-adapted object recognition, exploration area coverage, and loop closure detection.

CCM-SLAM, a collaborative framework for a centralized server and several agents, such as tiny UAVs or robots, was presented by [12]. The server detects redundancy, optimizes globally, and stores agents' estimates. CCM-SLAM's two-way communication improves onboard estimates and exploits many agents in the mission region. A study [13] introduces a simple visual semantic slam technique using planar surfaces for high-level semantics. VO/VIO, object-based, and geometric SLAM algorithms are outperformed by the algorithm. Indoor and 2 m/s flying experiments indicate its efficacy. According to Liang et al. [14], an ORB-SLAM-integrated Semantic SLAM system may generate semantic maps using object-level entities. The system adds object detection, keyframe-object connections, and feature point filtering to ORB-SLAM2.

2.5D mapping has been utilized for robot localization and mapping in many settings. Vinicio et al. [15] created a multi-robot localization system for multi-level exploration employing a leader and follower robot. The leader employs a Kinect camera for RGB photos and a 3D LIDAR for 2.5D mapping, while the follower uses a 2D LIDAR for even exploration. The research [16] introduced a graph optimization-based SLAM system with low-cost LiDAR and vision sensors, updated cost-function, Bag of Words model visual features, and 2.5D map including obstacles and vision characteristics. Nam et al. [17] used low-cost RGB-D sensors and 2D Laser sensors to produce a 2.5D elevation map for ground robots in indoor localization employing aerial and ground robotics.

This work introduces 2.5D mapping, which adds information to 2D mapping to overcome 3D mapping's information restrictions. Intel RealSense D435 RGBD frames are required for 2.5D mapping. These frames generate hitmaps for items of interest. Next, we update the Hitmap based on the robot's self-location and the relatively accurate center locations of the people. Our robot uses ocular odometry and inertial measurement to locate itself. Hitmaps are three-dimensional maps made up of points of interest and self-placements. A 3D map to 2.5D conversion takes two steps. A two-dimensional map is created. Each map point gets a feature vector array with relevant data. Rasterize the two-dimensional map to create a three-dimensional representation. The 2.5D mapping approach employs two-dimensional feature vectors. We provide 2.5D maps of merging robots. Map transmission shows user landscapes. Visual information decreases when 3D images become 2.5D. Operators receive the aggregated map. Humans generate 3D maps from 2.5D maps. We made successful components to prove the design's practicality.

The subsequent sections of the document are organized in the following manner: The methodology of the system is explicated in Sect. 2. Section 3 presents the implementation of the system and the resulting output, while Sect. 4 provides a comprehensive analysis. The final section of the study presents the conclusion and provides recommendations for further measures.

3 Proposed Methodology

The 2.5 mapping method presented in this study has three unique phases, as indicated in Fig. 1. Stage 1 generates basic 3D mapping. In Stage 2, three-dimensional maps become two-and-a-half-dimensional. It must also include notable entities, navigation strategies, and other interactive elements. Finally, 2.5D maps are converted to 3D in stage 3. The technology will be able to create three-dimensional maps from 2.5D maps and vector data.

3.1 Stage 1

Stage 1 consists of three modules. Module 1 is responsible for generating the Hitmap, Module 2 serves as the localizer, and lastly, Module 3 is dedicated to generating the 3D map.

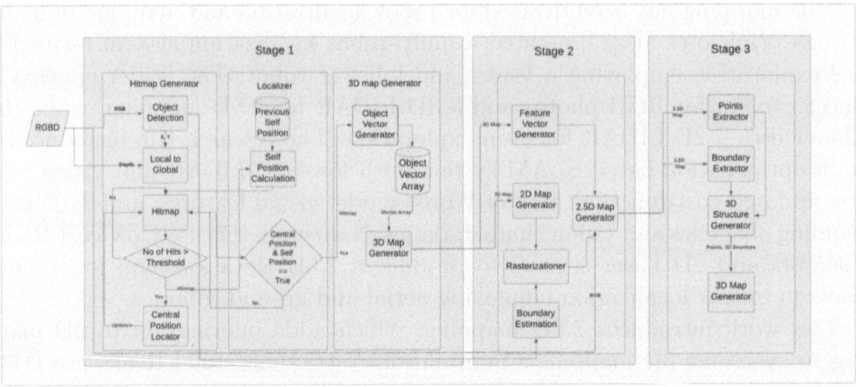

Fig. 1. Basic Architecture of the Proposed System

Hitmap Generator. Initial Hitmap generator input is RGBD frames. The Intel Realsense D435 offers RGBD. The RGB frames and depth frames make up RGBD frames. Figure 2 shows how the YOLOv8 model works with RGB input to identify and classify objects of interest and interactable objects. Once the model detects these objects, it sends their centroid coordinates from the RGB frame to the depth frames for depth computation. After then, Hitmap receives these positions. People commonly accept that Hitmap accurately determines real-time global positions. Raw Intel Realsense D435 data is error-prone. YOLOv8's center calculation is likewise inaccurate. We fix this by processing the Hitmap. For every object detection, the Hitmap records a hit. Hitmap analysis shows a clustering of hits near each object. To achieve our goal, we keep 10 cm between objects. We consider any hits within 10 cm as part of the object. The two entities are different and separate when the distance exceeds 10 cm, the two entities are distinct.

This process consolidates the dispersed data points associated with each object into a single data point using the Hitmap algorithm. Only the item's core locations will remain in the Hitmap. Current location data is local-only. Local viewpoint to global position is achieved with the Local to Worldwide Position Module. This module uses the Intel Realsense D435 camera's advanced settings to convert local to global location. It also warps the local vision into the global view. To calculate 3D global position following formulas:

$$X = \frac{x - \text{depth\_intrin.ppx}}{\text{depth\_intrin.fx}} \cdot \text{depth} \tag{1}$$

$$Y = \frac{y - \text{depth\_intrin.ppy}}{\text{depth\_intrin.fy}} \cdot \text{depth} \tag{2}$$

$$Z = \text{depth} \tag{3}$$

Where: X, Y, and Z represent the coordinates of the 3D point. x and y are the pixel's X and Y coordinates in the depth image, respec-

Fig. 2. Human and Chair Detection using YOLOv8.

tively. depth_intrin.ppx and depth_intrin.ppy denote the principal point X and Y coordinates, usually half of the image width and height, respectively. depth_intrin.fx and depth_intrin.fy are the focal lengths along the X and Y axes. Finally, depth represents the depth value at pixel $[x, y]$.

Localizer. After creating the Hitmap, the robot's internal location must be updated. Localizers are responsible for localizing. The suggested system assumes the robot starts from (0, 0, 0). Then, using visual odometry and an IMU, the system updates its position. The Hitmap vector array receives the altered position. Our robot's position is rendered when our 3D map is generated.

3D Map Generator. After transforming the view from local to global, the points will be appended to a vector with the global position, object class, and object sequence. Each object gets a vector. Array vector addition will occur. Our Hitmap will be regenerated using RGBD frames. If the detected object is already in the array, the software will check if its position has changed. Update the vector in the array. After creating the Hitmap and vector array, the 3D map generator receives it. Within the framework of the 3D map generator, the generation of 3D points is carried out for each vector present in the array. Scaling and visualizing 3D points aids observation.

3.2 Stage 2

Within the framework of our proposed system, the second stage assumes the responsibility of transforming three-dimensional maps into two-and-a-half-dimensional representations. Stage 2 comprises five distinct modules, namely a feature vector generator, a 2D map generator, a rasterizationer, a boundary estimator, and a 2.5D map generator.

Feature Vector Generator. The feature vector is crucial to our 2.5D mapping system. Using a three-dimensional map as input, feature vectors are generated. The feature vectors and two-dimensional mapping technique form the basis for our 2.5D mapping approach. The feature vectors in our proposed system are adaptable. Based on user needs, change is possible. In this case, the feature vector was a 10-dimensional vector.

$$FeatureVector = [X, Y, Z, P, P_x, P_y, P_{xx}, P_{yy}, P_h, P_c] \tag{4}$$

where:

$$\begin{aligned}
X &: \text{Position in } x\text{-axis} \\
Y &: \text{Position in } y\text{-axis} \\
Z &: \text{Position in } z\text{-axis} \\
P &: \text{Occupancy at } (X, Y, Z) \\
P_x &: \text{Occupancy from robot's lowest point to camera in } z \\
P_y &: \text{Occupancy from robot's highest point to camera in } z \\
P_{xx} &: \text{Robot pass rate at } X, Y \text{ from lowest to camera in } z \\
P_{yy} &: \text{Robot pass rate at } X, Y \text{ from highest to camera in } z \\
P_h &: \text{Human presence at } (X, Y, Z) \\
P_c &: \text{Chair presence at } (X, Y, Z)
\end{aligned}$$

The feature vector can be used for object recognition, path detection, and other activities depending on its dimensions, according to the analysis. 2.5D mapping is more versatile than 3D mapping, making it better.

2D Map Generator. The 2D map generator utilizes the conversion of a 3D map into a 2D map by considering only the (X, Y) coordinates of the objects of interest at a given elevation. The camera's height is being utilized for our intended purpose, as it serves as the primary reference point in our suggested system. Subsequently, the aforementioned 2D map is presented as input for the Rasterizationer.

Boundary Estimator. The boundary estimator estimates item height and width. The submitted information is essential for three-dimensional reconstruction of stage 3. Information is lost when mapping from 3D to 2.5D. However, height and breadth data can reduce this loss. Height and width are not directly used. This is for processing in the sterilizer.

Rasterizationer. From 2.5D representations, the Rasterizationer module generates the additional data points needed to recreate 3D structures. This method uses height-width information from the boundary estimator and the 2D map generator's 2D map. Creating a three-dimensional structure from one point is impossible. This method mostly creates skeletal representations for intriguing objects on the two-dimensional map. The 2.5D map generator will use the rasterized 2D map. See the rasterized 2D map in Fig. 3.

2.5D Map Generator. The 2.5D map generator utilizes a combination of rasterized 2D maps and feature vectors to construct a 2D array of vectors. In addition to the cartographic data, a visual representation of the surroundings is obtained by the camera's capture. Subsequently, the 2.5-dimensional map and the snapshot are integrated and subsequently disseminated to other robotic entities for the purpose of merging.

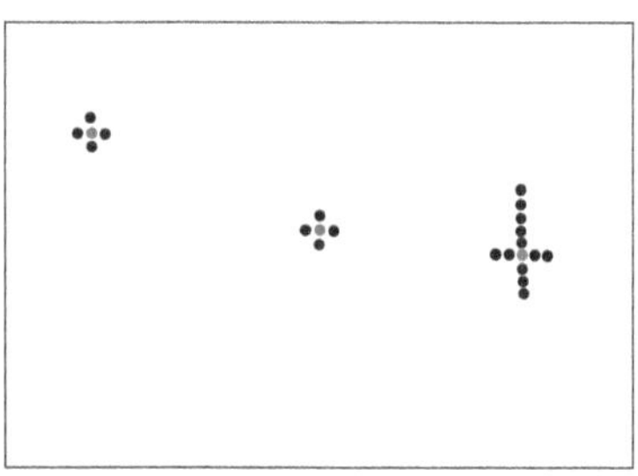

Fig. 3. Rasterized 2D Map.

Fig. 4. Reconstructed 3D map.

3.3 Stage 3

During the third stage of the process, the 2.5D map undergoes a conversion to restore its original three-dimensional representation. The current stage is comprised of four distinct modules, namely the Points Extractor, Boundary Extractor, 3D Structure Generator, and the 3D Map Generator.

Points Extractor. The points extractor module uses the 2.5D map to find the centroid points of items of interest. Using feature vectors, the points are converted from 2D to 3D. The 3D coordinates are sent to the 3D Structure Generator to determine the structure's centroid.

Boundary Extractor. From a 2.5D map, the boundary extractor module estimates the height and breadth of each object of interest. Using the feature vector and centroid, the object's height and breadth are determined. Then, the height and width data are sent to the 3D Structure Generator to determine the structure's height and width.

3D Structure Generator. This module utilizes the centroid and dimensions (height and breadth) of the item as input parameters, which are then integrated to generate a three-dimensional structure. The 3D structures are amalgamated and afterwards transmitted to the 3D Map Generator.

3D Map Generator. The 3D Map Generator creates 3D maps from 3D points and structure. Humans can understand the three-dimensional map. When 3D mapping fails, environmental images can be used. See Fig. 4 for the created 3D map.

3.4 Significance of the Proposed System

The recommended 2.5D mapping method can recreate three-dimensional maps while retaining enough information. 2.5D maps are hard to understand in human touch, therefore this is vital. 2.5D-to-3D mapping is needed. In the proposed method, each robot uses 2.5D mapping to construct spatial linkages and integrates them to represent the environment. The operator then converts the 2.5D map into a 3D map for human perception. Robots use a 2.5D map and a picture of the surroundings to capture information lost during cloud point conversion. Also, 2.5D mapping provides more data than 3D mapping. Vectors in 2.5D mapping. Highlighted objects, occupancy, and pathfinding are provided via feature vectors. The feature vectors alter. Their flexibility shows. The system can be tailored to any user or situation. Adaptability makes our proposed SLAM system better than others.

Additional customization options are available for our suggested system. Use a two-dimensional mapping to deploy this system on cheap computational equipment. Systems adapt to human demands. 2.5D mapping recognizes one or more items. In perilous situations, search and rescue robots can help. This strategy is excellent for drone scanning unexplored locations. While 2.5D mapping lacks fine detail, it often outperforms 3D mapping. The suggested method simplifies computing, provides additional data not available in current 3D mapping techniques, may be effortlessly transformed into a 3D mapping representation, improves data transfer rates, simplifies integration, and has many other advantages.

4 Implementation and Results Analysis

4.1 Implementation

In this investigation, we started 2.5D mapping. Stage 1 includes the Hitmap Generator, Localizer, and 3D Map Generator. The Hitmap Generator marks centroids to show item placements. The Localizer then adds localization data to the Hitmap. The Hitmap was used to create a three-dimensional map. The next discussion will analyze implementation in detail.

The Intel RealSense D435 camera provides Hitmap Generator's initial RGBD frames. RGB frames are used for object detection in RGBD imaging. A cup was used to prove work during implementation. The YOLOv8 model [18] detects objects and returns centroid and bounding box coordinates. Using depth frames, the centroid is used to calculate the precise global position, as shown in Fig. 5. In Stage 2, the Bounding Estimator needs bounding box data. The global position is used to create a 1000-by-1000 Hitmap.

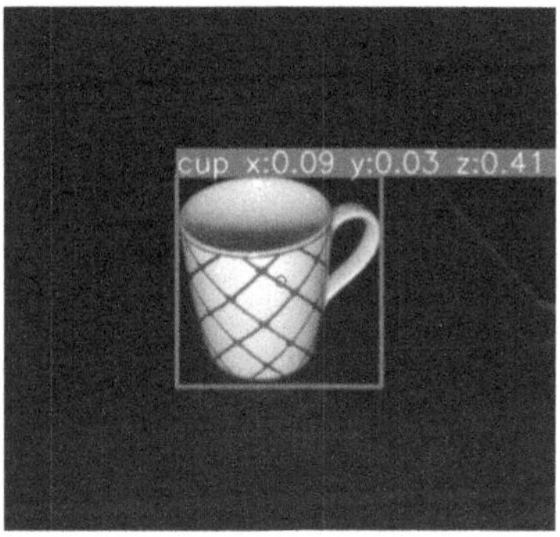

Fig. 5. Detection of Cup using YOLOv8.

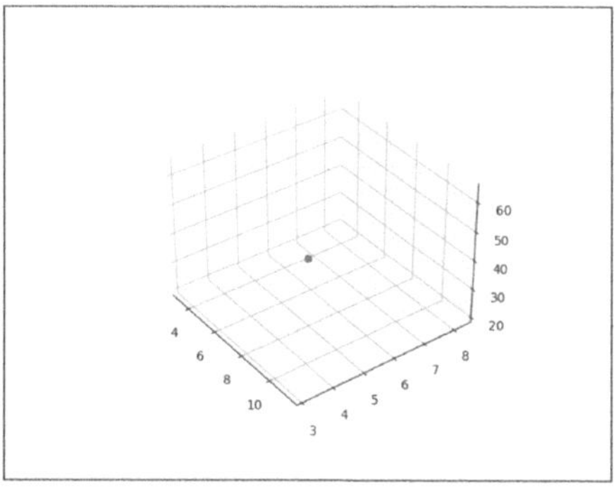

Fig. 6. 3D Generated map using Hitmap.

After recording enough hits, the self-position module computes the robot's location and updates the Hitmap. A maximum function ensures that the Hitmap retains only the highest hit value for each object. Hitmap generating problems in Yolov8 harm the local to global position module. Thus, object hit positions can vary. Max function reduces value magnitude.

The final phase is creating a 3D map after Hitmap development and optimization. The first step in 3D map building is vector generation. Each vector in this array represents an object and contains its class, sequence number, and global position. Since the array can only have non-negative values, we place the origin at (500, 500). A 3D map is created using global locations, as seen in Fig. 6.

4.2 Results

The system proposed in this paper will serve to provide a new way to create and share mapping data between multiple actors in a system. One benefit of the system is the decreased bandwidth usage when transferring map data from one actor to another. As the map size increases, a 3D map's size increases in a cubic fashion. That is, the total number of bytes used to represent a 3D map can be described as X * Y * Z, where X is the width of the map, Y is the length of the map, and Z is the height of the map. By utilizing the proposed 2.5D mapping system, a map's size increases in a quadratic manner instead. The total number of bytes for a 2.5D map can be described as X * Y * V, where X and Y remain the same, but Z is replaced by V which represents the size of the vector. Since for a map of any given size, V remains the same, it can be seen that the size of the 3D map quickly exceeds the size of the 2.5D map, as shown in Fig. 7.

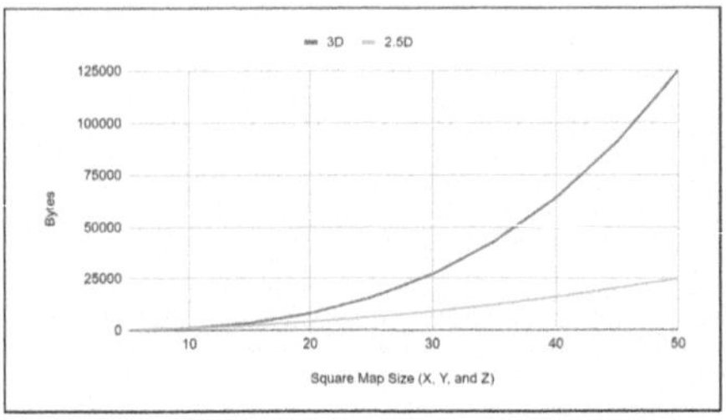

Fig. 7. A comparison of the growth rate in bytes of a 2.5D map with a 10D feature vector and a 3D map.

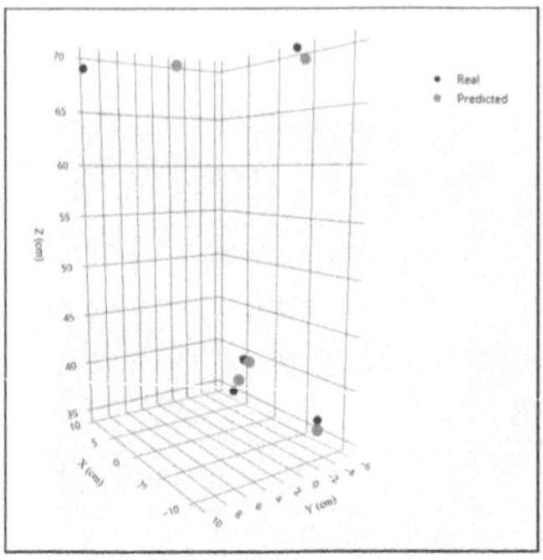

Fig. 8. A 3D representation of the values predicted by the Hitmap Generator.

Table 1. Real and Predicted Positions of Objects using Hitmap Generator

Real X (cm)	Real Y (cm)	Real Z (cm)	Predicted X (cm)	Predicted Y (cm)	Predicted Z (cm)	Error (cm)
10	0	69	9	0	70	1.41
−10	0	69	−11	0	68	2.10
5	−6	39	4	−6	39	1.00
−7	−6	36	−7	−6	35	1.00
−5	4	41	−5	2	41	2.00

This study focuses primarily on the implementation of the "Stage 1" module, which acts as the foundation for the system. Two parts of this system, the Hitmap Generator and Localizer, were tested together to predict the location of objects in a scene. The tests resulted in very accurate 3D position generation, which can be seen in Table 1. The data, further represented in Fig. 8, shows the high performance of the Hitmap Generator. The measurements, taken in centimeters, show that the Hitmap Generator was able to generate the location of the objects within nearly 2 cm for every trial.

5 Discussion

We employ 2.5D mapping in this investigation. It's better than 3D mapping. This approach outperforms low-resource mapping methods. Real-time 3D SLAM matching using adaptive feature vectors. Robot route finding and localization can be reduced by several prediction methods and high-dimensional feature vectors. It uses feature vectors to search. This reduces computing and communication bandwidths, allowing smaller robots to scan new areas. As processing resources decrease, robots can calculate most jobs without additional resources. At the conclusion, we suggested making our 2.5D map 3D.

This 3D map is human-readable. Very well-defined 2.5D maps can produce accurate 3D maps. 2.5D map possibilities abound. This article shows that 3D maps with reconstruction lose important information. Confuses 3D map interpretations. This is handled by full-feature vectors. Our 2.5D mapping permits two-way robot-to-robot communication, unlike robot-human communication. Robots learn from humans. Connotative 3D maps enable robot and user feedback. Share 2.5D mapping feedback with resource-constrained robots. Both humans and robots learn faster, speeding map creation and merging.

6 Conclusion and Future Works

This work suggests 2.5D mapping, which adds information to 2D mapping, to solve 3D mapping's information restrictions. Intel RealSense D435 RGBD frames for 2.5D mapping. Hitmaps are built from frame highlights. The Hitmap is updated by the robot's self-position and people's mostly error-free center positions. Inertial measuring unit and visual odometry measure robot self-position. Hitmaps show 3D self-placements and attractions. 2-step 3D-to-2.5D map conversion. Making 2D map. We create a data-filled feature vector array for each map point. Rasterization transforms 2.5D maps into 3D models. 2.5D mapping uses feature vectors and 2D maps. Integrating robots get 2.5D maps. People can see their surroundings on the map. 2.5D 3D photo graphics lose detail. Humans get combined maps. Humans create 3D maps from 2.5D. We tested the concept with numerous components and got good results.

This effort implemented the first level of our planned design. First-stage improvements are conceivable. First, localization was done manually. Ocular odometry and IMU will possibly detect localization. We just built the first step in this study, but we could build stages 2 and 3 to use the complete architecture. 2.5D and 3D maps work with stage 2 of our architecture. This study examined stage 2's 10-dimensional feature vector. We can add more parameters to the feature vector with this adaptable vector, expanding 2.5D mapping. This study detailed phase 3's 3D map reconstruction. Implementing this level proves our system can employ standard mapping methods. Stage 3 can be improved. Reconstructing the 3D map with 10D feature vectors is suggested in Stage 3. The 3D map can gain detail from future feature vector dimensions.

References

1. Mahmud, S., Ferdous, M., Sourave, R., Rahman Shuvo, M.I., Kim, J.-H.: An essential robot vision system for robot assisted plant disaster prevention and response missions. In: 2022 25th International Conference on Computer and Information Technology (ICCIT), Cox's Bazar, Bangladesh, pp. 1116–1121 (2022). https://doi.org/10.1109/ICCIT57492.2022.10056111
2. Mahmud, S., Dannemiller, J., Sourave, R., Lin, X., Kim, J.-H.: Smart robot vision system for plant inspection for disaster prevention. In: 2022 Sixth IEEE International Conference on Robotic Computing (IRC), Italy, pp. 416–420 (2022). https://doi.org/10.1109/IRC55401.2022.00079
3. Aguiar, A.S., dos Santos, F.N., Cunha, J.B., Sobreira, H., Sousa, A.J.: Localization and mapping for robots in agriculture and forestry: a survey. Robotics **9**(4), 97 (2020)
4. Kim, P., Chen, J., Cho, Y.K.: Autonomous mobile robot localization and mapping for unknown construction environments. In: Construction Research Congress 2018, pp. 147–156 (2018)
5. Kohlbrecher, S., Meyer, J., Graber, T., Petersen, K., Klingauf, U., von Stryk, O.: Hector open source modules for autonomous mapping and navigation with rescue robots. In: Behnke, S., Veloso, M., Visser, A., Xiong, R. (eds.) RoboCup 2013. LNCS (LNAI), vol. 8371, pp. 624–631. Springer, Heidelberg (2014). https://doi.org/10.1007/978-3-662-44468-9_58
6. Wang, H., Zhang, C., Song, Y., Pang, B., Zhang, G.: Three-dimensional reconstruction based on visual SLAM of mobile robot in search and rescue disaster scenarios. Robotica **38**(2), 350–373 (2020)
7. Xiao, L., Wang, J., Qiu, X., Rong, Z., Zou, X.: Dynamic-SLAM: semantic monocular visual localization and mapping based on deep learning in dynamic environment. Robot. Auton. Syst. **117**, 1–16 (2019)
8. Li, X., Du, S., Li, G., Li, H.: Integrate point-cloud segmentation with 3D LiDAR scan-matching for mobile robot localization and mapping. Sensors **20**(1), 237 (2019)
9. Rosas-Cervantes, V., Lee, S.G.: 3D localization of a mobile robot by using Monte Carlo algorithm and 2D features of 3D point cloud. Int. J. Control Autom. Syst. **18**, 2955–2965 (2020)
10. Luo, R.C., Lin, S.-C., Lai, C.C.: Indoor autonomous mobile robot localization using natural landmark. In: 34th Annual Conference of IEEE Industrial Electronics, Orlando, FL, USA 2008, pp. 1626–1631 (2008). https://doi.org/10.1109/IECON.2008.4758197
11. Wang, H., Zhang, C., Song, Y., Pang, B., Zhang, G.: Three-dimensional reconstruction based on visual SLAM of mobile robot in search and rescue disaster scenarios. Robotica **38**(2), 350–373 (2020)
12. Schmuck, P., Chli, M.: CCM-SLAM: robust and efficient centralized collaborative monocular simultaneous localization and mapping for robotic teams. J. Field Robot. **36**(4), 763–781 (2019)
13. Bavle, H., De La Puente, P., How, J.P., Campoy, P.: VPS-SLAM: visual planar semantic SLAM for aerial robotic systems. IEEE Access **8**, 60704–60718 (2020). https://doi.org/10.1109/ACCESS.2020.2983121
14. Zhang, L., Wei, L., Shen, P., Wei, W., Zhu, G., Song, J.: Semantic SLAM based on object detection and improved Octomap. IEEE Access **6**, 75545–75559 (2018). https://doi.org/10.1109/ACCESS.2018.2873617

15. Rosas-Cervantes, V.A., Hoang, Q.D., Lee, S.G., Choi, J.H.: Multi-robot 2.5 d localization and mapping using a Monte Carlo algorithm on a multi-level surface. Sensors **21**(13), 4588 (2021)
16. Jiang, G., Yin, L., Jin, S., Tian, C., Ma, X., Ou, Y.: A simultaneous localization and mapping (SLAM) framework for 2.5 D map building based on low-cost LiDAR and vision fusion. Appl. Sci. **9**(10), 2105 (2019)
17. Nam, T.H., Shim, J.H., Cho, Y.I.: A 2.5 D map-based mobile robot localization via cooperation of aerial and ground robots. Sensors **17**(12), 2730 (2017)
18. Home - Ultralytics YOLOv8 Docs. https://docs.ultralytics.com/

Context-Aware Robotic Assistance for Workers Using Intention Recognition and Semantic Digital Twin

Snehal Walunj[1(✉)], Parsha Pahlevannejad[1], Michael Sintek[2], Juan Carlos Saborio[3], Prerna Garg[1], Hooman Tavakoli[1], Christiane Plociennik[1], Nicolas Grossmann[1], Ansgar Bernardi[1], Martin Ruskowski[1], Joachim Hertzberg[3], and Andreas Dengel[1]

[1] German Research Center for Artificial Intelligence, Kaiserslautern, Germany
snehal.walunj@dfki.de
[2] German Research Center for Artificial Intelligence, Berlin, Germany
[3] German Research Center for Artificial Intelligence, Osnabrück, Germany

Abstract. Context-aware assistance systems are becoming essential for aiding decision-making and task management for the workers. However, integrating these systems with various other entities such that they interact seamlessly, poses a challenge. This paper proposes a solution that leverages advanced interaction protocols and context-aware robotic assistance to ensure smooth and effective collaboration in the factory environment supported by three pillars: 1) an ontology-based Digital Twin (DT) representing the semantic dimension and providing object-specific information, 2) an egocentric perception system using HoloLens2 to observe the worker with individually tailored dataset for context recognition, and 3) a robotic assistant guided by an intention-recognition module that responds to emerging worker needs, based on behavioral and environmental observations. In a manual assembly scenario, our framework provides useful contextual information as well as automated and adaptive robotic assistance. We present the results from a user experiment which was conducted to compare the stress level of participants with and without the use of the assistance system in a manual assembly scenario. Physiological sensors from the Empatica E4 wrist-watch were used to collect physiological data on which the stress levels were detected using a deep learning model. Additionally, the user ratings on the System Usability Scale (SUS) were used to assess the usability and acceptance of robotic-assistance technology.

Keywords: Smart Factory · Intention Recognition · Context Recognition · Egocentric Perception · Object Detection · Ontology-based DT · Robotic Assistance · Worker Assistance

1 Introduction

Unlike machines, humans contribute to the modern manufacturing system, through their creativity, dexterity, cognition, and decision-making prowess. The

D. Singh et al. (Eds.): IHCI 2024, LNCS 15557, pp. 92–104, 2025.
https://doi.org/10.1007/978-3-031-88705-5_8

manual workforce serves as a wellspring of innovation, representing a significant asset for industries. In factories, workers are required to perform various manual tasks, which means their hands are often not free and voice commands are not practical due to noisy factory environment. Hence, a worker assistance system must provide the necessary situation-specific, intuitive and prompt support in new, adaptive and innovative ways.

In this paper we present a framework that encompasses the core topics of semantic technologies, intention recognition and mobile robots, which together provide the means to integrate a variety of principled and applied innovations that address the problem of automated worker assistance in smart factories. Semantic technologies enhance the DT model, which represents the physical factory and its processes in a virtual world. The DT can simulate different scenarios, test changes to the factory layout or equipment, and optimize overall performance. This DT also provides uniform access to software components that benefit from object-specific information, such as planners for actions, tasks, and (worker) intention recognition [25]. Information can be provided on-demand by introducing *zoom levels* (explained in Sect. 3.1).

By utilizing a head-mounted camera, one can effectively monitor the worker's Field-of-View (FOV) which serves as crucial contextual information of work status. A YOLOv5 object-detection model was trained on custom synthetic image datasets and deployed. To respond and adapt to worker activity, we introduce an intention recognition module that uses external observations to estimate the worker's current plan, and consequently their upcoming actions and overall goals. This information allows the robot to pursue tasks and actions that fulfill worker needs, by maximizing the expected return in an underlying Partially-Observable Markov Decision Process (POMDP) where rewards represent the fulfillment of goal conditions shared between the robot and the worker [33]. Finally, a robot equipped with an arm and several onboard sensors, is controlled by multiple AI methods that allow the robot to perform a variety of actions in response to the worker's intention. The robot can navigate the factory setting, locate and grasp objects, and perform smooth hand-over motions [22].

Furthermore, a user study was conducted, with a focus on the acceptance and perception of automated assistance during an assembly task, by measuring the physiological parameters and estimating the stress of a group 20 of participants. Participant feedback was recorded through self-reported questionnaires on the System Usability Scale (SUS) [4] and NASA Task Load Index (NASA-TLX) [13].

2 Related Work

Industries are actively pursuing mass customization of products to attain reduced unit costs per customized item. As product variety increases, the support for human workers involved in production becomes increasingly crucial. A worker assistance system, is designed to aid its users in their tasks, offering relevant information, and addressing both physical and cognitive aspects of the manual work [20], while aiming to prevent stress from arising due to physical and mental activities. Since this paper addresses both the technical platform as well as its evaluation, we need to review the topics of robotic assistance, intention recognition on robots and digital twins as well as system usability.

The usage of robots in industry was proposed in 1961. [15] deals with the technology of autonomous mobile robots (AMR) and their implementation on a production line. Robots can also be used to assist humans in situations such as a manual assembly process or a maintenance task [5]. Mobile robots in general require different types of planners for navigation, manipulation or grasping; in our robot platform we use the Unified Planning (UP) library together with the Embedded Systems Bridge [34], both part of the project AIPlan4EU. UP provides access to multiple planners but we primarily rely on Fast Downward for deterministic planning [14].

For goal and intention recognition we use the Partially Observable Markov Decision Process (POMDP) model and a suitable POMDP planner. POMDP planning possesses a large body of work but we focus on generative sampling methods (c.f. [37,39]) capable of handling comparatively large and complex problems. Our intention recognition system is based on the RAGE planner, an MCTS approach that uses relevance-based planning improvements from [31,32].

Work in intention recognition (IR) addresses the core problem of correctly identifying the behavior of an external agent. An *intention* may be seen as a *motivation*, a subgoal or a set of possible future actions [9]. The challenge is estimating the probability that another agent is pursuing some given plan, based on activity observations. Existing work often focuses on merely observing and classifying intentions and there is no direct interaction [10,28,29,40], and relies on pre-computed worker plans obtained from a library or constructed from a grammar [10,27]. We instead use the *reverse* planning approach where a sequence of actions or activities is observed to guide the search for the current goal or plan [8,28,38]. Most of this existing work assumes determinism and full observability, but efforts also exist to address missing information [38] and to extend goal recognition to probabilistic and partially observable domains [29]. Our approach addresses these last two dimensions.

We follow the approach known as Active Goal Recognition (AGR), since our robot must support and not just observe the worker. Previous work in AGR combines the robot and worker variables in POMDP states using factored or mixed structures [1,23]. Our method also uses extended POMDP states with worker variables but no explicit factorization is required. An additional strength is that our proposal uses an anytime, online generative POMDP planner that may adapt to changes quickly by interleaving planning and acting.

In previous years, Digital Twins were combined with semantic technologies like RDF [19], RDFS [11], OWL [16] ontologies, and other semantic web technologies like reasoning. In "A Holistic Digital Twin Based on Semantic Web Technologies to Accelerate Digitalization", Moder et al. deal with a Digital Twin using Semantic Web representations of supply chains deploying and manufacturing semiconductors, where the ontology is based on existing W3C recommendations like a sensor ontology [24]. In "A design framework for adaptive digital twins", Erkoyuncu et al. developed a new DT design framework that uses ontologies to enable co-evolution with complex engineering systems by capturing data in terms of variety, velocity, and volume across the asset life-cycle [6].

The perceived usability of the assistance system can be evaluated using System Usability Scale (SUS) [4] which is a standardized questionnaire-based approach to work out system usability which is subjective in nature [21]. NASA-TLX [13] is another questionnaire-based method which is widely used to find out workload.

3 Implementation

3.1 Ontology-Based Environment Model for DT

Our DT is modeled within Unity. Since CAD models are available for industrial components as they are an integral part of the product development process [35], they are used for modeling the factory environment in 3D. This helps to model and visualize every component of the system and its 3D correspondences with other features of our assistance system. For example, the 1:1 scaled 3D environment is useful for simulating the robot navigation and identifying obstacles in its path.

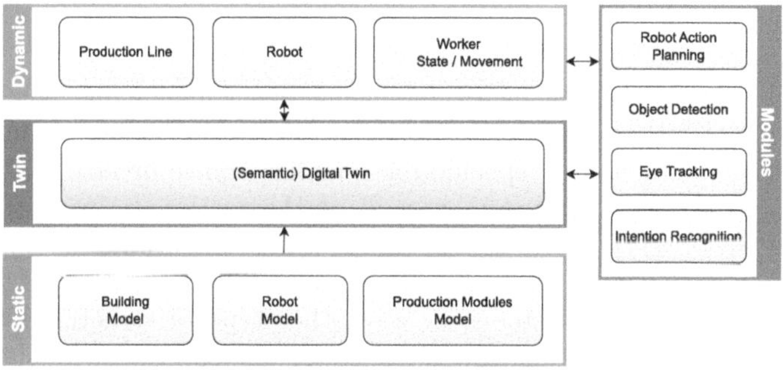

Fig. 1. System architecture for Smart Factories

Figure 1 shows our system architecture for DTs. The lower lane contains the static data sources used to create the DT, like the models of the building and the production modules, and the robot model. The upper lane contains the dynamic sources that are fed into the DT at run time, and some software modules that interact both with the dynamic sources and the DT. The semantic DT is built by reusing existing, standardized ontologies that are extended to represent factories and their DTs. Using ontologies for DTs is beneficial since ontologies guarantee a common understanding of the DT. Secondly, query and rule languages building on these ontologies improve designing APIs for DTs. The ontology-based model of the DT thus constitutes a semantic environment model. API access to knowledge represented in this semantic environment model and stored in standard triple stores is facilitated by using the SPARQL query language [12] and SPARQL protocol [7].

The environment model has to aggregate information about buildings, rooms, and factory objects like production lines, tools, humans, robots, etc. We therefore identified the Building Information Modeling (BIM) [3] as a solid source for all building-related aspects of a factory setting, ready to be automatically mapped and integrated into our model of the environment [41]. We chose the "Building Topology Ontology" (BOT) as the base of our ontology as it is a W3C standardized ontology and widely accepted in industry [30], see Fig. 2.

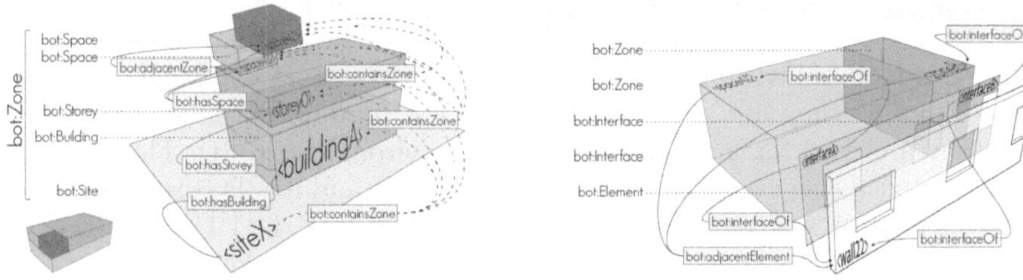

(a) Zones

(b) Interfaces and elements

Fig. 2. BOT ontology

An extension for general factory elements was added to be able to represent all kinds of structures in a factory, like production lines, modules, tools, etc. Since these elements are specific for a factory, quite diverse and dynamic, we simply introduced a means to tag all of these elements with a tag hierarchy using SKOS [17]. SKOS is the Simple Knowledge Organization System recommended by the W3C for representing thesauri, classification schemes, taxonomies, subject-heading systems, or any other type of structured controlled vocabulary. The tags we use here are a taxonomy of factory elements like demonstrators, production modules, tools etc. We also extended various tools from the DT tool chain to allow the user to add, edit, and view these tags for all factory elements. Their further extensions include several properties for spatial relations and also *zoom levels* [26], which can be used for spatial inferences which are used, e.g., for robot planning. Refer to full paper for details of our ontology based DT [41].

3.2 Context Recognition

To be able to anticipate the worker's future actions and cater to their needs, it is essential to observe the worker for context. We use the head-mounted camera from HoloLens-2 to observe the worker's field-of-view for objects as in Fig. 3, which is the necessary context in the assembly process. These observations serve as inputs for the intention recognition module.

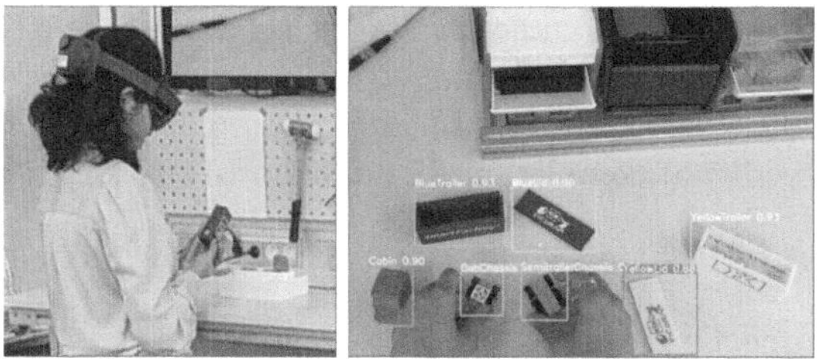

Fig. 3. Manual assembly (left); egocentric object detection (right)

Object detection is performed on egocentric videos using YOLOv5 object detection model [18]. For training the model, a synthetic dataset was generated using Unity's Perception Package [2]. It is often difficult and time consuming to collect real data, and label it manually. 3D CAD model of the industrial objects can be used to model simulation scene in Unity. various data augmentations were deployed using scene randomizers. Additionally, the head mounted camera is in constant motion hence the noise on account of motion blur and changing views is immensely high compared to a fixed camera. Such data augmentations were included during dataset generation. Figure 3 (right) shows the results of multi-object detection on an image.

3.3 Intention and Goal Recognition

Intention recognition is a process that integrates perception, representation and planning. We assume workers follow plans in order to achieve goals, so the task of the robot is to identify those goals in advance. This is inherently a probabilistic and partially observable process, and we approach it using reverse planning: given a sequence of actions, determine the current worker's plan.

We modeled these dynamics into our AGR-POMDP which represents the robot's decision-making process, with the ability to simulate the worker dynamics. This becomes an action-value maximization problem, so we also rely on the well-known convergence properties (as well as limitations) of POMDPs. The AGR-POMDP model was formally presented in [33].

In very broad terms, the AGR problem is solved by planning over a POMDP where states represent the robot and the worker, and robot actions may allow the worker to complete goals. A stochastic worker model represents the solution (or policy) to a worker MDP, and is used as a simulation of their behavior. This worker model is part of the POMDP states, such that the robot can simulate the results of its own actions in combination with the worker behavior (out of the robot's control). The robot actions can, however, provide or enable conditions that help the worker achieve goals. Because the robot behavior follows a POMDP, at each point in time we are interleaving actions and (real-world) observations, and this information is used to update the robot's own estimate of the actual state of the environment. This also means the robot can respond to emerging worker needs, environment changes, uncertain action outcomes, etc.

The resulting system can handle many of the challenges of AGR by exploiting the features of relevance-based planning [31,32]. Among such challenges are the inherent lack of information, large state-, action- and observation-spaces, delayed rewards and delayed goal completion, in addition to the long planning horizons in complex POMDPs.

3.4 Performance Overview

The resulting AGR-POMDP was evaluated in simulation with the RAGE planner and with POMCP [37] as a performance baseline. RAGE achieves much better returns with lower error, which suggests more stable solutions focusing on goal completion (i.e. assistance). For further details refer to the full paper [33].

The better performance of the RAGE planner (based on relevance estimation) highlights the fact that solving intention recognition problems in real-time, in practical applications such as factory assistance, demands a more sophisticated approach to partially-observable planning. Intention recognition by planning and acting online requires high-quality plans interleaved with an efficient use of information from observations, in order to quickly find useful actions and avoid unnecessary steps. This is an important lesson when transferring AI methods to physical robots, particularly in challenging scenarios involving humans and other machines. The resulting IR module (AGR-POMDP with RAGE) was integrated onboard a real robot platform with very successful results. In our controlled tests, the robot seeks to collect observations about worker activity and part availability, and attempts to bring missing tools and parts. For example, is the truck being assembled needs glue, it will proceed to bring the glue. If a lid is missing, it brings an additional set. This non-trivial integration is illustrated in Fig. 4, where the IR-planner receives information from sensors and delegates

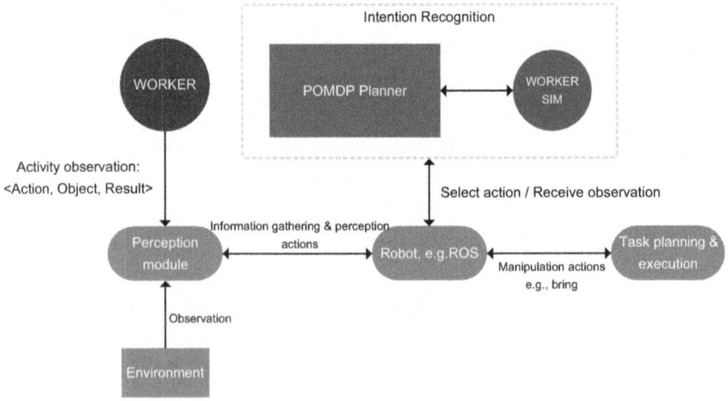

Fig. 4. Intention Recognition framework

high-level manipulation tasks. Further details about the robot and the different supporting technologies can be found in [22].

4 Evaluation

To evaluate our intent-based robotic assistance system, a user test was conducted with 20 participants, quantitatively and qualitatively. Participants underwent assembly tasks in two cases: Case 1 without robot assistance and Case 2 with robot assistance. Total stress level of participants in both cases was computed to investigate if robotic assistance helps reduce stress levels of the participants. We also measured the perceived usability of the system through reported questionnaires. The hypothesis is that workers working in case 2 would have lower stress compared to workers in case 1.

Stress Detection Model: We developed a stress detection model using an open source dataset WESAD [36] that maps physiological signals from Empatica E4. For binary classification, baseline and amusement classes were combined into a single class (non-stress) with label zero and stress as another class with label one in WESAD. A deep learning model based on Convolutional Neural Network (CNN) – Long Short-Term Memory (LSTM) for binary classification was trained on this data. This model analyzes physiological signals from four sensors of E4 wristband. The physiological data includes Electrodermal Activity (EDA), Blood Volume Pulse (BVP), Temperature (TEMP) and Acceleration (ACC) for inferring the persons stress levels. We used all sensor data separately as input to the model.

The signals were pre-processed, normalized, and segmented using a sliding window of length 1 s with no overlap. The CNN component extracted spatial patterns from the sensor data, while the LSTM component captured temporal dependencies within the sequential data. After the CNN and LSTM layers, the features extracted from different signals are concatenated into single feature vector and then passed through multiple dense layers for binary classification. User independent model was trained using Leave-One-Participant out cross validation technique for generalization purpose. Final accuracy of CNN LSTM model was 0.84 with F1 score of 0.82. We saved the model weights after training for inference on user test data.

User Study: Participants were graduate students as well as professionals with mean age of 32 $\pm$ 2 years. They were asked to performed an assembly task in case 1 (Fig. 5a) and case 2 (Fig. 5b). To balance the ordering effect, we split the participants into two groups with opposite orders in which they underwent the two cases. With HoloLens mounted on head, Empatica E4 was worn on their non-dominant hand. A timer was set, during which they had to assemble the parts and their physiological signals were recorded. They were asked to fetch parts/tools from inventory themselves in case 1, whereas intent recognition based robot assisted the participants in fetching the parts from inventory in case 2.

Further, we used our trained stress-detection model to predict stress on physiological data collected in case 1 and case 2. On average, participants were stressed

(a) Case 1: Without Robot Assistance

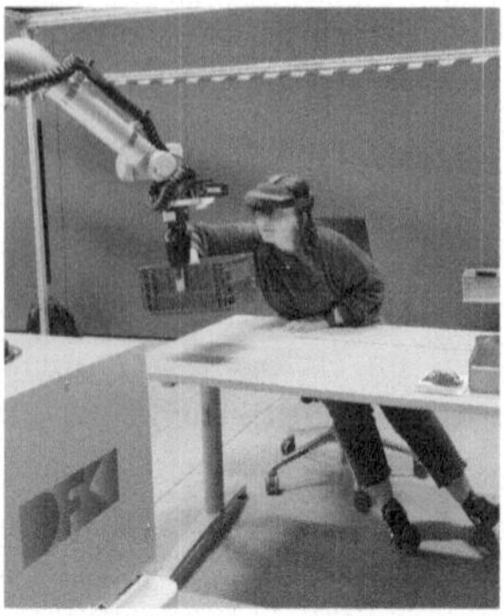

(b) Case 2: With Robot Assistance

Fig. 5. Participants undergoing assembly in both cases with worn sensors

around 63.25% of the total experimental duration during case 2 compared to nearly 76.24% in case 1. Hence, as per the hypothesis, it can be seen that stress levels of participants were slightly lower in case 2 with robotic assistance. Moreover, the effect of robotic assistance on the percentage stress duration was found out to be statistically significant ($p < 0.05$). Percentage stress duration ($p = 0.018$) exhibits a significant disparity between two cases, indicating that the assistance provided by the robot in case 2 indeed had a substantial impact on stress duration. Post experiment, participants were asked to fill in questionnaires and provide feedback to determine the usability of the worker assistance system. In our study, we utilized the System Usability Scale (SUS) and NASA-Task Load Index questionnaire. **System Usability Scale** serves as a standard tool to assess a range of critical factors when evaluating a system for its efficiency, intuitiveness, ease of use, and user satisfaction. SUS provides valuable insights into the usability and user experience of a given system or product. SUS scores above 68 are considered above average [4]. Scores below 50 are considered 'Not Acceptable' and Scores above 71 are considered 'Acceptable'. Figure 6 depicts individual SUS score ratings given by participants. On average the participants ratings fall above the acceptance level with an average score of 75.375 which validates the overall worker assistance system to be acceptable based on usability and user-friendliness. **NASA-Task Load Index** was used to assess perceived workload following each task on six dimensions (i.e., mental, physical, and temporal demands, performance, effort, and frustration) and the overall weighted workload. Figure 7a shows the mean score given by participants for each dimension in each condition. The effect of robotic assistance on the NASA-TLX overall weighted score dimension was found out to be statistically significant ($p < 0.05$). Weighted scores (p value $= 0.002$) exhibit a significant disparity between two cases, indicating that the assistance provided by the robot in case 2 had a substantial impact on perceived workload.

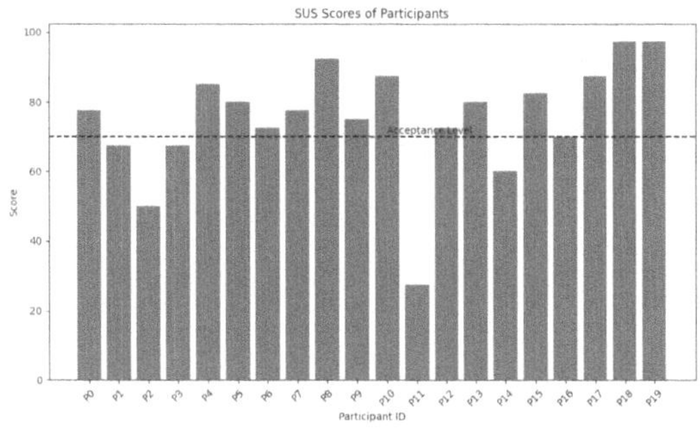

Fig. 6. System Usability score in case 1 and case 2

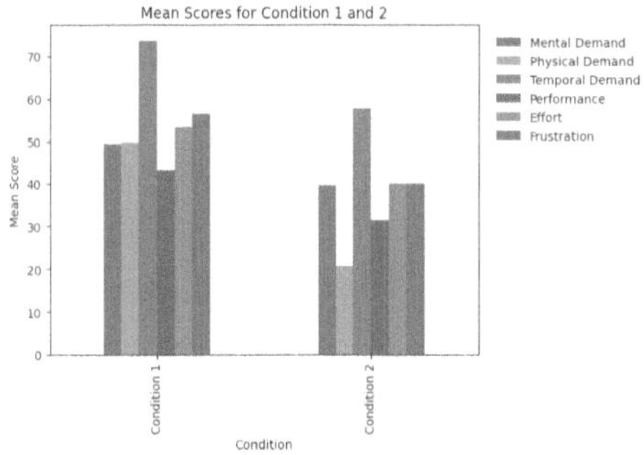

(a) Mean Score for each dimension

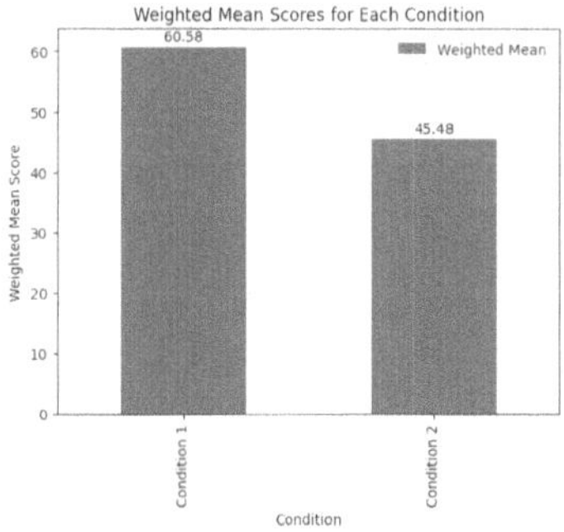

(b) Weighted mean score

Fig. 7. Mean score and weighted mean score for condition 1 and 2

5 Conclusion and Future Work

In this paper, we have described a worker assistance system for smart factories, built around a robot that bases its assistance functionality on the worker's intentions and an HRC approach enabled by a semantic, ontology-based digital twin. The technical elements of our proposal are theoretically sound, and we also demonstrated the feasibility of our platform in a manual assembly scenario which included its evaluation in a user study. Results show that the participants were more efficient and less stressed with robot assistance, and that they perceived the robot assistance as useful. In their feedback participants addressed well-known issues, such as the robot being perceived as slow, and that robot assistance could have been faster, both of which respond to technical and safety limitations of our specific robot platform. Other suggestions were concerned with verbal interaction between human and robot, and another one mentioned feedback from the robot while assisting workers. All of these comments highlight valuable research avenues for the future, as they directly connect with open problems in robot planning and acting and in human-robot interaction.

Acknowledgments. The project InCoRAP is supported by the German Federal Ministry of Education and Research under grant no. 01IW19002. We would also like to thank our student-assistants and all the participants of the user study.

References

1. Amato, C., Baisero, A.: Active goal recognition. arXiv preprint arXiv:1909.11173 (2019)
2. Borkman, S., et al.: Unity perception: generate synthetic data for computer vision. arXiv preprint arXiv:2107.04259 (2021)
3. Borrmann, A., König, M., Koch, C., Beetz, J.: Building information modeling: Why? What? How? In: Borrmann, A., König, M., Koch, C., Beetz, J. (eds.) Building Information Modeling: Technology Foundations and Industry Practice, pp. 1–24. Springer, Cham (2018). https://doi.org/10.1007/978-3-319-92862-3_1
4. Brooke, J.: SUS: a 'quick and dirty' usability. Usability Eval. Ind. **189**(3), 189–194 (1996)
5. Costa, G.d.M., Petry, M.R., Moreira, A.P.: Augmented reality for human-robot collaboration and cooperation in industrial applications: a systematic literature review. Sensors **22**(7), 2725 (2022)
6. Erkoyuncu, J.A., del Amo, I.F., Ariansyah, D., Bulka, D., Vrabič, R., Roy, R.: A design framework for adaptive digital twins. CIRP Ann. **69**(1), 145–148 (2020). https://doi.org/10.1016/j.cirp.2020.04.086. https://www.sciencedirect.com/science/article/pii/S0007850620301086
7. Feigenbaum, L., Williams, G.T., Clark, K.G., Torres, E.: SPARQL 1.1 protocol. W3C Recommendation (2013). https://www.w3.org/TR/2013/REC-sparql11-protocol-20130321/
8. Freedman, R., Zilberstein, S.: Integration of planning with recognition for responsive interaction using classical planners. In: Proceedings of the AAAI Conference on Artificial Intelligence, vol. 31 (2017)
9. Freedman, R.G., Zilberstein, S.: A unifying perspective of plan, activity, and intent recognition. In: Proceedings of the AAAI Workshops: Plan, Activity, Intent Recognition (Honolulu, HI), pp. 1–8 (2019)
10. Geib, C.: Lexicalized reasoning. In: Proceedings of the 3rd Annual Conference on Advances in Cognitive Systems ACS, p. 19 (2015)
11. Guha, R., Brickley, D.: RDF schema 1.1. W3C Recommendation (2014). http://www.w3.org/TR/2014/REC-rdf-schema-20140225/
12. Harris, S., Seaborne, A.: SPARQL 1.1 query language. W3C Recommendation (2013). https://www.w3.org/TR/2013/REC-sparql11-query-20130321/
13. Hart, S.G.: Nasa-task load index (NASA-TLX); 20 years later. In: Proceedings of the Human Factors and Ergonomics Society Annual Meeting, vol. 50, pp. 904–908. Sage publications, Los Angeles (2006)
14. Helmert, M.: The fast downward planning system. J. Artif. Intell. Res. **26**, 191–246 (2006)
15. Hercik, R., Byrtus, R., Jaros, R., Koziorek, J.: Implementation of autonomous mobile robot in smartfactory. Appl. Sci. **12**(17) (2022). https://doi.org/10.3390/app12178912. https://www.mdpi.com/2076-3417/12/17/8912
16. Hitzler, P., Krötzsch, M., Parsia, B., Patel-Schneider, P.F., Rudolph, S.: OWL 2 web ontology language primer. W3C Recommendation (2009). https://www.w3.org/TR/owl2-primer/

17. Isaac, A., Summers, E.: SKOS simple knowledge organization system primer. W3C Recommendation (2009). https://www.w3.org/TR/skos-primer/
18. Jocher, G., et al.: ultralytics/yolov5: v3.1 - Bug Fixes and Performance Improvements (2020). https://doi.org/10.5281/zenodo.4154370
19. Klyne, G., Carroll, J.J.: Resource description framework (RDF): Concepts and abstract syntax. W3C Recommendation (2004). http://www.w3.org/TR/2004/REC-rdf-concepts-20040210/
20. König, M., et al.: MA2RA – manual assembly augmented reality assistant. In: 2019 IEEE International Conference on Industrial Engineering and Engineering Management (IEEM), pp. 501–505 (2019). https://doi.org/10.1109/IEEM44572.2019.8978844
21. Lewis, J.R.: The system usability scale: past, present, and future. Int. J. Hum.-Comput. Interact. **34**(7), 577–590 (2018)
22. Lima, O., et al.: A physics-based simulated robotics testbed for planning and acting research. In: ICAPS Workshop on Planning and Robotics (PlanRob 2023) (2023)
23. Massardi, J., Beaudry, É.: Active goal recognition using intention aware motion planning. In: Plan, Activity and Intent Recognition (PAIR) 2021 (2021)
24. Moder, P., Ehm, H., Jofer, E.: A holistic digital twin based on semantic web technologies to accelerate digitalization. In: Keil, S., Lasch, R., Lindner, F., Lohmer, J. (eds.) EADTC 2018-2019. LNEE, vol. 670, pp. 3–13. Springer, Cham (2020). https://doi.org/10.1007/978-3-030-48602-0_1
25. Orsag, L., Stipancic, T., Koren, L.: Towards a safe human-robot collaboration using information on human worker activity. Sensors **23**(3) (2023). https://doi.org/10.3390/s23031283. https://www.mdpi.com/1424-8220/23/3/1283
26. Perlin, K., Fox, D.: PAD: an alternative approach to the computer interface. In: Proceedings of the 20th Annual Conference on Computer Graphics and Interactive Techniques, SIGGRAPH 1993, pp. 57–64. Association for Computing Machinery, New York (1993). https://doi.org/10.1145/166117.166125
27. Rafferty, J., Nugent, C.D., Liu, J., Chen, L.: From activity recognition to intention recognition for assisted living within smart homes. IEEE Trans. Hum.-Mach. Syst. **47**(3), 368–379 (2017)
28. Ramírez, M., Geffner, H.: Probabilistic plan recognition using off-the-shelf classical planners. In: Proceedings of the Conference of the Association for the Advancement of Artificial Intelligence (AAAI 2010), pp. 1121–1126. Citeseer (2010)
29. Ramírez, M., Geffner, H.: Goal recognition over POMDPs: inferring the intention of a POMDP agent. In: IJCAI, pp. 2009–2014. IJCAI/AAAI (2011)
30. Rasmussen, M.H., Lefrançois, M., Schneider, G.F., Pauwels, P.: BOT: the building topology ontology of the W3C linked building data group. Semant. Web J. (2020). http://www.semantic-web-journal.net/system/files/swj2279.pdf
31. Saborío, J.C., Hertzberg, J.: Efficient planning under uncertainty with incremental refinement. In: Proceedings of the 35th Conference on Uncertainty in Artificial Intelligence, Tel Aviv, Israel, 22–25 July 2019, UAI 2019, p. 112 (2019)
32. Saborío, J.C., Hertzberg, J.: Planning under uncertainty through goal-driven action selection. In: van den Herik, J., Rocha, A.P. (eds.) Agents and Artificial Intelligence, pp. 182–201. Springer, Cham (2019). https://doi.org/10.1007/11829263_12
33. Saborío, J.C., Hertzberg, J.: Towards intention recognition for robotic assistants through online POMDP planning. In: ICAPS 2023 Workshop on Plan, Activity and Intent Recognition (PAIR 2023) (2023)
34. Sadanandam, S.H.S.S., et al.: A closed-loop framework-independent bridge from AIPlan4EU's unified planning platform to embedded systems. In: 33rd Interna-

tional Conference on Automated Planning and Scheduling (ICAPS 2023) PlanRob Workshop (2023)
35. Schäper, T., Foege, J.N., Nüesch, S.: Toolkits for innovation: how digital technologies empower users in new product development. R&D Manag. **54**(1), 95–117 (2024)
36. Schmidt, P., Reiss, A., Duerichen, R., Marberger, C., Van Laerhoven, K.: Introducing WESAD, a multimodal dataset for wearable stress and affect detection, pp. 400–408 (2018). https://doi.org/10.1145/3242969.3242985
37. Silver, D., Veness, J.: Monte-Carlo planning in large POMDPs. In: Advances in Neural Information Processing Systems, vol. 23, pp. 2164–2172 (2010)
38. Sohrabi, S., Riabov, A.V., Udrea, O.: Plan recognition as planning revisited. In: IJCAI, New York, NY, pp. 3258–3264 (2016)
39. Somani, A., Ye, N., Hsu, D., Lee, W.S.: DESPOT: online POMDP planning with regularization. In: Burges, C.J.C., Bottou, L., Welling, M., Ghahramani, Z., Weinberger, K.Q. (eds.) Advances in Neural Information Processing Systems, vol. 26, pp. 1772–1780. Curran Associates, Inc. (2013). http://papers.nips.cc/paper/5189-despot-online-pomdp-planning-with-regularization.pdf
40. Sukthankar, G., Geib, C., Bui, H., Pynadath, D., Goldman, R.P.: Plan, Activity, and Intent Recognition: Theory and Practice. Newnes, London (2014)
41. Walunj, S., Sintek, M., Pahlevannejad, P., Plociennik, C., Ruskowski, M.: Ontology-based digital twin framework for smart factories. In: Information Systems Development, Organizational Aspects and Societal Trends (ISD2023 Proceedings) (2023)

Conversational Agent Design: A Comprehensive Analysis of Research from Leading Conferences

Neşe Baz Aktaş[1,2](✉) and Adem Akbıyık[1]

[1] Sakarya University, Sakarya, NJ 08544, Turkey
nese.aktas@bogazici.edu.tr, nese.aktas1@ogr.sakarya.edu.tr
[2] Bogazici University, 17, 69121 Istanbul, Turkey

1 Introduction

Lately, the industry has witnessed a significant increase in using natural language in software applications. Specifically, chatbots have gained widespread use, allowing interactions through natural written language in various application fields. In addition, the adaptive capabilities of systems like ChatGPT 3.5 have influenced the CA design strategies, utilizing natural language understanding and generation [1, 2].

With the advancements in technology, it's no wonder that investments in CAs have significantly increased. The expected growth of the conversational agent market, anticipated to reach US$86.42 billion by 2032 [3], highlights its expanding footprint in global markets. Scholars have noted these trends and begun researching how people interact with this technology. These studies are crucial for understanding the dissatisfaction and negative emotions caused by agents that fail to meet user expectations or possess limited capabilities, potentially driving improvements in agent design.

While the extensive research on CA design focuses on user perceptions during verbal interactions, there remains a gap in our understanding of the broader human-agent interactions—specifically, what individuals anticipate, feel, and experience when engaging with CAs [4]. Despite this, numerous studies have provided insights into user experiences with these agents, particularly in relation to their design. For example, integrating human-like linguistic cues in online shopping chatbots has been shown to enhance trust [5]. Similarly, the use of female agents in service robots has been linked to increased user satisfaction [6], and distinct explanatory features in recommender agents significantly influence users' willingness to disclose personal information [7].

Given this context, our study seeks to address the following research question: "How do current CA design practices and trends address human-agent interactions and user expectations?" We conduct a systematic literature review to explore how CA design research examines human-agent interactions. Our focus is on scholarly works from conferences spanning the last fourteen years (2010–2024), a period during which technology has not only advanced to meet user expectations but has also seen a significant increase in CA research. Throughout this review, we apply an HCI perspective, prioritizing the human aspects of interactions rather than the technical performance or capabilities of the systems.

D. Singh et al. (Eds.): IHCI 2024, LNCS 15557, pp. 105–121, 2025.
https://doi.org/10.1007/978-3-031-88705-5_9

2 Research Background

The development of conversational agents began with ELIZA in 1960, a simple program capable of mimicking human conversation by reformulating user input into questions [8]. This innovation marked the beginning of significant advancements in natural language processing. The field has since evolved to include more sophisticated systems, such as voice-activated assistants like Siri and Google Assistant in the 2010s. A notable recent milestone was the introduction of ChatGPT 3.5 in 2022, renowned for its ability to generate human-like responses and adapt to user interactions [1, 2].

Due to the significant improvements of CA, scholars have employed diverse methodologies to investigate how different characteristics of CA affect user perceptions and experiences. These research often involve multidimensional analyses that consider factors such as the context of use, the nature of the task, and demographic differences among users [9]. For example, experimental studies can measure user satisfaction, emotional response, and task efficiency in scenarios where agents display varying levels of anthropomorphism [10–12]. Qualitative approaches, such as interviews or focus groups, provide insights into how users feel about the authenticity and utility of these agents [13].

Furthermore, empirical studies have shown that anthropomorphic features like expressive voices or facial expressions significantly increase user trust and engagement [14]. Such designs have been linked to positive outcomes, including enhanced trust [15–17], improved attitudes [18], greater enjoyment of interactions [19], increased willingness to make e-commerce payments [20], higher acceptance of technology [21], and overall user satisfaction [6, 22]. Specific design elements like naturalness, which can be enhanced by reducing cognitive effort and ambiguity while increasing physiological arousal, have also been shown to improve the interaction quality [23]. Other studies suggest that non-verbal behaviors, casual conversation styles, and physical embodiments further build trust [15, 24, 25].

3 Research Method

3.1 Phase 1: Selection of Sources

We began our literature review in January 2024 by identifying relevant outlets from prominent IS and HCI conferences, chosen for their credibility and current reflection of trends in the field. The search utilized major databases such as Web of Science, AISeL, ACM Digital Library, and websites of key organizations. We conducted a keyword-based search with terms related to conversational agents, including synonyms such as "chatbot" and "virtual agent" [26, 27]. This initial phase focused on aligning with our study's objectives to synthesize pivotal works within CA design. This effort resulted in 463 studies from IS and HCI conferences.

(((Conversational OR Virtual OR Interactive OR Intelligent) AND AGENT) OR Chatbot OR "Digital Assistant") AND Design (Table 1).

Table 1. Extracted Literature

Journals and Conferences	Search Results	Related Papers After Abstract Review	Related Papers After Full-Text Review
ACM CHI Conference on Human Factors in Computing Systems	325	70	19
European Conference on Information Systems	33	16	6
Americas Conference on Information Systems	43	9	2
International Conference on Information Systems	40	25	12
Pacific Asia Conference on Information Systems	22	9	1
Total	463	129	40

3.2 Phase 2: Screening and Categorization

Two of the authors conducted an initial screening of the first 50 abstracts to refine the inclusion and exclusion criteria further. The abstracts of the identified papers were then screened using these refined criteria to ensure the relevance and integration of high-quality studies [28]. We included studies that focused on any type of CA, investigated agent design, empirically tested design and their effects, and were published in reputable conferences. Exclusions were based on lack of relevance to CA design, insufficient design implications, or absence of empirical evaluation. Subsequently, one of the authors took the lead in coding the studies using predefined dimensions (Table 2) related to communication modes, agent types, and research methods using MaxQDA24. The extraction of information from full texts was rendered straightforward due to the well-established nature of the concepts under investigation [29, 30].

3.3 Phase 3: Synthesis and Analysis

In the final phase of our review, we integrated the coded data into a concept matrix. This matrix helped us focus on the main ideas and themes from the literature rather than just summarizing the content. Such an approach enhanced our understanding of the underlying concepts and allowed us to see how different characteristics spread across the studies [28]. This deep analysis revealed key insights into conversational agent design and pointed out areas needing further investigation. By following this structured method, we conducted a thorough and detailed review of the field, using well-established literature review techniques to uncover meaningful trends and patterns in conversational agent research [31].

Table 2. CA Design Research Coding Framework

Communication Mode	Voice-based	Text-based	Text and voice-based
Agent Type	Virtual Humans	Robot	Intelligent Agent
	Digital Assistant	Recommender Agent	Chatbot
	Virtual Agent		
Perception and Outcome	Performance	Learning	Relationship
	Trust	Emotion	Acceptance
	Perception	Attitude	Ethics
	Other		
Research Method	Focus Group	Interview	Survey
	Design Science Research	Laboratory Experiment	Online Experiment

4 Results

In this section, we present the concept matrix derived from the evaluation of 40 conference papers. Our comprehensive literature review, as depicted in Table 3, highlights an emerging trend in the field of conversational agent design research, particularly over the past five years. Table 3 provides a detailed breakdown of the research methods, communication modes, and agent types utilized in these studies, further categorizing the diverse approaches and thematic focuses observed in contemporary research.

First of all, the studies cover two primary communication modes, text-based and voice-based, with no research exclusively focusing on combined modes. Notably, voice-based interactions were examined in 11 studies, exploring elements like expressive auditory **gestures** in pedagogical virtual agents to enhance learning outcomes [32], user responses to digital assistants' **verbal expressions** [33], and the effects of **female and male voices** on anthropomorphism [34]. Additionally, text-based communication modes dominated our findings, with 30 out of 40 studies investigating various aspects, such as the **communication patterns** (helper, victim, and persecutor) that affect users' perceptions, frustrations, and behaviors when interacting with error-producing CAs [35], the influence of anthropomorphic designs featuring varying levels of **verbal expressiveness** on perceived empathy and competency in therapeutic healthcare services [36], and the impact of human-like design elements, including **greetings, self-references, and polite phrases**, on enhancing service satisfaction [37].

In terms of agent types, most research concentrated on chatbot design, highlighted by 21 studies, followed by virtual agent design with 11 studies. Regarding chatbots Gnewuch et al. (2019) studied using **typing indicators** in chatbot design that affect the perceived social presence of chatbots and Sun and Sundar (2022) study chatbot design with the design of **different explanations** in chatbots. Fewer studies focused on the design of recommender agents and digital assistants. Additionally, Gambino et al.

Table 3. Research Stream Over the Year.

	2010	2012	2015	2016	2017	2018	2019	2020	2021	2022	2023	SUM
Communication Mode												
Voice-based	1	–	1	–	–	–	3	2	1	2	1	**11**
Text-based	–	–	–	–	–	2	3	5	7	5	8	**30**
Text&voice-based	–	–	–	–	–	–	–	–	–	–	–	**0**
Agent Type												
Virtual Humans	1	–	–	–	–	–	–	–	–	–	–	**1**
Robot	–	–	–	–	–	–	1	–	–	–	–	**1**
Intelligent Agent	–	–	–	–	–	–	–	1	–	–	–	**1**
Digital Assistant	–	–	–	–	–	–	–	2	–	1	–	**3**
Recommender Agent	–	–	–	–	–	–	–	–	–	1	1	**2**
Chatbot	–	–	–	–	–	2	3	2	5	4	5	**21**
Virtual Agent	–	–	1	–	–	–	2	2	2	1	3	**11**
Perception and Outcome												
Performance	–	–	–	–	–	–	–	–	1	2	2	**5**
Learning	–	–	–	–	–	–	–	–	–	1	1	**2**
Relationship	–	–	–	–	–	–	–	–	–	–	–	**0**
Trust	–	–	1	–	–	–	–	1	3	4	1	**10**
Emotion	–	–	1	–	–	–	1	3	1	1	–	**7**
Acceptance	–	–	–	–	–	–	2	1	1	2	3	**9**
Perception	1	–	–	–	–	1	–	6	3	–	–	**11**
Attitude	–	–	–	–	–	–	2	2	2	–	2	**8**
Ethics	–	–	–	–	–	–	–	–	–	–	–	**0**
Other	–	–	–	–	–	1	1	1	2	1	1	**7**
Research Method												
Focus Group	–	–	–	–	–	–	1	–	–	–	–	**1**
Interview	–	–	–	–	–	–	–	–		–	1	**1**
Survey	–	–	–	–	–	1	–	–	–	–	–	**1**
Design Science Research	–	–	–	–	–	–	1	–	–	–	1	**2**
Laboratory Experiment	1	–	1	–	–	–	4	6	4	4	3	**20**
Online Experiment	–	–	–	–	–	1	1	3	4	2	4	**16**

(The 40 analyzed conference papers may contain multiple dimensions, agent types, or communication modes, resulting in total counts that occasionally exceed 40).

(2019) explore the psychology behind user acceptance of robot receptionists in healthcare settings, while Wang and Gratch (2010) investigate user perceptions of virtual humans with behaviors like **gazing, head nods, and posture mimicking**. Lastly, Hass and Moussawi (2020) examine users' perceptions of the intelligence and anthropomorphism of intelligent agents.

In the reviewed studies, experiments are the primary research method, with laboratory experiments (20 of 40) being the most common, followed by online experiments (16 of 40). In laboratory settings, between-subject designs featuring randomized tasks were used to facilitate agent-user interactions [32, 42, 43], and factorial designs were employed to manipulate various design elements within agent construction [33, 36, 44, 45]. In the online experiment settings, participants were asked to engage with an online agent within a specified task and subsequently complete an online survey [35, 46–48]. Some treatments implemented a between-subject design with a predefined scenario [10, 49, 50]. Furthermore, researchers employed DSR, focus group, and interview methods to examine CA design. For instance, Feine et al. (2019) applied DSR to leverage empirical research knowledge for creating a prototype, followed by focus group workshops to evaluate it. Posser and Bittner (2023) defined meta requirements and gathered expert opinions on designed chatbot snippets through DSR. Harrington and Egede (2023) created agent utterances for health information seeking and presented participants with related photos, conducting interviews to explore users' trust perceptions toward chatbots. Lastly, Hu et al. (2018) employed formative research to design various **tones** for a chatbot, asking survey participants to evaluate these tones.

In our corpus, studies explore different perception and outcome dimensions related to CAs. Perception constructs are dominant, with a total of 11 studies indicating a strong focus in recent research on how users perceive CAs. For instance, a considerable number of these studies examine the extent to which agents can emulate human traits and behaviors, termed as perceived humanness (5 out of 11 studies) [55], or how an agent conveys its presence as a real and interactive entity, known as perceived social presence (4 out of 11 studies) [56, 57]. The emphasis on these constructs likely arises from their impact on user engagement and acceptance. Agents perceived as more human-like and present tend to foster a more natural and comfortable interaction environment, enhancing user acceptance and interaction [21, 58, 59]. Scholars also intend to explain other perception constructs, such as perceived empathy, aggression, rapport, intelligence, anthropomorphism, and uncanniness, receive less attention (see Appendix 2).

Concerning the trust and acceptance dimensions, they had considerable attention (respectively, 10 and 9 of 40 studies) (see Appendix 2 for construct details). For example, studies have explored the relationship between trust and various design elements, such as human-like or anthropomorphic design and the agent's ability to provide explanations [36, 39]. Research on acceptance has examined how different designs are received in areas like education and healthcare [50, 60, 61]. This aligns with prevailing trends in IS research that emphasize the importance of developing systems that are both trustworthy and acceptable [62, 63].

The following scholars examine the constructs of attitude (8) and emotion (7). These investigations aim to elucidate the dynamics between users and agents by exploring constructs such as satisfaction/dissatisfaction [37, 51], service satisfaction [46], enjoyment [44], or frustration [35]. Additionally, only two studies specifically address design aspects that influence learning outcomes with virtual agents for educational purposes [32, 64]. Our framework for perception and outcome also includes constructs related to relationships and ethics. However, our review found that no scholars have yet focused on these dimensions in the context of conversational agent design and interaction.

Furthermore, our findings indicate that scholars are exploring constructs such as motivation, external regulation, introjected regulation, persuasiveness, fidelity, teamwork outcomes, appropriateness, helpfulness, and negative word-of-mouth. These constructs, which do not currently feature within our established framework, may need to be investigated as they offer insights into human-like interactions that differ from traditional IS constructs typically applied to agent design.

5 Discussion

Our literature review aims to provide a comprehensive overview of the CA design research and offers directions for future research. In this section, we interpret research screams from our results to offer avenues for researchers and practitioners.

5.1 Current Trends and Practices

It is seen that the studies predominantly used text-based and voice-based communication modes. Voice-based modes typically focus on **auditory elements** [32], while text-based modes concentrate on the **expressiveness** of the agent [65, 66]. The use of **anthropomorphic elements** has been found to significantly impact the user experience for both text-based and voice-based communication modes [25, 37, 67, 68]. However, it appears that studies focusing on the integration of these modes and their combined usage have not been addressed in any of the research that this study sampled. Given these insights, future studies should thoroughly explore less examined topics and elements specific to communication modes.

Considering the agent types in the studies within the scope of this research, it is observed that chatbot studies are prevalent. This prevalence may be attributed to the relative ease and lower cost of creating chatbots, alongside their versatility in a wide array of everyday applications. However, the user perception of virtual humans, service robots, digital assistants, and intelligent assistants, which can occupy various roles, remains unclear. This ambiguity represents a critical area that requires further exploration by researchers.

Additionally, experimental studies, especially those that are scenario-based, are frequently employed in CA design research. This frequent use may be attributed to the simpler design and greater accessibility of participants that scenario-based studies offer. Additionally, the experiments have investigated user experiences using prototypes or product visuals rather than products in actual use. Examining the designs of agents, which are becoming increasingly common in daily life, through real usage will provide a more comprehensive perspective on users' intentions to continue using these products and their usability [69].

In the context of perception and outcomes, while studies on perception (11) are notably frequent, the recent expansion of CA usage has also significantly heightened the focus on technology acceptance (9), attitudes (8), and trust in technology (10), making these areas increasingly prominent in current research. As underscored by other scholars, this trend indicates the increasing interest in conversational agents, marking a shift from theoretical constructs to functional components of everyday life. Incorporating technologies into our lives initially influences perceptions and, subsequently, through usage, naturally leads to the acceptance and establishment of trust in these technologies [62, 63, 70].

Lastly, despite the significance of AI ethics, our sample lacks studies that specifically explore ethical considerations in CA design. Furthermore, although the deployment of CA systems in education has expanded to enhance the learning experience as engaging and enjoyable, the actual contribution to learning has been scarcely investigated (only 2 studies).

5.2 Research Avenues

The commonly used dimensions and subdimensions based on the 40 conference papers we reviewed are presented in Appendix 1. Findings show that, instead of using the perception and outcome framework proposed by Zhang & Li (2005) and Diederich et al.

(2022), researchers have started employing new dimensions (which were categories under the 'other' category) to evaluate CA user experience. For example, scholars have explored **humanlike design elements**, demonstrating that such features can reduce negative word-of-mouth [10]. These findings suggest a need for further exploration of these emerging dimensions across various types of agents and communication modes to better understand their implications for user experience.

While most studies focus on aspects such as **anthropomorphic design and verbal and non-verbal cues** [25, 36, 37], there is a lack of research addressing the improvements in communication. Brendel et al. (2020) study is among the few that explore the differentiation of **communication patterns** in chatbot interactions, specifically for appointment offers, and investigate their effect on service satisfaction. This study highlights the importance of incorporating specific elements into agent design to enhance communication. For this purpose, components such as **explanations and expressive speech acts,** which have been used by Al-Natour et al. (2021) to increase the effectiveness of recommendations in recommender agents, could be examined across different agent types.

Lastly, despite the critical role of ethical considerations in impacting user perception and experience, none of the studies in our corpus directly addressed the ethical dimensions of conversational agents (CAs). Future research must prioritize the integration of ethical evaluations to thoroughly assess how biases, privacy concerns, and the overall integrity of AI interactions influence users.

6 Conclusion

This literature review presents current trends and practices in conversational agent design research across the 40 papers we analyzed. It highlights that researchers studied predominantly text-based CAs, and there is growing interest in voice-based agents within specific fields such as education. The research also points out that laboratory and online experiments have played pivotal roles in developing interaction designs for CAs but emphasizes the need for further research on their real-world application, including ethical and integration challenges. In addition, we notice that scholars focused on anthropomorphic and multimodal features within user-centric design frameworks for CAs; also, identifying these features has been shown to significantly improve empathy, trust, user satisfaction, and even service quality. Based on our review, further studies should investigate how conversational agents can meet user expectations in sectors with complex communication needs. Future research should also focus on the detailed examination of successful design practices for conversational agents across various sectors. This effort will guide the enhancement of conversational agent technology, ensuring it is both innovative and applicable to real-world demands.

Appendix 1

Concept Matrix for Evaluated Research.

Concept Matrix for Evaluated Research.

		C. M.			Agent Type							Research Method						Performance and Outcome									
Research	Conference	Voice-based	Text and voice based	Text-based	Virtual Humans	Robot	Intelligent Agent	Digital Assistant	Recommender Agent	Chatbot	Virtual Agent	Focus Group	Interview	Survey	Design Science Re-	Laboratory Experi-	Online Experiment	Performance	Other	Ethics	Learning	Relationship	Trust	Emotion	Acceptance	Perception	Attitude
[35]	1	-	-	X	-	-	-	X	-	-	-	-	-	-	-	-	X	-	-	-	-	-	-	X	-	X	X
[47]	2	-	-	X	-	-	-	-	-	X	-	-	-	-	-	-	X	X	-	-	-	-	-	-	-	X	-
[42]	3	-	-	X	-	-	-	-	X	-	-	-	-	-	-	X	-	-	-	-	-	-	X	-	X	-	-
[43]	3	-	-	X	-	-	-	-	X	-	-	-	-	-	-	X	-	-	-	-	-	-	-	-	X	-	-
[32]	3	X	-	-	-	-	-	-	-	-	X	-	-	-	-	X	-	-	-	-	X	-	-	-	-	-	-
[44]	3	-	-	X	-	-	-	-	-	X	-	-	-	-	-	X	-	-	-	-	-	-	-	X	-	-	-
[33]	3	X	-	-	-	-	-	X	-	-	-	-	-	-	-	X	-	-	-	-	-	-	-	X	-	X	-
[45]	4	-	-	X	-	-	-	-	-	-	X	-	-	-	-	X	-	-	-	-	-	-	-	-	X	-	-
[36]	5	-	-	X	-	-	-	-	-	-	X	-	-	-	-	X	-	-	-	-	-	-	X	-	-	X	-
[72]	4	-	-	X	-	-	-	-	-	X	-	-	-	-	-	-	X	-	-	-	-	-	-	-	-	-	X
[46]	5	-	-	X	-	-	-	-	-	-	X	-	-	-	-	-	X	-	-	-	-	-	-	-	-	X	X
[51]	4	-	-	X	-	-	-	-	-	X	-	X	-	-	X	-	-	-	-	-	-	-	-	-	-	-	X
[40]	3	X	-	-	-	X	-	-	-	-	-	-	-	-	-	X	-	-	-	-	-	-	-	-	X	-	-
[38]	4	-	-	X	-	-	-	-	-	X	-	-	-	-	-	-	X	-	-	-	-	-	-	-	-	X	-
[61]	4	-	-	X	-	-	-	-	-	-	X	-	-	-	-	-	X	-	-	-	-	-	-	-	X	-	-
[60]	4	-	-	X	-	-	-	-	-	X	-	-	-	-	-	-	X	-	X	-	-	-	X	-	X	-	-
[66]	3	X	-	-	-	-	-	-	-	-	X	-	-	-	-	-	X	-	-	-	-	-	X	-	-	-	-
[34]	1	X	-	-	-	-	X	-	-	-	-	-	-	-	-	X	-	-	-	-	-	-	-	-	-	X	-
[53]	3	-	-	X	-	-	-	-	-	X	-	-	X	-	-	-	-	-	-	-	-	-	X	-	-	-	-
[10]	4	-	-	X	-	-	-	-	-	X	-	-	-	-	-	-	X	-	X	-	-	-	-	-	-	-	-
[73]	4	-	-	X	-	-	-	-	-	X	-	-	-	-	-	-	X	-	-	-	-	-	-	-	-	-	X
[54]	3	-	-	X	-	-	-	-	-	X	-	-	-	X	-	-	-	-	X	-	-	-	-	-	-	-	-
[48]	3	-	-	X	-	-	-	-	-	X	-	-	-	X	-	-	-	-	X	-	-	-	-	-	-	-	-
[74]	3	X	-	-	-	-	-	-	-	-	X	-	-	-	-	X	-	-	-	-	-	-	X	X	-	-	-
[75]	3	-	-	X	-	-	-	-	-	X	-	-	-	-	-	X	-	X	-	-	-	-	X	X	-	-	-
[76]	5	-	-	X	-	-	-	-	-	X	-	-	-	-	-	-	X	-	X	-	-	-	-	X	-	X	-
[50]	4	-	-	X	-	-	-	-	-	X	-	-	-	-	-	-	X	-	-	-	-	-	-	-	X	-	-

(continued)

(*continued*)

[68]	5	-	-	X	-	-	-	-	-	X	-	-	-	-	-	X	-	-	-	-	-	-	X	-	-	X	-
[49]	4	-	-	X	-	-	-	-	-	X	-	-	-	-	-	-	X	-	-	-	-	-	-	-	X	-	-
[52]	5	-	-	X	-	-	-	-	-	-	X	-	-	-	X	-	-	X	-	-	-	-	-	-	-	-	-
[77]	4	-	-	X	-	-	-	-	-	-	-	-	-	-	-	-	X	-	-	-	-	-	-	X	-	X	X
[37]	4	-	-	X	-	-	-	-	-	X	-	-	-	-	-	-	X	-	-	-	-	-	-	-	-	-	X
[25]	5	-	-	X	-	-	-	-	-	X	-	-	-	-	-	X	-	-	X	-	-	-	X	-	-	-	-
[78]	3	X	-	X	-	-	-	X	-	X	-	-	-	-	-	X	-	X	-	-	-	-	-	-	-	-	-
[67]	3	X	-	-	-	-	-	-	-	-	X	-	-	-	-	X	-	-	X	-	-	-	-	-	-	-	-
[39]	3	-	-	X	-	-	-	-	-	X	-	-	-	-	-	-	X	-	-	-	-	-	X	-	-	-	-
[79]	3	X	-	-	-	-	-	-	-	-	X	-	-	-	-	X	-	-	-	-	-	-	-	-	X	-	-
[41]	3	X	-	-	X	-	-	-	-	-	-	-	-	-	-	X	-	-	-	-	-	-	-	-	-	X	-
[64]	3	X	-	-	-	-	-	-	-	-	X	-	-	-	-	X	-	X	-	-	X	-	-	-	-	-	-
[65]	6	-	-	X	-	-	-	-	-	X	-	-	-	-	-	X	-	-	-	-	-	-	-	-	-	-	X

1: AMCIS, 2: HICSS, 3: CHI, 4: ICIS, 5: ECIS, 6: PACIS

Appendix 2

Perception and Outcome	Construct	Total Research	Perception and Outcome	Construct	Total Research
Performance	Service Quality	2	**Perception**	Perceived Empathy	1
	Information Quality	1		Aggression	1
	Perceived Effectiveness	1		Perceived Intelligence	1
	Perceived Usability	1		Anthropomorphism	2
Learning	Learning Outcome			Perceived Humanness	5
Trust	Trust	5		Uncanniness	1
	Risk Awareness	1		Social Presence	4
	Qualification Trust	1	**Attitude**	Dissatisfaction	1
	Goodwill Trust	1		Positive/Negative	1
	Competence	2		Sustainability Belief	1
	Trusting Belief	1		Satisfaction	1
	Privacy Concern	1		Familiarity	1
	Trustworthiness	1		Service satisfaction	5

(*continued*)

(*continued*)

Perception and Outcome	Construct	Total Research	Perception and Outcome	Construct	Total Research
	Perceived Competence	1	**Others**	Amotivation	1
Emotion	Intrinsic Motivation	2		External Regulation	1
	Warmth	1		Introjected Regulation	1
	Enjoyment	1		Persuasiveness	1
	Frustration	2		Fidelity	1
Acceptance	Behavioral Intention	5		Teamworking	1
	Continued Use	1		Appropriateness	1
	Usage Intention	1		Helpfulness	1
	Intention to Use	2		Negative WoM	2

References

1. Rajamaran, V.: From ELIZA to ChatGPT history of human-computer conversation. Resonance **28**, 889–905 (2023)
2. Rudolph, J., Tan, S., Tan, S.: War of the chatbots: bard, bing chat, ChatGPT, ernie and beyond: the new AI gold rush and its impact on higher education. J. Appl. Learn. Teach. **6**(1), 364–389 (2023). https://doi.org/10.37074/jalt.2023.6.1.23
3. Conversational AI Market Size, Share, and Trends 2024 to 2034. Precedence Research (2023). https://www.precedenceresearch.com/conversational-ai-market
4. Luger, E., Sellen, A.: "Like having a really bad pa": the gulf between user expectation and experience of conversational agents. In: Conference on Human Factors in Computing Systems - Proceedings, pp. 5286–5297 (2016). https://doi.org/10.1145/2858036.2858288
5. Konya-Baumbach, E., Biller, M., von Janda, S.: Someone out there? A study on the social presence of anthropomorphized chatbots. Comput. Hum. Behav. **139**, 107513 (2023). https://doi.org/10.1016/j.chb.2022.107513
6. Seo, S.: When female (male) robot is talking to me: effect of service robots' gender and anthropomorphism on customer satisfaction. Int. J. Hosp. Manag. **102**, 103166 (2022). https://doi.org/10.1016/j.ijhm.2022.103166
7. Al-Natour, S., Benbasat, I., Cenfetelli, R.: Designing online virtual advisors to encourage customer self-disclosure: a theoretical model and an empirical test. J. Manag. Inf. Syst. **38**(3), 798–827 (2021). https://doi.org/10.1080/07421222.2021.1962595
8. Weizenbaum, J.: ELIZA—a computer program for the study of natural language communication between man and machine. Commun. ACM **9**(1), 36–45 (1966)
9. Rapp, A., Curti, L., Boldi, A.: The human side of human-chatbot interaction: a systematic literature review of ten years of research on text-based chatbots. Int. J. Hum. Comput. Stud. **151**, 102630 (2021). https://doi.org/10.1016/j.ijhcs.2021.102630

10. Hildebrandt, F., Lichtenberg, S., Brendel, A.B., Landmann, E.: Who's bad?-the influence of perceived humanness on users' intention to complain about conversational agent errors to others. In: ICIS 2023, Hyderabad, India (2023). https://www.researchgate.net/publication/374388134
11. Ho, A., Hancock, J., Miner, A.S.: Psychological, relational, and emotional effects of self-disclosure after conversations with a chatbot. J. Commun. **68**(4), 712–733 (2018). https://doi.org/10.1093/joc/jqy026
12. Fast, E., Chen, B., Mendelsohn, J., Bassen, J., Bernstein, M.S.: Iris: a conversational agent for complex tasks. In: Conference on Human Factors in Computing Systems – Proceedings. 2018-April (2018). https://doi.org/10.1145/3173574.3174047
13. Følstad, A., Nordheim, C.B., Bjørkli, C.A.: What makes users trust a chatbot for customer service? An exploratory interview study. In: Bodrunova, S.S. (ed.) Internet Science: 5th International Conference, INSCI 2018, St. Petersburg, Russia, October 24–26, 2018, Proceedings, pp. 194–208. Springer International Publishing, Cham (2018). https://doi.org/10.1007/978-3-030-01437-7_16
14. Magyar, G., et al.: Anthropomorphic virtual assistant to support self-care of type 2 diabetes in older people: a perspective on the role of artificial intelligence. In: VISIGRAPP 2019 - 14th International Joint Conference on Computer Vision, Imaging and Computer Graphics Theory and Applications, pp. 323–331 (2019). https://doi.org/10.5220/0007572403230331
15. Cassell, J., Nakano, Y.I., Bickmore, T.W., Sidner, C.L., Rich, C.: External manifestations of trustworthiness in the interface. Commun. ACM **43**(12), 50–56 (2001)
16. Waytz, A., Heafner, J., Epley, N.: The mind in the machine: anthropomorphism increases trust in an autonomous vehicle. J. Exp. Soc. Psychol. **52**, 113–117 (2014). https://doi.org/10.1016/j.jesp.2014.01.005
17. Seeger, A.M., Pfeiffer, J., Heinzl, A.: Texting with humanlike conversational agents: designing for anthropomorphism. J. Assoc. Inf. Syst. **22**(4), 931–967 (2021). https://doi.org/10.17705/1jais.00685
18. Balakrishnan, J., Dwivedi, Y.K.: Conversational commerce: entering the next stage of AI-powered digital assistants. Ann. Oper. Res. **333**(2–3), 653–687 (2021). https://doi.org/10.1007/s10479-021-04049-5
19. Haugeland, I.K.F., Følstad, A., Taylor, C., Alexander, C.: Understanding the user experience of customer service chatbots: an experimental study of chatbot interaction design. Int. J. Hum. Comput. Stud. **161**, 102788 (2022). https://doi.org/10.1016/j.ijhcs.2022.102788
20. Yuan, L., Dennis, A.R.: Acting like humans? Anthropomorphism and consumer's willingness to pay in electronic commerce. J. Manag. Inf. Syst. **36**(2), 450–477 (2019). https://doi.org/10.1080/07421222.2019.1598691
21. Gursoy, D., Chi, O.H., Lu, L., Nunkoo, R.: Consumers acceptance of artificially intelligent (AI) device use in service delivery. Int. J. Inf. Manag. **49**(March), 157–169 (2019). https://doi.org/10.1016/j.ijinfomgt.2019.03.008
22. Xie, Y., Zhu, K., Zhou, P., Liang, C.: How does anthropomorphism improve human-AI interaction satisfaction: a dual-path model. Comput. Hum. Behav. **148**, 107878 (2023). https://doi.org/10.1016/j.chb.2023.107878
23. Chandra, S., Shirish, A., Srivastava, S.C.: To be or not to be …human? Theorizing the role of human-like competencies in conversational artificial intelligence agents. J. Manag. Inf. Syst. **39**(4), 969–1005 (2022). https://doi.org/10.1080/07421222.2022.2127441
24. Chi, O.H., Jia, S., Li, Y., Gursoy, D.: Developing a formative scale to measure consumers' trust toward interaction with artificially intelligent (AI) social robots in service delivery. Comput. Hum. Behav. **118**, 106700 (2021). https://doi.org/10.1016/j.chb.2021.106700
25. Seeger, A.-M., Heinzl, A.: Association for information systems association for information systems chatbots often fail! can anthropomorphic design mitigate trust chatbots often fail!

can anthropomorphic design mitigate trust loss in conversational agents for customer service? loss.In: ECIS 2021 (2021). https://aisel.aisnet.org/ecis2021_rp

26. Diederich, S., Brendel, A.B., Morana, S., Kolbe, L.: On the design of and interaction with conversational agents: an organizing and assessing review of human-computer interaction research. J. Assoc. Inf. Syst. **23**(1), 96–138 (2022). https://doi.org/10.17705/1jais.00724
27. Rheu, M., Shin, J.Y., Peng, W., Huh-Yoo, J.: Systematic review: trust-building factors and implications for conversational agent design. Int. J. Hum.-Comput. Interact. **37**(1), 81–96 (2021). https://doi.org/10.1080/10447318.2020.1807710
28. Levy, Y., Ellis, T.J.E.: A systems approach to conduct an effective literature review in support of information systems research. Inf. Sci. J. **9**(2006), 181–212 (2006)
29. Miles, M.B., Hubermen, A.M., Saldana, J.: Qualitative Data Analysis: A Methods Sourcebook. SAGE Publications, Thousand Oaks (2014)
30. Leavy, P.: The Oxford Handbook of Qualitative Research (Second). Oxford University Press, Cambridge (2020)
31. Bandara, W., et al.: Achieving rigor in literature reviews: insights from qualitative data analysis and tool-support. Commun. Assoc. Inf. Syst. **37**, 154–204 (2015)
32. Ceha, J., Law, E.: Expressive auditory gestures in a voice-based pedagogical agent. In: Conference on Human Factors in Computing Systems - Proceedings (2022). https://doi.org/10.1145/3491102.3517599
33. Chin, H., Molefi, L.W., Yi, M.Y.: Empathy is all you need: how a conversational agent should respond to verbal abuse. In: Conference on Human Factors in Computing Systems - Proceedings (2020). https://doi.org/10.1145/3313831.3376461
34. Haas, C., Moussawi, S.: Are anthropomorphic intelligent agents more intelligent? In: AMCIS 2020 (2020). https://aisel.aisnet.org/amcis2020
35. Brendel, A.B., Greve, M., Diederich, S., Riquel, J.: 'You are an Idiot!'-how conversational agent communication patterns influence frustration and harassment. In: AMCIS 2020 (2020). www.uni-goettingen.de/im
36. Cui, T., Wang, X., Qi, J.: Designing anthropomorphic therapeutic conversational agents: an uncanny valley perspective. In: ECIS 2021 Research-in-Progress Papers (2021). https://aisel.aisnet.org/ecis2021_rip/30
37. Riquel, J., Brendel, A.B., Hildebrandt, F., Greve, M.: "Even the Wisest Machine Makes Errors"-an experimental investigation of human-like designed and flawed conversational agents. In: ICIS 2021, Austin (2021). https://tu-dresden.de/bu/wirtschaft/winf/isd
38. Gnewuch, U., Morana, S., Adam, M.T.P., Maedche, A.: "The Chatbot is typing ..." – the role of typing indicators in human-chatbot interaction. In: SIGHCI 2018, San Francisco, CA (2018). https://aisel.aisnet.org/sighci2018
39. Sun, Y., Sundar, S.S.: Exploring the effects of interactive dialogue in improving user control for explainable online symptom checkers. In: CHI Conference on Human Factors in Computing Systems (2022). https://doi.org/10.1145/3491101
40. Gambino, A., Shyam Sundar, S., Kim, J.: Digital doctors and robot receptionists: user attributes that predict acceptance of automation in healthcare facilities. In: Conference on Human Factors in Computing Systems - Proceedings (2019. https://doi.org/10.1145/3290607.3312916
41. Wang, N., Gratch, J.: Don't just stare at me!. In: Proceedings of the SIGCHI Conference on Human Factors in Computing Systems, pp. 1241–1250 (2010). https://doi.org/10.1145/1753326.1753513
42. Cai, W., Jin, Y., Chen, L.: Impacts of personal characteristics on user trust in conversational recommender systems. In: Conference on Human Factors in Computing Systems - Proceedings (2022. https://doi.org/10.1145/3491102.3517471

43. Cai, W., Jin, Y., Zhao, X., Chen, L.: "Listen to Music, Listen to Yourself": design of a conversational agent to support self-awareness while listening to music. In: Proceedings of the 2023 CHI Conference on Human Factors in Computing Systems, pp. 1–19 (2023). https://doi.org/10.1145/3544548.3581427
44. Chin, H., Yi, M.Y.: Should an agent be ignoring it? A study of verbal abuse types and conversational agents' response styles. In: Conference on Human Factors in Computing Systems - Proceedings (2019. https://doi.org/10.1145/3290607.3312826
45. Cui, T., Peng, X., Wang, X.: Understanding the effect of anthropomorphic design: towards more persuasive conversational agents. In: ICIS 2020 (2020). https://aisel.aisnet.org/icis2020/user_behaviors/user_behaviors/9
46. Diederich, S., Lembcke, T.-B., Brendel, A.B., Kolbe, L.: not human after all: exploring the impact of response failure on user perception of anthropomorphic conversational service agents. In: ECIS 2020, Marrakech, Mororcco (2020). https://www.researchgate.net/publication/341189578
47. Bührke, J., Brendel, A.B., Lichtenberg, S., Greve, M., Mirbabaie, M.: Is making mistakes human? On the perception of typing errors in chatbot communication. In: HICSS 2021, pp. 4456–4465 (2021)
48. Hwang, A.H.C., Won, A.S.: Ideabot: investigating social facilitation in human-machine team creativity. In: Conference on Human Factors in Computing Systems - Proceedings (2021. https://doi.org/10.1145/3411764.3445270
49. Mozafari, N., Weiger, W.H., Hammerschmidt, M.: That's so embarrassing! when not to design for social presence in human-chatbot interactions. In: ECIS 2021, Austin (2021). https://ssrn.com/abstract=4305288
50. Lichtenberg, S., Morana, S.: Become a lifesaver-how to design conversational agents to increase users' intention to donate blood. In: ICIS 2023, Hyderabad, India (2023). https://tu-dresden.de/bu/wirtschaft/winf/isd
51. Feine, J., Morana, S., Maedche, A.: Designing a chatbot social cue configuration system. In: ICIS 2019, Munich, Germany (2019). http://ksri.kit.edu
52. Poser, M., Bittner, E.: May the guide be with you: ca-facilitated information elicitation to prevent service failure. In: ECIS 2023 (2023). https://www.researchgate.net/publication/370745703
53. Harrington, C.N., Egede, L.: Trust, comfort and relatability: understanding black older adults' perceptions of chatbot design for health information seeking. In: Proceedings of the 2023 CHI Conference on Human Factors in Computing Systems, pp. 1–18 (2023). https://doi.org/10.1145/3544548.3580719
54. Hu, T., et al.: Touch your heart: a tone-aware chatbot for customer care on social media. In: Proceedings of the 2018 CHI Conference on Human Factors in Computing Systems, pp. 1–12 (2018). https://doi.org/10.1145/3173574.3173989
55. Nass, C., Fogg, B.J., Moon, Y.: Can computers be teammates? Int. J. Hum. Comput. Stud. **45**(6), 669–678 (1996). https://doi.org/10.1006/ijhc.1996.0073
56. Biocca, F., Harms, C., Burgoon, J.K.: Toward a more robust theory and measure of social presence: review and suggested criteria. Pres. Teleoper. Virtual Environ. **12**(5), 456–480 (2003). https://doi.org/10.1162/105474603322761270
57. Lombard, M., Ditton, T.: At the heart of it all: the concept of presence. J. Comput.-Mediated Commun. **3**(2) (1997). https://doi.org/10.1111/j.1083-6101.1997.tb00072.x
58. Lee, S.C., Nadri, C., Sanghavi, H., Jeon, M.: Eliciting user needs and design requirements for user experience in fully automated vehicles. Int. J. Hum.-Comput. Interact. **38**(3), 227–239 (2022). https://doi.org/10.1080/10447318.2021.1937875
59. Jin, S.V., Youn, S.: Social presence and imagery processing as predictors of chatbot continuance intention in human-AI-interaction. Int. J. Hum.-Comput. Interact. **39**(9), 1874–1886 (2023). https://doi.org/10.1080/10447318.2022.2129277

60. Pietrantoni, N., Brendel, A.B., Greulich, R.S., Hildebrandt, F.: Follow me if you want to live—understanding the influence of human-like design on users' perception and intention to comply with COVID-19 education chatbots (2022)
61. Greulich, R.S., Morana, S.: Be a miracle-designing conversational agents to influence users' intention regarding organ donation. In: ICIS 2023, Hyderabad, India (2023). https://tu-dresden.de/bu/wirtschaft/winf/isd
62. Gefen, D., Straub, D.W.: Consumer trust in B2C e-Commerce and the importance of social presence: experiments in e-products and e-services. Omega **32**(6), 407–424 (2004). https://doi.org/10.1016/j.omega.2004.01.006
63. Venkatesh, V., Thong, J.Y.L., Xu, X.: Consumer acceptance and use of information technology: extending the unified theory of acceptance and use of technology. MIS Q. **36**(1), 157–178 (2012)
64. Xu, W., et al.: MathKingdom: teaching children mathematical language through speaking at home via a voice-guided game. In: Conference on Human Factors in Computing Systems - Proceedings (2023. https://doi.org/10.1145/3544548.3581043
65. Yang, Y., Kankanhalli, A.: Humor touch: should conversational agents express humor when they fail? In: PACIS 2023 (2023). https://aisel.aisnet.org/pacis2023
66. Guo, J., Guo, J., Yang, C., Wu, Y., Yang, W., Sun, L.: Shing: a conversational agent to alert customers of suspected online-payment fraud with empathetical communication skills. In: Conference on Human Factors in Computing Systems - Proceedings (2021). https://doi.org/10.1145/3411764.3445129
67. Sin, J., Munteanu, C.: A preliminary investigation of the role of anthropomorphism in designing telehealth bots for older adults. In: Conference on Human Factors in Computing Systems - Proceedings (2019. https://doi.org/10.1145/3290607.3312941
68. Morana, S., Gnewuch, U., Jung, D.: The effect of anthropomorphism on investment decision-making with robo-advisor chatbots. In: ECIS 2020, Marrakech, Morocco (2020). https://www.researchgate.net/publication/341277570
69. Sun, X., May, A.: A comparison of field-based and lab-based experiments to evaluate user experience of personalised mobile devices. Adv. Hum.-Comput. Interact. **2013**, 1–9 (2013). https://doi.org/10.1155/2013/619767
70. Hernández-Ortega, B.: The role of post-use trust in the acceptance of a technology: drivers and consequences. Technovation **31**(10–11), 523–538 (2011). https://doi.org/10.1016/j.technovation.2011.07.001
71. Zhang, P., Li, N.: The Intellectual development of human-computer interaction research: a critical assessment of the MIS literature (1990–2002). J. Assoc. Inf. Syst. **6**(11), 227–292 (2005)
72. Diederich, S., Lichtenberg, S., Brendel, A.B., Trang, S.T.N.: Promoting sustainable mobility beliefs with persuasive and anthropomorphic design: insights from an experiment with a conversational agent. In: ICIS 2019, Munich, Germany (2019). www.uni-goettingen.de/im
73. Hildebrandt, F., Brendel, A.B., Dennis, A.R., Sachdeva, A.: New bots-the influence of a conversational agent's rookie personality on users' satisfaction. In: ICIS 2023 (2023). https://www.researchgate.net/publication/374388320
74. Hyde, J., Carter, E.J., Kiesler, S., Hodgins, J.K.: Using an interactive avatar's facial expressiveness to increase persuasiveness and socialness. In: Proceedings of the 33rd Annual ACM Conference on Human Factors in Computing Systems, pp. 1719–1728 (2015). https://doi.org/10.1145/2702123.2702465
75. Jung, J.-Y., Qiu, S., Bozzon, A., Gadiraju, U.: Great chain of agents: the role of metaphorical representation of agents in conversational crowdsourcing. In: Proceedings of the 2022 CHI Conference on Human Factors in Computing Systems (2022). https://doi.org/10.1145/3491102.3517653

76. Lembcke, T.-B., Diederich, S., Brendel, A.B.: Supporting design thinking through creative and inclusive education facilitation: the case of anthropomorphic conversational agents for persona building. In: ECIS 2020 (2020). www.uni-goettingen.de/im
77. Riquel, J., Brendel, A.B., Hildebrandt, F., Greve, M.: "F*** You!"-an investigation of humanness, frustration, and aggression in conversational agent communication. In: (2021)ICIS 2021. https://tu-dresden.de/bu/wirtschaft/winf/isd
78. Setlur, V., Tory, M.: How do you converse with an analytical chatbot? revisiting gricean maxims for designing analytical conversational behavior. In: CHI Conference on Human Factors in Computing Systems, pp. 1–17 (2022). https://doi.org/10.1145/3491102.3501972
79. Wang, I., Smith, J., Ruiz, J.: Exploring virtual agents for augmented reality. In: Proceedings of the 2019 CHI Conference on Human Factors in Computing Systems, pp. 1–12 (2019). https://doi.org/10.1145/3290605.3300511

Enhancing Firefighting Robot's Capability: A Comprehensive Implementation for Enhancing Firefighting Robot's Operations

Ovishek Pal[1], Arpita Das[2], Saifuddin Mahmud[3](✉), and Awal Ahmed Fime[4]

[1] Department of Computer Science and Engineering, Southeast University, Dhaka 1208, Bangladesh
[2] Department of Computer Science and Engineering, Khulna University of Engineering and Technology, Khulna 9203, Bangladesh
[3] Computer Science and Information Systems, Bradley University, Peoria, IL 61625, USA
smahmud@bradley.edu
[4] Artificial Intelligence, Kent State University, Kent, OH 44240, USA
afime@kent.edu

Abstract. Fire accidents have severe consequences in terms of life loss, injuries and damages to the property. In hazardous areas, firefighters risk their life to safeguard others. In search of minimizing these risks has contributed to the development of small firefighting robots that can respond quickly. This work aims to develop a state-of-the-art firefighting robot where the robot is designed to help firefighters and save lives. In the paper, the firefighting robot is designed with the combination of hardware devices and deep learning algorithms. The system has different hardware devices like Raspberry Pi, Arduino Uno, Pi camera, 3 flame sensors, servo motor, motor driver, DC motors, water pump etc. The three flame sensor cover almost 180° area and each flame sensor can detect fire within 30 cm. The system moves to three different directions, like left, right and front, based on which flame sensor detected the fire. After detecting fire by the flame sensor, the Pi camera captures images and send to the YOLOv8 model for identifying the exact location of the fire in the image. The system calculates the distance and angle of fire from the water pipe using Projectile Motion formulas. Finally, the servo motor moves according to the calculated angle to point the water pipe in the exact position of the fire and water pump activates to give water for putting down the fire. The constructed YOLOv8 model for the system is trained with 6520 online images collected from Roboflow Universe and tested with 371 real time images captured by the Pi camera of the proposed system. The advanced firefighting robot achieves accuracy of 76.4%, F1 Score of 0.73 for the test dataset. The Mean Absolute Error (MAE) value for test dataset of the system is equal to 1.14.

Keywords: Firefighting Robot · YOLOv8 Model · Roboflow Universe · Arduino Uno · Raspberry Pi · Fire-Detection

D. Singh et al. (Eds.): IHCI 2024, LNCS 15557, pp. 122–133, 2025.
https://doi.org/10.1007/978-3-031-88705-5_10

1 Introduction

Fire accidents are one of the most dangerous destruction in the whole world. Fire accidents can be caused by human or it can take place naturally. Global warming is a big factor for fire accidents in the nature. A survey indicates that almost 20,000 people die because of fire accidents in the world every year [1]. As human can not move to places in fire, fire fighting robot can be a rescuer in this matter. As of late, many mechanical technology has been developed as a solution for putting out fires, and the development of putting out fire robots has become a significant exploration matter [2].

Automatic fire fighting robot can be a blessing for saving human life as well as different resources. The fire fighting robot must have the ability to detect fire's exact location and take actions to put down the fire. There are different works that have taken place to design fire fighting robots like IoT based system [3], Smart Sensor Network based system [4], etc. Most of the system use Arduino or Raspberry Pi models for implementing the systems. Different deep learning algorithms and fuzzy logic methods are implemented to detect fire locations in these systems.

In this paper, an automatic fire fighting robot is developed using the combination of Arduino Uno, Raspberry Pi and different sensors. The fire robot can detect the fire from images capture using Pi camera using the trained YOLOv8 model. Three flame sensors (left, front and right) are used to detect the initial location of the fire. The robot moves based on which flame sensor detected the fire. Then Pi camera captures the image of the fire location. A YOLOv8 model takes the image as input and identifies the exact position of the fire in the input image. Then, the distance and angle of the detected fire from the water pipe line is calculated. Finally, a servo motor is used for moving the water pipe to the calculated angle for putting water in the fire location. The implemented YOLOv8 model in the proposed system is trained using multiple datasets collected from Roboflow Universe. The model is tested using real life images collected using the Raspberry Pi and Pi camera of the proposed system. The performance of the model is evaluated using accuracy, f1 score and Mean Absolute Error (MAE) value.

So our contributions to this paper are:

1) Designing an automatic fire fighting robot using Raspberry pi, Arduino Uno and different sensors.
2) Implementing YOLOv8 model by 5 datasets available in online for detecting fire location in the input images.
3) Detecting fire location using YOLOv8 model and testing the model's performance using real time images captured using Pi camera.
4) Calculating distance and angle of the fire location for moving the servo motor, attached with the water pipe, for putting down the fire.

The following sections provide more details about the design of the fire fighting robot proposed in the paper. Section 2 explains some of the works related to this field done previously. Section 3 explains the flowchart of the whole system, the components used for developing the system, and the implementation of hardware and software parts of the system. Section 4 explains the hardware setup of the system and the performance analysis of the trained machine learning model. Finally, Sect. 5 concludes the overall processes of the system and future plan for this research.

2 Literature Review

Many research works have taken place for designing an effective and faster fire fighting robots in recent years.

In paper [5], the authors developed a fire fighting robot using a ESP32 micro-controller. The system is established by combining flame sensor, servo motor, DC motor, water pump and motor driver. The communication between hardware and the app is established using Internet of Things (IoT) technology. The authors also developed a Blynk mobile app for controlling the robot remotely. But the proposed system can not move and detect fire automatically because it always needs human supervisor. Nagaraja et al. [6] developed a robot that uses Arduino Uno as the main control device. This developed system uses flame sensor, GSM module, servo motor, DC motor, ultrasonic sensor, water pump and motor driver. Flame sensor detects the fire and ultrasonic sensor can detect if any obstacle comes in front of the robot. But the system also can not detect the exact location of the fire, also can not calculate distance between the robot and fire.

In the paper [7], Patil et al. developed a fire fighting robot car with the NodeMCU micro-controller. The micro-controller has builtin wi-fi capabilities. The micro-controller also controls the movement of the motor driver used in the system. The whole system is controlled using a mobile application which is connected with the NodeMCU micro-controller using wi-fi. Dubey et al. [8] designed an Arduino based fire fighting robot that can notify about the location of the affected area to the user. The system uses both smoke detector and IR sensor for identifying fire location. GSM module and GPS module is used for notifying about the affected area in the developed system. The movement of the robot is also controlled by the developed mobile application. Krishan et al. [9] proposed a fire fighting robot system that can detect fire and obstacle controlled by Arduino Uno. For fire detection, the system uses IR flame sensor and for obstacle detection, ultrasonic sonar sensor is used for autonomous movement. The system has water tank and a spray mechanism for putting down fire in the affected area.

Monica et al. [10] developed a fire fighting robot car that moves using motor drivers in enclosed spaces. The system also uses fire or flame sensor for detecting fire. The system is developed using Arduino Uno as the controlling device. The system has 4 flame sensors to cover 360° circular area. Rofid et al. [11] proposed a fire fighting robot system using fuzzy logic control. The system uses a flame detector KY -026 ultrasonic sensor for detecting fire and Arduino MEGA 2560 is used as the micro-controller of the system. For robot movement and control, fuzzy logic control method is implemented. The system can detect fire within the range of 20 cm for the ultrasonic sensor and the ADC value of 300 for the flame detector.

The authors in [12] designed a four-wheeled mobile robot that can detect and extinguishes ground-level emergency fire flames in the industrial environment. The system applies Cartesian Coordinate Mapping technique and Mathematical Operations for location identification. MATLAB Simulink models with image processing techniques are used for detecting and calculating distance of fire from the robot. The authors in paper [13] proposed a fire fighting robot system that uses temperature sensor for detecting fire. The robot can also detect obstacle using ultrasonic sensor and move to other directions using the techniques of fuzzy logic control. The system uses Arduino as the micro-controller. If it detects any fire, it also can takes pictures of the scene and send to the user so that the user can understand the situation properly.

In the paper [14], the authors used a AVR ATMega-32 micro-controller to design a robotic vehicle for fire extinguishing operation. The system is controlled using an android application developed in Android studio. The system uses bluetooth device for communication between the micro-controller and android application. It has a water tanker and a pump which is controlled over wireless communication to sprinkle water to the fire.

3 Proposed Methodology

The proposed system includes the following two main components: i) Hardware including devices with physical entities and ii) Software, including programs and applications. Flowchart of Proposed system is shown in Fig. 1.

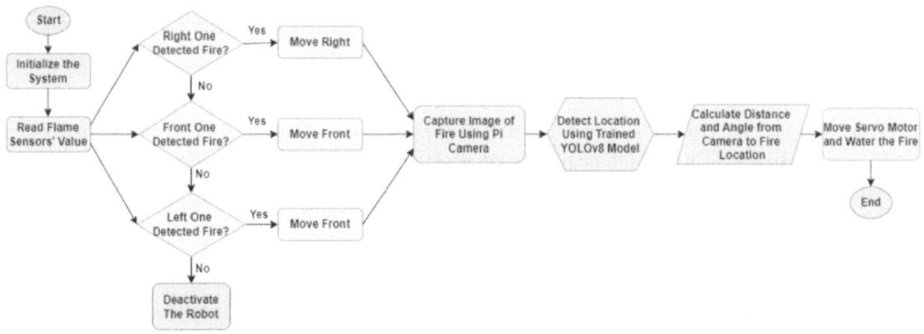

Fig. 1. Flowchart of the proposed system.

3.1 Hardware

The hardware is used as the basic structure of the system shown in physical context. Hardware component can be divided into two parts.

Hardware Components. In the proposed system, there are seven primary components. Additionally, different supporting components like 4WD Car Chassis, resistors, capacitors, two types batteries, different types of wires etc., contribute to the overall completion and functionality of the system. The major components are:

- **Arduino Uno:** A Arduino Uno R3 model is used for combining the sensors of the system.
- **Raspberry Pi:** A Raspberry Pi B model is utilized for capturing images and implementing YOLOv8 model.
- **Pi Camera:** Pi camera is used for capturing pictures of real time for detecting fire.
- **Flame Sensor:** Three flame sensors (left, front and right) are used for finding initial location of the fire and movement of the robot accordingly. AB007 flame sensor model is used in the proposed system. The sensor can detect fire in rage of 30 cm.
- **Servo Motor:** A Servo SG-90 motor is used for moving the pipe of the water tank.
- **Motor Drive:** A L298N Motor Driver is working for the movement of the robot.
- **Water Pump:** A 5V DC Water Pump is used for developing the system.

Connection. Arduino Uno is the main micro-controller of the whole system. All sensors and devices interface with the Arduino Uno. Arduino Uno communicates with Raspberry Pi via serial communication. Raspberry Pi is used for developing the deep learning model for image detecting and a Pi-Camera is attached to the Raspberry Pi for capturing real life images. Figure 2 shows the circuit diagram of the proposed system which describes all components connections.

3.2 Software

Software serves as the brain of the proposed system which controls and manages all its capabilities. This part consists of four components: Dataset used for training, model construction, testing with real life collected data images and finally location detection.

Collected Dataset. The custom fire detection dataset is compiled from multiple datasets via Roboflow Universe that are put together in one single training set available to Roboflow Universe, an online site for computer vision datasets and algorithms. The dataset is created by combining 5 datasets [15–19]. The final combined dataset is used for prepare the training dataset and validation dataset. Figure 3 shows a training dataset sample used in the system. Total number of images in the training dataset are 6520, where 84% data is used for the train set, 16% data is used for the validation set.

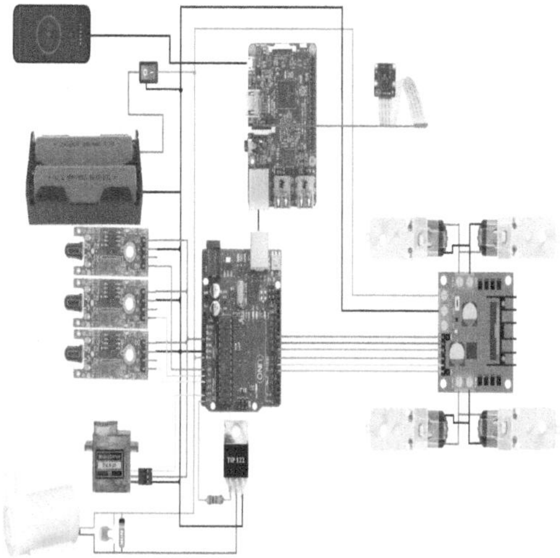

Fig. 2. Circuit Diagram of the Proposed System.

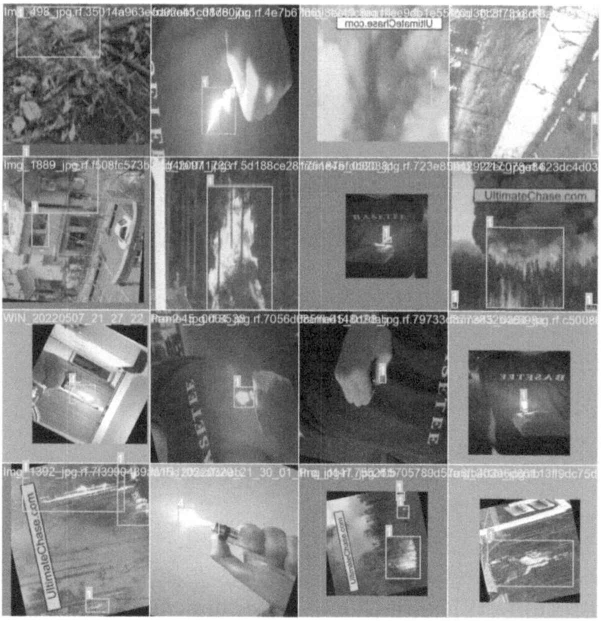

Fig. 3. Train Dataset.

Model Construction. The model is built with YOLOv8 by Ultralytics, a powerful object detection model implemented in PyTorch. It boasts of an effective feature extraction backbone, a strong detection system at multiple scales with PANet and BiFPN, as well as a state-of-the-art detection head for accurate localization and classification [20]. For constructing the YOLOv8 model, Ultralytics and Roboflow packages are used. The home directory is defined in the shell language script. After fine-tuning the hyper-parameters of the model, the model is trained for total 75 epochs to get the maximum accuracy.

Model Training and Testing. After creating the model, the model is trained using the online training dataset collected from Roboflow. Real life images, collected from the Pi camera in the system, is used for testing the developed model. Total 120 images are captured using Pi camera and image augmentation is applied to create total 371 images for test dataset. Figure 4 shows the data collected in real time using the Pi camera.

Fig. 4. Test Dataset Sample.

Location Detection. When the flame sensor and yoloV8 model detect fire, the appropriate angle from the water pump needs to be calculated to throw water. YoloV8 provides a box with the left, top, right, bottom values. From these values, the center of the fire is detected. Also from the coordination of the center, it is possible to decide whether it is on the left or right of the water pump. From the ratio of the image frame width and position of the fire, we can decide the angle. We need to turn the motor in horizontal direction through Eq. 1.

$$\theta_x = sin^{-1}(\frac{opp}{hyp}) \tag{1}$$

where *opp* is the opposite of the angle, that will come from the ratio of the frame width and the x directional position of the fire center. *hyp* is the hypotenuse of the triangle, this is constant for the flame sensor.

To decide y angle for throwing water, Eq. 2 is used.

$$\theta_y = \frac{1}{2} sin^{-1}(\frac{gR}{v^2}) \tag{2}$$

where g is gravity which is fixed. R is the distance from the pump to the flame it is fixed by the flame sensor. The velocity is also constant for the proposed case.

4 Experimental Results

4.1 Experimental Setup

The connection established in the methodology is implemented successfully. Different devices are connected with both analog and digital pins of the Arduino Uno for developing the proposed system. The right and left motors are operated using motor driver L298. The right motor is connected with pin 8, 9 and 10, whereas the left motor is connected with pin 5, 6 and 7. The right, front and left flame sensors are connected in analog pins A0, A1 and A2 respectively. Analog Pin A4 is being used to control the servo motor and Analog pin A5 is used to turn on/off the water pump. Extra batteries are used for giving extra power to the whole system. Raspberry Pi is connected with the Arduino Uno with serial communication. Figure 5 shows left view, front view and right view setup of the proposed system.

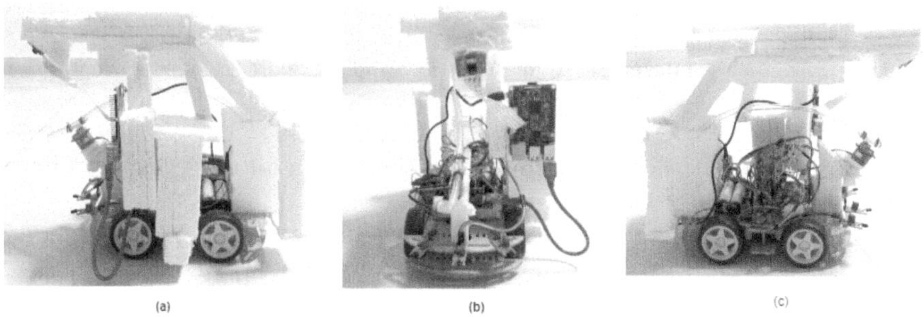

Fig. 5. Setup Views of the System: a) Left View, b) Front View and c) Right View.

4.2 Experimental Results

The custom YOLOv8 object detection model continues to improve over epochs 1–75. The YOLOv8 model is evaluated based on accuracy, f1 score and Mean Absolute Error (MAE) value for the test dataset. Table 1 shows the performance evaluation matrix values of the proposed system model.

Table 1. Performance Analysis of the Model with Accuracy, F1 Score, and MAE.

Variable	Result
Accuracy(%)	76.4
F1 Score	0.73
Mean Absolute Error(MAE)	1.41

Figure 6 Performance Analysis Curve for accuracy and f1 score in each epoch for the trained YOLOv8 model in the system. The green line shows the accuracy value and yellow line shows the f1 score in each epoch in the curve.

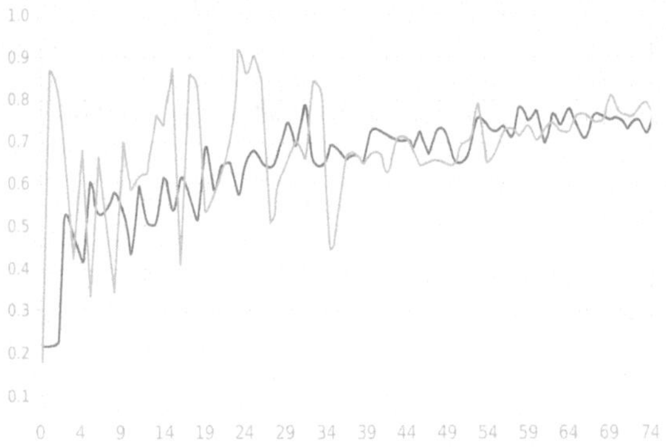

Fig. 6. Performance Analysis Curve for Accuracy and F1 Score. (Color figure online)

Figure 7 displays the confusion matrix of the system for detecting fire in test images.

Figure 8 shows the Label Correlogram graph for the trained YOLOv8 model in the system.

A sample test dataset along with the predicted bounding boxes showing the fire instances by the model is represented in Fig. 9.

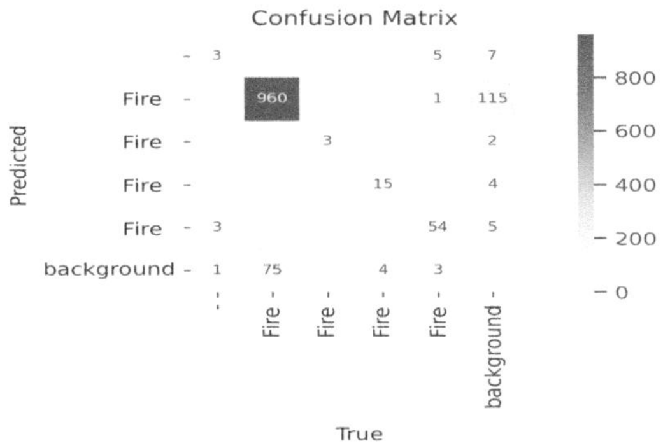

Fig. 7. Confusion Matrix of the Proposed System.

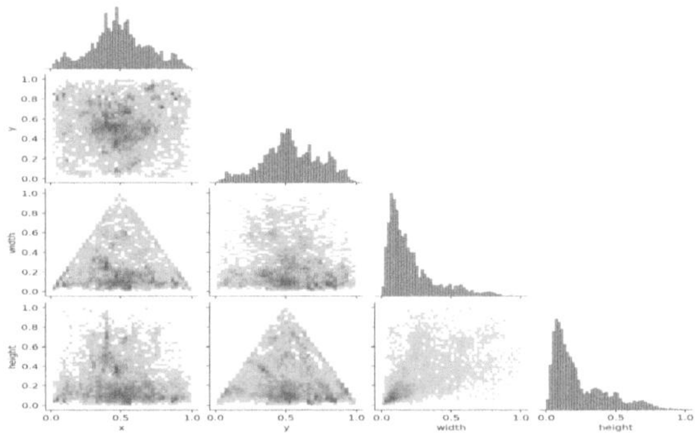

Fig. 8. Label Correlogram Graph.

Fig. 9. Test image Predicted-Sample

5 Conclusions

Fire accidents are increasing day by day. Global warming and human actions are mainly responsible for frequent fire accidents. Human can not enter every place that is affected because of fire. That is why, an autonomous fire fighting robot can be a great solution here. In the paper, an autonomous firefighting robot system is established using the combination of Raspberry Pi with Pi camera, Arduino Uno and different sensors. The proposed system can detect fire using flame sensor and Pi camera. The system uses 3 flame sensor to cover almost 180° area. YOLOv8 model is constructed to identify the exact location of the fire in the image captured by the camera. The model is trained with multiple online datasets having 6520 images and tested with 371 real time images. The distance and angle of the fire from the water pipe is calculated using the mathematical formulas of Projectile Motion. Finally, based on the angle and distance, the servo motor is used to move the water pipe to give water to the fire using the water pump.

In the future, more real time images will be collected and image processing will be be implemented to increase the accuracy of the model. More deep learning algorithms will be constructed for fire detection in the proposed system. For now, the system can detect fire in the ground level or the same level as the robot. But in the future, the robot will be updated to detect and give water to the higher level of fire location. Obstacle detection ability using ultrasonic sensor can also be added.

Acknowledgement. The authors wish to acknowledge the department of Computer Science and Engineering of Southeast University, Bangladesh (@SEU, Bangladesh) for providing their support and necessary equipment in developing the system and conducting the research.

References

1. Imran, I.M., Ramesh, R.K., Abineshwar, S.S., Pandyaraj, V.: Intelligent firefighting robot with deep learning. In: 2022 International Conference on Communication, Computing and Internet of Things (IC3IoT), pp. 1–5 (2022)
2. Vishwakarma, D., Srivastava, A., Singh, A.K., Tripathi, A.M., Yadav, S., Dubey, A.K.: Design and fabrication of automated fire fighting robot (2023)
3. Imteaj, A., Rahman, T., Hossain, M.K., Alam, M.S., Rahat, S.A.: An iot based fire alarming and authentication system for workhouse using raspberry pi 3. In: 2017 International Conference on Electrical, Computer and Communication Engineering (ECCE), pp. 899–904 (2017)
4. Radha, D., Arun Kumar, M., Telagam, N., Sabarimuthu, M.: Smart sensor network-based autonomous fire extinguish robot using iot (2021)
5. Dalvi, K., Dhapate, A., Adik, R., Jadhav, P.: Automatic fire fighting robot using iot. SSRN 4841278 (2024)
6. Kumar, N., Mounika, P., Ramani, D., Saketh, M., Sujith, M.S.: Arduino fire fighting robot. **04**, 1377–1381 (2024)

7. Patil, A.K., Mohite, T., Mulani, A., Borage, A., Jadhav, S.: Fire fighting robot car. **07**, 5476–5481 (2024)
8. Dubey, A., Singh, B., Kumar, N., Alkhafaji, M.A., Isam, M.: Fire fighting robot with more accuracy and implementation using arduino uno. In: 2023 3rd International Conference on Advance Computing and Innovative Technologies in Engineering (ICACITE), pp. 20–23 (2023)
9. Arora, K., Kumar, H., Singh, R.R.: Autonomous fire fighting robot. In: 2023 International Conference on Computational Intelligence, Communication Technology and Networking (CICTN), pp. 431–435 (2023)
10. Suresh, M.P., Rhythesh, V.R.V., Dinesh, J., Deepak, K., Manikandan, J.: An arduino uno controlled fire fighting robot for fires in enclosed spaces. In: 2022 Sixth International Conference on I-SMAC (IoT in Social, Mobile, Analytics and Cloud) (I-SMAC), pp. 398–402 (2022)
11. Ikbar, R.K., Mulyana, E., Mardiati, R., Nurmalasari, R.R.: Fire fighting robot using flame detector and ultrasonic based on fuzzy logic control. In: 2022 16th International Conference on Telecommunication Systems, Services, and Applications (TSSA), pp. 1–6 (2022)
12. Perera, R.K.K., Priyankara, H.D.N.S., Madhusanka, B.G.D.A.: A fire detection and extinguishing mobile robot for industrial environments. In: 2021 10th International Conference on Information and Automation for Sustainability (ICIAfS), pp. 393–398 (2021)
13. Hasan, M., Sonet, R.I., Sen, A., Das, R.P., Ahmed, J.: Design and development of a fire fighting mobile surveillance robot with autonomous collision avoidance. In: 2021 International Conference on Automation, Control and Mechatronics for Industry 4.0 (ACMI), pp. 1–5 (2021)
14. Patel, K., Pancholi, B.K.: A novel fire extinguishing robotic vehicle controlled by android application. In: 2017 IEEE International Conference on Smart Technologies and Management for Computing, Communication, Controls, Energy and Materials (ICSTM), pp. 417–422 (2017)
15. LTTTNT. Fire dataset (2023). https://universe.roboflow.com/ltttnt/fire-vqbia. Accessed 26 June 2024
16. REALLYai. Fire detection dataset (2023). https://universe.roboflow.com/reallyai-78cm0/fire_detection-lwcsv. Accessed 26 June 2024
17. fire. fire custom dataset (2024). https://universe.roboflow.com/fire-8luym/fire-custom-wv7aq. Accessed 26 June 2024
18. fabritalk. Fire dataset (2024). https://universe.roboflow.com/fabritalk/fire-soybb. Accessed 26 June 2024
19. Kuumoneko. Fire detection for khkt dataset (2023). https://universe.roboflow.com/kuumoneko/fire-detection-for-khkt. Accessed 26 June 2024
20. Tan, M., Pang, R., Le, Q.V.: Efficientdet: scalable and efficient object detection. In: Proceedings of the IEEE/CVF Conference on Computer Vision and Pattern Recognition, pp. 10781–10790 (2020)

A Novel Framework for Collaborative Knowledge Sharing and Learning for Disaster Response Robots

M. I. R. Shuvo, Nafiul Alam, Awal Ahmed Fime, Ahmed Aziz Jaidane, Saifuddin Mahmud, and Jong-Hoon Kim(✉)

Advanced Telerobotic Research Lab, Computer Science, Kent State University, Kent, OH, USA
smahmud@fsmail.bradley.edu
https://www.atr.cs.kent.edu/

Abstract. Collaborative robotics research shows multi-robot systems' utility in various applications. This study proposes a novel approach to knowledge sharing and learning across heterogeneous robots. The suggested architecture allows robotic agents to easily share vision, control, and decision-making algorithms. We solved object detection, global position estimation, and reconstruction using a stage-wise modular method, demonstrating good performance even in changing situations. The system's initial findings show that RealSense technology's rigorous calibration allows for minimum global position estimate errors. The PCA-based orientation module was highly accurate. The paper examines the bounding box estimation module's complexity and Z-axis orientation flexibility. The reconstruction module needs improvement due to intrinsic conversion problems between pixel data and 3D points, however the system estimates height and width accurately. This research improves adaptability and interoperability in complex multi-agent circumstances by fostering collaborative intelligence in different robot teams.

Keywords: Collaborative Robotics · Knowledge Sharing · Object Detection · Global Position Estimation · Principal Component Analysis · Disaster Response

1 Introduction

Every aspect of modern life uses deep learning and machine learning. Additionally, they understand robots [1–3]. Robots used in disaster response and search and rescue must have a machine learning model to assess different types of information utilizing their robotic vision. Object identification, location, appearance, and visibility are examples. With this knowledge, robots can make real-time decisions and act accordingly. Diverse robots may be at the disaster site. There will be terrestrial and maybe UAV/drone robots. Robots will encounter many scenarios and things as they explore the tragedy scene. They may encounter the

D. Singh et al. (Eds.): IHCI 2024, LNCS 15557, pp. 134–146, 2025.
https://doi.org/10.1007/978-3-031-88705-5_11

same thing from various perspectives. This could be due to eyesight, obstacles, or robot position. Due to the conditions stated, robots with machine learning technology may have trouble understanding and responding to information.

Different robots are commonly used in Search and Rescue (SAR) missions during disasters [4–6]. Disaster zones include many objects and robots to find. Each robot's fundamental machine learning model can understand its surroundings and objects, but may struggle to perceive anything beyond its field of vision. If robots share information, they can explore and restore the catastrophe site more easily. Two robots may examine artifacts and recognized items from different angles at a disaster site. They can detect unknown objects using their fundamental machine learning model's class, global position, and orientation, even with minimal information or the image's perspective from their current location. Data from all local robots can be combined to help robots navigate and retrain models. This scenario uses autonomous robots to classify and retrain.

Transferring knowledge between robots or agents improves their performance and helps the system achieve its goals. Chen et al. [7] described a method for quadrotor UAVs to learn from tracking errors of other UAVs with different dynamics. Learning algorithm communicates with target UAV feedforward loop without affecting closed-loop stability. Tatiya et al. [8] adopted KEMA to facilitate haptic information transmission between robots. Identification recognition models were trained on a single latent space, not each robot's sensorimotor feature space. The Sarma et al. [9] KT-BT system was used in this study to represent and convey information in a novel way. Local Procrustes Analysis enhances linear Procrustes Analysis to estimate non-linear relationships between manifolds, as stated by Makondo et al. [10].

Automatic annotation and tagging are done with machine learning in computer vision, natural language processing, robotics, and other fields. Classifying robot vision images and videos takes time and money. Annotating or labeling the robot's AI system may help. Wang et al. [11] modified CaffeNet to create AIA, as described in reference eleven. Raw photos are annotated superficially with clear category naming. Sarangi and Sekhar used T-DSN and K-DCN convex deep learning models to evaluate images using deep convolutional network features [12].

This paper presents an approach for search and rescue robots to communicate and learn autonomously during a crisis by combining their experience and information from other robots in the same environment without human intervention. A rudimentary machine learning model, Yolov8 [13], was developed to fulfill this goal, assuming all robots now had it. Yolov8 detected full-view objects. We trained the basic machine learning model Yolov8 with a unique dataset of water bottles. Our proof of concept test contains two robots and one object. One robot sees the thing completely, while the other sees it partially. The robots' mobility can move the bottle. Robot 1 uses Yolov8 to identify the bottle and automatically determine its position, shape, and orientation. A real-time principal component analysis (PCA) [14] was employed to establish the bottle orientation. After that, Robot 2 received this information. After collect-

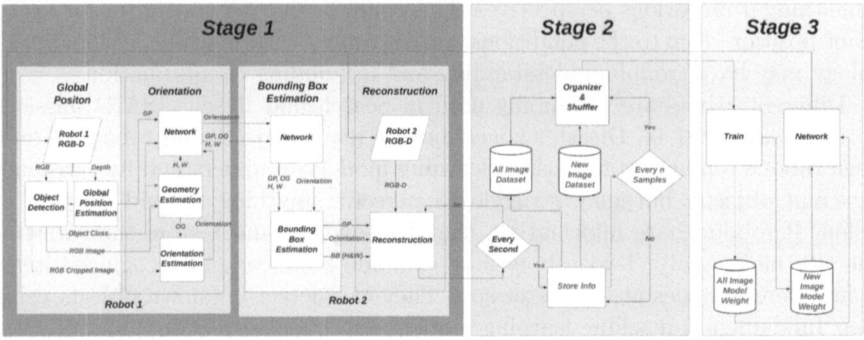

Fig. 1. Basic Architecture of the Proposed System.

ing data, Robot 2 locates and aligns the object. Based on the object's distance, orientation, and geometric features, the algorithm estimates the bounding box (BB)'s height (H) and width (W). The bounding box information detected the bottle without a machine learning model. Automatically tag scene objects and produce a dataset for machine learning model retraining with this knowledge. We successfully created and evaluated the system till bounding box generation using data from other robots, yielding satisfactory results.

The following portions of the document are arranged as such: The system's method is detailed in Sect. 2. Section 3 presents the implementation and system's output, while Sect. 4 provides a comprehensive analysis. The study's conclusion and suggested follow-up actions are detailed in the final section.

2 Collaborative Intelligence and Transfer Learning Approach

The subsequent part will discuss the strategies that support the proposed system and facilitate intelligent collaboration among different entities. We will start by discussing the system's autonomous Information Sharing and Learning by Robots capabilities. Then we will discuss the system's implementation aspects. This section concludes with an examination of the significance of the proposed system.

2.1 Information Sharing and Learning by Robots Without Human Intervention

For simplicity, we will consider that the disastrous scene involves two robots. Robotic vision fields can contain numerous objects. Robot 1 has a complete view of an object while Robot 2 has a partial or complete view.

Figure 1 shows system workflow. Only two robots are viewing the same object from different angles. Since Robot 1 has a full view, its object detection module will detect it. The pixel-to-global position module uses Robot 1's RGB frame and Intel RealSense camera depth information to determine the object's global position in the scene. X and Y represent the object's centroid, whereas Z represents the robot's distance. The geometry estimation module analyzes RGB frame and object class data to estimate object geometry (OG) such shape, height, and width. Bottle identification uses the geometry estimate module to recognize a cylindrical shape with height and diameter measurements. This data helps maintain Robot 2's bounding box reconstruction aspect ratio.

The orientation module will utilize the cropped image of the identified object to provide its orientation. The dimensions of the bounding box are influenced by the object's orientation, making this information crucial. Robot 1 transmits the object's class, global position, geometric shape, height, width, and orientation to Robot 2's bounding box estimation module. The bounding box estimation module will provide the dimensions of the bounding box using Robot 2 vision's Z, geometry information, and orientation. The reconstruction module creates the bounding box using the position, height, width, and orientation.

Robot 2 can recognize scene items without an object detection model utilizing input from another robot. A dataset for model training is created by storing Robot 2's picture view and bounding box information from the bounding box generating module. The store info module receives one-second data. The store info module organizes and saves images and labels in the all image dataset, which includes training data, and the new image dataset, which contains created data. It also counts added samples. The organizer and shuffler module activates when samples exceed n. The latter manages all image and fresh image databases. Once triggered, it shuffles datasets and feeds the train module training data. The train module will train Yolov8 models with that data. One Yolov8 model will use original data, one will use original plus synthetic data, and one will use exclusively synthetic data in Robot 2. Robot 1 receives the models after training so both robots can detect the thing from different angles.

SC module processes stored data to acquire further information. Robot 1 may know an object's class name and owner. Robot 1 records an object's position, orientation, and owner in the GK after detecting it, such as a bottle at (10, 10, 10). Robot 2 can then use its Object Authentication module to verify its findings using the shared data from Robot 1. Robot 2 can ask the SC module for more information after verifying the information and object. Since GK (Global Knowledge) includes the owner's information, SC (Semantic Classification) can merge the basic class name with the owner's information to create a more comprehensive class name. This improved class name can be used to train the object detection model later.

Thus, Robot 2 can distinguish scene objects without a model using input from the other robot. Robot 2's image view and bounding box information are stored combined to create a dataset for model training by the bounding box creation module. Data enters the store info module every second. Store info maps and saves images and labels in the all image dataset, which contains initial training data, and the new image dataset, which contains intentionally generated data.

The organiser and shuffler module counts new samples from all image databases and fresh datasets. The organizer and shuffler modules shuffle and store data in both databases after collecting a predetermined number of samples. The train module will get data for training again. To train Yolov8 models with scrambled data, the train module requests all image databases and new image datasets. Three Yolov8 models will be in Robot 2. One with initial data, one with initial and artificially generated data, AND one with artificial data. The models are sent to Robot 1 after training so both robots can detect the thing from any viewpoint.

2.2 Detailed Description of the Modules

This section offers a comprehensive overview of the modules utilized by a robot to calculate the bounding box based on data gathered from other robots in the environment.

Object Detection. The architecture centers on the ODM. YoloV8, trained on a custom dataset, fulfilled our demands. Input Intel BRG color from Real Sense depth camera. YoloV8 converts RGB with OpenCV. This model gives bounding box, item kind, and detection confidence. Bounding box shows item centroid. The global position estimation module receives this center coordinate. The global position estimation module receives this center coordinate.

Global Position Estimation. The Global Position Estimation module translates 2D coordinates to 3D. The previous module identified the object's center. GPEM receives this center position. GPEM finds depth from the Depth frame for this 2D coordinate. Intel RealSense supplies the Depth frame. To reduce distortion and inaccuracy, the RGB and Depth frames are aligned. Additionally, RealSense's profile can be improved. Shooting parameters like focus point and lens curvature are in this profile. After alignment, 2D coordinates become 3D global positions. To calculate 3D global position following formulas:

$$X = \frac{x - \text{depth\_intrin.ppx}}{\text{depth\_intrin.fx}} \cdot \text{depth} \tag{1}$$

$$Y = \frac{y - \text{depth\_intrin.ppy}}{\text{depth\_intrin.fy}} \cdot \text{depth} \tag{2}$$

$$Z = \text{depth} \tag{3}$$

where: X, Y, and Z represent the coordinates of the 3D point. x and y are the pixel's X and Y coordinates in the depth image, respectively. depth_intrin.ppx and depth_intrin.ppy denote the principal point X and Y coordinates, usually half of the image width and height, respectively. depth_intrin.fx and depth_intrin.fy are the focal lengths along the X and Y axes. Finally, depth represents the depth value at pixel $[x, y]$.

Geometry Estimation. The Geometry Estimation Module (GEM) estimates object geometry from ODM class information. The mapping function in GEM converts geometric structure to object class. Look for the mapping in Table 1. It shows the object's height and width. Second robot's bounding box estimation module gets object height and width from GEM. Height and width are needed. In Table 1, the Bottle and Fire Extinguisher have similar geometric shapes but different height and width. Based on the geometric design, the second robot cannot calculate height and width. Fix this by specifying height and width. Maps will feature new items with tangible height and breadth.

Table 1. Objects, Shapes, Height, and Width

Object	Shape	Height (m)	Width (m)
Bottle	Cylinder	0.22	0.06
Cup	Cube	0.09	0.09
Ball	Sphere	0.08	0.08
Fire Extinguisher	Cylinder	0.60	0.18
Apple	Sphere	0.10	0.10

Orientation Estimation. The Object Orientation Estimation Module (OEM) utilizes the RGB frame to determine the direction of an object. We used the following technique to determine the principal axis and its orientation in regard to the x, y, and z axes of a set of three-dimensional points. The approach initially centers the data around its mean, as depicted in Eq. 4. This is the initial step in the algorithm.

$$\bar{\mathbf{p}} = \frac{1}{N}\sum_{i=1}^{N}\mathbf{p}_i, \mathbf{p}_i' = \mathbf{p}_i - \bar{\mathbf{p}} \tag{4}$$

The covariance matrix, crucial for understanding the relationships between different dimensions of the dataset, is calculated using Eq. 5 after that.

$$\mathbf{C} = \frac{1}{N}\sum_{i=1}^{N}(\mathbf{p}_i' - \bar{\mathbf{p}})(\mathbf{p}_i' - \bar{\mathbf{p}})^T \tag{5}$$

We may identify the principal axis as the eigenvector corresponding to the largest eigenvalue by utilizing the eigenvalue decomposition outlined in Eq. 5.

$$\mathbf{C}\mathbf{v}_i = \lambda_i\mathbf{v}_i \tag{6}$$

Trigonometric functions are utilized to calculate the angles between the primary axis and the coordinate axes, as demonstrated in Eqs. 7 through 9.

$$\theta_x = \arccos\left(\frac{\mathbf{v}_p \cdot \mathbf{e}_x}{|\mathbf{v}_p||\mathbf{e}_x|}\right) \tag{7}$$

$$\theta_y = \arccos\left(\frac{\mathbf{v}_p \cdot \mathbf{e}_y}{|\mathbf{v}_p||\mathbf{e}_y|}\right) \tag{8}$$

$$\theta_z = \arccos\left(\frac{\mathbf{v}_p \cdot \mathbf{e}_z}{|\mathbf{v}_p||\mathbf{e}_z|}\right) \tag{9}$$

This strategy not only offers valuable insights into the dataset's orientation but also simplifies the implementation of applications in computer vision, robotics, and geometry processing techniques.

Bounding Box Estimation. Bounding Box Estimation (BBEM) is in the second robot. The second robot can approximate the bounding box's dimensions and direction using the first robot's global coordinates and orientation. Estimate bounding box size using linear regression. Object distance from the camera determines bounding box height in the model. Input layer node of the model is object distance from robot camera. Dense 32- and 16-node layers of the neural network. One output layer node holds the bounding box height. Robot 1 sends geometry, height, breadth, global coordinates, and orientation. After considering geometry, height, breadth, global coordinates, and orientation, the BBEM model calculates bounding box dimensions. Model uses real height and width data and geometric structure to forecast bounding box height and breadth.

Reconstruction Module. The second machine has the Reconstruction Module. This module maps an object's global position to its local position in a second machine's field of vision. This translation is easy because we know all the machines' global positions. Subtracting the second camera position from the object's global location converts global to local coordinates. BBEM height and width are used to draw the object's bounding box in the second machine's view. The bounding box center is local. Rotate the bounding box if the item is oriented differently. The bounding box is rotated according to BBEM's orientation. Following for calculating the Local coordinate.

$$\begin{aligned}
X_2 &= [x_{2_1}, x_{2_2}, x_{2_3}] \text{ (object's position)} \\
X_1 &= [x_{1_1}, x_{1_2}, x_{1_3}] \text{ (robot's position)} \\
\Delta x &= x_{2_1} - x_{1_1} \\
\Delta y &= x_{2_2} - x_{1_2} \\
\Delta z &= |x_{2_3} - x_{1_3}| \\
\Delta X &= [\Delta x, \Delta y, \Delta z]
\end{aligned}$$

Following for calculating the 3D coordinate to Pixel coordinate.

$$u = \frac{f_x \cdot x}{\text{depth}} + p_x \tag{10}$$

$$v = \frac{f_y \cdot y}{\text{depth}} + p_y \tag{11}$$

where u and v are the pixel coordinates in the image, f_x and f_y are the focal lengths in x and y directions respectively, p_x and p_y are the principal points, typically near the center of the image, and depth is the depth value of the 3D point. Following for estimating the orientation of the bounding box. 1. Rotation Matrices: Convert the Euler angles into rotation matrices:

$$R_{\text{object}} = f(r1, r2, r3) \tag{12}$$

$$R_{\text{cam2}} = f(r11, r12, r13) \tag{13}$$

where f is a function that converts Euler angles to a rotation matrix. 2. Transformation Matrices: The transformation matrix for the object in the global frame is given by:

$$T_{\text{object}} = \begin{bmatrix} R_{\text{object}} & [x, y, z]^T \\ \mathbf{0} & 1 \end{bmatrix} \tag{14}$$

The transformation matrix for Camera 2 in the global frame is:

$$T_{\text{cam2}} = \begin{bmatrix} R_{\text{cam2}} & [x1, y1, z1]^T \\ \mathbf{0} & 1 \end{bmatrix} \tag{15}$$

3. Inverse Transformation: The inverse transformation matrix for Camera 2 is:

$$T_{\text{cam2}}^{-1} = T_{\text{cam2}}^{-1} \tag{16}$$

4. Object's Pose Relative to Camera 2: Multiply the inverse transformation of Camera 2 by the object's transformation to get:

$$T_{\text{relative}} = T_{\text{cam2}}^{-1} \cdot T_{\text{object}} \tag{17}$$

Auto-Annotation and Retraining. Robot 2's bounding box information is computed by the reconstruction module and recorded with the image view to build a dataset for model training. One-second intervals are used to retrieve store info. After mapping, the store info module saves the mapped images and labels in the new image dataset with purposely generated data and the training image dataset. Additionally, it tracks new samples. If additional samples exceed n, The organizer/shuffler module activates. Shuffles both datasets and sends to train module. The Yolov8 model is trained using that data. Robot 2 will have three Yolov8 model versions: one with the original data, one with the original data and artificially generated data, and one with the intentionally generated data solely. After training, Robot 1 receives the models to let both robots identify the object from any angle.

Table 2. Global Position Estimation

Actual (m)	Estimated (m)	Error
(0.1, 0.0, 0.69)	(0.09, 0.0, 0.7)	0.014
(–0.1, 0.0, 0.69)	(–0.11, 0.0, 0.68)	0.014
(0.05, –0.06, 0.39)	(0.04, –0.06, 0.39)	0.01
(–0.07, –0.06, 0.36)	(–0.07, –0.06, 0.35)	0.01
(–0.05, 0.04, 0.41)	(–0.05, 0.02, 0.41)	0.02

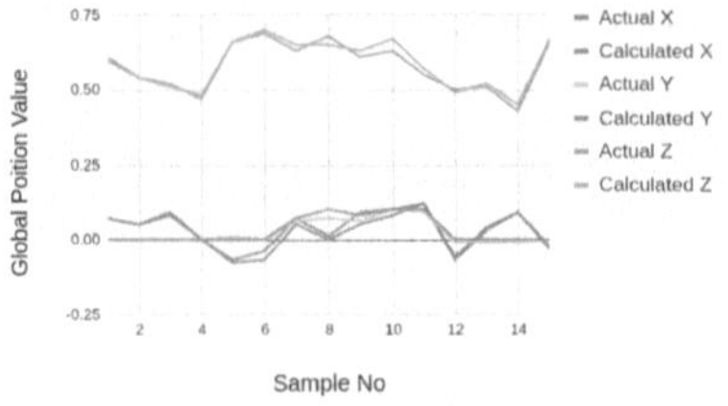

Fig. 2. Performance of Global Position Estimation Module

3 Implementation and Results Analysis

This section covers the execution of the experiment and its results. We showcased the performance of the stage one module during its implementation. The results are categorized into four distinct groups: global position estimate, orientation, bounding box size, and coordinate reconstruction. Finally, we display the sample output produced by all of the operations. For the purpose of training Yolov8's fundamental object detection model ahead of its deployment, a customized bottle dataset consisting of 500 images was deployed. The model learned at a rate of 0.01 during the course of 300 epochs. The 2D coordinate of the object's center, as well as the height and breadth of the bounding box, may be calculated by this module. Immediately after the detection of the object, the global location was determined by employing the technique described in the reference Sect. 2.2.

The Table 2 displays the performance of the module responsible for calculating the global position. We have calculated the Euclidean distance, or L2 norm, between the two points using the equation provided. Next, we use this distance as a measure to calculate the level of inaccuracy. This mistake provides the straight-line distance between the two points in three-dimensional space. This inaccuracy provides a precise quantification of the discrepancy between the estimated points and the actual points.

$$\text{Error} = \sqrt{(x_a - x_e)^2 + (y_a - y_e)^2 + (z_a - z_e)^2} \tag{18}$$

A manual calculation was performed in order to determine the real global position of the object on the scene. In the following step, this module determines the object's position on the global scale. Based on the data presented in the table, it is evident that this module produces outstanding outcomes, as the error rate is really close to zero. With the help of this module, we are able to obtain a location of the item that is nearly correct. Additionally, the same properties are depicted in Fig. 2.

Figure 3 depicts the performance of the orientation estimation module. The bottle is split in Fig. 3(a) and 3(b) using depth information from Intel Realsense. A bounding box was generated for the object of interest using Yolov8. To determine the major axis and orientation, the bottle must be segmented and the backdrop eliminated. This is achieved by applying a threshold to the depth

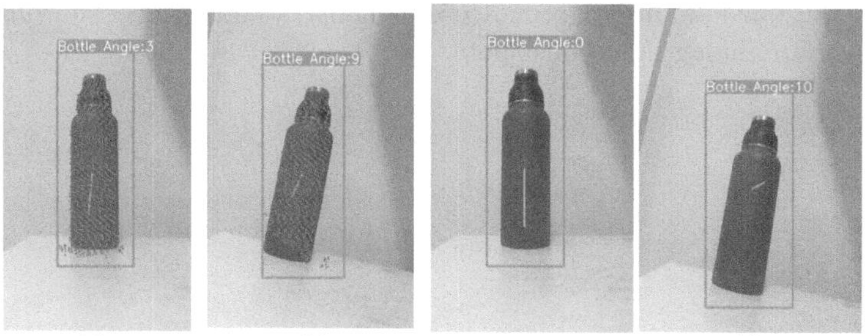

(a) (b) (c) (d)

Fig. 3. Performance of Orientation Estimation Module.

Table 3. Bounding Box Height-Width Estimation

Actual (pixels)	Estimated (pixels)	Error
(280, 96)	(293, 97)	13.03
(177, 58)	(193, 64)	17.08
(329, 121)	(347, 115)	18.97
(264, 83)	(276, 91)	14.42
(277, 90)	(285, 94)	8.94

Table 4. Reconstruction Coordinate Estimation

Actual (pixels)	Estimated (pixels)	Error
(315, 240)	(319, 231)	9.85
(256, 242)	(258, 231)	11.18
(272, 241)	(276, 231)	10.77
(127, 247)	(126, 242)	5.10
(183, 242)	(231, 235)	11.00

value. Typically, RealSense provides high values for the background. The depth data along with their corresponding 2D positions are inputted into primary Components Analysis (PCA) to compute the primary axis. Figure 3(c) represents the original angle (0°C) and Fig. 3(d) represents the orientated angle. This module demonstrates excellent performance. We utilized YoloV8's bounding box to determine the true height and width of the bounding box. We utilized the bounding box estimation module to determine the estimated height and width of the bounding box. The bounding box estimation relies on geometric data, the object's true dimensions, Global Coordinates, and orientation. For our purposes, bounding box estimate relies solely on geometric information, specifically the bounding box ratio (Width/Height). BBEM use its regression model to predict the height and width. Table 3 displays the height, width, and inaccuracy of both the real and estimated bounding boxes. The table indicates that BBEM's performance is outstanding.

Figure 4 illustrates the difference between actual and predicted height and width, highlighting their imprecision. The estimated and real values are similar. There are a few oversights and undersights, but they are mainly accurate. The reconstruction module receives global coordinates and orientation data following bounding box estimate. Converts global coordinates to local coordinates. Points are converted into pixels.

We can align our bounding box along the z-axis using orientation data. Table 4 displays a slight discrepancy between the actual and computed centers of the object. The module performance is demonstrated in Fig. 5, showing a high level of overlap among numerous spots.

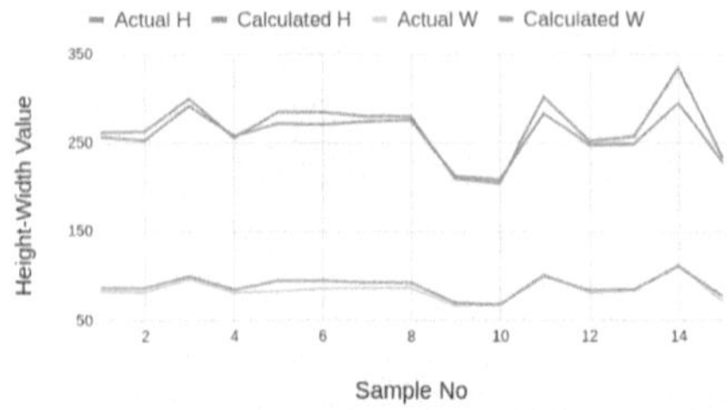

Fig. 4. Performance of Bounding Box Height-Width Estimation Module.

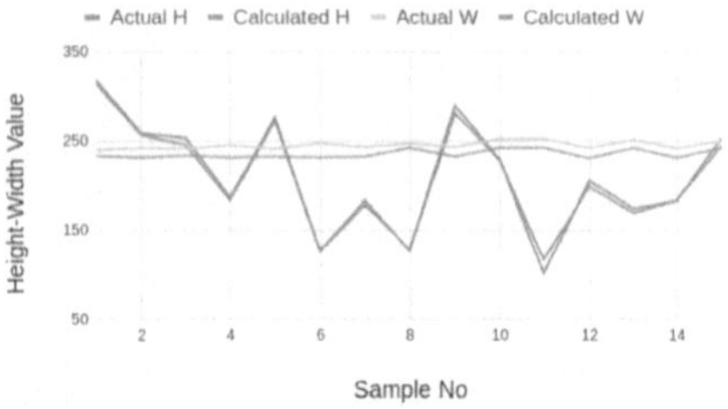

Fig. 5. Performance of Reconstruction Coordinate Estimation.

Figure 6 depicts the end result of the first stage of the proposed technique. Figure 6(a) depicts Robot 1's perspective of identifying the bottle. We determined that Robot 1 was located at position (0,0,0). Figure 6(b) displays the bounding box used to detect the bottle, created with data obtained from Robot 1. We assumed that Robot 2 is in the same location as Robot 1. It excelled in this scenario. We assess the technique by employing two distinct positions for Robot 1 and 2. Figure 6(c) illustrates the performance, showing that the size and position of the bounding box formed by Robot 2 are accurately right. The implemented system performs well under different settings, as seen from this.

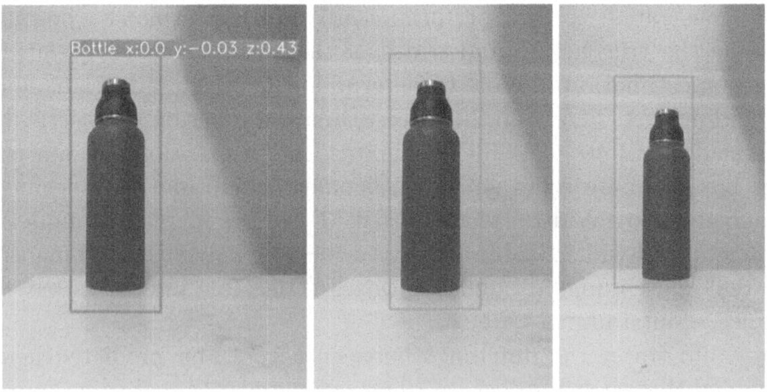

(a) Robot 1's View

(b) Robot 2's View from the same position of Robot1

(c) Robot 2's View from the different position of Robot1

Fig. 6. Final Outcome of Stage 1.

4 Discussion

This study examined the first phase of our suggested design, which contains five components (Fig. 1). Table 2 and Fig. 2 show negligible GPEM error. The study used orientation, geometry, and global position. This study uses the Z-axis, although the bounding box estimation module in the proposed architecture can create orientations in any direction. BBEM linear model size depend on camera distance. Table 3 and Fig. 4 show that height and weight measurements match estimates.

Even after implementation, the BBEM Reconstruction module (RM) is underutilized. The RM's extra functionalities were disabled. Table 4 and Fig. 5 show that BBEM inaccuracies exceed height and width inconsistencies. From pixel data, GPEM developed 3D points. Reversing the Reconstruction Module amplifies mistakes since pixels are distinct. The center moves left (Fig. 6(c)). When two cameras are in the same place, their coordinates match. It reduces rounding errors. A RealSense depth measurement error may affect the Reconstruction module.

5 Conclusion and Future Works

This research introduces a new framework to enhance the collaboration capabilities of diverse groups of robots. The system addresses key challenges in object recognition, global positioning estimates, and orientation by facilitating efficient exchange of perception, control strategies, and decision-making algorithms. The global position estimation and orientation tasks demonstrate promising accuracy, although there is potential for enhancement in the pixel-to-3D point conversion within the reconstruction module.

Possible enhancements can be implemented in the future. We restricted the object's position in the scene and focused on a single object for the proof of concept in this study. We aim to incorporate several objects into the test environment and dynamically determine their positions to accurately represent the real world. We currently only take into account the X-axis value for orientation, but we want to include the Z-axis in the future. We will utilize several robots to collect information from the exploration area and combine their knowledge to construct a comprehensive knowledge repository.

References

1. Mahmud, S., Ferdous, M., Sourave, R., Shuvo, M.I.R., Kim, J.: HAn essential robot vision system for robot assisted plant disaster prevention and response missions. In: 2022 25th International Conference on Computer and Information Technology (ICCIT), Cox's Bazar, Bangladesh, pp. 1116–1121 (2022). https://doi.org/10.1109/ICCIT57492.2022.10056111
2. Mahmud, S., Dannemiller, J., Sourave, R., Lin, X., Kim, J.-H.: Smart robot vision system for plant inspection for disaster prevention. In: 2022 Sixth IEEE International Conference on Robotic Computing (IRC), Italy, pp. 416–420 (2022). https://doi.org/10.1109/IRC55401.2022.00079

3. Shuvo, M.I.R., Wimer, B., Mahmud S., Kim, J.H.: A novel collaborative knowledge sharing and self-learning framework for robotic systems in search and rescue operations. In: IECON 2023- 49th Annual Conference of the IEEE Industrial Electronics Society, Singapore, Singapore, pp. 1–6 (2023). https://doi.org/10.1109/IECON51785.2023.10312013
4. Firozjae, H.M., Moghaddam, J.Z., Ardebilipour, M.: Performance analysis of an UAV-assisted cognitive D2D communication-based disaster response network. In: 2022 30th International Conference on Electrical Engineering (ICEE). IEEE (2022)
5. Qu, C., et al.: Environmentally-aware and energy-efficient multi-drone coordination and networking for disaster response. IEEE Trans. Netw. Serv. Manag. **20**, 1093–1109 (2023)
6. Krenn, D., Eder, M.J., Steinbauer-Wagner, G.: Ground robots in disaster response-why are they still not there?. In: Disaster Research Days (2022)
7. Chen, Z., Liang, X., Zheng, M.: Knowledge transfer between different UAVs for trajectory tracking. IEEE Rob. Autom. Lett. **5**(3), 4939–4946 (2020)
8. Tatiya, G., et al.: Haptic knowledge transfer between heterogeneous robots using kernel manifold alignment. In: 2020 IEEE/RSJ International Conference on Intelligent Robots and Systems (IROS). IEEE (2020)
9. Venkata, S.S.O., Parasuraman, R., Pidaparti, R.: KT-BT: a framework for knowledge transfer through behavior trees in multi-robot systems. arXiv preprint arXiv:2209.02886 (2022)
10. Makondo, N., Rosman, B., Hasegawa, O.: Knowledge transfer for learning robot models via local procrustes analysis. In: 2015 IEEE-RAS 15th International Conference on Humanoid Robots (Humanoids). IEEE (2015)
11. Wang, R., et al.: Large scale automatic image annotation based on convolutional neural network. J. Visual Commun. Image Represent. **49**, 213–224 (2017)
12. Sarangi, N., Chandra Sekhar, C.: Automatic image annotation using convex deep learning models. In: ICPRAM, no. 2 (2015)
13. Home - Ultralytics YOLOv8 Docs. https://docs.ultralytics.com/
14. Maćkiewicz, A., Ratajczak, W.: Principal components analysis (PCA). Comput. Geosci. (1993). https://doi.org/10.1016/0098-3004(93)90090-r

More Human, Less Bot: How Anthropomorphism Impacts User Experience in Customer Support Chatbots

Natalia Amat-Lefort(✉) and Christiaan van Driel

Leiden Institute of Advanced Computer Science. Leiden University, Einsteinweg 55, 2333 CC Leiden, The Netherlands
n.amat.lefort@liacs.leidenuniv.nl, christiaanvandriel@live.nl

Abstract. Understanding how to design engaging and trustworthy chatbots is crucial for businesses. While research has explored the impact of anthropomorphism on user experience, the specific contributions of different anthropomorphic dimensions (i.e., human-likeness, naturalness, lifelikeness, professionalism) remain unclear. This study investigates how various dimensions of chatbot anthropomorphism, influenced by visual design cues, affect user satisfaction and trust in customer support interactions. An online experiment was conducted where participants were randomly assigned to view one of three pre-recorded customer support scenarios. The scenarios featured identical conversations but varying chatbot designs (no avatar, static human-like avatar, dynamic human-like avatar with facial expressions). Interestingly, static human-like avatars did not significantly impact satisfaction or trust. However, the presence of an avatar with dynamic facial expressions did have a positive impact on trust. Additionally, user perceptions of the chatbot's human-likeness, naturalness, and professionalism were linked to higher satisfaction and trust. These findings suggest that while not all anthropomorphic design cues might have an impact on user perceptions, anthropomorphism dimensions positively influence user experience with chatbots. This highlights the importance of considering user experience in chatbot design, with a focus on user satisfaction and trust.

Keywords: Customer Support Chatbots · Anthropomorphism · User Experience

1 Introduction

Customer service is undergoing a digital revolution, with chatbots – computer programs designed to simulate human-like conversations – rapidly becoming widespread (Adamopoulos and Moussiades 2020). While chatbots offer benefits such as 24/7 availability and faster response times (Adam et al. 2020), limitations in handling complex requests and expressing empathy can lead to frustration. Gartner predicts that by 2027, a quarter of organizations will rely on chatbots as their primary customer service channel (Gartner 2022). This underscores the critical need to design chatbots that deliver positive user experiences.

N. Amat-Lefort and C. van Driel—Equal contribution/.

D. Singh et al. (Eds.): IHCI 2024, LNCS 15557, pp. 147–158, 2025.
https://doi.org/10.1007/978-3-031-88705-5_12

One key factor influencing user experience is chatbot anthropomorphism – attributing human qualities to a non-human entity (in this case, a chatbot). While previous research has explored the impact of anthropomorphism on user perceptions, the specific contributions of different anthropomorphic dimensions remain unclear. This study aims to answer the following research questions: How do different dimensions of anthropomorphism influence user satisfaction and trust in the context of customer support chatbots? By examining the impact of naturalness, human-likeness, lifelikeness, and professionalism on user satisfaction and trust, this study offers a nuanced understanding of how anthropomorphic design can be optimized.

We conducted a randomized experiment where participants were exposed to one of three customer support chatbots featuring identical conversations but varying levels of anthropomorphism (no avatar, static human-like avatar, dynamic avatar with facial expressions). Drawing upon Social Presence Theory (Short et al. 1976), we hypothesized that chatbots with anthropomorphic design cues (e.g., human-like avatars and dynamic facial expressions) would lead consumers to perceive the interaction as more natural and engaging, positively influencing satisfaction and trust.

Interestingly, the experiment found no significant impact of the human-like avatar on user satisfaction nor trust. On the other hand, the results revealed a positive effect of the avatar with dynamic facial expressions on trust. Moreover, the study showed positive relationships between user ratings of anthropomorphism dimensions and user satisfaction and trust in the chatbot. These findings suggest that while users may not consciously notice certain subtle differences in the chatbot's design, anthropomorphism dimensions do influence user experience with chatbots. This demonstrates the importance of considering user experience in chatbot design, with a focus on user satisfaction and trust.

The results offer valuable insights that bridge the gap between theory and real-world application. From a theoretical standpoint, this study extends the Social Presence Theory by testing its applicability in the context of human-computer interaction with customer support chatbots. By establishing the link between visual design cues, anthropomorphism dimensions, and consumer outcomes, this research sheds light on how chatbots can be designed to foster more positive consumer experiences.

In practical terms, the findings can guide the development and implementation of customer support chatbots that leverage anthropomorphism to enhance consumer experience. Understanding the design elements that influence consumer perceptions of anthropomorphism allows companies to optimize their chatbots for a more natural and engaging customer service experience. This can lead to improved consumer satisfaction, trust, and, overall, a more positive brand image.

2 Literature Review

2.1 Use of Chatbots for Customer Support

Businesses are increasingly seeking innovative ways to enhance their services and provide accessible communication channels for their users. One such innovation is the use of chatbots, computer programs designed to simulate human conversation.

Despite the increasing popularity of chatbots, challenges persist in meeting user expectations. While businesses benefit from cost savings and efficiency gains, chatbots often fall short in handling complex queries and providing empathetic responses to customers (Adam et al. 2020). To address these limitations and improve user satisfaction, it is crucial to design chatbots that can effectively bridge the gap between human and machine interaction.

2.2 Chatbot Anthropomorphism and User Experience

A core aspect of chatbot design is anthropomorphism, the attribution of human-like qualities to non-human entities. Chatbot anthropomorphism has emerged as a critical area of research, examining the extent to which designing chatbots with human-like qualities influences user perceptions and behaviors. Existing studies have explored a range of anthropomorphic features and their impact on diverse user outcomes.

A predominant theme in the literature is the potential for anthropomorphism to enhance user experience. For instance, Han (2021) demonstrates that anthropomorphic chatbots can increase consumer perceptions of social presence and enjoyment, thereby positively influencing purchase intentions. Similarly, Cai et al. (2022) highlight the role of emotional message cues in driving user adoption of chatbots in the online travel industry. These studies suggest that designing chatbots with human-like qualities can foster positive affective responses and enhance user engagement.

However, the relationship between anthropomorphism and user experience is not uniformly positive. Crolic et al. (2022) found that anthropomorphic chatbots can negatively impact customer satisfaction, firm evaluation, and purchase intentions when customers are angry. This finding underscores the importance of considering the emotional context of user interactions when designing anthropomorphic chatbots. Moreover, Song and Shin (2024) highlight the potential for the uncanny valley effect, where excessive human-likeness can lead to eeriness and negatively impact user trust and purchase intentions.

While these studies offer valuable insights, they primarily focus on specific anthropomorphic features or outcomes. A meta-analysis by Yanxia et al. (2024) provides a broader perspective, demonstrating that anthropomorphism positively influences various aspects of the customer journey, including purchasing attitudes, perceived interest, social presence, and trust. However, the study also highlights the need for further research to examine the moderating effects of factors such as user demographics.

In summary, the literature on chatbot anthropomorphism reveals a complex and multifaceted relationship between anthropomorphic design and user experience. While anthropomorphism can enhance user perceptions and behaviors, its effectiveness is contingent upon various factors, including the specific anthropomorphic features employed, the context of the interaction, and user characteristics.

Despite the growing body of research on chatbot anthropomorphism, there remains a need for studies that delve deeper into the specific dimensions of anthropomorphism and their differential impacts on user experience. Existing research often relies on broad conceptualizations of anthropomorphism, limiting our understanding of the precise elements that contribute to positive user outcomes. Additionally, while some studies have examined the role of user characteristics, further research is needed to explore how individual differences interact with anthropomorphic cues to shape user experiences.

The present study addresses these gaps by investigating the influence of specific chatbot design cues on user perceptions of anthropomorphism, and how different dimensions of anthropomorphism impact user experience in customer support interactions. By examining the impact of naturalness, human-likeness, lifelikeness, and professionalism on user satisfaction and trust, this research offers a more nuanced understanding of how anthropomorphic design can be optimized to enhance user experience. Additionally, by focusing on the customer support context, the study contributes to the practical application of anthropomorphism in real-world settings. To account for potential individual differences, the model includes several control variables: Gender, Age, ICT familiarity, and Personality (Extraversion, Agreeableness, Conscientiousness, Emotional Stability, and Openness to new experiences).

2.3 Assessment of Anthropomorphism Dimensions

Human-likeness, Naturalness, and Lifelikeness

To assess chatbot anthropomorphism, this study focuses on three key dimensions: Human-likeness, Naturalness, and Lifelikeness (Powers and Kiesler 2006; Kim and Sundar 2012). We asked participants to assess whether the chatbot looks human-like/machine-like, natural/unnatural, or lifelike/artificial.

- Human-likeness: Refers to the extent to which a chatbot exhibits human-like attributes.
- Naturalness: Focuses on how natural and effortless the interaction with the chatbot feels, resembling a human conversation.
- Lifelikeness: This dimension goes beyond visual appearance to capture the overall sense of a chatbot being alive and responsive.

These dimensions encompass both visual appearance and language characteristics. Studies have shown that incorporating human-like features, such as facial expressions and polite language, can enhance user experience and perceptions of humanness (Meyer et al. 2016). Moreover, Haugeland et al. (2022) emphasize the role of human-likeness in fostering anthropomorphism and social presence.

Professionalism and Visual Appearance

To further enrich the assessment of chatbot anthropomorphism, two additional dimensions are considered: Professionalism and Perceived Visual appearance (moderated by Importance of appearance: how important is the chatbot's appearance for the user). These dimensions collectively contribute to the overall perception of a chatbot as a genuine and competent conversational partner.

- Professionalism: In customer service interactions, professionalism is essential, reflecting the chatbot's competence, skill, and seriousness (Goodwin and Smith, 1990).
- Visual Appearance: We asked participants to rate (on a 5-point Likert scale) how much they liked the chatbot's appearance.

- Importance of Appearance: We acknowledge that the chatbot's appearance might not be important to everyone. Thus, we included a moderating variable to moderate the impact of Visual Appearance on Satisfaction and Trust.

By considering these five dimensions, we aim to provide a comprehensive assessment of Anthropomorphism in chatbots and its influence on user experience.

3 Hypotheses Development and Research Model

3.1 Anthropomorphic Design Cues and Social Presence

Social Presence Theory (SPT) (Short et al. 1976) notes that consumers perceive interactions with entities exhibiting human-like qualities as more natural and engaging, leading to a stronger sense of Social Presence. In the context of chatbots, this translates to consumers feeling as if they are interacting with another person rather than a machine. Drawing upon SPT (Short et al. 1976), we suggest that chatbots with anthropomorphic design cues (e.g., human-like avatars) can trigger these associations. As a result, we expect chatbots with such design elements to be perceived as having a higher degree of Social Presence compared to those without, leading to higher user satisfaction and trust scores.

Impact of Chatbot Design on Satisfaction

The presence of a human-like avatar in a chatbot can function as a powerful visual cue, suggesting to the consumer that they are interacting with another person rather than a machine. This, in turn, can increase the perceived human likeness of the interaction. Furthermore, a well-designed avatar can be crafted to appear natural and lifelike, further enhancing the overall sense of Social Presence. The avatar can also project professionalism, leading users to perceive the chatbot as a more trustworthy and reliable source of information. In essence, the human-like qualities embedded within the avatar can contribute to a more positive consumer experience on multiple levels. Therefore, we propose the following hypothesis:

H1a: The presence of a (static) human-like avatar will have a significant positive impact on Satisfaction scores in customer support chatbots.

Facial expressions are a fundamental aspect of human communication, conveying emotions, intentions, and even personality traits without a single word spoken (Hess 2020). When interacting with others, humans tend to pay close attention to nonverbal cues, including facial expressions (Hess 2020). Research shows that recognizing emotions in facial expressions is not only automatic but also plays a crucial role in social interactions, fostering empathy and understanding (Hess 2020).

Media Richness Theory (MRT) (Daft and Lengel 1984) suggests that communication channels with richer nonverbal cues enhance user understanding and satisfaction. Facial expressions are a key form of nonverbal communication. Aligning with Media Richness Theory, we propose that chatbot avatars with dynamic facial expressions will be perceived as richer and more natural compared to those with static expressions or those without an avatar. By providing nonverbal cues that mirror human interaction, this increased richness can contribute to a stronger sense of Social Presence. This leads us to the following hypothesis.

H1b: The presence of a human-like avatar featuring dynamic facial expressions will have a significant positive impact on Satisfaction scores in customer support chatbots.

Impact of Chatbot Design on Trust

Building trust with consumers is essential for successful technology adoption (Lukyanenko et al. 2022), and Social Presence can play a crucial role in this process. When interacting with a machine, consumers may experience feelings of uncertainty or apprehension (Dekkal et al. 2024). Social Presence can foster a more trusting relationship by creating a sense of familiarity and reducing this uncertainty. This is supported by Social Identity Theory (SIT) (Turner et al. 1979), which suggests that individuals are more likely to trust those they perceive as similar to themselves. By exhibiting human-like characteristics, a chatbot can trigger a sense of in-group membership, increasing user trust. Furthermore, interpersonal trust often relies on nonverbal cues that signal sincerity and competence. Leveraging anthropomorphic design cues, chatbots can increase perceived Social Presence and project trustworthiness and reliability. A chatbot that exhibits human-like qualities through design cues may be perceived as more relatable and trustworthy compared to one lacking such features. Therefore, we propose the following hypotheses:

H2a: The presence of a (static) human-like avatar will have a significant positive impact on Trust scores in customer support chatbots.

H2b: The presence of a human-like avatar featuring dynamic facial expressions will have a significant positive impact on Trust scores in customer support chatbots.

3.2 Anthropomorphism Dimensions, User Satisfaction, and Trust

In the context of chatbots, Anthropomorphism can be promoted through design elements that evoke human-like qualities. The Elaboration Likelihood Model (ELM) (Petty and Cacioppo 1986) provides a valuable framework for understanding how anthropomorphism might influence consumer experience. ELM suggests that individuals process information through two primary routes: the central route and the peripheral route. The central route involves thoughtful consideration of the information itself, while the peripheral route relies on simpler cues to form judgments.

Customer support chatbot interactions often involve the peripheral route, as consumers may not need to engage in deep analysis of the information exchanged. In these situations, peripheral cues like a chatbot's appearance can play a significant role in shaping consumer perception. Building on ELM, we suggest that a chatbot with a high degree of anthropomorphism, achieved through anthropomorphic design cues, can trigger positive associations with human interaction. These positive associations, even without extensive information processing, can lead to a more positive consumer experience. Thus, we propose the following hypotheses:

H3a: Perceived anthropomorphism in a customer support chatbot will be positively associated with user satisfaction.

H3b: Perceived anthropomorphism in a customer support chatbot will be positively associated with user trust.

3.3 Research Model

The proposed research model visually depicts the relationships between perceived Social Presence dimensions and consumer outcomes (satisfaction, trust, reuse intention). The Social Presence dimensions are the independent variables that influence consumer outcomes. These variables include Human-likeness, Naturalness, Lifelikeness, Professionalism, and Appearance, where Appearance has the moderator variable 'Importance of Appearance.' The model also includes several control variables: Gender, Age, ICT familiarity, and Personality (Extraversion, Agreeableness, Conscientiousness, Emotional Stability, and Openness to new experiences) (Fig. 1).

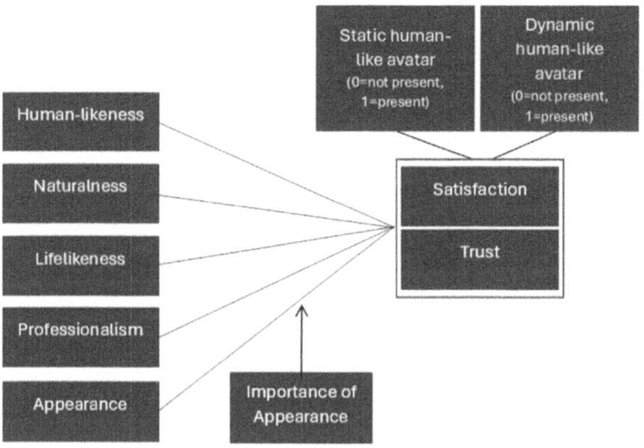

Fig. 1. Research model.

4 Methodology

4.1 Experimental Design

This study investigated consumer responses towards the use of chatbots as customer support agents. To achieve this, a between-subjects experiment was conducted. Participants were randomly assigned to watch one of three variations of a pre-recorded customer support scenario. The scenario depicted a customer contacting a chatbot named "Eva" about damaged sunglasses received from an online store. Three animations were created, each featuring an identical conversation but with various levels of anthropomorphism:

- Faceless Avatar: Conversation featuring a faceless avatar.
- Human-like Avatar: Conversation featuring a static, human-like avatar image.
- Human-like Avatar with Dynamic Facial Expressions: Conversation featuring a human-like avatar that displays a range of emotions (e.g., smiling for positive messages, frowning for negative messages) to reflect the conversation's tone.

The decision to use a female persona ("Eva") for the chatbot was informed by previous research suggesting that users tend to rate female virtual agents more favorably

(Toader et al. 2019). Using pre-recorded animations ensured minimal variability in the chatbot experience across participants, allowing for reliable comparisons of consumer responses. After watching the animation, participants were asked to rate various aspects of the chatbot.

Chatbot Design

The first variation of the chatbot presented the conversation featuring a faceless avatar representing the chatbot "Eva". This design served as the baseline condition with the lowest level of anthropomorphism. The second variation of the chatbot introduced a static human-like avatar image for "Eva" throughout the conversation. A stock image was used for this purpose (iStock 2018). This design increased the level of anthropomorphism compared to the baseline condition. See all chatbot avatar variations in Fig. 2. The third animation featured a human-like avatar with dynamic facial expressions that responded to the emotional tone of the conversation (see Fig. 3).

Fig. 2. Avatar variations.

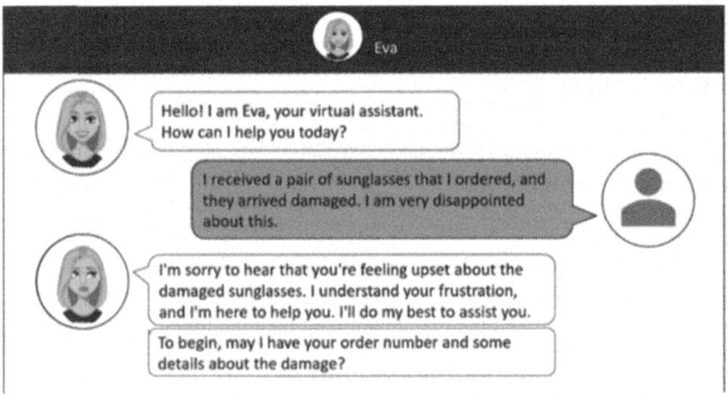

Fig. 3. Screenshot of the conversation in which the avatar's facial expressions match the emotions expressed in the textual messages.

By manipulating the level of anthropomorphism across these animations, the experiment aimed to investigate how users perceive and respond to chatbots with varying degrees of human-like characteristics.

4.2 Questionnaire Development

To assess participant responses to the chatbot animations, a survey instrument was developed and distributed online. The questionnaire included several key sections. First, an initial set of demographic questions gathered background information on participants, including age, gender, and ICT knowledge. Next, an introductory section presented participants with a brief description of the customer support scenario. Participants were instructed to imagine themselves as the customer experiencing this situation.

Following the scenario introduction, participants viewed a pre-recorded animation depicting the customer interacting with the chatbot. A timer measured viewing time to assess if participants watched the entire video before proceeding with the questionnaire. Finally, the survey concluded with a chatbot evaluation section. In this section, participants answered multiple Likert-scale questions evaluating various aspects of the chatbot. A total of 93 responses were collected. However, only 73 responses were considered complete and unique. Incomplete surveys or those from participants who did not watch the entire video were excluded. Participants were comprised of 42 males (57.5%) and 31 females (42.5%). The average age of participants was 39.6 years.

4.3 Cumulative Link Model Analysis

The survey data consisted of Likert-scale responses. Since Likert-scale data is ordinal (structured into ranked categories), Cumulative Link Models (CLMs) were chosen to conduct the analysis.

A critical step in CLM analysis is selecting the link function that best describes the relationship between the ordinal responses and the predictor variables. Several link functions (logit, probit, cloglog, loglog, and cauchit) were evaluated to determine the best fit based on the Akaike Information Criterion (AIC) score.

For each of the dependent variables (satisfaction and trust), a separate CLM analysis was conducted.

The models explored how independent variables (Human Likeness, Naturalness, Lifelikeness, Professionalism, and Appearance) influenced each of the dependent response variables. Additionally, the models included an interaction term (Appearance*Importance of Appearance), and several control variables representing participant characteristics (Gender, Age, ICT knowledge, Personality).

Following the construction of the Cumulative Link Models for Satisfaction and Trust, post-hoc testing was conducted via the Likelihood ratio test. For each of the CLM models a null model (containing only the control variables) was tested. The full models were then compared to their null models using a likelihood ratio test (R: Likelihood Ratio Test), which follows a chi-squared distribution, to determine if the models were statistically significant. The results indicate that all models have significantly better AIC scores compared to their respective null models ($p < 0.001$).

5 Results

The Cumulative Link Models of Satisfaction and Trust highlighted multiple significant relationships. The models returned significant results for Human-likeness, Naturalness, and Professionalism, which are shown in Table 1. The estimates of these variables were all positive, which suggests that a higher rating for one of these variables leads to a higher Satisfaction and Trust score. While Lifelikeness and Appearance had no significant impact on consumer responses, Professionalism showed the greatest positive effect on Satisfaction and Trust. These findings partially support H3a and H3b.

Moreover, the presence of a dynamic avatar had a positive impact on Trust (supporting H2b), but no significant impact on Satisfaction (rejecting H1b). On the other hand, the presence of a static human-like avatar had no significant impact on Satisfaction nor Trust, rejecting H1a and H2a. Finally, there were also significant results for the Control variables, including some of the Personality traits. While Openness to Experience and Extraversion had a significant negative effect on Satisfaction, Emotional Stability positively impact Satisfaction. Interestingly, age and gender (male) showed a significant positive effect on Trust.

Table 1. CLM analysis results. Note Significant effects are marked in bold. (*p < 0.05; **p < 0.01; *** < 0.001; N = 73).

Variable	Effect on Satisfaction (SE)	Effect on Trust (SE)
Main effects		
Chatbot types		
Static Human-like avatar	0.228 (1.215)	0.154 (0.420)
Dynamic Human-like avatar	0.406 (1.734)	**1.047* (0.435)**
Anthrop. Dimensions		
Human-likeness	**5.221* (2.422)**	**0.569* (0.290)**
Naturalness	**3.874* (1.926)**	**0.598* (0.292)**
Lifelikeness	−1.637 (1.016)	−0.157 (−0.595)
Professionalism	**9.897* (4.495)**	**0.641* (0.268)**
Appearance	−2.574 (1.402)	0.382 (0.324)
Control variables		
Gender: Male	4.978 (2.955)	**0.806* (0.366)**
Age	−0.193 (0.105)	**0.038** (0.013)**
ICT	3.705 (2.227)	−0.640 (0.415)
P1; Extraversion	**−3.291* (1.578)**	−0.056 (0.214)
P2; Agreeableness	1.842 (1.080)	**0.425* (0.176)**
P3; Conscientiousness	−0.870 (0.557)	**−0.404* (0.161)**
P4; Emotional Stability	**4.496* (2.028)**	0.210 (0.161)
P5; Openness to Exp	**-2.977* (1.383)**	0.096 (0.137)
Interaction effect		
Appearance: Importance	0.661 (0.342)	0.024 (0.056)

6 Discussion

6.1 Theoretical and Practical Implications

This study contributes to the existing body of knowledge on chatbot anthropomorphism and user experience. By demonstrating the differential impact of various anthropomorphic dimensions on user satisfaction and trust, the research extends the understanding of how to effectively design engaging and trustworthy chatbots. The results also highlight the importance of social presence theory in explaining the relationship between anthropomorphic cues and user outcomes.

Moreover, the findings offer valuable insights for chatbot developers and organizations seeking to optimize customer service interactions. By prioritizing the incorporation of human-like qualities, such as naturalness and professionalism, businesses can enhance user satisfaction and trust. Additionally, the study emphasizes the importance of visual design elements, particularly dynamic facial expressions, in fostering user trust. These findings can inform the development of more effective and user-centric chatbots, ultimately leading to improved customer satisfaction and loyalty.

6.2 Limitations and Opportunities for Future Research

It is important to recognize that this research has several limitations, which could present opportunities for future research. Firstly, the study used chatbot animations instead of using a fully functioning chatbot interface. Future research could benefit from developing a real chatbot with dynamic expressions. This would allow users to interact with the chatbot in unique ways, which better imitates the way chatbots are used in business settings.

Secondly, the survey population was small, with the data of 73 respondents which could be used for the CLM. Future studies could validate these models and analyze them with a larger survey population to improve the reliability and generalizations of the findings.

Lastly, the study did not further explore the effects of personality dimensions on Satisfaction and Trust scores. Since the personality questions seem to have an impact on either Satisfaction or Trust scores it could be interesting to further study how users' personalities influence these scores. Chatbot practitioners should consider that users' personalities can influence the Satisfaction and Trust scores both positively and negatively.

References

Adam, M., Wessel, M., Benlian, A.: AI-based chatbots in customer service and their effects on user compliance. Electron. Mark. **31**(2), 427–445 (2020)

Adamopoulou, E., Moussiades, L.: An overview of chatbot technology. In: IFIP International Conference on Artificial Intelligence Applications and Innovations, pp. 373–383. Springer, Cham (2020)

Cai, D., Li, H., Law, R.: Anthropomorphism and OTA chatbot adoption: a mixed methods study. J. Travel Tour. Mark. **39**(2), 228–255 (2022)

Crolic, C., Thomaz, F., Hadi, R., Stephen, A.T.: Blame the bot: anthropomorphism and anger in customer–chatbot interactions. J. Mark. **86**(1), 132–148 (2022)

Daft, R.L., Lengel, R.H.: Information richness: a new approach to managerial decision making. Acad. Manag. Rev. **9**(2), 199–213 (1984)

Dekkal, M., Arcand, M., Prom Tep, S., Rajaobelina, L., Ricard, L.: Factors affecting user trust and intention in adopting chatbots. J. Finan. Serv. Mark. **29**, 699–728 (2024)

Gartner. Gartner Predicts Chatbots Will Become a Primary Customer Service Channel Within Five Years (2022). https://www.gartner.com/en/newsroom/press-releases/2022-07-27-gartner-predicts-chatbots-will-become-a-primary-customer-service-channel-within-five-years

Goodwin, C., Smith, K.L.: Courtesy and friendliness: conflicting goals for the service providers. J. Serv. Mark. **4**(1), 5–20 (1990)

Han, M.C.: The impact of anthropomorphism on consumers' purchase decision in chatbot commerce. J. Internet Commer. **20**(1), 46–65 (2021)

Haugeland, I.K.F., Følstad, A., Taylor, C., Bjørkli, C.A.: Understanding the user experience of customer service chatbots: an experimental study of chatbot interaction design. Int. J. Hum Comput Stud. **161**, 102788 (2022)

Hess, U.: Body Language. Springer, Heidelberg (2020)

iStock (2018). https://www.istockphoto.com/nl/vector/gezicht-uitingen-van-vrouw-met-blond-haar-gm1081125798-289884774

Kim, Y., Sundar, S.S.: Anthropomorphism of computers: is it mindful or mindless? Comput. Hum. Behav. **28**(1), 241–250 (2012)

Lukyanenko, R., Maass, W., Storey, V.C.: Trust in artificial intelligence: from a foundational trust framework to emerging research opportunities. Electron. Mark. **32**(4), 1993–2020 (2022)

Meyer, J., Miller, C., Hancock, P., De Visser, E.J., Dorneich, M.: Politeness in machine-human and human-human interaction. In: Proceedings of the Human Factors and Ergonomics Society Annual Meeting, vol. 60, No. 1, pp. 279–283, SAGE Publications, Sage CA: Los Angeles, CA (2016)

Petty, R.E., Cacioppo, J.T.: The elaboration likelihood model of persuasion. In: Communication and Persuasion. Springer, New York (1986)

Powers, A., Kiesler, S.: The advisor robot: tracing people's mental model from a robot's physical attributes. In: Proceedings of the 1st ACM SIGCHI/SIGART Conference on Human-Robot Interaction, pp. 218–225 (2006)

Short, J., Williams, E., Christie, B.: The Social Psychology of Telecommunications. John Wiley and Sons Ltd., Hoboken (1976)

Toader, D.C., et al.: The effect of social presence and chatbot errors on trust. Sustainability **12**(1), 256 (2019)

Song, S.W., Shin, M.: Uncanny valley effects on chatbot trust, purchase intention, and adoption intention in the context of e-commerce. Int. J. Hum.-Comput. Interact. **40**(2), 441–456 (2024)

Turner, J.C., Brown, R.J., Tajfel, H.: Social comparison and group interest in ingroup favouritism. Eur. J. Soc. Psychol. **9**(2), 187–204 (1979)

Yanxia, C., Shijia, Z., Yuyang, X.: A meta-analysis of the effect of chatbot anthropomorphism on the customer journey. Mark. Intell. Plan. **42**(1), 1–22 (2024)

'I Prefer Robot Cats!' Reflections on Robot Animal-Like Morphology from an in-the-Wild Child-Robot Interaction Workshop

Alicja Wróbel[1], Paulina Zguda[2](✉), Karolina Źróbek[3], Marie-Monique Schaper[4], Paweł Gajewski[3], and Bipin Indurkhya[1]

[1] Jagiellonian University, ul. Gołębia 24, Kraków, Poland
[2] Doctoral School in the Humanities, Jagiellonian University, Kraków, Poland
paulina.zguda@doctoral.uj.edu.pl
[3] AGH University of Krakow, al. Adama Mickiewicza 30, Kraków, Poland
[4] University of Lleida, Lleida, Spain

Abstract. This study is based on an in-the-wild child-robot interaction event involving 100 children aged 7 to 9. The children interacted with four distinct robots: Pepper, Nao, Misty, and MiRo-E, during a workshop designed to explore their engagement with these robots through various activities. Audio and video recordings captured these interactions, providing rich data for analysis. This study focuses on the MiRo-E robot and investigates how specific animal-like features, such as appearance, lifelike behaviors, and sounds, influence children's engagement and emotional connections with robots. The findings reveal that children tend to favor robots resembling domestic animals, such as dogs or cats, due to their approachable appearance. Furthermore, robots designed with ambiguous, creature-like features, evoking multiple animal species, elicit heightened interest and fascination. This suggests that morphing features blending both animal and robotic characteristics are particularly appealing to children. These preferences underscore the significance of robot morphology in fostering deeper engagement and emotional connections. The study offers design recommendations, emphasizing the critical role of animal-like features in enhancing user experience, particularly for interactive and educational robotic technologies designed for children.

Keywords: Social Robotics · Child-Robot Interaction · In-the-wild Methodology

1 Introduction

Children's perceptions of a social robot's morphology (physical appearance and structure) are a subject of significant interest in the field of Child-Robot Interaction (CRI). A robot's physical features like appearance, size, shape, and sensory

D. Singh et al. (Eds.): IHCI 2024, LNCS 15557, pp. 159–170, 2025.
https://doi.org/10.1007/978-3-031-88705-5_13

modalities determine how children interact with the robot in a meaningful way [18,22].

The physical design of the robot also plays a crucial role in facilitating interactions that involve multiple sensory experiences in education, particularly in CRI [5,9]. A robot's form, movement capabilities, and sensory input systems must be designed to effectively engage with children through touch, proximity, and verbal communication [11].

Another important factor is the robot's visual resemblance to living creatures, such as animals or humans [11,16,21]. For instance, a robot that mimics the appearance of a beloved pet or favorite animal tends to evoke positive responses and a sense of familiarity in children, which may facilitate easier interaction [16].

In this study, we examined how children respond to various modes of interaction with a robot, including conversation (engaging in verbal activities with the robot) and movement (engaging in motor activities with the robot). We explored the impact of animal-like features, appearance, lifelike behaviors, and sounds on children's engagement and emotional connections with robots.

2 Children's Perceptions of Animal-Like Robots

Previous work on the perception of animal-like robots has highlighted the significance of appearance in determining a robot's likability [19,28,34]. Human and animal appearances are generally preferred, with prior expectations that stem from familiarity aiding in interactions [17,19,27]. A preference for animal-like robots has been noted, which may be driven by concerns over human-like robots causing uncanny feelings or failing to meet expectations [19]. However, pet robots offer limited possibilities in comparison to real pets, leading to more frequent termination of interactions due to inconsistencies in appearance [19]. Additionally, previous research on animal-like robots has shown that children perceive these robots as living beings, attributing them with emotional and moral characteristics, even though they distinguish them from real animals [14,20]. AIBO, a robot dog, has been the subject of many studies, with findings indicating that children display more positive attitudes toward AIBO than adults, assigning cognitive abilities and emotions to it [14]. Furthermore, the robot's ability to evolve from a 'puppy' to a 'dog' enhances its perceived social engagement, resulting in higher levels of emotional attachment from users [15]. These studies also suggest that children construct new conceptual categories for personified artifacts, challenging conventional ontological genres [15]. A study using the baby dinosaur robot Pleo indicated that participants had difficulty engaging in typical play activities (other than petting), indicating a lack of understanding of how to play with such a robot, leading to the toy being put down. Participants reported problems creating a long-term relationship with the robot [10]. While several studies have delved into children's perceptions of animal-like robots, these investigations have encountered limitations that deserve consideration. Notably, there has been a dearth of explicit focus on the design of a robot's form and structure, a critical aspect of industrial design [19]. Many prior studies examining children's perceptions of animal-like robots have predominantly concentrated on robots resembling dogs and tend to compare them to those animals [1,2,20]. However, these studies often overlook the potential of a hybrid approach towards a robot's appearance, limiting a comprehensive understanding of children's responses to the ambiguous appearance of robotic companions. Furthermore, these studies have yielded incompatible outcomes regarding the impact of inconsistencies in robots' appearance on people's impressions, like adding fur to robotic dogs, with some suggesting it enhances interactive engagement. In contrast, others assert it evokes uncanny feelings in individuals [11,19]. Additionally, the presence of uncanny valley effects in animal-like robots calls for further research to mitigate these disconcerting responses through improved robot design. Enhancing human-robot interactions and attachment bonds necessitates a deeper exploration of a robot's physical and socio-cognitive developmental capabilities, as well as a consideration of demographic factors such as age, gender, and participants' levels of technology experience in the design of social robots. This paper will study how specific animal-like features, such as their appearance, lifelike behaviors and sounds, influence children's engagement and emotional connections with robots, and what the implications are for designing interactive and educational robotic technologies.

3 Method

3.1 In-the-Wild Child-Robot Interaction in Educational Settings

We employed the in-the-wild methodology, where a phenomenon is studied in its natural setting, rather than in a laboratory [30,31,33]. Examining children's perceptions of robots in their familiar environments is crucial for designing more effective real-life child-robot interactions [7]. These natural settings include homes, schools, and hospitals, each significantly affecting how children perceive and engage with robots. For example, children tend to feel at ease and comfortable interacting with robots in their homes, where the environment is known and comforting [6]. Conversely, the school environment, while rich in educational opportunities, may introduce stress and unfamiliarity, potentially impacting how children perceive and interact with robots. For our study, this obstacle was addressed beforehand. The pupils who participated in the workshop with four robots mostly had the opportunity to participate in previous workshops conducted by our research group (with only one robot and with involvement in other activities). The school chosen is an educational institution attended by the children of one of our researchers, which allowed for better contact with the school, as well as enabling mental preparation in a less stressful atmosphere. The choice of school as a workshop venue allowed a safe and familiar place where children know how to behave in a given context.

3.2 Participants

One hundred children (age: 7–9 yrs; M: 43, F: 52, unspecified: 5) from a local primary school (three first grades and a second grade) participated in our study.

The study was approved by the Ethics Committee of our university, and informed consent was obtained from the children's parents.

The children were divided into four groups of 25 each: each group was one section of the 1st or 2nd grade class. The same interaction pattern (described in the method section below) was followed for each group. The study was conducted over two days: on the first day, Group 1 followed by Group 2; and on the second day Group 3 followed by Group 4. Each interaction session lasted for about one hour.

Eleven researchers (ten university students and one faculty) were involved in the design and implementation of the study. All interaction sessions were closely supervised.

3.3 Activities

The analysis presented here is based on a larger study during which children could interact with four different robots: Pepper, NAO, Misty and Miro-E. The study focused on children's interactions with these robots in a school setting to see how they respond to different robots.

The whole study involved groups of 25 children participating in six activities. Initially, all children joined together for two activities: 'Brainstorming', where they drew and discussed robot designs, and 'Introduction', where they interacted with the Pepper robot in a mindfulness exercise, both lasting 20 min to gauge their design ideas and group interaction with robots.

Afterwards, the children were split into smaller groups (6–7 per group) for four more activities, each with a different robot, to observe their interactions in a more intimate setting. Each activity lasted 10 min, allowing every group to experience all four robots within 40 min. Nine researchers facilitated the sessions, with one leading the activity and another controlling the robot. Children were divided into small groups to ensure that each child had ample opportunity for direct interaction with the robot. In larger group settings, individual engagement would have been limited, reducing the chances for each child to interact as fully as they did in smaller groups.

We focus here on the interactions children had with the Miro-E robot. All children mentioned took part in this activity, in which the children were asked to program the MiRo-E robot's movements. We incorporated this activity as such form of control enhances children's engagement with robots and provides feedback regarding children's preferences for specific robot features [4,13,25].

After introducing the task and the robot, the children were given a set of possible commands on little sheets of paper, which consisted of turning left/right, going forward backwards, moving ears/head and wagging tail. Children could arrange the commands in a sequence they liked, and the researcher helped them enter those commands into the robot via its built-in programming interface called MiRoCODE. The robot movements corresponding to the command sheets given to the children were pre-programmed. The transformation of algorithms from the paper version into the digital one was performed by the researcher openly so the children could observe how it was done. Children interacting with the robot during the activity are shown in Fig. 1.

Fig. 1. Children interacting with Miro-E

4 Data Collection and Analysis

The interaction was recorded by three cameras set up in different parts of the room, and there was a voice recorder at each station to record the children's conversations while interacting with robots.

We used qualitative analysis methods, which involves examining non-numerical data, such as behaviors, interactions, and perceptions, to understand underlying patterns and meanings [8,32]. In a study exploring how children perceive and engage with robots, qualitative analysis is crucial because it allows researchers to capture the nuances of children's experiences, such as how they describe and interact with the robots [4,12]. This approach helps reveal deeper insights into the children's thoughts, feelings, and behaviors that might not be evident through quantitative methods alone [4,9]. Due to the nature of the data and analysis, it was not possible to conduct a statistical analysis showing the quantitative occurrence of specific behaviors, such as in the context of gender-related dynamics.

During the analysis of the data, one researcher analysed videos and transcriptions of children's responses to the animal-like robot Miro. All analyzes were performed manually without the use of software. We present below the details of each of these analyses.

Table 1 shows the coding scheme for analyzing children's perceptions of animal-like robots. This scheme was based on children's statements about the

Miro-E robot. Five characteristic categories were identified: animal-like appearance (e.g. children describing robot's appearance in animal-like terms), animal-like behavior (e.g. children describing robot's behavior in animal-like terms), ambiguous appearance (e.g. children describing robot's appearance as ambiguous), machine-like characteristics (e.g. children describing robot's appearance and behavior in machine-like terms), and preferences for robot morphology (animal-like vs. human-like). The researcher developed these codes while watching the videos and basing on specific traits in children's behavior towards the robot. Then the codes were applied in a second review.

Table 1. Coding scheme of children's perceptions of animal-like robots analysis

Children's perceptions of animal-like robots analysis	
Behavior	Description
Animal-like appearance	Children describe the robot using words related to animal-like appearance. Examples would be "I like his ears" or "He is a cute puppy".
Animal-like behavior	Children describe the robot using words related to animal-like behavior. An example would be "He's purring" or "He's wagging his tail."
Ambiguous appearance	Children describe the robot as having features of different animals indicating that the robot's appearance is ambiguous. An example would be children not agreeing with each other on what animal the robot looks like.
Machine-like characteristics	Children describing the robot as having machine-like characteristics. An example would be "He has cameras in his eyes."
Preference for robot morphology	Children express their preference, whether they prefer animal-like robots or human-like robots.

5 Results

5.1 Animal-Like Morphology

We present here our observations regarding children's reactions to the animal-like robot Miro-E. We focused on children's perceptions of the robot's animal-like appearance and behavior, ambiguities in its appearance, a robot's machine-like features and human-like vs. animal-like robots. During the interaction, children exhibited different types of behaviors towards Miro-E.

Animal-Like Appearance. Children often commented on the robot's animal features and behavior. They seemed to like its animal-like features. Some of them

referred to those characteristics as 'cute' or 'adorable'. They said that the robot's movements like wagging tail or moving its ears remind them of animals like dogs or bunnies.
Child: I like his ears!
Child: *Miro wags its tail* He's definitely a doggie.
Child: He's so cute!
Child: Miro, you are a cute puppy!

Ambiguous Appearance. Children seemed concerned with guessing what animal the robot resembled the most. They cheered when they were asked what animal the robot reminds them of. Children shouted over each other to express their ideas and they listed many different animals as their guesses. They discussed their ideas of the robot's similarities to animals during the whole interaction.
Researcher: Hello! This is our robot, named Miro. What kind of animal do you think it is? Child: Rabbit!
Child: Donkey!
Child: Puppy?
Researcher: Why a puppy?
Child: Rabbit!
Child: Because it has a little tail, and big ears, but I don't know the breed.
Child: It's definitely a donkey.
Child: Donkey!
Researcher: Hi, this robot's name is Miro. What kind of animal do you think it is?
Child: Little bull!
Child: Rabbit!
Child: Cow!

Animal-Like Behavior. Children complained when the robot was not moving. They cheered when it wagged its tail and asked it to wag the tail again and again. Children cheered when the robot made sounds and seemed to relate its whir to animal-like sounds, e.g. cat's purring or a dog's barking. They said that they prefer animals that make sounds.
Child: Why doesn't he wag his tail?
Child: Make him wag his tail!
Child: Look he wagged! Like a dog!
Child: He said something! I heard! He said ''tytytyty''.
Child: He's purring!
Child: Yes, purring!

Machine-Like Characteristics. Despite the animal appearance, children seemed interested in the robot's machine-like appearance as well. They often commented on cameras in robots' eyes, plastic ears or back glowing with colors. They seemed to enjoy interpreting its behavior by adding new machine-like meanings, e.g. robot closing and opening its eyes must be taking photos.

`Child: It took a photo of us! I heard!`
`Child: He has cameras in his eyes!`
`Child: Miro has cameras in his eyes, speakers in his ears and something weird in his nose.`
`Researcher: What ears does Miro have?`
`Children: Plastic!`
`Researcher: And eyes?`
`Children: Cameras!`
`Child: *pointing to Miro's glowing back* It looks like a turtle!`

Animal-Like Robots vs Human-Like Robots. Despite children's interest in animal-like robots, it is not clear if they prefer them over different kinds of robots. During the study, children disagreed amongst each other over which robots they preferred: animal-like or human-like.

`Researcher: Do you like these robotic animals or do you prefer little human-like robots?`
`Child: Yes.`
`Child: I prefer robot cats.`
`Child: And I prefer hedgehogs.`
`Child: I prefer the other one. *points at NAO, a humanoid robot*`
`Child: They are all cool.`
`Child: I saw a robot like that at my mom's work that helps with tasks. *points at Pepper, a humanoid robot*`

6 Discussion

6.1 Animal-Like Morphology and Its Design Implications

Our study offers a glimpse into the factors that drive children's attraction to animal-like robots. It is evident that children possess a fondness for robotic companions that embody the qualities of cute animals, as it has been stated before [5]. Basing on the outcomes of our study, children tend to favor robots that resemble domestic animals, such as dogs or cats, likely because of their approachable appearance. We suggest sticking with that design, since appropriate robot design may evoke feelings of comfort and emotional support, leading to a higher level of trust [3]. Basing on that, interacting with such robots could bring both playfulness and the potential for emotional connections, particularly when robots respond to affectionate petting with lifelike animal behaviors such as tail wagging or purring. Furthermore, these results expand the previous knowledge on the preferred robotic designs [19,28,34] by the addition of certain animal-like features. The incorporation of features deemed 'cute' by children into robot design holds promise for enhancing the effectiveness of educational, therapeutic, and entertainment applications. We suggest that by leveraging characteristics such as specific animal sounds (e.g., purring, barking, or squeaking) and animal behaviors (e.g., tail-wagging, sniffing, or seeking cuddles from children), designers can craft robots that not only elicit more pleasant and engaging interactions but also deepen children's engagement and participation. This study underscores the potential of cute animal-like features in the development of robotic technologies tailored to enrich children's experiences and foster meaningful connections.

Interestingly, our findings highlight a captivating phenomenon in children's interactions with robots: when a robot's appearance is intentionally crafted to be ambiguous, encompassing features that simultaneously evoke multiple animal species, it tends to elicit heightened interest and fascination among young users. This stands as contrary to the majority of dog-like or certain animal-like designs in the area of animal-like social robots [1,2,20]. Instead of conforming to a single animal archetype, ambiguous robots may be an answer to how to ignite curiosity and imagination in children, encouraging them to adapt and explore. Embracing ambiguity in robotic design could enhance children's engagement and learning by promoting dynamic, open-ended interactions that align with their innate curiosity and creativity.

This study also underscores the fascinating attraction children exhibit toward robots with morphing features that blend both animal and robotic characteristics. This tendency to imagine robotic functionalities opens up a rich realm of creative play and exploration. Designers can leverage this insight by crafting robots that intentionally blur the boundaries between animal and machine, thereby encouraging children to engage in imaginative scenarios where these robots possess extraordinary, technology-driven capabilities. This approach may not only enrich the play experience but also align with the educational potential of fostering curiosity, problem-solving, and an understanding of technology from a young age.

6.2 Further Application of the Study Results

The findings presented in this study can contribute significantly to improving the design of robots in several ways:

Designing and introducing robots that cater specifically to the needs of Polish children in a similar age group, such as providing entertainment (e.g., a robot as a play companion or daily life companion).

Enhancing our understanding of the specific robot features that capture children's attention, which can be leveraged to introduce robots as valuable educational tools. For example, integrating robots into classroom settings to assist in developing storytelling and creativity skills (as suggested by [24,29]), or facilitating discussions on difficult topics (as explored by [26]).

Familiarizing children with robots as artificial agents, which are likely to become increasingly prevalent in the environments of both children and adults.

These insights pave the way for more effective and engaging child-robot interactions, ensuring that robots can serve both educational and recreational purposes in the lives of young users. Our study suggests that there is a potential to use some characteristics of animal-like robots to encourage children to interact with robots or even deepen the bond between them. However, further research is needed to delve deeper into the cognitive and emotional mechanisms underlying CRI for the design of future educational and entertainment robots.

7 Limitations and Future Work

Due to the one-time nature of the event (a single-day workshop designed to allow children to freely interact with social robots in a safe, structured, yet non-restrictive environment), the study did not extensively explore the long-term effects of child-robot interactions on cognitive development or social skills. However, conducting multiple workshops with the same children, for instance, at monthly intervals, would provide a more comprehensive understanding. Additionally, the study primarily focused on describing the mental states of children as they explored the behavior and appearance of different social robots, rather than inferring causal relationships regarding the effects of social robots on cognitive or social development. This does not preclude future research from addressing these aspects, although it would require adopting a different methodological paradigm (for example, focusing on long-term engagement in children using methodologies already tested for autism therapy [23]).

Furthermore, an interesting avenue for future research would be to replicate a similar study design (with appropriately adjusted activities to engage participants) with older children interacting with social robots. This would help determine whether the preferences observed in younger children align with those exhibited by older children.

Acknowledgments. This research was supported in part by a grant from the National Science Centre, Poland, under the OPUS call in the Weave Programme under the project number 2021/43/I/ST6/02489. The authors would like to thank Zofia Samsel, Maria Kiraga, Maria Halesiak, Rebecca Fontanarosa, Marcin Wierzbowski and Bartłomiej Kozyra for their help with conducting the study.

Disclosure of Interests. The authors have no competing interests to declare that are relevant to the content of this article.

References

1. Barber, O., Somogyi, E., McBride, A.E., Proops, L.: Children's evaluations of a therapy dog and biomimetic robot: influences of animistic beliefs and social interaction. Int. J. Soc. Robot. **13**, 1411–1425 (2021)
2. Barber, O., Somogyi, E., McBride, E.A., Proops, L.: Exploring the role of aliveness in children's responses to a dog, biomimetic robot, and toy dog. Comput. Hum. Behav. **142**, 107660 (2023)
3. Bartneck, C., Suzuki, T., Kanda, T., Nomura, T.: The influence of people's culture and prior experiences with Aibo on their attitude towards robots. AI Soc. **21**, 217–230 (2006)

4. Belpaeme, T., Kennedy, J., Ramachandran, A., Scassellati, B., Tanaka, F.: Social robots for education: a review. Sci. Robot. **3**(21), eaat5954 (2018)
5. Breazeal, C.: Emotion and sociable humanoid robots. Int. J. Hum.-Comput. Stud. **59**(1), 119–155 (2003). Applications of Affective Computing in Human-Computer Interaction
6. Causo, A., Vo, G.T., Chen, I.-M., Yeo, S.H.: Design of robots used as education companion and tutor. In: Zeghloul, S., Laribi, M.A., Gazeau, J.-P. (eds.) Robotics and Mechatronics. MMS, vol. 37, pp. 75–84. Springer, Cham (2016). https://doi.org/10.1007/978-3-319-22368-1_8
7. Crabtree, A., Tolmie, P., Chamberlain, A.: "Research in the wild": approaches to understanding the unremarkable as a resource for design. In: Chamberlain, A., Crabtree, A. (eds.) Into the Wild: Beyond the Design Research Lab. SAPERE, vol. 48, pp. 31–53. Springer, Cham (2020). https://doi.org/10.1007/978-3-030-18020-1_3
8. Creswell, J.W., Poth, C.N.: Qualitative Inquiry and Research Design: Choosing Among Five Approaches. Sage Publications (2016)
9. Dautenhahn, K.: Socially intelligent robots: dimensions of human-robot interaction. Philos. Trans. Roy. Soc. B Biol. Sci. **362**, 679–704 (2007)
10. Fernaeus, Y., Håkansson, M., Jacobsson, M., Ljungblad, S.: How do you play with a robotic toy animal? A long-term study of Pleo. In: Proceedings of the 9th International Conference on Interaction Design and Children, pp. 39–48 (2010)
11. Fong, T., Nourbakhsh, I., Dautenhahn, K.: A survey of socially interactive robots. Robot. Auton. Syst. **42**, 143–166 (2003)
12. Kahn Jr, P.H., Friedman, B., Perez-Granados, D.R., Freier, N.G.: Robotic pets in the lives of preschool children. In: CHI 2004 Extended Abstracts on Human Factors in Computing Systems, pp. 1449–1452 (2004)
13. Kahn, P.H., Jr., Gary, H.E., Shen, S.: Children's social relationships with current and near-future robots. Child Dev. Perspect. **7**(1), 32–37 (2013)
14. Kerepesi, A., Kubinyi, E., Jonsson, G.K., Magnusson, M.S., Miklósi, Á.: Behavioural comparison of human-animal (dog) and human-robot (AIBO) interactions. Behav. Process. **73**, 92–99 (2006)
15. Krueger, F., Mitchell, K.C., Deshpande, G., Katz, J.S.: Human-dog relationships as a working framework for exploring human-robot attachment: a multidisciplinary review. Anim. Cogn. **24**, 371–385 (2021)
16. Kühnlenz, K., Sosnowski, S., Buss, M.: Impact of animal-like features on emotion expression of robot head EDDIE. Adv. Robot. **24**, 1239–1255 (2010)
17. Lin, Y.-C., Liu, T.-C., Chang, M., Yeh, S.-P.: Exploring Children's Perceptions of the Robots, pp. 512–517 (2009)
18. Lytridis, C., Bazinas, C., Papakostas, G.A., Kaburlasos, V.: On measuring engagement level during child-robot interaction in education. In: Merdan, M., Lepuschitz, W., Koppensteiner, G., Balogh, R., Obdržálek, D. (eds.) RiE 2019. AISC, vol. 1023, pp. 3–13. Springer, Cham (2020). https://doi.org/10.1007/978-3-030-26945-6_1
19. Löffler, D., Dörrenbächer, J., Hassenzahl, M.: The uncanny valley effect in zoomorphic robots, pp. 261–270. ACM (2020)
20. Melson, G.F., Kahn, P.H., Jr., Beck, A., Friedman, B.: Robotic pets in human lives: implications for the human-animal bond and for human relationships with personified technologies. J. Soc. Issues **65**, 545–567 (2009)
21. Mutlu, B., Forlizzi, J., Hodgins, J.: A storytelling robot: modeling and evaluation of human-like gaze behavior, pp. 518–523. IEEE (2006)
22. Podpečan, V.: Can you dance? A study of child-robot interaction and emotional response using the NAO robot. Multimodal Technol. Interact. **7**(9), 85 (2023)

23. Rakhymbayeva, N., Amirova, A., Sandygulova, A.: A long-term engagement with a social robot for autism therapy. Front. Robot. AI **8**, 669972 (2021)
24. Ryokai, K., Lee, M.J., Breitbart, J.M.: Children's storytelling and programming with robotic characters. In: Proceedings of the Seventh ACM Conference on Creativity and Cognition, pp. 19–28 (2009)
25. Šabanović, S., Bennett, C.C., Chang, W.L., Huber, L.: Paro robot affects diverse interaction modalities in group sensory therapy for older adults with dementia. In: 2013 IEEE 13th International Conference on Rehabilitation Robotics (ICORR), pp. 1–6. IEEE (2013)
26. Sanoubari, E., Muñoz Cardona, J.E., Mahdi, H., Young, J.E., Houston, A., Dautenhahn, K.: Robots, bullies and stories: a remote co-design study with children. In: Proceedings of the 20th Annual ACM Interaction Design and Children Conference, pp. 171–182 (2021)
27. Song, Y., Luximon, A., Luximon, Y.: The effect of facial features on facial anthropomorphic trustworthiness in social robots. Appl. Ergon. **94**, 103420 (2021)
28. Stower, R., Calvo-Barajas, N., Castellano, G., Kappas, A.: A meta-analysis on children's trust in social robots. Int. J. Soc. Robot. **13**, 1979–2001 (2021)
29. Tamura, Y., Shiomi, M., Kimoto, M., Iio, T., Shimohara, K., Hagita, N.: Robots as an interactive-social medium in storytelling to multiple children. Interact. Stud. **22**(1), 110–140 (2021)
30. Venture, G., Indurkhya, B., Izui, T.: Dance with me! child-robot interaction in the wild. In: Social Robotics: 9th International Conference, ICSR 2017, Tsukuba, Japan, 22–24 November 2017, Proceedings 9, pp. 375–382. Springer (2017)
31. Wróbel, A., Źróbek, K., Indurkhya, B., Schaper, M.M., Gunia, A., Zguda, P.M.: Are robots vegan? Unexpected behaviours in child-robot interactions and their design implications. In: Extended Abstracts of the 2023 CHI Conference on Human Factors in Computing Systems, pp. 1–7 (2023)
32. Yin, R.K.: Qualitative Research from Start to Finish. Guilford Publications (2015)
33. Zguda, P., Kołota, A., Venture, G., Sniezynski, B., Indurkhya, B.: Exploring the role of trust and expectations in cri using in-the-wild studies. Electronics **10**(3), 347 (2021)
34. Złotowski, J., Sumioka, H., Nishio, S., Glas, D.F., Bartneck, C., Ishiguro, H.: Appearance of a robot affects the impact of its behaviour on perceived trustworthiness and empathy. Paladyn J. Behav. Robot. **7**, 1 (2016)

Human Factors

Pipeline Leak Detection: Leveraging Acoustic Emission Signal Processing and Machine Learning

Muhammad Farooq Siddique[1], Wasim Zaman[1], Niamat Ullah[1], Saif Ullah[1], and Jong-Myon Kim[1,2(✉)]

[1] Department of Electrical, Electronics, and Computer Engineering, University of Ulsan, Ulsan 44610, South Korea
jmkim07@ulsan.ac.kr
[2] PD Technology Co. Ltd., Ulsan 44610, South Korea

1 Introduction

Pipelines are the most cost-effective way to transport substances like gas, oil, and water. However, prolonged exposure to harsh conditions can cause damage such as corrosion, leaks, and ruptures, leading to significant financial and environmental impacts [1]. To address these issues, advanced leak detection techniques are becoming a priority. Increased corrosion, aging infrastructure, and external pressures are making pipeline leaks more common, posing serious risks to public safety and the environment [2]. Therefore, continuous leak monitoring is crucial, especially for gathering pipelines that connect production wells to processing facilities. These pipelines face more challenging environments and cover larger areas than transmission and distribution pipelines, making timely and accurate leak detection both urgent and complex [3].

In recent years, various leak detection methods have been developed for pipelines carrying water, oil, and gas. These methods include transient-based models, negative pressure wave techniques, hardware-focused strategies, and data-driven approaches [3]. Transient techniques, which analyze the hydraulic behavior of pipelines, are known for their clarity and accuracy, but are typically tested in controlled environments like simulations or labs, limiting their real-world applicability. Negative pressure wave methods detect leaks through pressure drops but may miss certain leaks, especially in pipelines with controlled end pressures, such as natural gas pipelines. Hardware-based strategies employ specialized sensors, like fiber Bragg grating pressure sensors [4] nd hydrophones, and can include soil monitoring systems. While these methods are effective, they often require additional equipment, making them costly and pipeline-specific.

Data-driven methods for pipeline leak detection typically analyze operational data, such as pressure readings, flow and pressure combinations [4], acoustic and concentration signals. With advancements in sensor technology and the adoption of SCADA systems, there has been a shift towards using multi-sensor approaches that gather a broader range of pipeline operational data [5]. Despite this, research on handling multivariate time series data in leak detection is limited. Existing data-driven models often employ intelligent algorithms like support vector machines (SVMs) [6], least square

D. Singh et al. (Eds.): IHCI 2024, LNCS 15557, pp. 173–184, 2025.
https://doi.org/10.1007/978-3-031-88705-5_14

SVMs [7], and ANNs [8]. These models rely on advanced feature extraction techniques, including time-domain and frequency-domain methods, statistical and wavelet-based approaches [9], and statistical and Markov-based methods.

Deep learning techniques, in particular, allow for automated extraction of leak-related information and patterns, by processing time–frequency data representations [10]. However, the challenge remains in accurately identifying sudden changes indicative of unexpected leaks. While DL algorithms are adept at feature extraction, their accuracy is not infallible. CNNs have shown effectiveness in leak and fault detection, identifying relevant information from acoustic images and aiding in pipeline condition classification [11]. Deep coupled dense convolutional networks [12] and networks utilizing residual joint adaptation through adversarial learning have been developed for intelligent fault diagnosis. Convolutional Autoencoders (CAEs) also play a role, compressing images and extracting meaningful features from the input data [8]. AE sensors detect energy spikes, or AE hits, which can be masked by background noise. The CWT is used to analyze significant AE detections, converting time-domain signals into time-frequency representations. The resulting scalogram visually represents AE signal impacts on a three-dimensional surface across various time-frequency scales, with the third dimension indicating magnitude.

Pipeline leaks can severely compromise structural stability, leading to stress cracks, fatigue ruptures, and corrosion [13]. These damages affect fluid or gas flow in pipelines. AE events, caused by structural changes like leaks, are detected as 'hits' by AE sensors. However, distinguishing these signals from background noise is challenging. Expertise is needed to set proper thresholds for filtering AE signals. Traditional methods, which rely on AE scalogram images from continuous wavelet transforms and deep learning, struggle due to noise masking signal variations, reducing diagnostic reliability. This study introduces a novel method that addresses these challenges by creating new 2D images from AE signals. These filtered scalograms are processed using CNN for feature extraction and classified using ANN and t-SNE. The approach shows superior performance on real-time datasets. The key contributions are:

- Introduction of Leak-Enhanced Scalograms (LES), which process CWT images using Laplacian filtering, NLM for noise reduction, and AHE for contrast enhancement.
- CNN is used for feature extraction from LES.
- ANN and t-SNE effectively classify leak and normal conditions.
- Validation is performed on data from a laboratory-scale testbed simulating industrial pipelines.

The paper is organized as follows: Sect. 2 describes the experimental testbed, Sect. 3 outlines the proposed framework, Sect. 4 presents results and discussion, and Sect. 5 concludes and suggests future work.

2 Experimental Design

The experimental setup, as shown in Fig. 2, used a stainless-steel water pipeline with a diameter of 114 mm and a thickness of 6 mm. AE R15I-AST sensors (Mistras Group, Inc., USA) were employed to capture AE signals, recorded at 1 MHz using the National

Instruments NI-9223 system. To simulate leaks, an electrical drill was used to create holes of varying sizes, and water flow through these holes was regulated by a fluid control valve. Water was chosen for its safety and environmental suitability. The pipeline was kept at a controlled pressure of 13 bars, even when the valve was fully closed. Data was collected both before and after leaks of 1 mm, 0.5 mm, and 0.7 mm were introduced, under different pressure levels. For each leak size, 240 samples were gathered, evenly split between normal and leak conditions. AE signals captured under both conditions are shown in Figs. 3 and 4. To ensure safety, the collected fluid was safely diverted using a hose (Fig. 1).

Table 1. Pipeline dataset.

Experiment ID	Pressure (bars)-Substance	Leak Size (mm)	Time	Samples (Leak - Normal)
1	13-Water	1.00	6 min	240–120
2	13-Gas	0.50	6 min	240–120
3	18-Water	0.70	6 min	240–120
4	18-Gas	0.50	6 min	240–120

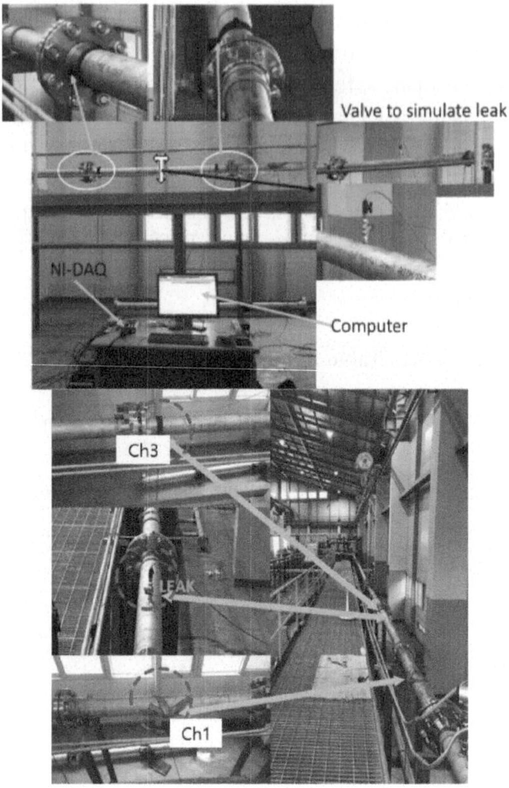

Fig. 1. Testing Environment for Data Acquisition Experiments

3 Proposed Method

The proposed approach is illustrated in Fig. 2. The method consist of the following steps:

Step 1. AE signals are collection is carried out from the pipeline setup.

Step 2. The AE signals are filtered using a Gaussian filter and converted into scalograms via CWT, showing energy variations with color intensities.

Step 3. NLM filtering reduces noise in the CWT scalograms, and AHE is applied for contrast enhancement.

Step 4. To evaluate the pipeline's condition, fault information is extracted from the scalograms and classified into leak and non-leak categories using a CNN. The table outlines the CNN's layer-by-layer architecture, detailing the input dimensions, filter sizes, kernel sizes, output dimensions, activation functions, and parameter counts for each layer. This progression illustrates how raw data is processed into classification outputs (Table 2).

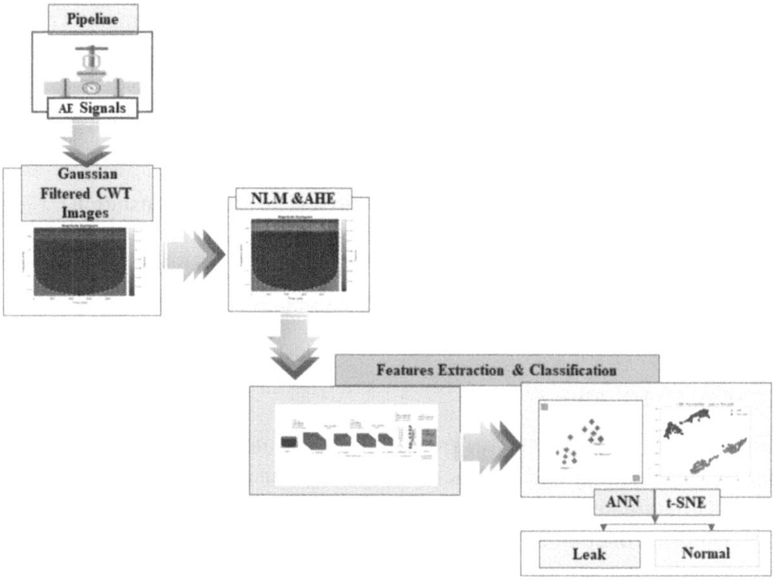

Fig. 2. Visual Representation of the Proposed Methodology Flow

Table 2. CNN Architecture.

Layer name	Input	Filters	Size of Kernel	Output	Activation	Parameters
Convolution 1	(3,224,224)	(16)	(3 × 3)	(16,224,224)	ReLU/-	(448)
Max Pooling	(16,224,224)	X	(2 × 2)	(16,112,112)	X	0
Convolution 2	(16,112,112)	(32)	(3 × 3)	(32,112,112)	ReLU/-	(4640)
Max Pooling	(32,112,112)	X	(2 × 2)	(32,56,56)	X	0
Convolution 3	(32, 56, 56)	(64)	(3 × 3)	(64,56,56)	ReLU/-	(18,496)
Max Pooling	(64 56,56)	X	(2 × 2)	(64,28,28)	X	0
Reshape	(64 28,28)	X	X	(50176)	X	0
FC1	(50176)	X	X	(128)	Sigmoid	(6,407,040)
FC2	(128)	X	X	(10)	X	(1290)

4 Technical Background

4.1 Continuous Wavelet Transform

CWT is used to transform the time-domain Acoustic Emission (AE) signal into three-dimensional scalograms. These graphical displays use varying shades of colors to show changes in energy levels at different time-frequency scales. In this study, the scalograms are particularly effective in highlighting energy shifts through the use of a wide range of colors to denote different energy intensities [14]. The CWT is formulated by aggregating wavelet functions that are scaled and shifted over the entire time domain. This method transforms the time-domain AE signal into 2D scalograms, adhering to the definition outlined by Hübner et al. [15], as shown in the subsequent Eq. (1).

$$CWT_X(\tau, S) = \frac{1}{\sqrt{|S|}} \int_{-\infty}^{+\infty} X(t).\psi * \left(\frac{t-\tau}{s}\right).dt \tag{1}$$

CWT produces coefficients determined by the scale ('s') and translation ('τ') parameters. Here, 'τ' represents the temporal location of the wavelet, and 's' controls its frequency and the size of the observation window. Higher scales provide a broader overview at lower frequencies, whereas lower scales reveal finer details of the signal at higher frequencies.

4.2 NLM and AHE Filters

The proposed method improves the quality of Gaussian filtered CWT scalograms by reducing unwanted noise and enhancing brightness, using the NLM technique [16]. This approach facilitates easier leak detection, segmentation, and classification due to its effective noise reduction and clarity enhancement. NLM leverages the inherent self-similarity found in mammographic images in the spatial domain to efficiently eliminate noise. It does this by comparing the gray levels at a specific point with the geometrical patterns of neighboring pixels, then calculating a weighted average for each pixel. This approach optimizes the quality of discrete noisy mammogram images, thus improving their characteristics for more effective subsequent analysis.

$$NL[m](a) = \sum_{b \in I} w(a, b)m(b) \tag{2}$$

Equation (2) determines the pixel weight w (a, b) by assessing the similarity between pixels 'a' and 'b' in the image. This calculation adheres to the standard constraints that $0 \leq$ w (a, b) ≤ 1, and it guarantees that the total sum of weights for a specific pixel 'a' ($\sum b$ w (a, b)) is equal to 1. These weights, used to define each pixel in the medical image, are based on the observed level of resemblance between neighboring pixels 'a' and 'b'. Additionally,

$$w(x) = \frac{1}{R_{sim}} \sum_{a=q}^{R_{sim}} \frac{1}{(2a+1)^2} \tag{3}$$

Equation (3) calculates the weights for filter neighbors based on their proximity to the filter's center. The weight of each pixel is determined by its similarity, represented as "sim," and the influence of the control parameter "h," which governs the distribution of weights. The NL-means algorithm operates under the assumption that image noise is similar to white noise, which aids in enhancing images that have high redundancy. In this process, NLM smoothens the scalograms, followed by contrast enhancement using limited AHE. This selective histogram equalization technique focuses on specific areas of the image, improving contrast. Bilinear interpolation further refines these results, ultimately improving the visibility and processability of mammographic images. Figure 3 illustrates the CWT images both with and without the application of NLM and AHE functions.

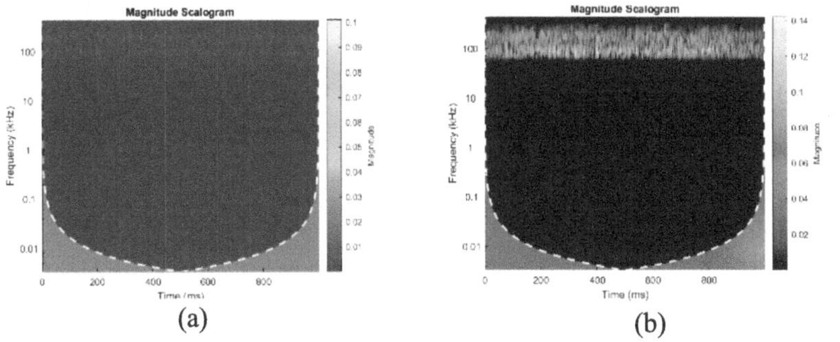

Fig. 3. CWT scalograms (a) Without filters (b) with NLM and AHE

4.3 CNN

CNNs are specialized deep learning models designed for processing visual data, particularly images. They use local receptive fields and shared weight parameters to efficiently extract hierarchical features. With multiple convolutional, pooling, and fully connected layers, CNNs autonomously learn intricate patterns and structures. By applying convolutional filters to specific input regions, CNNs excel in discerning spatial relationships and constructing representations from basic low-level features to complex high-level features. The following figure illustrates the whole working of CNN in this study.

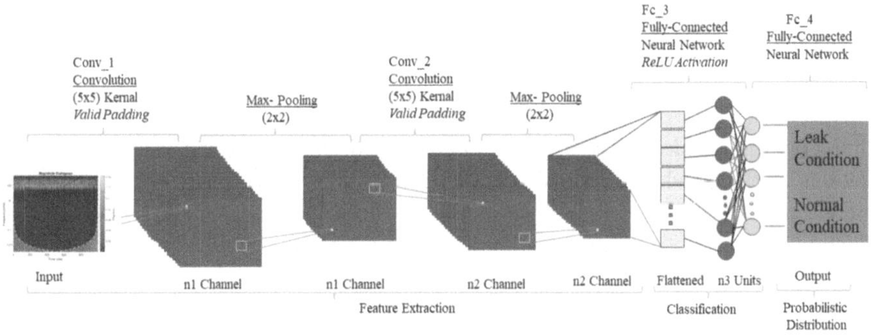

Fig. 4. Architetecture of CNN

4.4 t-SNE

The t-SNE method is an effective tool for creating 2D or 3D scatter plots, visually representing complex data sets while preserving their local structure [17]. Using the obtained feature set, t-SNE significantly enhances the efficiency and effectiveness of subsequent classification tasks. This is especially relevant for detecting anomalies in pipelines, where t-SNE excels at distinguishing subtle feature variations. Visual representations clearly depict this distinction, with leak-related features marked in red and non-leak features in blue, as shown in Fig. 2.

5 Compartive Analysis

To assess the effectiveness of the proposed methodology, a k-fold cross-validation with k = 3 was employed. To enhance result reliability, the experiments were repeated 15 times. In each iteration, 80% of the samples were randomly selected for model training, while the remaining samples were used for testing. A detailed description of the dataset is provided in Table 1.

The study includes a comparison between the proposed method and a reference technique introduced by Prosvirin et al. [18]. For evaluation, metrics such as average accuracy (AA), precision (Pm), recall (Rm), and F-1 score are calculated from the confusion matrices shown in Fig. 3. These metrics are computed using Eqs. 4–7.

$$Avg.Accuracy(AA) = \frac{\sum_{k}^{K} TP_k}{N} \times 100, \tag{4}$$

$$Precision(Pm) = \frac{1}{K}\left(\sum\nolimits_{k=1}^{K} \frac{TP_k}{TP_k + FP_k}\right) \times 100, \tag{5}$$

$$Recall\ R_m = \frac{1}{K}\left(\sum_{k=1}^{K} \frac{TP_k}{TP_k + FN_k}\right) \times 100, \tag{6}$$

$$F - 1\,score = 2 \times (Pr\,Pr\,e\,c_m \times c_m)/(Pr\,Pr\,e\,c_m + c_m), \tag{7}$$

TP and TN represent true positives and true negatives, respectively. False negatives of the classifier are denoted by FN, and N indicates the total count of samples within each class.

The performance of the proposed model is evaluated through a comparative analysis with recent models designed for similar tasks, highlighting its superiority. Unlike conventional methods, which often rely on 1D signal analysis, this approach transforms AE signals into 2D images, representing a significant shift in methodology. By using CNN for efficient feature extraction and combining it with ANN and t-SNE for classification, the model achieves more accurate feature representation and better classification outcomes. A critical innovation is the introduction of Leak-Enhanced Scalograms (LES), which improve image contrast and reduce noise, significantly aiding in the precise detection of subtle leak patterns that other methods might miss.

A comparative analysis using the same dataset was conducted against the approach by Prosvirin et al. [18], which employs kurtograms processed from vibration signals and a CNN to detect mechanical flaws in centrifugal pumps. While their method achieved an accuracy of 96.88%, precision of 96.82%, recall of 95.86%, and an F1 score of 95.86%, it is limited by the nature of the signals it processes, focusing primarily on mechanical flaws and vibration data. In contrast, the proposed model excels with a higher accuracy of 98.44%, precision of 100%, recall of 97.00%, and an F1 score of 98.00%. This improvement is due to the LES's ability to enhance critical leak-related features in the transformed 2D images and the combination of CNN with t-SNE for more refined classification. These enhancements enable the model to detect even minute leak patterns with greater precision, making it more suitable for pipeline leak detection in complex, real-world conditions. The superior performance reflects the model's ability to handle noise, improve image clarity, and deliver highly accurate classification results, as detailed in Table 3.

Table 3. Proposed method comparison with Prosvirin et al. [18] in performance metrics.

Approaches	Performance Metrics			Accuracy (Average)%
	Precision	Recall	F-1 Score	
Proposed	**100**	**98.25**	**98.50**	**98.40%**
Prosvirin et al. [18]	95.88	96.82	95.86	96.88%

The demonstrated leak detection efficacy of the proposed approach in pipeline diagnostics indicates that the framework is highly effective for identifying leaks. Its key strength lies in the advanced preprocessing of AE signals, which enhances the accuracy of leak detection. A comprehensive comparative analysis is presented in Figs. 5, 6, and 7, where both the proposed method and the contemporary model by Prosvirin et al. [18] are evaluated using the same dataset under identical conditions. Figure 5 compares the confusion matrices, highlighting the superior classification performance of the proposed model.

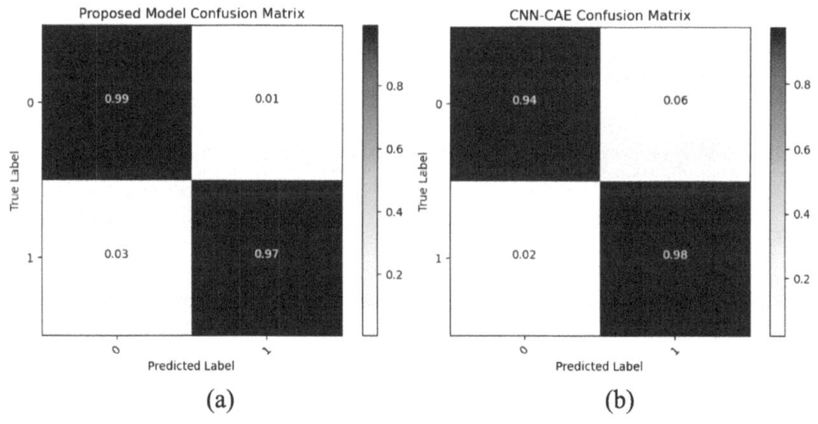

Fig. 5. Comparison of confusion matrices of (a) Proposed model and (b) Prosvirin et al. [18]

In similar manner, Fig. 6, the ROC curves show a higher true positive rate for the proposed approach, indicating better overall detection capability. Finally, Fig. 7 presents t-SNE plots, where the clearer separation of leak and non-leak conditions in the proposed model illustrates its effectiveness in feature space. These results underscore the robustness and precision of the proposed framework, establishing it as a significant improvement over existing methods for pipeline leak detection.

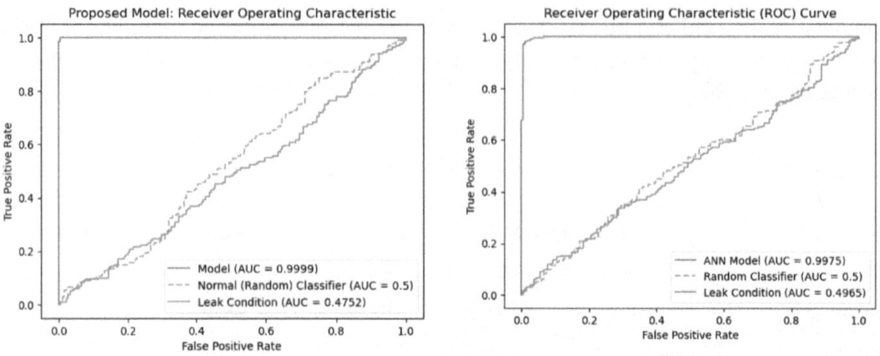

(a) (b)

Fig. 6. Comparison of ROC curves of (a) Proposed model and (b) Prosvirin et al. [18]

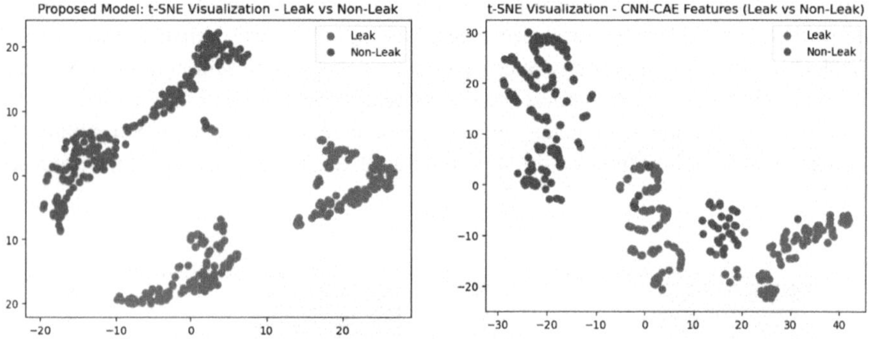

(a) (b)

Fig7. t-SNE plots comparison of (a) Proposed model and (b) Prosvirin et al. [18].

6 Conclusions

In conclusion, this research significantly improves pipeline leak detection by addressing noise issues in AE analysis. The enhancement of CWT scalograms using Laplacian filtering, NLM, and AHE improves image clarity and noise reduction, leading to more accurate feature extraction.

By employing a CNN for feature extraction and ANN with t-SNE for classification, our method demonstrated superior performance on real-world datasets and in an industrial-scale pipeline testbed. It outperformed existing techniques in classifying centrifugal pump vibration signals. The integration of advanced image processing with cutting-edge machine learning techniques not only enhances pipeline safety but also sets a new standard in the field, paving the way for future innovations in pipeline monitoring and maintenance.

Acknowledgements. This work was supported by the Korea Institute of Energy Technology Evaluation and Planning (KETEP) grant funded by the Korea government (MOTIE) ('RS-2023-00232515', 'Development of life prediction safety technology and hydrogen embrittlement estimation of LNG pipe mixed hydrogen'). This work was also supported by the Korea Institute of Energy Technology Evaluation and Planning (KETEP) grant funded by the Korea government (MOTIE) ('RS-2024-00449107', 'Development of Flexible Pipe and Connector for Hydrogen gas').

References

1. Siddique, M.F., Ahmad, Z., Ullah, N., Ullah, S., Kim, J.-M.: Pipeline leak detection: a comprehensive deep learning model using CWT image analysis and an optimized DBN-GA-LSSVM framework. Sensors **24**(12), 4009 (2024). https://doi.org/10.3390/s24124009
2. Rahman, I.U., et al.: Application of membrane technology in the treatment of waste liquid containing radioactive materials. J. Radioanal. Nucl. Chem. (2023). https://doi.org/10.1007/s10967-023-09169-9
3. Latif, J., Shakir, M.Z., Edwards, N., Jaszczykowski, M., Ramzan, N., Edwards, V.: Review on condition monitoring techniques for water pipelines. Measurement **193**, 110895 (2022). https://doi.org/10.1016/j.measurement.2022.110895
4. Zhou, P., Peng, Z., Chen, S., Yang, Y., Zhang, W.: Non-stationary signal analysis based on general parameterized time–frequency transform and its application in the feature extraction of a rotary machine. Front. Mech. Eng. **13**, 292–300 (2018)
5. Bakhtawar, B., Zayed, T.: Review of water leak detection and localization methods through hydrophone technology. J. Pipeline Syst. Eng. Pract. **12**(4) (2021). https://doi.org/10.1061/(ASCE)PS.1949-1204.0000574
6. Piltan, F., Prosvirin, A.E., Sohaib, M., Saldivar, B., Kim, J.-M.: An SVM-based neural adaptive variable structure observer for fault diagnosis and fault-tolerant control of a robot manipulator. Appl. Sci. **10**(4), 1344 (2020)
7. Sun, J., Xiao, Q., Wen, J., Wang, F.: Natural gas pipeline small leakage feature extraction and recognition based on LMD envelope spectrum entropy and SVM. Measurement **55**, 434–443 (2014). https://doi.org/10.1016/j.measurement.2014.05.012
8. Siddique, M.F., Ahmad, Z., Ullah, N., Kim, J.: A hybrid deep learning approach: integrating short-time fourier transform and continuous wavelet transform for improved pipeline leak detection. Sensors **23**(19), 8079 (2023). https://doi.org/10.3390/s23198079
9. Lilly, J.M.: Element analysis: a wavelet-based method for analysing time-localized events in noisy time series. Proc. Roy. Soc. A Math. Phys. Eng. Sci. **473**(2200), 20160776 (2017). https://doi.org/10.1098/rspa.2016.0776
10. Zaman, W., Ahmad, Z., Siddique, M.F., Ullah, N., Kim, J.-M.: Centrifugal pump fault diagnosis based on a novel SobelEdge scalogram and CNN. Sensors **23**(11) (2023). https://doi.org/10.3390/s23115255
11. Goodfellow, I., Bengio, Y., Courville, A.: Deep Learning. MIT Press (2016)
12. Wang, F., Lin, W., Liu, Z., Wu, S., Qiu, X.: Pipeline leak detection by using time-domain statistical features. IEEE Sens. J. **17**(19), 6431–6442 (2017). https://doi.org/10.1109/JSEN.2017.2740220
13. Kim, H., Lee, J., Kim, T., Jin Park, S., Kim, H., Doo Jung, I.: Advanced thermal fluid leakage detection system with machine learning algorithm for pipe-in-pipe structure. https://ssrn.com/abstract=4147041

14. Chen, J., et al.: Wavelet transform based on inner product in fault diagnosis of rotating machinery: a review. Mech. Syst. Signal Process. **70–71**, 1–35 (2016). https://doi.org/10.1016/j.ymssp.2015.08.023
15. Hübner, H.B., Duarte, M.A.V., da Silva, R.B.: Automatic grinding burn recognition based on time-frequency analysis and convolutional neural networks. Int. J. Adv. Manuf. Technol. **110**(7–8), 1833–1849 (2020). https://doi.org/10.1007/s00170-020-05902-w
16. Siddique, M.F., Ullah, N., Kim, J.-M.: Comprehensive pipeline leak detection using induced-leak enhanced scalogram analysis and deep learning. In: 2023 IEEE International Conference on High Performance Computing & Communications, Data Science & Systems, Smart City & Dependability in Sensor, Cloud & Big Data Systems & Application (HPCC/DSS/SmartCity/DependSys), pp. 877–883. IEEE (2023). https://doi.org/10.1109/HPCC-DSS-SmartCity-DependSys60770.2023.00127
17. Van der Maaten, L., Hinton, G.: Visualizing data using t-SNE. J. Mach. Learn. Res. **9**(11) (2008)
18. Prosvirin, A.E., Ahmad, Z., Kim, J.M.: Global and local feature extraction using a convolutional autoencoder and neural networks for diagnosing centrifugal pump mechanical faults. IEEE Access **9**, 65838–65854 (2021). https://doi.org/10.1109/ACCESS.2021.3076571

Application of Machine Learning Techniques for Leak Detection in a Horizontal Pipeline Transporting a Water-Glycerol Mixture

Adalberto Gámez-De-León[1], Javier Jiménez-Cabas[1], Luis Díaz-Charris[1](✉), and Jorge Herrera Cuartas[2]

[1] Universidad de la Costa, Barranquilla, Atlántico, Colombia
ldiaz28@cuc.edu.co

[2] Facultad de Ciencias Naturales e Ingeniería, Universidad de Bogotá Jorge Tadeo Lozano, Bogotá, Colombia

1 Introduction

Timely detection of leaks is crucial in preventing water loss and avoiding significant economic and environmental disasters. Pipeline leaks can occur due to a variety of factors, including poor pipe materials, weak joints, ground movement, internal corrosion, corrosive soils, excavation activities, temperature fluctuations, heavy traffic, tidal forces, water hammer, and air entrapment, among others. When leaks occur, liquid is lost from the pipeline, leading to inefficiency and waste. In the context of drinking water distribution, leaks are particularly costly on a global scale, as they not only waste a critical natural resource but also lead to significant economic losses due to the energy-intensive process of water purification [1].

Timely detection of leaks in pipe networks is essential for several reasons. Firstly, it prevents water loss and mitigates negative economic and environmental impacts by preserving valuable resources and reducing the strain on water treatment systems. Secondly, early leak detection significantly reduces the financial consequences associated with material losses and the costs of repairing extensive damage, making it a critical factor in maintaining the efficiency and sustainability of water distribution systems [2].

In this work, two machine learning techniques are applied to systematically detect leaks by analyzing process data from the pipelines under study. By leveraging data such as flow rates and pressure measurements, the implemented algorithms can effectively identify the presence of leaks in real-time, offering a more efficient and accurate approach to leak detection.

In Colombia, at present, the companies that operate the drinking water treatment and distribution service do not have a tool for real-time detection of leaks in pipes with high water flow and high pressure [3]. This presents a significant operational and economic challenge, as part of the water treated is wasted, leading to both resource and financial losses. Additionally, the water purification process is highly resource-intensive, requiring substantial input and energy, further exacerbating the costs associated with leaks. However, it is evident that the operating companies choose to have good leakage control policies through the reporting of damage by customers and have detection

D. Singh et al. (Eds.): IHCI 2024, LNCS 15557, pp. 185–192, 2025.
https://doi.org/10.1007/978-3-031-88705-5_15

systems in the Control Centers and brigades of operators who locate leaks directly in the field [4]. However, these strategies do not allow for real-time leak detection.

Timely detection of leaks in pipe networks is important because, on the one hand, it prevents water loss and helps to avoid negative economic and environmental impacts. On the other hand, timely leak detection will substantially reduce the economic impacts caused by material losses due to leakage [2].

It should also be noted that any practical leak detection solution should detect multiple leaks in pipeline networks, as well as identify in which pipe the leaks have occurred. In addition, the solution should be easily scalable, and its implementation and maintenance costs should be reasonable. This is because, during the last decades, the problem of finding leaks in water distribution networks (WDN) has been a challenge due to the complexity and extent of the leaks [2]. Currently, leak detection in fluid transport pipes is performed using specialized hardware such as Thermal Imaging Cameras, Moisture detectors, and Geophone. However, due to the extent of WDNs, real-time detection with these instruments is complex, as they are required to be installed in the field.

2 Methodology

The methodology described outlines the process of developing and implementing machine learning algorithms to detect leaks in a horizontal pipe carrying a mixture of water and glycerol. The primary goal is to predict whether a leak is occurring based on flow rate and pressure measurements at various points along the pipe (P1–P5). Here is a detailed explanation of the methodology and the architecture that supports it.

1. ***Problem definition:definition:*** the task is a classification problem, where the objective is to classify instances as "Leak" or "No Leak" based on the given features: flow rate (Q) and pressure readings (P1–P5).
2. ***Data preparation:*** the dataset contains measurements for each instance, including flow rate (Q) and pressure readings (P1, P2, P3, P4, P5). The dataset is divided into two parts: 70% for training the model (training set), and 30% for testing the model (test set).
3. ***Scatter lot visualization:*** scatter diagrams are generated to visualize the relationship between the features (e.g., Q and P4) and the target labels (leak or no leak). This helps to understand how the features are distributed and whether they show any clear separation between leak and no-leak instances.
4. ***Application of machine learning techniques:***
   a. ***Decision Tree Algorithm.*** This algorithm is implemented to classify the instances as leak or no leak. The decision tree works by choosing the most informative features and splitting the data based on specific thresholds.
   b. ***Support Vector Machine (SVM) Algorithm.*** This algorithm is also implemented to solve the same classification problem. SVM works by mapping the data to a higher-dimensional space where it can draw a decision boundary that separates the leak from the no-leak data points.
5. ***Performance evaluation of techniques:*** several important metrics were used to measure how well each model classified the data as "Leak" or "No Leak." Here's what happened during the evaluation.

a. In the *Decision Tree evaluation*, the main metric used was accuracy. This measures the proportion of correct predictions made by the model out of all the predictions made. Additionally, the evaluation provided some derived metrics such as: Hit rate and Misclassification rate.
b. In the *SVM evaluation*, three metrics were used: the Confusion Matrix, which provided a detailed breakdown of correct and incorrect predictions; the f1_score, which focused on the balance between precision and recall; and the Jaccard Index, which evaluates the similarity between predicted and actual classes."

2.1 Experiment

The data set presented in this paper is part of the research paper "Water and glycerol flow data set in a leaky and non-leaky horizontal pipe" developed by researchers in a flow loop laboratory, which was designed to investigate high viscosity flows. The data set consists of 1200 s (equivalent to 12,000 samples) of pressure and mass flow measurements taken at five points along the pipe. The first 300 s were recorded when the flow in the loop was composed of only glycerol. The remaining data were acquired when the flow was composed of a mixture of water and glycerol.

The prototype, shown in Fig. 1, is equipped with: a 0.0762 m (3 in) diameter, 54 m long steel pipe, and a 40 HP progressing cavity pump (Seeped Mod. BN35-24), five intermediate pressure measurement points (P1 to P5), a mass flow sensor installed at the pipe inlet, and a 1.5 m3 separator tank.

A valve is used to emulate the appearance of leakage. Therefore, the data set can be used to validate similar leakage diagnostic approaches [5].

The Experimental Data. The data starts to be extracted when the first leak is induced, at time t = 567 s. When a valve is opened, the valve is closed at time t = 668 s. Then, at time t = 1045 s, a second leak is induced until time t = 1136 s when the valve is closed. It should be noted that data were recorded from the first opening of the valve until the end of the experiment [5].

Considering the previously collected data, the supervised learning algorithms for classification were selected: Decision Trees and Support Vector Machines. The data was taken from the article dataset; the data was cleaned. The physical setup of the experiment can be seen in Fig. 1.

3 Algorithm Implementations

The task at hand is to detect potential leaks in a horizontal pipe carrying a mixture of water and glycerol using machine learning algorithms. The problem is framed as a classification challenge where the algorithm must determine whether a leak is present based on specific features such as flow rate (Q) and pressure measurements taken at various points (P1–P5). To solve this, a machine learning model is trained on a dataset that contains both flow and pressure readings, along with the corresponding labels indicating whether a leak is occurring or not.

The dataset consists of two main components: the features (X), which include flow rate and pressure measurements, and the labels (y), which indicate whether a leak is

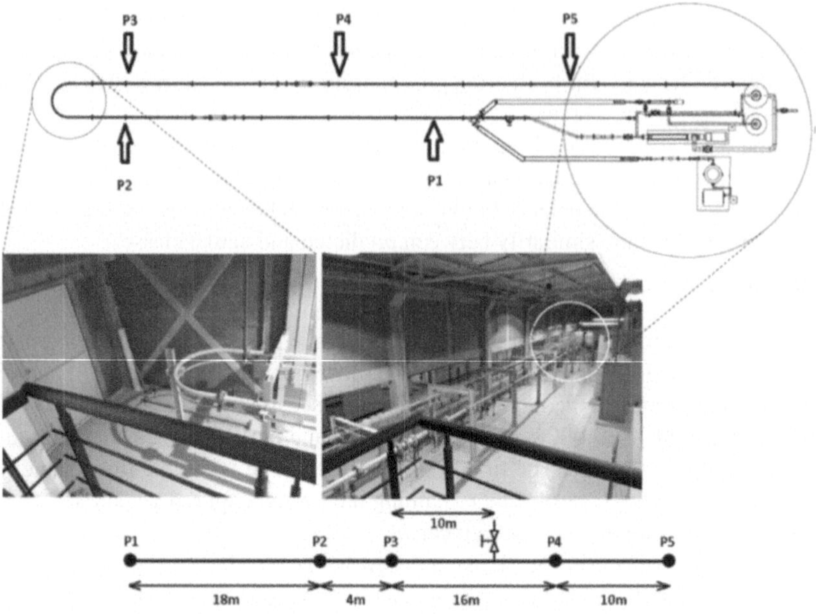

Fig. 1. The physical setup of the experiment [5].

present. The Dataset is divided in two: 70% training data (or training set) and 30% test data (or test set). The scatter diagram of the classes according to flow rate "Q" and pressure "P4" is generated (Figs. 2 and 3).

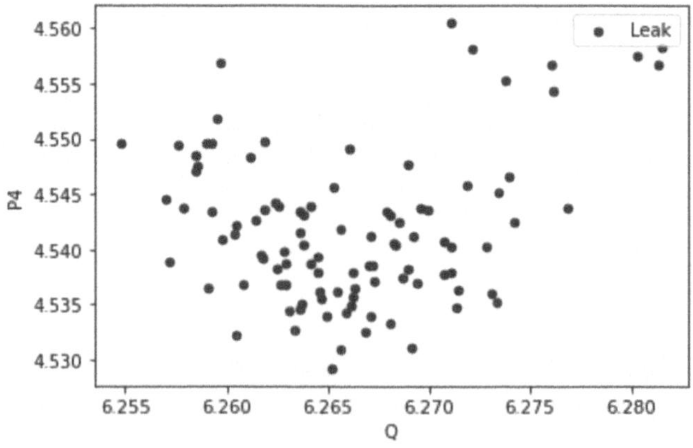

Fig. 2. Dispersion of data with leaky P4 pressures.

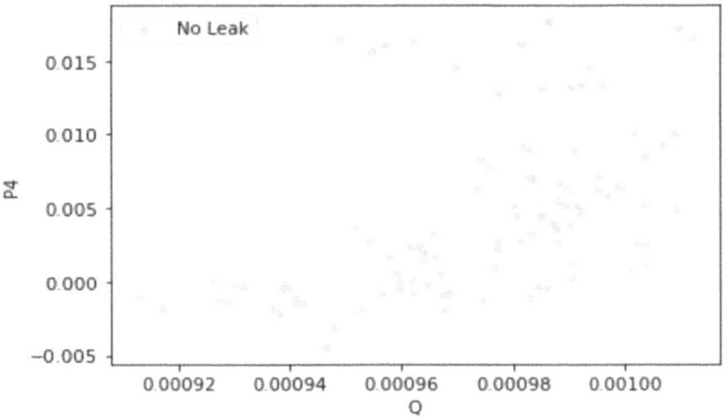

Fig. 3. Dispersion of data with no leakage of P4 pressures.

3.1 Decision Trees

The algorithm chooses the most predictive feature to split the data, and the decision tree is displayed in Fig. 4. If the pressure P3 is less than 10.159 psi, then look when the pressure P3 is less than or equal to 9.639 psi; if the condition is met, no leak; if, on the contrary, the pressure P3 is greater than 9.639 psi, look if the pressure P2 is less than or equal to 11.572 psi; if it is true, no leak; otherwise, it seems that there would be a leak; and finally, if the pressure P3 is greater than 10.159 psi, no leak with a high probability.

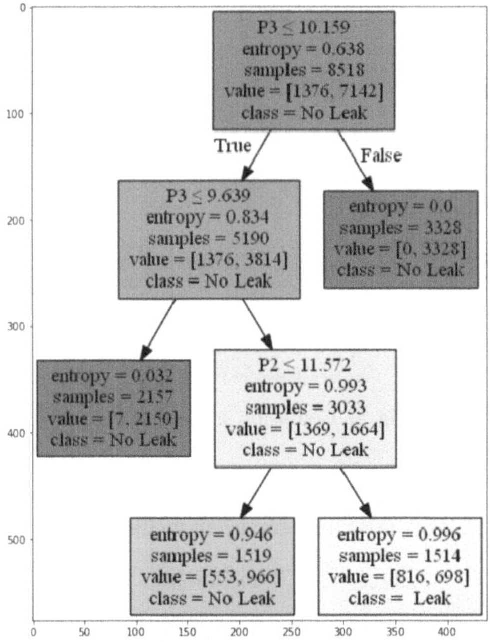

Fig. 4. Visualization of the Decision Tree.

Data Performance: Test examples: 657; examples correctly classified: 581; misclassified examples: 76; hit rate: 0.884; and failure rate: 0.116.

Among the metrics used to evaluate the performance of the algorithm, we have the accuracy score metric, which consists of evaluating the accuracy score, i.e. how many correct predictions the model made with respect to the test solutions. According to this metric, the model has 97% training effectiveness. The algorithm was trained with 1533 data points, of which it managed to correctly classify 1490 data points and misclassify 43 data points from the dataset, which is equivalent to a failure rate of 28%. As for the test data, it has an effectiveness of 88%. The algorithm was tested with 657 data points, of which it managed to correctly classify 581 data points and misclassify 76 data points, equivalent to a failure rate of 11.6%.

3.2 Support Vector Machines

SVM works by mapping data to a high-dimensional feature space so that data points can be categorized, even when the data is not linearly separable. The algorithm predicts solutions for the data set, and then evaluates its performance. The performance is evaluated using the confusion matrix. The confusion matrix is adjusted by defining the labels to be displayed, the accuracy and other parameters (Fig. 5).

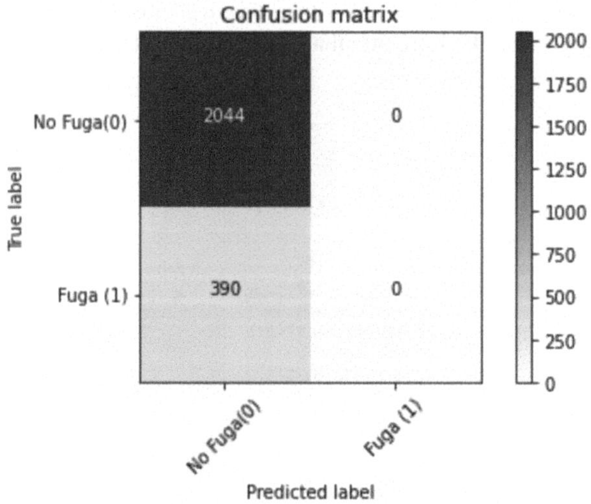

Fig. 5. Confusion or error matrix.

The evaluation metric f1_score is used to measure the accuracy of the model on the dataset. The Jaccard index is used for more accurate model performance evolution.

$$f1_{score} = 0.7666323040828108$$

$$I_J = 0.8397699260476582$$

3.3 Comparison

The Decision Tree performed well on both the training and test data. It had a relatively high training accuracy (97%) but showed signs of overfitting, as the test accuracy was lower (88%). The model was good at making simple, interpretable decisions based on feature thresholds but might have struggled with more complex patterns in the data.

The novelty of this approach lies in how the Decision Tree selects feature thresholds based on the pressure readings from multiple sensors (P1–P5). For example, the algorithm identifies that when P3 $<$ 10.159 psi, it should check additional conditions on P2 to further refine the prediction. This layered decision-making process ensures that the model provides interpretable results that align with physical conditions in the pipeline.

The SVM provided a more nuanced classification, which is reflected in the f1_score and Jaccard Index. The f1_score (0.7666) indicates that the model balanced precision and recall, but its performance was slightly lower in capturing all true positives compared to the Decision Tree. However, the Jaccard Index suggests that the overall prediction alignment with true outcomes was better than the Decision Tree.

Moreover, the SVM model demonstrates how mapping the input features (flow rate and pressures) into a higher-dimensional space enables it to better differentiate between leak and no-leak conditions, even when these conditions are not easily separable in the original feature space.

4 Conclusions

The combination of these two techniques offers a comprehensive solution to the leak detection problem. The Decision Tree provides an interpretable, rule-based approach that can be easily implemented in industrial settings, while the SVM introduces a more nuanced classification capability by mapping data into a higher-dimensional space.

This work utilizes dual-model approach that leverages both interpretable and high-performance machine learning techniques. Additionally, the use of advanced evaluation metrics such as f1_score and Jaccard Index for model evaluation represents a significant advancement over traditional accuracy-based assessments. This methodology provides a more holistic evaluation of model performance, ensuring robustness in real-world applications where the cost of misclassification is high.

By employing these machine learning techniques, this work improves the accuracy of leak detection in pipeline systems and offers new insights into how such techniques can be effectively evaluated and applied in industrial settings.

References

1. Adedeji, K.B., Hamam, Y., Abe, B.T., Abu-Mahfouz, A.M.: Towards achieving a reliable leakage detection and localization algorithm for application in water piping networks: an overview. IEEE Access **5**, 20272–20285 (2017). https://doi.org/10.1109/ACCESS.2017.2752802
2. Fereidooni, Z., Tahayori, H., Bahadori-Jahromi, A.: A hybrid model-based method for leak detection in large scale water distribution networks. J. Ambient Intell. Humaniz. Comput. 1–17 (2020). https://doi.org/10.1007/s12652-020-02233-2

3. Subdirección General de Agua Potable Drenaje y Saneamiento, Volumen 42 Mantenimiento y Reparación de Tuberías y Piezas Especiales (2015)
4. Cardelus, D., Lorenzo, L.: Localización de fugas en terreno con machine learning. XXXV Jornadas Técnicas de AEAS, pp. 925–926 (2019)
5. Noguera-Polania, J.F., et al.: Dataset on water–glycerol flow in a horizontal pipeline with and without leaks. Data Briefef **31**, 5 (2020). https://doi.org/10.1016/j.dib.2020.105950This

Connected Lives: Revolutionizing Health and Social Interaction with Wearable Technology in India

M. Anees(✉) and Madhavi Ravikumar

Department of Communication, University of Hyderabad, Gachibowli 500046, Hyderabad, India
{21snpc04,mrksns}@uohyd.ac.in

Abstract. This paper investigates the impact of wearable technology on everyday life in India, focusing on topographical wearable devices like smartwatches. It examines how these devices enable users to monitor and track their bodily activities, creating new dynamics in personal health management and social interactions. Through fieldwork and interviews, this study explores the relationship between wearable devices and users, analyzing how these technologies influence behavior, provide agency, and integrate into socio-technical systems.

Keywords: Wearable Technology · Health Management · Ubiquitous Interactions

1 Introduction

Wearable computing devices, particularly smartwatches and health trackers, have witnessed unprecedented growth in India, underscoring a transformative shift in how individuals engage with technology for personal health management and daily life enhancement. The wearable technology market in India expanded by 34% in 2023 alone, reaching a milestone of 134 million units sold, according to the International Data Corporation (IDC, 2023). Initialy conceptualized by Claude Shannon in 1961, wearable devices have evolved significantly to encompass a range of formats, including clothing, wristwatches, and implantable devices, demonstrating major technological advancements (Raad, 2020);(Pedersern & Iliadis, 2020);(Dierkes & Hoffmann, 1992). This study focuses on topographical wearable devices—those that maintain direct contact with the user's skin—facilitating real-time monitoring of health metrics. By offering continuous access to personal health data, these devices are redefining health management practices, enabling users to gain insights into heart rate, sleep patterns, physical activity, and other critical indicators. The adoption of wearable technology in India not only marks a shift towards self-monitoring and proactive health management but also indicates a broader societal transformation in everyday interactions with digital tools. Wearable devices increasingly influence social dynamics, subtly shaping how individuals interact in social contexts by signaling health consciousness, technological affinity, or lifestyle preferences. Additionally, the unique socio-technical systems that emerge around wearable

D. Singh et al. (Eds.): IHCI 2024, LNCS 15557, pp. 193–202, 2025.
https://doi.org/10.1007/978-3-031-88705-5_16

technology reflect a blend of individual, social, and technological interactions within India's distinctive socio-cultural landscape. These systems comprise the devices themselves, the networks they connect to, the users, and the societal norms surrounding technology use, creating new modes of engagement and community functionality.

The primary research objectives are designed to investigate these shifts comprehensively. First, the study aims to explore how wearable technology enhances personal health monitoring, enabling users to manage their health more actively. Second, it examines the influence of wearable devices on everyday social interactions and personal agency, considering how wearables impact social relationships, perceptions, and communication. The third objective is to analyze the socio-technical systems formed around wearable technology in India, focusing on the interplay between cultural, social, and technical factors unique to this context. Finally, the study seeks to identify the challenges and limitations faced by wearable technology users, particularly issues related to data privacy, usability, and economic accessibility that may affect user adoption and experience. To achieve these objectives, a mixed-methods approach will be employed, including qualitative interviews, quantitative surveys, and data analysis.

Despite the rapid growth and adoption of wearable technology in India, there is a substantial research gap in understanding its comprehensive impact on health management, social interactions, and the socio-technical systems unique to the Indian context. Existing research has often focused on wearable technology's general health benefits or use in Western contexts. In contrast, studies on the cultural, social, and economic landscape of wearable adoption in India remain sparse. Additionally, previous literature has not sufficiently addressed how wearable technology influences social dynamics, personal agency, or evolving societal norms within India. A notable gap also exists in examining the socio-technical systems surrounding wearables, particularly the complexities introduced by mobile networks, cloud platforms, and data-sharing applications that operate within India's regulatory and privacy frameworks. These frameworks present distinct challenges for data security and user trust, particularly relevant in a rapidly digitizing society with diverse digital literacy levels. Furthermore, while India's wearable technology market is expanding, there is limited research on the barriers users encounter when integrating these devices into daily life. Factors such as cost, accessibility, usability in resource-constrained environments, and potential disparities in device functionality across regions have not been adequately explored. This study addresses these gaps by focusing explicitly on India's wearable technology landscape, investigating not only the health benefits but also the socio-cultural implications of wearable use. By analyzing how wearables shape and are shaped by Indian social norms and behaviours while also identifying critical limitations and user challenges, this research contributes to a nuanced understanding of wearable technology's role in India.

The research questions guiding this study are structured to address these multifaceted interactions and gaps. The first question examines how wearable devices influence users' personal health management, exploring how these devices support or enhance health-related practices. The second question investigates how wearable technologies impact social interactions and relationships, aiming to uncover shifts in communication, behavior, and social dynamics resulting from wearable usage. The third question focuses on the socio-technical systems that emerge around wearable devices in India, exploring the mutual influence between technology and social factors unique to the Indian context. Lastly, the fourth question addresses the challenges and limitations users encounter, acknowledging potential barriers to the seamless integration of wearable technology into daily routines, such as privacy concerns, technological reliability, and accessibility. Overall, this study aims to provide a comprehensive understanding of wearable technology's impact on health management, social interactions, and socio-technical systems in India while identifying key barriers to its effective adoption. By addressing these research gaps, the study offers insights that contribute to the global discourse on wearable technology, expanding it beyond predominantly western perspectives and highlighting areas that require further academic and industry focus for the effective development and implementation of wearables tailored to India's needs.

2 Methodology

The methodology of this study is structured around a socio-technical systems framework to explore the complex interplay between wearable technology and its users, particularly within the Indian context. Grounded in cybernetics and theories of embodiment, this framework emphasizes the interconnectedness of information, materiality, and agency as facilitated by wearable devices. Drawing on the theoretical perspectives of Hayles (1999) and Bennett (2004), the study investigates how wearable technology generates affective forces and agential capacities, thereby shaping user behavior and influencing social dynamics through continuous data feedback and interactive engagement. A mixed-methods approach was employed, integrating both qualitative and quantitative data to achieve a comprehensive analysis. The qualitative component involved fieldwork and 15 in-depth, semi-structured interviews conducted with a diverse sample of wearable device users across various regions in India, comprising nine men and six women. These interviews were designed to capture users' experiences, usage patterns, and perceptions of how wearable technology affects their daily lives and decision-making processes. To complement the qualitative insights, quantitative analysis of market data and industry reports, including statistics from the International Data Corporation (IDC), was conducted to understand the growth trajectory, user demographics, and market diversification of wearable technology in India. While existing literature on wearable technology often prioritizes technical specifications and general market trends, there remains a significant gap regarding its socio-cultural impact, particularly in the Indian setting. This study directly addresses this gap by examining how wearable technology affects individual health management, social interactions, and the formation of socio-technical systems within India's unique socio-cultural landscape. Focusing on topographical wearable devices—those in direct contact with the skin, enabling continuous health monitoring—this research delves into how these technologies extend users' agency over their bodies. By facilitating real-time tracking of health parameters, these devices empower users with actionable health data, influencing behavioral decisions and promoting a more informed approach to personal health (Lupton, 2017). Findings from the interviews reveal that users actively utilize wearable devices to monitor various health metrics, such as heart rate, Electrocardiogram, and sleep patterns, enabling them to make informed health choices and exercise greater personal agency over their wellness routines. Through this multi-layered approach, the study contributes to a nuanced understanding of wearable technology's role in shaping personal agency, health management, and social dynamics within India's evolving technological landscape.

3 Findings and Analysis

Wearable technology has emerged as a pivotal tool in personal health management, bridging the gap between individual monitoring and broader socio-technical systems. In India, the adoption of smartwatches and health trackers has enabled real-time health monitoring, often linked to support networks that transcend geographical boundaries. This section examines the case of Raheem, an elderly man whose use of an Apple watch exemplifies the evolving interplay between wearable devices, personal agency, and remote health monitoring. By exploring this incident within the frameworks of socio-technical systems and affective interaction, the analysis highlights how wearable technology reshapes daily health practices, fosters connectivity, and subtly challenges traditional healthcare structures.

3.1 Wearable Technology, Health Monitoring, and the Socio-technical Linkage in Daily Life

During a group hiking expedition in Wayanad, a mountainous region requiring a steep and challenging ascent, an incident involving the eldest member of our group, Raheem, underscored the transformative role of wearable technology in health monitoring and socio-technical connectivity. When 20-member group was traversing the rocky trail Raheem experienced a sudden fall, making it impossible for him to continue the climb. A few individuals assisted him back to the base camp. En route, he received a call from his nephew, who was residing in Qatar, inquiring about the fall and subsequent details. This incident marked the starting point of this inquiry, sparking a keen interest in understanding the affective and functional relationship between wearable devices and human well-being. In 2020, upon the recommendation of his nephew, Raheem adopted the Apple watch to address health monitoring concerns due to his health conditions. Raheem, a former migrant worker, has family members, including siblings and nephews, who settled in Qatar. With limited immediate family in his home country, Raheem's wearable device served as a vital link, ensuring that his health metrics could be monitored remotely by family members. This wearable device is connected to his nephew's mobile phone, allowing him to receive timely updates regarding Raheem's health status. Such linkage enabled Raheem's nephew to be informed about the hiking incident immediately after it occurred. Raheem regularly uses his Apple watch to monitor key health parameters, such as his ECG, heart rate, and sleep patterns. When irregularities arise, he consults his family doctor, providing a new layer of agency in his health management. Notably, he also uses the device to check his wife's health metrics, such as ECG and heart rate, in instances of sudden fatigue, thereby extending the device's utility beyond individual monitoring to family health oversight.

Historically, accessing real-time data on metrics such as heart rate, ECG, or blood oxygen levels required a visit to a medical facility or practitioner. Although mobile devices have incorporated some monitoring functions, wearable technology has made such information more accessible, affordable, and integrated into daily routines. De Certeau (2011) theorizes technological shifts as "tactics," where individuals employ tools to manoeuvre within and partially outside established structures. In contrast, he defines the reliance on institutional systems for self-knowledge as "strategies," underscoring the power dynamics that structure societal norms. Cyborg theory, as explored by Haraway and Wolfe (2016), examines how these strategies are potentially circumvented by wearable technologies, where users can temporarily sidestep traditional "medical gaze" systems that exert authority over health-related data and intervention. De Certeau (2011) suggests that although wearables provide an alternative to institutionalized strategies, they cannot entirely replace traditional frameworks, given the embedded power structures in society.

Analyzing Raheem's case through a socio-technical systems framework, the Apple watch serves as an intermediary node connecting him to both personal and broader medical networks. The device facilitates real-time data transmission, allowing for swift, remote intervention, if necessary, thereby aligning with Bennett's (2004) concept of affective force and agential capacity, where the device empowers the user with crucial health information, thereby enhancing self-governance over personal health. Furthermore, the Apple watch's autonomous operation in health monitoring exemplifies Bennett's notion of "thing-power," illustrating how non-human entities like wearable devices actively influence socio-technical systems. Hayles (1999) argues that information, when generated within an embodied and material context, is inseparable from the user's physicality. In Raheem's case, the data collected by the Apple watch effectively extends his physical presence, enabling his relatives across geographical boundaries to remain connected to his well-being. This socio-technical linkage reflects how wearable technology can bridge distances and reinforce support networks, affirming its potential to reshape traditional health and social frameworks within a digitizing world.

3.2 Early Detection and Remote Intervention: Wearable Technology in Health Monitoring and Socio-technical Support Systems

Wearable technology has shown significant potential in detecting early warning signs of health conditions, especially for older adults who may not actively monitor their own health metrics (Mirjalali et al., 2022). By integrating wearable devices into everyday life, users gain a safety net through remote monitoring capabilities, often supported by caregivers or family members (Lindeman et al., 2020). This section presents a case study from Malappuram, where wearable technology facilitated early medical intervention by detecting a critical drop in blood oxygen levels (SpO2), leading to timely hospitalization and likely preventing severe complications. The case underscores how wearable devices, even when operated by non-users, can act as effective intermediaries within a socio-technical system involving the user, caregiver, and medical institutions.

In this case, a 78-year-old individual from the village of Malappuram experienced sudden fatigue and low energy, which can often be subtle indicators of underlying medical issues. Following a previous incident of low SpO2 levels, his daughter-in-law acquired a smartwatch capable of monitoring blood oxygen saturation. Connected to her mobile device, the smartwatch monitored his SpO2 in real time. When his oxygen levels dropped below a critical threshold, the device sent an alert to the daughter-in-law's mobile phone, prompting her to take immediate action by admitting him to a hospital. The early detection provided by the smartwatch played a crucial role in promptly addressing medical emergencies, demonstrating the life-saving potential of wearable devices in health management. Interestingly, the elderly man was largely unaware of the smartwatch's capabilities and did not actively engage with the device himself. This case illustrates how socio-technical systems can compensate for user limitations in monitoring personal health. Here, the smartwatch operates autonomously, serving as an external agent that continuously tracks health metrics and communicates them to a remote caregiver. This arrangement contrasts with other wearables where users actively monitor, as the technology here is primarily managed by a secondary user—the caregiver. This dynamic reflects Dourish and Bell's (2011) concept of ubiquitous computing, wherein technology is seamlessly integrated into daily routines to provide invisible yet essential support. Additionally, this case aligns with the principles of cybernetics, where wearable devices blur the lines between information and materiality (Wiener, 2019), transforming health data into a key aspect of the wearer's embodied experience (Stark, 2018). Although the elderly man does not directly interact with the device, it becomes a part of his extended health system, connecting him to his caregiver and healthcare resources. This socio-technical linkage, where the device autonomously aids in health monitoring, exemplifies how wearable technology extends agency to both wearers and their support networks, enhancing proactive healthcare management in daily life.

3.3 Wearable Technology, Agency, and Health Empowerment

Wearable devices play an increasingly prominent role within broader socio-technical systems, integrating users, technology, and social structures to influence everyday health practices and empower individuals. In India, however, the digital divide creates disparities in technology adoption, affecting how users use and benefit from these devices. This section examines how wearable technology fosters agency and autonomy, particularly in the context of pregnancy, through concepts of "thing power" and relational agency as outlined by theorists like Dourish and Bell (2011), Bennett (2004), and Lupton (2017). Drawing on real-life experiences, this case study explores how wearable devices, as components of socio-technical assemblages, can become influential tools in personal health management. Wearable devices are embedded in expansive socio-technical systems where technology intersects with users' daily environments, shaping patterns of interaction and dependency. Advertising campaigns frequently emphasize the empowering nature of wearable devices with slogans like "*The watch knows you best*" and "*Do more, live more*," suggesting that these devices offer unprecedented agency over personal health and wellness. This concept of agency is understood as a relational force within socio-technical systems, continuously shaped by dynamic interactions among users, devices, and data (Lupton, 2017). Bennett (2004) describes this influence as "Thing Power," a form of affective force exerted by non-human agents, like wearable devices, within human experience. One example illustrating the impact of wearable technology is the case of Rajeswari Natarajan, a 27-year-old woman from Tamil Nadu who relied on a smartwatch to manage her health during her first pregnancy. Living away from her family and juggling professional responsibilities, Rajeswari faced challenges in adhering to her doctor's advice to maintain physical activity and stay hydrated. Recognizing her need for assistance, her husband purchased a smartwatch designed to support daily activity monitoring. The smartwatch provided automated reminders to walk, drink water, and avoid prolonged inactivity, helping her adhere to her doctor's recommendations. These reminders were critical in supporting her health and well-being, ultimately contributing to a successful pregnancy outcome. Through this socio-technical lens, Rajeswari's smartwatch functioned as a vital agent within her health management system, illustrating Bennett's (2004) notion of "thing power." The data generated by the watch, such as reminders and activity metrics, became an active force in shaping her daily habits and ensuring compliance with health guidelines. In this way, the smartwatch's data-driven influence exemplified Hayles' (1999) concept of embodied information, where data is inseparably linked to the user's physical experience. By continuously monitoring her actions and delivering reminders, the wearable device facilitated Rajeswari's autonomy and personal agency within her pregnancy management. Lupton's (2017) view of agency as relational is also pertinent, as the smartwatch's role in supporting Rajeswari underscores how wearable technology can empower users by integrating seamlessly into their lives, providing critical assistance, and enhancing health outcomes. This case underscores wearable technology's potential to support complex health needs, demonstrating its capacity to become an integral part of personal health management systems, especially for users navigating physical or situational limitations. Through its embedded role in socio-technical assemblages, wearable technology can contribute to relational agency, enhancing users' control over their health and fostering positive outcomes.

3.4 Wearable Devices as Diagnostic Tools: Enhancing Health Awareness and Agency Through Sleep Monitoring

Wearable devices are increasingly integrated into daily routines, functioning not only as health monitoring tools but also as diagnostic aids that enhance individual health awareness and agency (Dias et al., 2018). This case study highlights the experience of a 29-year-old IT professional from Bangalore who utilized an Apple watch to detect sleep abnormalities impacting his daily well-being. Through socio-technical theory, this section explores how the Apple watch functions within a system that connects the user, healthcare providers, and technological infrastructure, underscoring the capacity of wearable technology to empower users and improve health outcomes by offering continuous, actionable insights.

IT professional faced persistent fatigue and exhaustion during the day without any visible health issues. Upon consulting a doctor, he was advised to monitor his sleep patterns, as poor sleep quality can often lead to such symptoms. He began wearing his Apple watch at night, which revealed inadequate sleep patterns, providing critical insights into his condition. The data allowed him to recognize and address his sleep deficit, making the smartwatch an invaluable diagnostic tool in his health management. In this case, the Apple watch operates within a socio-technical system that includes the IT professional, his physician, and the broader healthcare network. The watch's capacity to autonomously monitor and record sleep patterns illustrates Dourish and Bell's (2011) concept of ubiquitous computing, where technology seamlessly integrates into daily life, enabling automated and continuous health observation. Additionally, the smartwatch's influence on the user's behavior reflects Bennett's (2004) notion of affective force and agential capacity, empowering the IT professional to make informed health decisions. By receiving real-time feedback on sleep quality, the device enhances his agency over well-being, aligning with Hayles' (1999) theory that information produced by wearable devices becomes an embodied extension of the user, shaping personal health practices within daily routines. Furthermore, many wearable devices offer users satisfaction through goal-setting features. Participants in this study reported that completing activity goals, such as walking a set number of steps or exercising daily, contributed to a sense of accomplishment. These socio-technical promissory narratives of wearables foster meaningful changes in lived experiences, as users engage with technology not only for health monitoring but also for enhancing personal fulfilment and achievement in everyday life.

4 Challenges and Limitations of the Study

While wearable technology offers numerous benefits, its adoption and effectiveness are accompanied by significant challenges and limitations. Key issues identified in this study include data privacy concerns, the persistence of the digital divide, and increasing dependency on technology.

Firstly, data privacy is a critical concern, as wearable devices continuously collect sensitive health data, often transmitting it over networks that may not be secure. This raises questions about data ownership, confidentiality, and the potential misuse of personal health information by third parties or organizations. Ensuring robust data protection mechanisms is essential to build user trust and promote the widespread adoption of wearable technology in health monitoring.

Secondly, the digital divide presents a barrier to the equitable integration of wearable technology. Socioeconomic disparities affect access to wearable devices, with cost and digital literacy influencing who can benefit from these advancements. In countries like India, where disparities in technological access remain significant, (Digital divide, 2022) wearable technology risks reinforcing existing inequalities unless affordability and accessibility are prioritized.

Lastly, the dependency on wearable technology poses a challenge as users may become increasingly reliant on devices for health insights and lifestyle choices, potentially reducing personal agency in managing one's own health. This dependency raises questions about the long-term impacts on users' autonomy and decision-making, especially if devices fail or data is misinterpreted without professional guidance.

Addressing these challenges is essential for ensuring that wearable technology can be seamlessly and equitably integrated into everyday life. Through focused efforts on privacy protection, equitable access, and fostering balanced use, wearable technology can more effectively fulfil its potential to enhance personal health management and well-being on a broad scale.

5 Conclusion

Wearable technology has profoundly impacted personal health management and social interactions in India, providing users with enhanced agency over their health through continuous monitoring and real-time feedback, as demonstrated by the case studies of Rahim, the elderly man from Malappuram, Rajeswari, and the IT professional from Bangalore. These devices facilitate proactive health management, bridging physical distances and fostering remote caregiving, thus transforming traditional healthcare dynamics. The socio-technical systems framework reveals how wearable devices seamlessly integrate into everyday life, becoming indispensable tools that enhance personal empowerment and create new patterns of interaction and dependency. However, the challenges of data privacy and the digital divide highlight the need for policies that ensure secure and equitable access to these technologies. As wearable technology evolves, future research must focus on making these devices more accessible and user-friendly, ensuring their benefits are inclusive and far-reaching. By addressing these challenges and leveraging interdisciplinary insights, wearable technology can continue to revolutionize personal health management and social interactions, making it a vital part of everyday life in India and beyond.

Disclosure of Interests. The authors have no competing interests to declare that are relevant to the content of this article. Authors have not received any grants for this research.

References

Bennett, J.: The force of things: steps toward an ecology of matter. Polit. Theory **32**(3), 347–372 (2004)

Raad, H.: Fundamentals of IoT and Wearable Technology Design. Wiley (2020)

de Certeau, M.: The Practice of Everyday Life (S. Rendall, Trans.; 3rd ed.) (2011)

Dias, D., Paulo Silva Cunha, J.: Wearable health devices—vital sign monitoring, systems and technologies. Sensors **18**(8), 2414 (2018)

Dierkes, M., Hoffmann, U.: New Technology at the Outset: Social Forces in the Shaping of Technological Innovations. Westview Press (1992)

Digital Divide: India Inequality Report 2022 (2022). https://ruralindiaonline.org/en/library/resource/digital-divide-india-inequality-report-2022/

Dourish, P., Bell, G.: Divining a digital future: mess and mythology in ubiquitous computing. MIT Press (2011)

Haraway, D.J., Wolfe, C.: Manifestly Haraway. University of Minnesota Press (2016). https://doi.org/10.5749/minnesota/9780816650477.001.0001

Hayles, N.K.: How we became posthuman: virtual bodies in cybernetics, literature, and informatics. University of Chicago Press (1999)

India's Wearable Device Market Grew 34% in 2023 to 134 Million Units, says IDC. IDC: The Premier Global Market Intelligence Company (n.d.). https://www.idc.com/getdoc.jsp?containerId=prAP51880624#:~:text=INDIA%2C%20February%2016%2C%202024%20%E2%80%93,over%2Dyear%20(YoY)

Iliadis, A.: Computer guts and swallowed sensors: ingestibles made palatable in an era of embodied computing. In: Pedersen, I., Iliadis, A. (eds.) Embodied Computing: Wearables, Implantables, Embeddables. The MIT Press (2020)

Lindeman, D.A., Kim, K.K., Gladstone, C., Apesoa-Varano, E.C.: Technology and caregiving: emerging interventions and directions for research. Gerontologist **60**(Suppl._1), S41–S49 (2020)

Lupton, D.: Wearable Devices: Sociotechnical Imaginaries and Agential Capacities (SSRN Scholarly Paper 3084419) (2017). https://papers.ssrn.com/abstract=3084419

Mirjalali, S., Peng, S., Fang, Z., Wang, C.H., Wu, S.: Wearable sensors for remote health monitoring: potential applications for early diagnosis of Covid-19. Adv. Mater. Technol. **7**(1), 2100545 (2022)

Pedersen, I., Iliadis, A.: Introduction. In Embodied computing: Wearables, Implantables, Embeddables. The MIT Press (2020)

Stark, L.: Visceral data. In: Sampson, T.D., Maddison, S., Ellis, D. (eds.) Affect and Social Media: Emotion, Mediation, Anxiety and Contagion, pp. 42–51. Rowman & Littlefield (2018)

Wiener, N.: Cybernetics: Control and communication in the animal and the machine (Second edition, 2019 reissue). The MIT Press (2019)

Online Monitoring System of Groundwater in the Aral Region Based on the Internet of Things

Islambek Saymanov[1,2(✉)], Sherzod Boltaev[1], and Nodirbek Urinov[2]

[1] National University of Uzbekistan, Tashkent 100007, Uzbekistan
islambeksaymanov@gmail.com

[2] Andijan State University, Andijan 170100, Uzbekistan

Abstract. Considering the opportunities that rapidly evolving technologies provide to current and future information technologies, including systems and processes implemented through technological tools, it is necessary to apply the unique aspects of IoT in an online groundwater monitoring system. Online monitoring of water levels in remote areas using technologies for remote control of groundwater regime parameters is becoming a requirement today.

This article develops an online groundwater monitoring system as a modern high-performance tool to support strategic and operational decision-making, providing direct support to users for analysis and management decisions based on the prompt provision of necessary information about groundwater parameters. This, in turn, makes it possible to quickly identify and assess changes in the state of groundwater, prevent negative consequences and monitor quality indicators. Thanks to this system, it is possible to carry out real-time operational monitoring of groundwater regime parameters and management capabilities at monitoring wells belonging to the "Reclamation Expedition of Karakalpakstan".

In this work, the implementation of a monitoring system for the Aral region based on the Internet of Things was carried out in order to realize the effectiveness of groundwater monitoring through radically new management methods. These methods rely on the use of telecommunications, Internet of Things and software tools to make decisions, analyze the status of groundwater regimes and reduce the time spent on data processing.

1 Introduction

The problem of the Aral Sea is a clear example of the negative anthropogenic impact on the ecological state of the natural environment around us. The international community defines the Aral crisis as a global environmental disaster of our time, the possible consequences of which are still not fully understood.

The relevance of this problem is caused not only by its global nature, but also by the fact that efforts to solve it are urgent.

D. Singh et al. (Eds.): IHCI 2024, LNCS 15557, pp. 203–217, 2025.
https://doi.org/10.1007/978-3-031-88705-5_17

The environmental situation in the Aral region, in particular in the Southern Aral Sea region on the territory of our republic, has an extremely negative impact on the natural environment and living conditions of the region's population (Fig. 1).

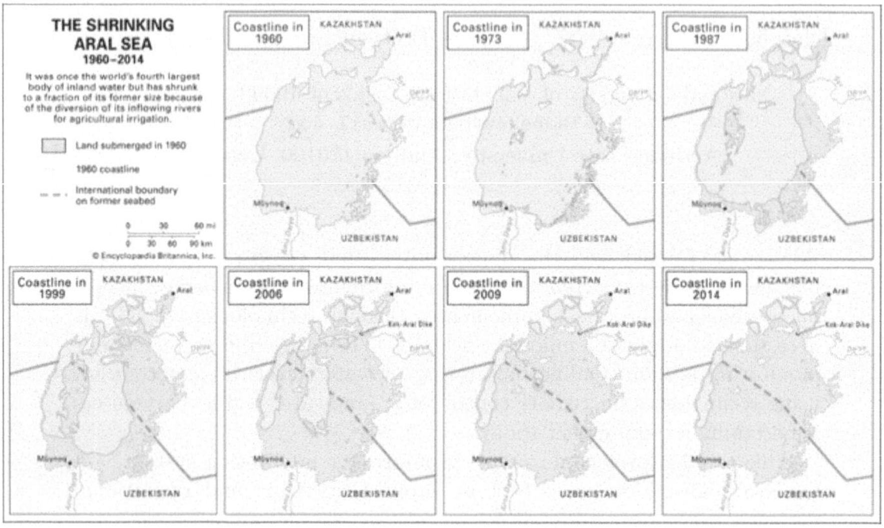

Fig. 1. State of the Aral Sea in 1960–2014.

Groundwater monitoring can be divided into three groups. The first is to measure groundwater levels at monitoring points using classical instruments, electronic and physical control methods. The second group involves the widespread use of satellite data to provide reliable and cost-effective real-time monitoring of groundwater level changes [1–5]. Most satellite data is used to indirectly determine the location of groundwater. However, GRACE (Gravity Recovery and Climate Experiment) satellite data, supported by computational tools, GAT (Geoinformation system) technology and hydroclimate models, have proven to be the most effective for groundwater management. The third group of monitoring is groundwater numerical modeling, which is a very effective tool for assessing and predicting future groundwater resources. GRACE allows for more efficient assessment of groundwater resources based on the integration of satellite modeling with numerical modeling of groundwater [6,7].

Automated measurement systems that can digitize data from devices directly on the network are widely used. Today, the most popular automated measuring instruments are diving measuring instruments from Schlumberger Water Services and Eijkelkamp [8–10].

These sensors measure water and air pressure, water temperature and electrical conductivity, and also, as a result of certain calculations, determine the level, temperature and general mineralization of groundwater. The process of moving from manual to automatic measurement provides the most efficient and reliable information. In addition, trips to hydrogeological monitoring wells are minimized, the number of sufficient measurements is increased and continuous data is provided [11–14].

In practice, it is possible to automatically measure the level, temperature and general mineralization of groundwater using sensors for water and air pressure, temperature and electrical conductivity of groundwater in monitoring wells and take measurements from the sensors every hour. The service life of the battery installed in the device is 3-4 years [15,16].

Developed by Australia's CSIRO, SENSEI Diver is the world's first automated real-time monitoring and reporting system that can operate in long-term and extreme conditions, including low pH (acidic) conditions and groundwater pressures of at least 20 bar (approx. operates at a depth of 200 m from the surface). The apparatus consists of robust solid-state electrochemical sensors to measure a range of chemical properties when immersed in water. The device includes sensors for pH, conductivity, pressure, depth and temperature. There are wired cables between the communication gateway and the sensor array to transmit data and power, and data from the gateway is transmitted to an online server using wireless connections. Measurements are taken every second and sent to an online data server. This data can be securely transferred and managed from anywhere in the world through a dashboard, facilitating analysis and quick decision-making [17,18].

2 Literature Review

The primary focus of this research was to explore and develop methods for processing and protecting data transmitted from monitoring wells located in the Aral Sea region, with the application of digital technologies in monitoring environmental conditions. A distinctive aspect of this work is its combination of environmental monitoring, data security, and the development of algorithmic models specific to an ecologically sensitive and isolated region.

The study introduces an innovative algorithmic model designed to enhance data reliability and ensure the security of transmitted information. While other works focus primarily on abstract or theoretical cryptographic models, particularly in the realm of Boolean functions [19–21], the practical application of such models to environmental monitoring is relatively underexplored. This research addresses that gap by applying data processing techniques and security measures to real-world environmental challenges, particularly in the Aral Sea region, an area affected by significant ecological degradation.

A key strength of the research lies in the formalization of the system used to monitor water and soil conditions, incorporating a microcontroller-sensor system for accurate data collection and interactive analysis. While previous studies [22,23] have explored environmental monitoring in different contexts, such as renewable energy or atmospheric pollution, the present study offers a comprehensive approach that not only monitors environmental data but also ensures its secure transmission and reliability over potentially unstable communication channels.

In comparison to the body of literature on Boolean functions and cryptographic models, this study goes beyond theoretical constructs by focusing on applied security in monitoring systems. The integration of information security into environmental monitoring systems is a notable advancement, particularly when compared to the abstract cryptographic challenges examined in works like those of [24] or [25]. The focus on the Aral Sea region adds further uniqueness, as no prior works address the specific challenges faced in this ecologically fragile area.

Another distinguishing feature of this research is the development of software for interactive analysis of water and soil data. This system allows for real-time monitoring of the environmental conditions during agricultural production in the region. The inclusion of harmful substance distribution models adds a layer of practicality that is absent in many other mathematical models for environmental monitoring [26,27]. These models can provide valuable insights into the environmental impact of agricultural practices, potentially guiding sustainable development efforts in the region.

In this research offers a novel and comprehensive approach to addressing the environmental challenges in the Aral Sea region through the integration of digital technologies, algorithmic modeling, and information security. It bridges the gap between theoretical studies in Boolean functions and cryptography and the practical need for secure and reliable environmental monitoring systems. This combination of approaches ensures that the data transmitted from monitoring wells remains protected, while providing stakeholders with accurate and actionable insights into the region's environmental health. Future research can build on this by further refining the models and expanding their applicability to other regions facing similar environmental challenges.

3 Analysis and Features of the Methodology for Managing Groundwater Monitoring Systems

The most effective way to ensure the rational use of groundwater, monitor the state of groundwater and its relationship with various components of the natural environment, and also obtain the necessary information for making management decisions is to conduct a monitoring system [28–31]. Taking into account the set goals and objectives, the groundwater monitoring system includes a number of main functions:

1. Conduct systematic monitoring in order to obtain information characterizing:

- water layers and groundwater in them;

- the amount of water taken from water intake structures and entering the groundwater layers during artificial replenishment of groundwater reserves;

- exogenous processes that arise or change during the exploitation of groundwater;

- technical condition of artificial filling of groundwater reserves and water intake structures;
- condition of sanitary protection zones of groundwater, sources of anthropogenic pollution of groundwater and surface water.

2. Collection of observation materials in monitoring systems.
3. Pre-processing and synthesis of groundwater monitoring and collection data.
4. Preparation and maintenance of information databases that provide an assessment of the state of groundwater, forecasting its area, time, short-term and long-term changes during the observation period.
5. Development of recommendations on the state of groundwater monitoring objects and changes in individual components of the environment and the prevention of their negative consequences.

Direct measurement of groundwater levels in observation wells can be carried out by continuous recording of observations, manual or automatic measurement technologies. The level, temperature and chemical composition of groundwater are monitored through monitoring wells in the study areas. Control of the chemical composition and physical properties is carried out by submitting samples to accredited laboratories. The quality of drinking water is usually determined by the following indicators: changes in bacteriological, organoleptic and chemical composition. The main indicators of the groundwater monitoring system include the following:

- chemical (increased general mineralization of groundwater, increase in the concentration of micro- and macrocomponents, the appearance in groundwater of mineral and organic compounds that are not characteristic of them);
- bacteriological (appearance of 19 microorganisms with pathogenic and sanitary-hygienic indicators);
- thermal (an increase in the temperature of groundwater can be caused, firstly, by the infiltration of heated wastewater from above, secondly, by the discharge of industrial wastewater into deep underground reservoirs, and thirdly, by the influence of the residential area as a whole);
- radioactive (increase in the content of radioactive substances in groundwater at different half-lives).

The sources of saturation of underground water bodies are surface waters; depending on their hydrogeological characteristics, their formation, accumulation and movement along the slope occurs in soil layers. Depending on their underground location, they can be in vapor, liquid and solid form.

Groundwater is divided into 5 classes according to the level of mineralization (Table 1).

Depending on the layer from which groundwater is extracted, it can be at different temperatures depending on the depth and characteristics of its location. According to temperature, groundwater is divided into 7 types (see Table 2).

Table 1. Classification of groundwater by level of mineralization.

Salinity type	HCO_3	Cl	SO_4	Ca
Unsalted	$\alpha < 0.02$	$\alpha < 0.03$	$\alpha < 0.05$	$\alpha < 0.01$
Slightly saline	$0.02 < \alpha < 0.03$	$0.03 < \alpha < 0.1$	$0.05 < \alpha < 0.12$	$0.01 < \alpha < 0.05$
Medium saline	$0.03 < \alpha < 0.05$	$0.1 < \alpha < 0.3$	$0.12 < \alpha < 0.35$	$0.05 < \alpha < 0.1$
Highly saline	$0.05 < \alpha < 0.1$	$0.3 < \alpha < 0.6$	$0.35 < \alpha < 0.7$	$0.1 < \alpha < 0.2$
Very highly saline	$\alpha > 0.1$	$\alpha > 0.6$	$\alpha > 0.7$	$\alpha > 0.2$

Table 2. Types of underground water according to temperature.

No.	Type of water	Temperature, C
1	Very cold	below 0
2	Very cold	0-4
3	Cold	4-20
4	Warm	20-37
5	Hot	37-42
6	Very hot	42-100
7	Too hot	above 100

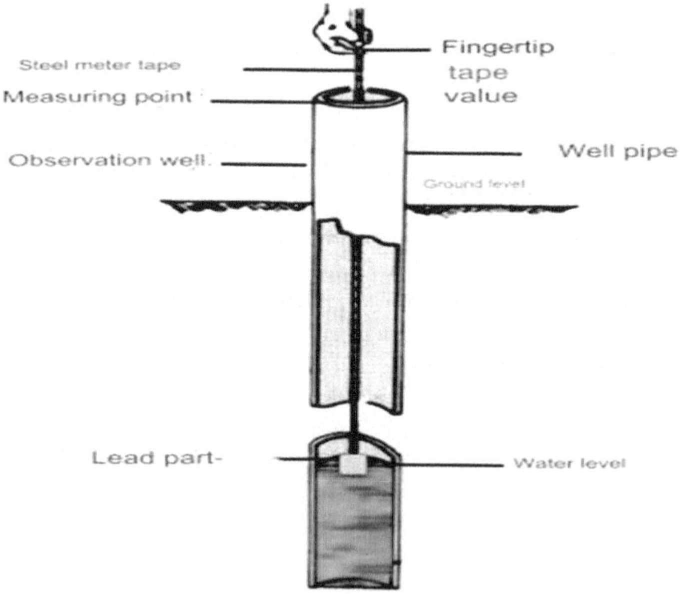

Fig. 2. Roulette method for measuring water level

4 Water Level Measurement Methods

4.1 Roulette

Tape measures allow you to measure the water level in a well if there is an open well or a large gap between the pipes. Tape reels are made of steel and range in length from 10 to 100 m (Fig. 2).

The steel tape and chalk method is labor intensive and is not suitable for use in deep wells. In addition, its weight can cause damage to the submersible pump in the well. Another disadvantage of the steel tape is the inaccuracy of placing the tape in the hole, that is, at an equal distance from the surface of the earth, which can lead to the fact that when determining the water level, the tape will be lowered into the well several times.

4.2 Hydrostatic Measurement Method

Based on a hydrostatic pressure sensor (also called hydrostatic sensing) in monitoring wells, this method allows continuous measurement of groundwater levels in real time. This method consists of an air pressure sensor and a water pressure sensor. They can be lowered into the well using a well pump or attached directly to the wellhead by running a cable down the well. The pressure sensor measures the equivalent hydrostatic pressure of the water above, converts the corresponding signal through the cable into a level and transmits the signal to the data processing and control system. Hydrostatic pressure sensors determine the water level in a well using the hydrostatic effect when there is no flow.

$$P_{static\ fluid} = \rho * g * p$$

Here ρ - fluid density; g - falling speed; p - liquid depth or filling height (Fig. 3).

4.3 Ultrasonic Method for Measuring Satellites

To measure the water level, I used the ultrasonic level measurement method. Ultrasonic devices are used in many measurement fields, especially for water level measurement. The ultrasonic satellite measurement method is considered a remote non-conductive satellite measurement technology. It emits a high frequency pulse, usually between 20 and 200 kHz, and is echoed. The impulse is usually transmitted in the form of a cone with an area of maximum 60 hectares. The pulse hits the surface of the water and returns to the receiving sensor.

To determine the water level, the total return time between the transmitted ultrasonic pulse and the reflected sound is measured. Paste ultrasonic sensors are used to measure water levels at great depths, and high frequency sensors are used to measure water levels at short distances. The distance along the water surface is calculated using the formula S.

$$S = \frac{V}{2 * t}$$

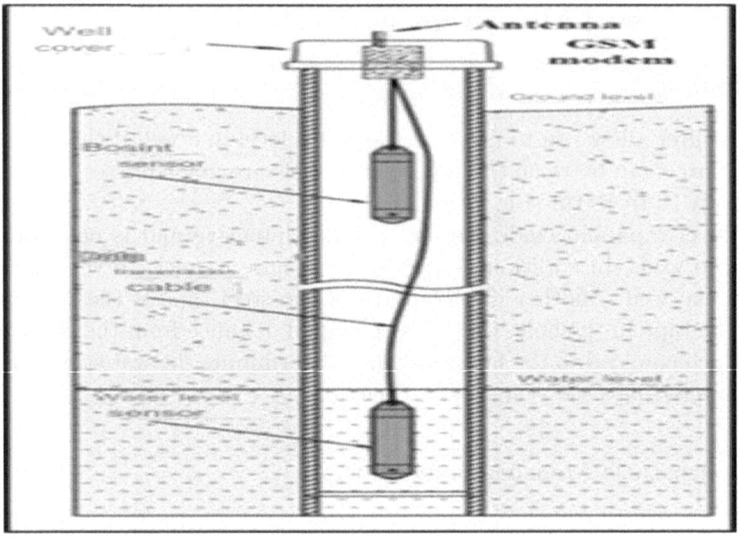

Fig. 3. Hydrostatic method for measuring water level

Here V - is the speed of sound in the medium; t - is the time between the transmitted pulse and the reflected echo (Fig. 4).

The main data required for water level measurement in an IoT environment are:

Sensors: Water level sensors are required to measure the water level in an IoT system. These sensors can be products that measure water levels, such as pH sensors, temperature sensors, oxygen sensors, salinity sensors, etc.

Communication Modules: Necessary communication modules such as Wi-Fi, Bluetooth, Zigbee or LoRa communications can be used to send sensor data directly to the IoT platform.

Internet of Things Platform: An IoT platform is needed to acquire, store, analyze, and perform other tasks. These platforms provide data viewing, analysis, and IoT device management capabilities.

Data Analytics: Data analytics is important for analyzing and taking action on water level data from IoT systems. This includes analyzing data on water levels, water quality, water flow, and troubleshooting.

Data Visualization: Data such as water levels, analysis results and passenger imaging systems are also important. This ensures that data can be displayed in the best possible way using graphs, pointers and other visual aids.

Payment and Offense: Fee and offense structures are needed to measure the water level of an IoT system. This is necessary to provide payment methods to customers, to ensure the protection and confidentiality of user data.

Internet Connection: An Internet connection is required to receive, analyze, and perform other IoT functions based on data received by the IoT system.

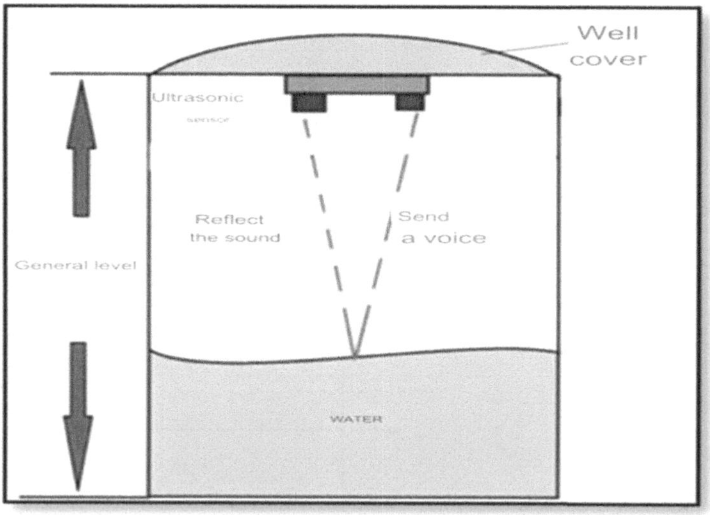

Fig. 4. Ultrasonic method for measuring water level

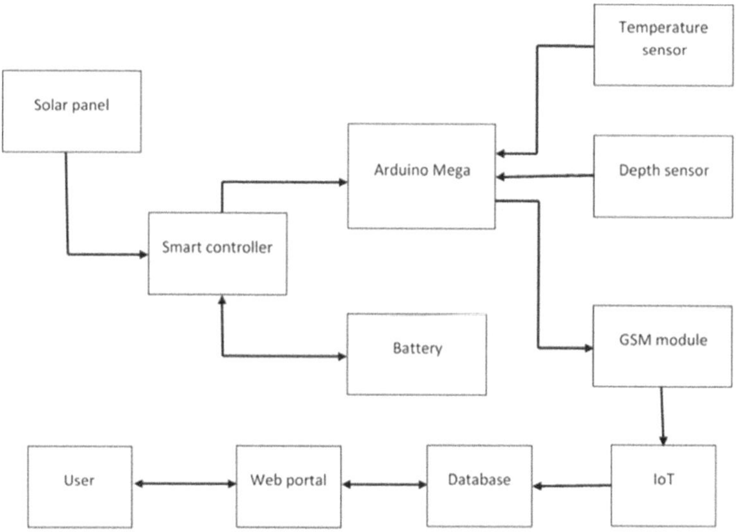

Fig. 5. Scheme of operation of the environmental monitoring system in interactive mode

It allows you to send data to an IoT platform and receive data from that platform (Figs. 5 and 6).

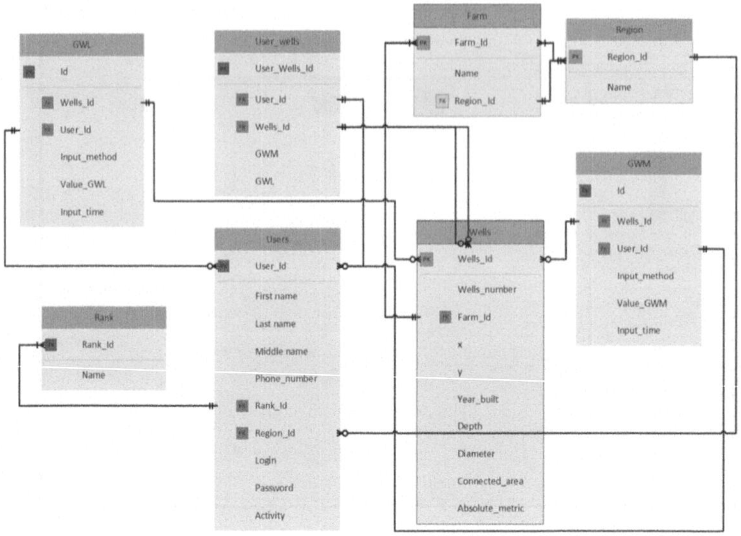

Fig. 6. Datalogical model of environmental monitoring information system

5 Results

In order to implement the above methods, in the presence of the Reclamation Expedition of the Republic of Karakalpakstan, to determine the state of groundwater from monitoring wells No. 2, 3, 7 and monitoring wells No. permissive, scientific and practical research work is carried out. aimed at improving the sys-

Fig. 7. Sensor power supply from solar panel

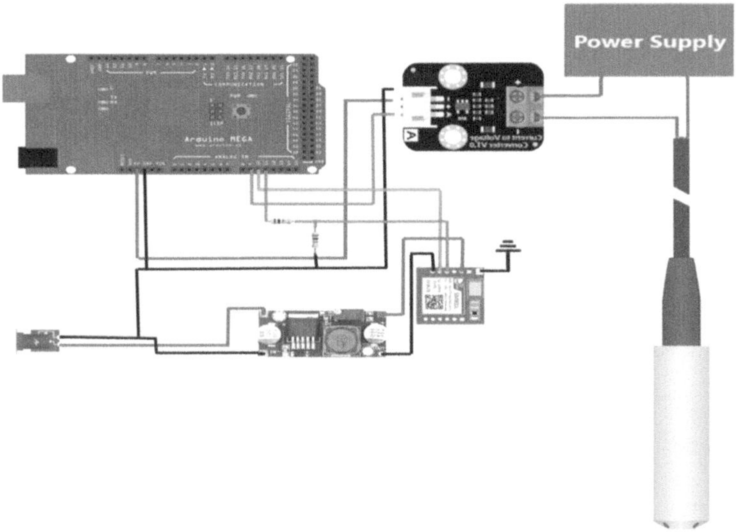

Fig. 8. Schematic diagram of environmental monitoring system tools that interactively transmit information about groundwater levels

Fig. 9. Tests with ultrasonic sensor

tem for monitoring groundwater parameters. As a result of scientific and practical research, the structure of a system for monitoring groundwater levels, temperature and electrical conductivity parameters has been developed. According to it, the structure of an online groundwater monitoring system consists of sys-

No.	Reference	Focus Area	Relevance to Data Processing & Protection in Monitoring Wells	Relevance to Digital Technologies in Environmental Monitoring	Relevance to Aral Sea Region	Algorithmic Models & System Formalization	Information Security
1	[19]	Boolean functions	-	-	-	+	+
2	[20]	Cryptographic functions	+	-	-	+	+
3	[21]	Reed-Muller codes	-	-	-	+	+
4	[22]	Same as above	-	-	-	+	+
5	[23]	Boolean semantics	-	-	-	+	-
6	[24]	Quadratic Boolean functions	+	-	-	+	+
7	[25]	Balanced Boolean functions	+	-	-	+	+
8	[26]	Air quality predictions	-	+	-	-	-
9	[27]	Agri-food supply chains	-	+	+	-	-
10	This Work		+	+	+	+	+

Fig. 10. Comparison table of works completed with other works

tems divided by functional or structural characteristics that meet specific goals and objectives (Figs. 7, 8, 9 and 10).

For field testing, the average groundwater level was 260 cm per month, based on sensor data from 3,583 wells. The cataloged minimum and maximum values are 137 sm and 345 sm respectively. The minimum and maximum salinity values in the Aral Sea region are 0.78 g/l and 18.560 g/l, respectively. Average salinity - 4.64 g/l.

6 Conclusion

General statistical analysis of groundwater level data obtained from observation wells No. 2, 3, 7 and observation wells No. 12, 17, 19 in the Muynak region farm during the Reclamation Expedition of the Republic of Karakalpakstan, in connection with the results of groundwater monitoring in the village. The Aral economy of the Muynak region has been completed. With the help of a groundwater monitoring system and the use of an effective method of extrapolation of data analysis, high efficiency has been achieved and the possibility of guaranteeing the results obtained has been created.

References

1. Kabulov, A.V., Normatov, I.H.: About problems of decoding and searching for the maximum upper zero of discrete monotone functions. J. Phys. Conf. Ser. **1260**(10), 1–8 (2019). https://doi.org/10.1088/1742-6596/1260/10/102006
2. Kabulov, A.V., Normatov, I.H., Ashurov, A.O.: Computational methods of minimization of multiple functions. J. Phys: Conf. Ser. **1260**(10), 1–11 (2019). https://doi.org/10.1088/1742-6596/1260/10/102007
3. Kabulov, A., Normatov, I., Seytov, A., Kudaybergenov, A.: Optimal management of water resources in large main canals with cascade pumping stations. In: 2020 IEEE International IOT, Electronics and Mechatronics Conference (IEMTRONICS), pp. 1–4. Vancouver, BC, Canada (2020). https://doi.org/10.1109/IEMTRONICS51293.2020.9216402
4. Kabulov, A., Saymanov, I., Yarashov, I., Muxammadiev, F.: Algorithmic method of security of the Internet of Things based on steganographic coding. In: 2021 IEEE International IOT, Electronics and Mechatronics Conference (IEMTRONICS), pp. 1–5. Toronto, ON, Canada (2021). https://doi.org/10.1109/IEMTRONICS52119.2021.9422588
5. Kabulov, A., Normatov, I., Urunbaev, E., Muhammadiev, F.: Invariant continuation of discrete multi-valued functions and their implementation. In: 2021 IEEE International IOT, Electronics and Mechatronics Conference (IEMTRONICS), pp. 1–5. Toronto, ON, Canada (2021). https://doi.org/10.1109/IEMTRONICS52119.2021.9422486
6. Kyllmar, K., Forsberg, L.S., Andersson, S., Mårtensson, K.: Small agricultural monitoring catchments in Sweden representing environmental impact. Agr. Ecosyst. Environ. **198**, 25–35 (2014). https://doi.org/10.1016/j.agee.2014.05.016
7. Fritz, S., See, L., Bayas, J., Waldner, F., Jacques, D., Becker-Reshef, I., McCallum, I.: A comparison of global agricultural monitoring systems and current gaps. Agric. Syst. **168**, 258–272 (2019)
8. Navruzov, E., Kabulov, A.: Detection and analysis types of DDoS attack. In: 2022 IEEE International IOT, Electronics and Mechatronics Conference (IEMTRONICS), pp. 1–7. Toronto, ON, Canada (2022). https://doi.org/10.1109/IEMTRONICS55184.2022.9795729
9. Kabulov, A., Yarashov, I., Otakhonov, A.: Algorithmic analysis of the system based on the functioning table and information security. In: 2022 IEEE International IOT, Electronics and Mechatronics Conference (IEMTRONICS), pp. 1–5. Toronto, ON, Canada (2022). https://doi.org/10.1109/IEMTRONICS55184.2022.9795746
10. Kabulov, A., Saymanov, I., Yarashov, I., Karimov, A.: Using algorithmic modeling to control user access based on functioning table. In: 2022 IEEE International IOT, Electronics and Mechatronics Conference (IEMTRONICS), pp. 1–5. Toronto, ON, Canada (2022). https://doi.org/10.1109/IEMTRONICS55184.2022.9795850
11. Kabulov, A., Baizhumanov, A., Berdimurodov, M.: On the minimization of k-valued logic functions in the class of disjunctive normal forms. J. Math. Mech. Comput. Sci. **121**(1), 37–45 (2024). https://doi.org/10.26577/JMMCS202412114
12. Kabulov, A., Saymanov, I., Babadjanov, A., Babadzhanov, A.: Algebraic recognition approach in IoT ecosystem. Mathematics **12**(1086), 1–26 (2024). https://doi.org/10.3390/math12071086
13. Saymanov, I.: Logical automatic implementation of steganographic coding algorithms. Journal of Mathematics, Mech. Comput. Sci. **121**(1), 122–131 (2024). https://doi.org/10.26577/JMMCS2024121112

14. Kabulov, A., Normatov, I., Saymanov, I.: On the Completeness of classes of correcting functions of heuristic algorithms. Azerbaijan J. Math. (accepted)
15. Xu, Y.: Envirotyping for deciphering environmental impacts on crop plants. Theor. Appl. Genet. **129**(4), 653–673 (2016). https://doi.org/10.1007/s00122-016-2691-5
16. Li, S., Huang, B., Zhao, F., Lu, Z., Wang, X., Chen, X., Chen, Y.: Environmental impact assessment of agricultural production in Chongming ecological island. Int. J. Life Cycle Assessment **24**(11), 1937–1947 (2019). https://doi.org/10.1007/s11367-019-01614-w
17. Sinisterra-Solís, N., Sanjuán, N., Ribal, J., Estruch, V., Clemente, G.: An approach to regionalise the life cycle inventories of Spanish agriculture: monitoring the environmental impacts of orange and tomato crops. Sci. Total Environ. **856**, 1–17 (2023). https://doi.org/10.1016/j.scitotenv.2022.158909
18. Mishra, S., Sharma, S.K.: Advanced contribution of IoT in agricultural production for the development of smart livestock environments. Internet Things **22**, 100724 (2023). https://doi.org/10.1016/j.iot.2023.100724
19. Cusick, T.W.: Using easy coefficients conjecture for rotation symmetric Boolean functions. Inf. Sci. **659**, 120075 (2024). https://doi.org/10.1016/j.ins.2023.120075
20. Guo, F., Wang, Z.: Balanced even-variable rotation symmetric boolean functions with optimal algebraic immunity, maximum algebraic degree and higher nonlinearity. Int. J. Found. Comput. Sci., 1–26 (2023). https://doi.org/10.1142/s0129054123500016
21. Carlet, C., Feukoua, S.: On those Boolean functions that are coset leaders of first order Reed-Muller codes. Ann. Math. Artif. Intell. **92**(1), 7–27 (2024)
22. Islam, M.S., Khatun, M.S., Biswas, M.: Mathematical modelling of using renewable energy in the power sectors for the sustainable environment. Math. Modell. Numerical Simul. Appl. **4**(2), 216–237 (2024)
23. Kabulov, A., Baizhumanov, A., Saymanov, I.: Synthesis of optimal correction functions in the class of disjunctive normal forms. Mathematics **12**(2120) (2024). https://doi.org/10.3390/math12132120
24. Chirvasitu, A., Cusick, T.W.: Quadratic rotation symmetric Boolean functions. Discret. Appl. Math. **343**, 91–105 (2024)
25. Lou, Y., Wang, Q.: New constructions of balanced Boolean functions with the maximum possible Walsh supports. Discret. Appl. Math. **355**, 262–267 (2024)
26. Katariya, N.K., Choudhary, B.S., Pandey, P.: Air quality predictions through mathematical modeling for iron ore mine project. Appl. Sci. **14**(13), 5922 (2024)
27. Fathi, M.R., Zamanian, A., Khosravi, A.: Mathematical modeling for sustainable agrifood supply chain. Environ. Dev. Sustain. **26**(3), 6879–6912 (2024)
28. Singh, D., Singh, M., Hakimjon, Z.:Multidimensional Polynomial Splines. In: Signal Processing Applications Using Multidimensional Polynomial Splines. SpringerBriefs in Applied Sciences and Technology. 2019. Springer, Singapore. https://doi.org/10.1007/978-981-13-2239-6_4
29. Singh, D., Singh, M., Hakimjon, Z.: Spline evaluation for railways. In: Signal Processing Applications Using Multidimensional Polynomial Splines. SpringerBriefs in Applied Sciences and Technology. Springer, Singapore (2019). https://doi.org/10.1007/978-981-13-2239-6_8

30. Zaynidinov, H.N., Yusupov, I., Juraev, J.U., Singh, D.: Digital processing of blood image by applying two-dimensional haar wavelets. In: Singh, M., Kang, D.K., Lee, J.H., Tiwary, U.S., Singh, D., Chung, W.Y. (eds.) Intelligent Human Computer Interaction. IHCI 2020. LNCS, 2021, vol. 12615. Springer, Cham. https://doi.org/10.1007/978-3-030-68449-5_9
31. Zaynidinov, H., Mallayev, O., Kuchkarov, M.: Parallel algorithm for modeling temperature fields using the splines method: IEEE International IOT, Electronics and Mechatronics Conference (IEMTRONICS). Toronto, ON, Canada 2021, pp. 1–6 (2021). https://doi.org/10.1109/IEMTRONICS52119.2021.9422645

Enhancing Meal Search Engine for Fitness Applications Using Semantic Search with Sentence Transformers

Kim Aleksandr and Sang-Joong Jung(✉)

Dongseo University, 47 Jurye-Ro, Sasang-Gu, Busan, South Korea
sjjung@dongseo.ac.kr

1 Introduction

The global emphasis on a healthy lifestyle has fueled substantial growth in the health and fitness sectors. The convergence of technology with diverse domains has inevitably led to its integration into the fitness industry. Modern fitness applications address various user needs, offering features like workout tracking and dietary management. Nutrition apps, crucial for calculating intake and analyzing well-being, generated a market revenue of US$4.79 billion in 2023, with a growth rate of 13.9%, according to Statista [1]. To optimize user experience, services provide extensive meal databases. However, traditional meal search engines in the most popular existing applications predominantly rely on keyword-based search systems, which, while functional, often struggle to capture the intricacies of user preferences and the nuances associated with meal-related queries. This research proposes a solution by leveraging semantic search powered by Language Models, addressing these limitations, and enhancing precision. By understanding user intent more accurately, this method aims to deliver more relevant and personalized search results.

2 Related Works

2.1 Semantic Search

The fundamental idea behind a semantic search engine is understanding the "semantics" or "meaning" behind the textual data. While the conventional method analyzes the lexical matches between the query and the documents, semantic search aims to comprehend the content of the query. Early studies [2] laid the groundwork for semantic search, emphasizing the importance of capturing user intent and context to improve the precision and relevance of search results. The corpus entries are first embedded into the vector space, whereas query embeddings are calculated during the search time. Once converted, the statistical similarity, such as cosine similarity is calculated between those embeddings, and the most relevant results are presented.

D. Singh et al. (Eds.): IHCI 2024, LNCS 15557, pp. 218–222, 2025.
https://doi.org/10.1007/978-3-031-88705-5_18

2.2 Sentence-BERT

To compute embeddings for text corpus and user queries, we employ Sentence-BERT (SBERT), a transformer-based model [3] known for capturing semantic meaning [4]. This modification of pre-trained BERT models [5] excels in tasks requiring semantic similarity, such as information retrieval and semantic search. SBERT's effectiveness lies in encoding sentences into fixed-size vectors.

2.3 Bi-encoder and Cross-encoder

Within the SBERT family, the Bi-Encoder utilizes two identical encoders to process pairs of sentences independently. It prioritizes capturing individual semantic meanings and determining pair similarity based on the embeddings' distances or similarities. In contrast, the Cross-Encoder processes sentence pairs comprehensively (Fig. 1), achieving higher results but lacking efficiency due to the absence of embeddings [3]. The next chapter discusses a method to leverage the Cross-Encoder's precision effectively.

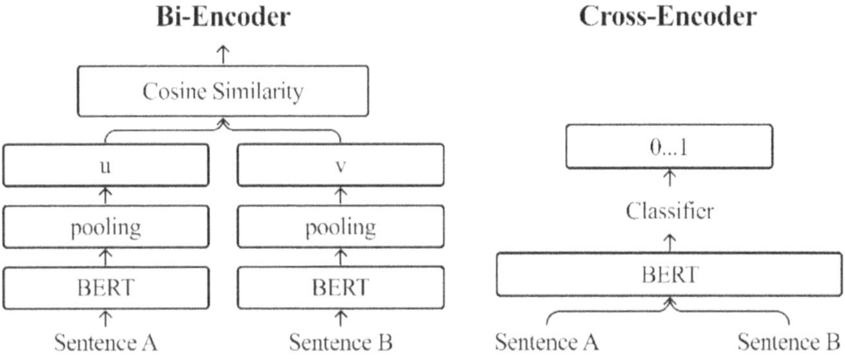

Fig. 1. Bi-Encoder vs Cross-Encoder architecture

3 Methodology

This section outlines the methodology employed to overcome limitations in conventional meal search engines within fitness applications. Beginning with data acquisition and corpus preparation, a food recipes dataset sourced from Kaggle [6, 7] underwent thorough processing, resulting in 14,113 refined rows. Synthetic Query Generation addressed the absence of labeled queries by utilizing a pre-trained T5 model [8], generating 70,575 query-recipe pairs [9] (Fig. 2).

Moving to model architecture, three SBERT variations were explored. First, the pre-trained baseline Bi-Encoder utilized "multi-qa-MiniLM-L6-cos-v1," creating 384-dimensional embeddings. Second, enhancing efficiency and accuracy, a Bi-Encoder Retriever fed predictions into a pre-trained Cross-Encoder Re-Ranker "ms-marco-MiniLM-L-6-v2". Finally, a fine-tuned Bi-Encoder Retriever, trained on the corpus with

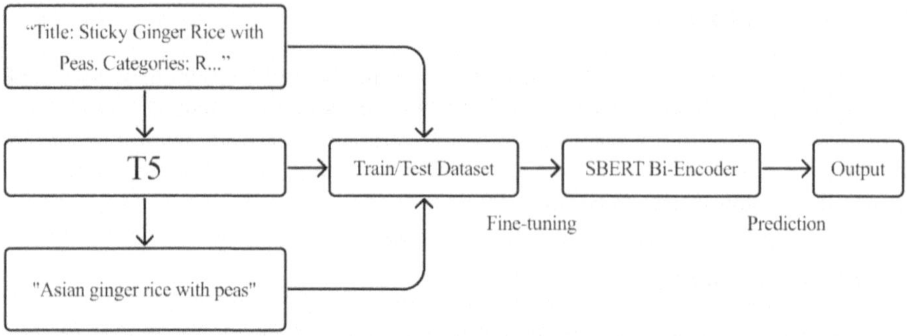

Fig. 2. Synthetic Query Generation pipeline

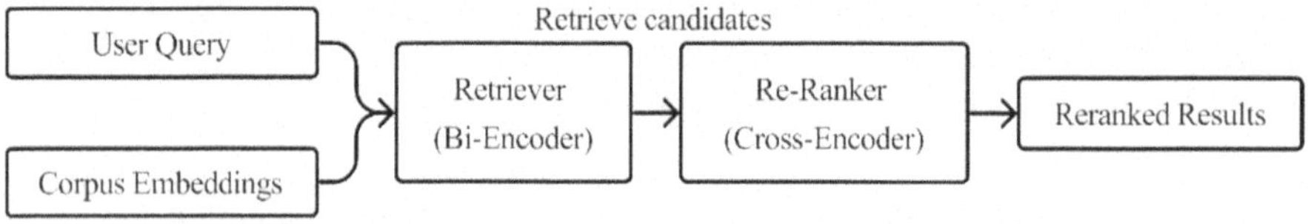

Fig. 3. Retrieve and Re-Rank pipeline

generated labels, further optimized performance for three epochs with warmup steps applied and the Multiple Negatives Ranking Loss [10] function used (Fig. 3).

The choice not to fine-tune the Cross-Encoder was informed by the absence of similarity labels in the dataset, making it challenging for the model to discern semantic relationships effectively. The methodology combines data refinement, synthetic query generation, and model architecture exploration to enhance semantic search within fitness applications.

4 Experiments and Results

The data for the training and test phases was split in an 80/20 ratio to preserve enough information for training and have sufficient data for performing evaluation. Moreover, to avoid data leakage, the data was split by rows, not the generated queries.

The data imperfections highlighted in preceding chapters have introduced a scenario requiring careful consideration of various trade-offs. The absence of ground truth in the data, originating from a lack of relevance labels, complicates the computation of relevance metrics. In addition, the limitation where each user query corresponds to a single recipe prevents the calculation of ranking performance metrics such as Precision at K or Recall at K. Therefore, only the top K predicted results generated by the models were considered the final output. Correct answers were determined based on the presence of the true label within the top K predicted values, while absence - an incorrect prediction. Taking into account those limitations, both the Accuracy and the F1 score were computed based on the outcomes generated by all three models during the evaluation on the test dataset. Accuracy is a metric calculated as the ratio of correctly predicted instances to the total number of instances in the dataset. While accuracy is a straightforward and intuitive measure, it may not be the most informative in some situations. Therefore, F1-score is additionally calculated. The F1-score is the harmonic mean of precision and recall, combining both measures into a single value. Precision represents the accuracy of positive predictions, while recall measures the ability of the model to capture all positive instances. The results are presented in Table 1.

Table 1. Models Performance Results

Architecture	Accuracy	F1-Score
BM25	0.7155	0.7913
Bi-Encoder	0.6524	0.7404
Bi-Encoder + Cross-Encoder	0.7634	0.8347
Fine-tuned Bi-Encoder + Cross-Encoder	**0.8359**	**0.8910**

The performance of keyword-based search was surprisingly high, even higher than the performance of the pre-trained Bi-Encoder. I suspect that it was caused by the abundance of same words presented in the queries and food recipes due to the work of query generator. Overall, the lexical search performance is far from perfect, as the generated user queries also included synonyms, which only can be spotted by the language model. That is why the performance of combined baseline pre-trained Bi- and Cross-encoders was higher. Finally, after training fine-tuning the pre-trained Bi-Encoder on the dataset and creating a new embedding, the model made big jump in performance, around 7.2% in accuracy and 5.6% in F1-score.

5 Conclusion

This study navigates the intersection of nutrition and technology, addressing limitations in conventional meal search engines within fitness applications. Leveraging semantic search powered by Language Models, specifically Sentence-BERT (SBERT), the methodology combines data refinement, synthetic query generation, and an exploration of SBERT model variations. Despite challenges posed by data imperfections, the proposed methodology showcases significant advancements. Notably, fine-tuning the pre-trained bi-encoder results in a substantial performance boost, namely, 7.2% improvement in accuracy and 5.6% in F1 score metrics. Furthermore, this project provided an environment for presentational purposes by building a mobile application prototype, a backend server, and the implementation of ElasticSearch indexing capabilities as a groundwork for further development in real-world conditions.

The future improvements to this study might include acquiring more suitable data, performing data labeling outsourcing, and adding a recommendation system to increase the relevancy of the search engine results. Moreover, the implementation of traditional methods, such as providing additional meta-data represented by categories, like calorie amount, dietary preferences, and meal type, can drastically improve the relevance of the search results. There is a vast space for experimentation with model architectures, yet the most crucial aspect of the project is appropriate data.

References

1. Statista Research Department: Nutrition Apps – Worldwide | Statista Market Forecast (2023). https://www.statista.com/study/40457/food-delivery
2. Guha, R., McCool, R., Miller, E.: Semantic search. In: Proceedings of the 12th International Conference on World Wide Web (2003)
3. Vaswani, A., et al.: Attention is all you need. In: Advances in Neural Information Processing Systems, vol. 30 (2017)
4. Reimers, N., Gurevych, I.: Sentence-BERT: Sentence embeddings using siamese BERT-networks. arXiv preprint arXiv:1908.10084 (2019)
5. Devlin, J., et al.: BERT: pre-training of deep bidirectional transformers for language understanding. arXiv preprint arXiv:1810.04805 (2018)
6. Hugo, D.: Epicurious - Recipes with Rating and Nutrition, Version 1 (2016). https://www.kaggle.com/rtatman/r-vs-python-the-kitchen-gadget-test
7. Epicurious – Recipes and Menus. http://www.epicurious.com/recipes-menus (2016)
8. Raffel, C., et al.: Exploring the limits of transfer learning with a unified text-to-text transformer. J. Mach. Learn. Res. **21**(140), 1–67 (2020)
9. Thakur, N., et al.: BEIR: a heterogeneous benchmark for zero-shot evaluation of information retrieval models. arXiv preprint arXiv:2104.08663 (2021)
10. Henderson, M., et al.: Efficient natural language response suggestion for smart reply. arXiv preprint arXiv:1705.00652 (2017)

Addressing Graphviz File Generation Issue in CPN Tools: A Java-Based Solution

Vikas Kumar Patel[1(✉)], Anshul Verma[1], and Pradeepika Verma[2]

[1] Banaras Hindu University, Varanasi, India
vikaspatelkomail@gmail.com
[2] TIH-Indian Institute of Technology, Patna, India

Abstract. This paper presents a novel approach to resolving a specific bug encountered in Colored Petri Nets (CPN) Tools during the generation of Graphviz files. CPN Tools, a powerful instrument for constructing and analyzing colored petri nets, has been identified to have limitations in graph visualization, particularly when dealing with token value strings or record/product with strings.The CPN Tools generate the Graphviz (dot) file, however it does not take into account color sets of strings, records, and products. If the color set is one of these, after generating the Graphviz file from CPN, Graphviz displays a syntax error. This issue results in a syntax mismatch with Graphviz, hindering the seamless visualization of graphs. To address this problem, the authors have developed a Java-based Graphical User Interface (GUI) program that takes the CPN Tools generated dot file as input, refactors it, and corrects the syntax, thereby enabling error-free graph visualization. The user merely needs to provide the CPN Tools created DOT file to the proposed system. It will make internal corrections and generate the graph. In addition, the user can download the revised version of the DOT file from the system. This solution not only resolves the identified bug but also enhances the overall user experience by simplifying the process of graph generation. The paper provides a detailed description of the bug, the development process of the Java GUI program, enhancing the page hierarchy graph of hierarchical colored petri nets, and its successful application in resolving the bug and generating correct graph. It also discusses the potential impact of this solution on the broader usability and efficiency of CPN Tools.

Keywords: CPN Tools · Graphviz · CPN · HCPN · State Space Graph · Dot File · Occurrences Graph · Syntax Based Error · SDLC Model

1 Introduction

Colored Petri Nets (CPN) have emerged as a powerful tool in the field of formal verification, providing a graphical and mathematical modeling language suitable

D. Singh et al. (Eds.): IHCI 2024, LNCS 15557, pp. 223–240, 2025.
https://doi.org/10.1007/978-3-031-88705-5_19

for the design, specification, validation, and verification of systems [1]. CPN extends the Petri net structure by introducing different colors (or tokens), which can be manipulated by associated functions, thereby allowing a more compact and comprehensible model representation. It comes with several built-in extensions. These extensions include Declare, Export Helper, Time Ranges, MSC [2], DCR Graphs, Debugger, PP-CPN, and Graphviz [3]. Each of these extensions serves a unique purpose and contributes to making the process of modeling and analyzing the net easier. For instance, the Declare extension aids in the declaration of variables and functions, while the Export Helper facilitates the export of models in various formats. The Time Ranges extension is useful for defining and managing time ranges in the model. Similarly, the MSC extension supports Message Sequence Charts, and the DCR Graphs extension is used for modeling dynamic condition response graphs. The Debugger helps in identifying and fixing issues in the model, and the PP-CPN extension is used for colored Petri nets [4]. Lastly, the Graphviz extension is used for creating and rendering graphs. These extensions collectively enhance the functionality of CPN Tools and simplify the task of net modeling and analysis.

The analysis of CPN models can be performed in several ways, including simulator reports [5], state space files [6], occurrence graphs, Graphviz, and state space queries. Simulator reports provide a detailed account of the system's behavior, while state space files offer a complete overview of all possible system states. Occurrence graphs represent the sequence of events in a system, and Graphviz provides a one-click solution for graph visualization. State space queries allow for the exploration of the state space in a structured manner.

Several verifications have been performed using state space graphs and Graphviz [7]. These verifications have proven the effectiveness of these tools in analyzing and understanding the behavior of systems modeled using CPN.

CPN Tools has its own method for generating Graphviz data. This data is produced according to the dot syntax [8]. However, in some instances, Graphviz fails to analyze the data generated by CPN Tools about the graph. This issue arises due to a syntax mismatch caused by the use of quotes for string tokens in CPN Tools. If the color set (colset) is defined to be only a string, one can rectify the issue from within CPN Tools itself; this is discussed in Sect. 3. This bug becomes evident when attempting to generate a graph with a token value string or a record/product with a string, resulting in a syntax mismatch with Graphviz.

This paper addresses this issue and proposes a solution in the form of a Java program with a Graphical User Interface (GUI) [9]. Swing, on the other hand, is a part of Java's foundation classes and is used for creating window-based applications. It provides a rich set of widgets and packages to make sophisticated GUI components for Java applications.

In our case, the Java program uses Swing to create a user-friendly interface [9]. The program takes the dot file generated by CPN Tools as input, refactors it, and corrects the syntax. This process enables error-free graph visualization, thereby addressing the bug effectively.

Additionally, to analyze the CPN page hierarchy in Hierarchical Colored Petri Nets (HCPN) [10], the tool generates a graph of the page hierarchy with multiple configurations. This feature allows users to visualize and understand the structure of their model better, aiding in more efficient analysis and debugging. The tool's ability to generate different configurations provides flexibility and caters to a variety of user needs.

2 Literature Surway

2.1 Graphviz

Formal verification, a cornerstone of ensuring the correctness and reliability of complex systems, relies heavily on meticulous analysis of system models. This analysis involves dissecting all possible configurations a system can adopt, a process often referred to as state-space exploration [11,12]. Graphviz [8,13] emerges as a powerful tool in this domain, empowering researchers to represent these configurations visually as nodes within a graph. Edges connecting these nodes symbolize transitions between states, triggered by specific events or actions. This graphical representation offers a multitude of benefits for formal verification.

2.2 CPN Tools and CPN IDE

A growing body of research has highlighted the effectiveness of Colored Petri Nets (CPNs) for system modeling and verification. However, CPN Tools, a widely used software platform for CPN analysis, has been shown to suffer from limitations in extensibility and its ability to adapt to evolving technologies [10]. In response to these shortcomings, researchers proposed CPN IDE as an extensible replacement for CPN Tools. This shift in focus towards a more adaptable platform for CPN modeling and analysis aligns with the need for future-proof software solutions.

The architecture of CPN IDE leverages Java Spring Boot technology for its backend [14], facilitating a robust and scalable foundation. This aligns with the current trends in software development that emphasize the utilization of web technologies. While CPN IDE currently focuses on core functionalities like model editing, it lacks some of the extensions present in CPN Tools, such as state space calculation. However, the modular design of CPN IDE, facilitated by the REST API communication between the frontend and backend, offers exciting possibilities for future development endeavors. This modularity paves the way for the potential integration of functionalities that were previously missing in CPN Tools, like state space analysis.

The work presented in this paper, which addresses a specific bug within CPN Tools related to Graphviz output, exemplifies the ongoing research efforts to refine and improve existing CPN software. By rectifying this software defect, the authors contribute to the overall enhancement of the CPN modeling ecosystem. Although CPN IDE may not currently encompass all functionalities of CPN Tools, its extensible architecture positions it as a promising platform for future development. The presented bug fix in CPN Tools, alongside the development of CPN IDE, exemplifies the continuous efforts undertaken by researchers to refine and improve CPN modeling tools.

3 Known Issues and Tool Limitations

3.1 Graphviz Features and Uses

One of the key strengths of CPN Tools lies in its seamless integration with Graphviz. CPN Tools provides functionalities to export the generated state space graph, the cornerstone of formal verification analysis, into a DOT file format readily understood by Graphviz. This integration empowers researchers to leverage the visualization capabilities of Graphviz to gain deeper insights into the behavior of their CPN models.

The export process is facilitated by an SML (Standard Markup Language) [15] code written in `OGtoGraphviz.sml` file. This code, typically located within the CPN Tools installation directory, acts as an extension and allows users to customize certain visual attributes of the exported graph. Functions like `OGSet.StringRepOptions'TI` and `OGSet.StringRepOptions'BE` enable modifications to token values, transitions, place names, and page names within the generated DOT file. Furthermore, CPN Tools offers granular control over the exported DOT file. Researchers can choose to export various elements, including individual nodes, node paths, arcs, arc paths, successors, predecessors, the entire state space, or even specific strongly connected components (SCCs) [16] within the model. This level of control allows for tailored visualizations that focus on specific aspects of the system behavior most relevant to the verification task at hand.

For instance, to generate a DOT file containing only a specific range of nodes (say, nodes 1, 5, and 8), the SML command `OGtoGraphviz.Export NodePath(outputpath, NodesInPath(1,8))` can be employed within CPN Tools. Similarly, the command `OGtoGraphviz.ExportStateSpace(outputpath)` can be used to export the entire state space as a DOT file. The sample of generated dot file is shown in Fig. 1(a).

By leveraging the combined strengths of CPN Tools and Graphviz, researchers gain a powerful and versatile toolkit for effectively modeling, analyzing, and verifying complex systems. CPN Tools provides the foundation for modeling system behavior and exploring the state space, while Graphviz offers the visual clarity needed to comprehend complex interactions and identify potential errors. This synergistic relationship empowers researchers to ensure the reliability and correctness of the systems they develop.

```
digraph cpn_tools_graph {
 N58238 -> N60750 [ label="A309815:58238-
>60750:InitPlatform'S_init_TT 2:
{station_name="Amausi"}" ];
 N60750 -> N63221 [ label="A323020:60750-
>63221:InitPlatform'Station_File 2:
{station_name="Amausi"}" ];
 N63221 -> N65686 [ label="A336129:63221-
>65686:InitPlatform'Initilize_Selector 2:
{t_no=1,train=("Metro1",80.0,16.4671577613,0.0,30,0)
}" ];
 N65686 -> N68104 [ label="A349054:65686-
>68104:InitPlatform'Station_Timetable_File 2:
{station_name="Amausi"}" ];
```

(a)

```
digraph cpn_tools_graph {
 N58238 -> N60750 [ label="A309815:58238-
>60750:InitPlatform'S_init_TT 2: {station_name=\
"Amausi\"}" ];
 N60750 -> N63221 [ label="A323020:60750-
>63221:InitPlatform'Station_File 2: {station_name=\
"Amausi\"}" ];
 N63221 -> N65686 [ label="A336129:63221-
>65686:InitPlatform'Initilize_Selector 2:
{t_no=1,train=(\"Metro1\
",80.0,16.4671577613,0.0,30,0)}" ];
 N65686 -> N68104 [ label="A349054:65686-
>68104:InitPlatform'Station_Timetable_File 2:
{station_name=\"Amausi\"}" ];
```

(b)

Fig. 1. CPN Tools (a) vs proposed work generated (b) dot file.

3.2 Known Issues

This section delves into a specific issue encountered during the generation of Graphviz DOT files from CPN Tools.

The challenge arises when dealing with colsets of type record, product or string (Fig. ??). Colsets are fundamental data structures in CPN Tools [16], and record colsets function similarly to structures in SML, enabling the storage of complex data with multiple attributes. As an example, consider a token value named `TRAIN` belonging to a record colset. This colset might hold attributes like: `train_number: Integer`, `train_name: String`, `distance_traveled: Real number`, `previous_station: String`, `next_station: String` [17]. Another colset, named `STATION`, might simply store string values representing station names.

The current mechanism for generating Graphviz DOT files in CPN Tools employs SML code to modify attributes before export. This approach works effectively for colsets containing only strings. For instance, quotation marks (`"`) within station names can be replaced with escaped double quotes (`\"`) using SML functions [18], ensuring proper representation in the DOT file. Alternatively, the easier way to get rid out of it is to remove the token values, for that we can use these functions:

```
OGSet.StringRepOptions'Arc(fn(ar,so,de)=>"");
OGSet.StringRepOptions'TI(fn(pg,tr,i)=>"");
OGSet.StringRepOptions'PI(fn(pg,pl,i)=>"");
OGSet.StringRepOptions'BE(fn(ti,b)=>"");
OGSet.StringRepOptions'Mark(fn(PI,mark)=>"");
```

However, adapting this method for record colsets proves more challenging. Modifying nested structures like records within SML code becomes cumbersome and lacks flexibility. A worst-case scenario would involve writing custom code to handle specific record types, resulting in code that only functions for a predefined record structure. Furthermore, any changes made to the record colset definition would necessitate corresponding modifications to the user-defined function, creating a maintenance burden.

3.3 Limitations

A significant limitation of CPN Tools lies in its inability to directly export a DOT file representing the page hierarchy (HCPN) for visualization purposes. While CPN Tools offers subpage generation [19], this method might not provide the desired level of control for visualizing the complete network structure.

To address these issues and limitations, this study proposes the development of a post-processing tool specifically designed to tackle the issue of syntax missmatch of colset representation within exported DOT files. This tool would operate after the initial DOT file generation by CPN Tools.

4 Proposed Work

In this research, the primary focus is on addressing a critical bug in CPN Tools while generating Graphviz files. The bug surfaces when trying to generate a graph with token value strings or record/product with strings, resulting in a syntax mismatch with Graphviz. This mismatch leads to errors in visualization, hindering the effective use of CPN Tools for graph generation and analysis. To rectify this issue, a Java program with a graphical user interface (GUI) has been developed. This program takes the CPN Tools generated dot file as input, refactors it, and fixes the error, enabling successful visualization of the graph.

4.1 Software Development Life Cycle

The Software Development Life Cycle (SDLC) [14,20] is a framework that defines the steps involved in the development of software at each phase, from inception to retirement. It's a roadmap for developers to create high-quality, reliable, and robust software. The SDLC consists of several stages, including requirement gathering, system analysis, system design, coding, testing, deployment, and maintenance. Each stage is essential and plays a pivotal role in the development of a software product. The SDLC not only ensures the production of high-quality software, but also improves the overall development process by making it more efficient and effective. Despite variations in the methodologies and processes, the fundamental principle of SDLC remains the same: to deliver high-quality software in a systematic and disciplined manner.

4.2 SDLC Model

developing a Java GUI program to address the syntax issue in DOT files generated by CPN Tools, a well-defined Software Development Life Cycle (SDLC) model is crucial for ensuring a successful project outcome. Here, we propose utilizing a Waterfall model [21] with iterative elements for this specific project.

The Waterfall model offers a structured approach, ideal for projects with well-defined requirements and a clear understanding of the desired functionalities. It follows a sequential flow, progressing through distinct stages:

Requirement analysis involves meticulously gathering and analyzing user requirements, including understanding syntax issues in DOT files, record colset types, and desired functionalities of the Java GUI tool. User interviews, surveys, and workshops aid in comprehensive requirement gathering. Based on established requirements, the system design phase focuses on defining the Java GUI tool's architecture, including modular components, user interface elements, and data structures for handling DOT file information. Java libraries or frameworks for GUI development and data processing are selected [22]. Leveraging the system design document, the development stage involves coding the Java application, building the GUI, implementing core functionalities for processing and correcting DOT file syntax, and integrating error handling mechanisms. Following this, the Java GUI tool can be deployed to users through download or integration into CPN Tools workflows. Ongoing maintenance addresses bugs, issues, and incorporates new features based on user feedback or evolving requirements.

While the Waterfall model provides a solid foundation, incorporating iterative elements can enhance the development process. After each stage (excluding deployment), conducting internal reviews or consultations with potential users can provide valuable feedback for refinement. This iterative approach allows for early identification and correction of potential issues, leading to a more robust and user-friendly final product.

By adhering to this structured yet adaptable SDLC model, the development of the Java GUI tool for fixing CPN Tools DOT file syntax can be effectively managed, resulting in a program that empowers researchers and developers working in the field of formal verification.

4.3 Feasibility Study

During the development of this project, a comprehensive feasibility study was undertaken to assess various factors that could impact the project's success and achieve the required user workflow (Fig. 3) smoothly. This study encompassed the following key aspects:

4.4 Requirement Analysis

The application provides a user interface for uploading CPN Tools-generated DOT files and importing CPN Tools saved XML files. Users select processing options, including fixing colset names or data types and choosing HCPN graph-related functionalities. Upon "Generate" button interaction, the application extracts relevant information from the imported file (CPU, memory, disk usage), displays it in a pop-up window, corrects DOT file syntax, generates a new corrected DOT file visualized in Fig. 1(b), and terminates the pop-up upon successful generation. The application offers saving the generated file and provides visualization capabilities. The application has a user-friendly Java Swing GUI, functions across various operating systems with compatible Java installations, exhibits acceptable performance, and adheres to error handling best practices with informative messages.

4.5 System Design

The application's architecture includes modular components for UI interactions, file processing, and multithreading, along with data structures for storing imported and processed data. Relevant Java libraries or frameworks are selected for GUI development, file processing, and optional graph visualization.

4.6 Development

The Java application is implemented adhering to system design specifications. The GUI is developed using Java Swing components [23], and functionalities for file import, processing (including colset token syntax correction), and corrected DOT file (visualized in Fig. 1 (b)) generation are programmed using Algorithm 1. Multi-threading for displaying CPU information during file processing is integrated, and error handling mechanisms for various scenarios are implemented.

4.7 Deployment and Maintenance

A deployment strategy is defined based on the target user base, with documentation including installation instructions and a user guide. A maintenance plan addresses post-deployment bugs and issues. User feedback is considered for future enhancements.

5 Implementation

5.1 Technologies Used

This section details the specific technologies chosen to implement the proposed Java GUI tool for addressing issues in CPN Tools-generated DOT files.

JFormDesigner. JFormDesigner is a valuable plugin for the IntelliJ IDEA integrated development environment (IDE) [24]. It facilitates the creation of user interfaces (UIs) using a visual drag-and-drop approach. This eliminates the need for manual Java code writing for UI component layout and structure. JFormDesigner empowers researchers to:

1. Design the application's UI elements visually, including buttons, text fields, and panels.

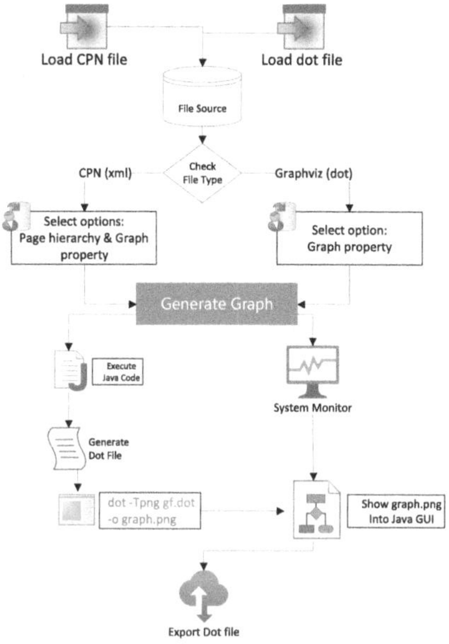

Fig. 2. Java program workflow.

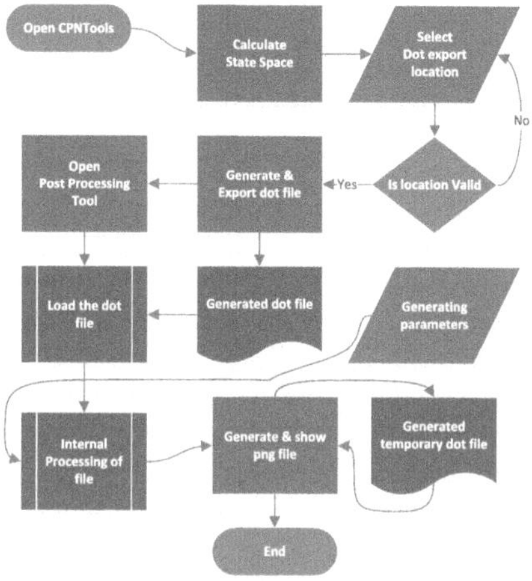

Fig. 3. Proposed user workflow.

2. Assign custom names to UI components for easy reference within the Java code.
3. Implement methods and callbacks associated with user interactions with UI elements. This allows researchers to program the desired behavior of the application in response to user actions.
4. By leveraging JFormDesigner, the development process becomes more efficient and visually intuitive, particularly for researchers who might not have extensive experience with low-level UI component manipulation in Java.

Java. Java [14,24] serves as the core programming language for developing the proposed Java GUI tool. Java's object-oriented nature and extensive libraries make it a well-suited choice for this project. Key advantages of using Java include: Java's platform independence, enabled by JRE compatibility, ensures the application's usability across different operating systems without requiring platform-specific modifications. The rich Java library ecosystem, including Java Swing for GUI development, Java concurrency libraries for multithreading, and potential XML processing libraries like JAXB or DOM, provides essential tools for the project. The large and active Java developer community offers valuable resources and support throughout the development process.

CPN Tools. Refer to the previously provided Sect. 2 for a detailed description of this tool.

Graphviz. Refer to the previously provided Sect. 2 for a detailed description of this tool.

The combination of these technologies provides a robust foundation for developing the proposed Java GUI tool. JFormDesigner streamlines UI creation, Java offers a versatile and platform-independent development environment, and CPN Tools, Graphviz serve as the core functionalities upon which the tool builds to address the identified record colset syntax issue.

5.2 Hardware and Software Requirements

System Specifications. The following minimum hardware specifications [25] are recommended for smooth operation of the application:

- Operating System: Windows 7 (64-bit) or later, macOS 10.10 (Yosemite) or later, Linux (x86-64 distribution)
- Processor: Intel Core i3 or equivalent AMD processor (multi-core processor recommended for improved performance)
- RAM: 4 GB RAM (8 GB RAM or more recommended for larger or more complex CPN models)
- Hard Disk Space: 500 MB free disk space
- Display: Resolution of 1024 × 768 pixels or higher

These are minimum recommendations. The actual hardware requirements might vary depending on the complexity of the CPN models being processed and the user's multitasking habits.

Software Interface. The application is designed as a standalone Java GUI application. To run the tool, users will need the following software:

- Java Runtime Environment (JRE): Java version 8 or later must be installed on the user's system. Most modern operating systems do not include JRE by default [26]. Users can download and install the latest JRE from the official Java website.
- Graphviz: The application leverages Graphviz for visualizing the processed DOT files. Users need to have Graphviz installed on their system and ensure the dot executable is accessible through the system's environment variables.

Algorithm 1. Refactor String and Remove Unwanted Characters

```
Require: Source dot file location (sourceDotFileLocation)
    Destination dot file location (destDotFileLocation)
Ensure: Exit code
labelStart ← "label=\""
labelEnd ← "]"
doubleQuote ← "\""
singleQuote ← "\\\\\""
lineSeparator ← System.lineSeparator()
writerFlag ← new AtomicBoolean(true)
for each line ∈ Files.lines(Paths.get(sourceDotFileLocation)) do
    startIndex ← line.indexOf(labelStart)
    endIndex ← line.indexOf(labelEnd)
    if startIndex ≠ −1 and endIndex ≠ −1 then
        textToReplace ← line.substring(startIndex+length(labelStart), endIndex − 2)
        newText ← textToReplace.replaceAll(doubleQuote, singleQuote)
        newLine ← line.substring(0, startIndex+length(labelStart)) + newText + line.substring(endIndex − 2)
        writer.write(newLine lineSeparator)
    else if startIndex == −1 and endIndex == −1 and writerFlag.get() then
        newLine ← line.replaceAll(doubleQuote, singleQuote)
        writer.write(newLine lineSeparator)
    else if startIndex ≠ −1 and endIndex == −1 then
        textToReplace ← line.substring(startIndex+length(labelStart))
        newLine ← textToReplace.replaceAll(doubleQuote, singleQuote)
        newLine ← newLine.substring(0, newLine.indexOf("@")).trim()
        writerFlag.set(false)
        writer.write(line.substring(0, startIndex+length(labelStart))+newLine)
    else if startIndex == −1 and endIndex ≠ −1 then
        textToReplace ← line.substring(0, endIndex − 1)
        newLine ← textToReplace.replaceAll(doubleQuote, singleQuote)
        writerFlag.set(true)
        writer.write(newLine + line.substring(endIndex − 1) + lineSeparator)
    else
        writer.write(line + lineSeparator)
    end if
end for
writer.close()
command ← "dot -Tpng "+destDotFileLocation+" -o hierarchy.png"
process ← Runtime.getRuntime().exec(command)
exitCode ← process.waitFor()
return exitCode
```

5.3 Methodology

The project progresses through several distinct phases. Here's an overview of the key stages involved.

1. Environment Setup:
   - Install Graphviz and configure the system environment variables to ensure the dot executable is accessible.
   - Install Java Runtime Environment (JRE) version 8 or later.
   - Install an Integrated Development Environment (IDE) of your choice (e.g., IntelliJ IDEA).
2. Project Initialization:
   - Launch the chosen IDE.
   - Start a new Java project.
     - Define a descriptive package name for your project.
     - Specify the desired project directory location.
3. IDE Plugin Installation:
   - Install the JFormDesigner plugin within your IDE. This plugin will facilitate visual UI design.
4. User Interface (UI) Design:
   - Utilize JFormDesigner to create the application's user interface visually. This includes designing buttons, text fields, panels, and other UI elements.
5. Code Implementation:
   - Implement Java classes and methods for functionalities related to:
     - CPN page hierarchy processing using xml file
     - DOT file processing, including correcting colset syntax using Algorithm 1.
   - Map the UI elements to the corresponding methods in your Java code. This ensures user interactions with the UI trigger the appropriate functionalities within your program.
6. Testing and Deployment:
   - Run the project within the IDE to test its functionality.

The workflow of resolving the issue internally is shown in Fig. 2. It leverages a Java-based post-processing tool to ensure accurate visualization of colset data.

6 Results

In this section, the results obtained by the proposed solution are discussed. Specifically, the focus is on the generation of a page hierarchy graph for a Colored Petri Net (CPN) model. To illustrate this, a sample CPN model representing the Indian Railway system in a moving block signaling context was created. The hierarchy of the system comprises several levels: at the topmost level, the Indian Railway Management Service (IRMS) is divided into zones, each containing a station section and a block section. Various railway operations, such as Train Dispatchers, Meet Conflict handlers, Train Receivers, and Moving Authorities, can interact with these zones and/or stations.

After modeling the CPN, the page hierarchy is visualized in Fig. 4. Notably, the Moving Authority is connected via the Zone, Station Section, and Block Section. Additionally, the Pass Conflict has three sub-pages, exclusively present in the station section. Interestingly, there are three zone pages: "Zone (1)", "Zone

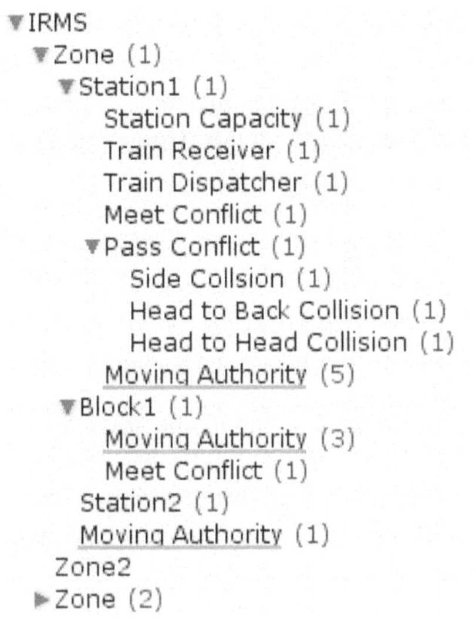

Fig. 4. Page hierarchy of testing CPN model

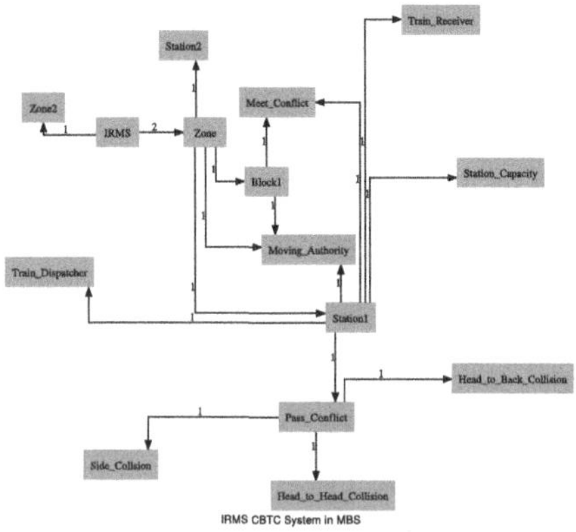

Fig. 5. Generated Graph of CPN model

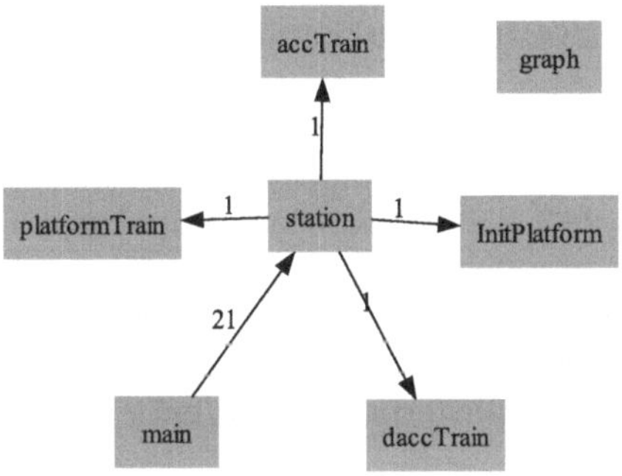

Fig. 6. Page hierarchy with custom arc length

Fig. 7. Part of state space graph generated by GUI tool

(2)", and "Zone2." Remarkably, "Zone (2)" is a replica of "Zone (1)", so any changes made in either of these will be reflected in the other. This behavior is indicated by the numbers on the edges, representing the number of replicas for each page.

The Fig. 5 illustrates the generated page hierarchy of a selected CPN file. Here's how the process works:

- Source Selection: The user selects the CPN Tools file from their disk using the provided graphical user interface (GUI).

- Configuration: Necessary options are chosen, configuring the graph generation process.
- Graph Generation: The user clicks the "Generate Graph" button.
- System Information Popup: After initiating the graph generation, a popup appears, displaying system information such as CPU and RAM usage.
- RAM Considerations: It's important to note that the amount of RAM required depends on the size of nodes and edges in the CPN model. Larger input files demand more RAM for graph generation. If insufficient RAM is available, the program may crash. Unfortunately, this limitation is inherent to the dot processing, and the proposed solution has no control over it

To address the objective of correcting string representations in the state space graphviz dot file, a CPN model with product and record color sets was created. This ensured that both types of color sets were included in the generated state space dot file. The SML code outlined in Sect. 3 was then used to generate the dot file, which initially contained syntax errors.

The GUI tool was utilized to import the state space dot file and customize graph options. Internally, the tool corrected and refactored the syntax code. Generating the graph resulted in Fig. 7, showcasing a graph with five nodes displaying page information, transaction names, and variable values at each state.

Variable 'a' held the value $(4, \text{``}DOCKER'')$, representing a product type variable. Variable '$x$' had the value $\text{``}DOCKER''$, indicating a string type variable. Variable 'c' was $[id = 4, name = \text{``}DOCKER'']$, representing a record type variable in SML.

No errors were encountered during graph generation. To validate the accuracy of the generated graph, a state space graph was manually obtained from CPN Tools (Fig. 8. Node numbers (1, 5, 31, 112, 330) from the generated graph were identified and used to plot nodes on a new page within CPN Tools' simulator. Successor nodes of nodes 1, 5, 21, and 112 were then generated to examine connectivity between them. The GUI-generated nodes and the CPN Tools-generated state space graph (Fig. 8) exhibited identical connectivity patterns for the selected five nodes.

Visual representations of the software are provided in Fig. 9 and 10. Figure 9 showcases the corrected dot file processing generated by CPN Tools. In some cases, CPN Tools might generate dot files with errors or inconsistencies that hinder proper visualization. This program addresses this issue by employing a correction mechanism that ensures the generated dot file is compatible with standard graph visualization tools. In contrast, Fig. 10 depicts the page hierarchy generation derived from a saved CPN Tools file (XML format). This functionality allows users to leverage existing CPN models stored in the XML format and visualize their page hierarchy structure within the program.

Figure 6, 11 illustrate the page hierarchy with varying user-defined configurations within the software tool. Figure 10. This configuration option allows users to control the visual complexity of the graph, making it easier to analyze for models with a large number of transitions or places. Conversely, Fig. 11 exhibits non-orthogonal (non-straight) arcs. This configuration might be useful

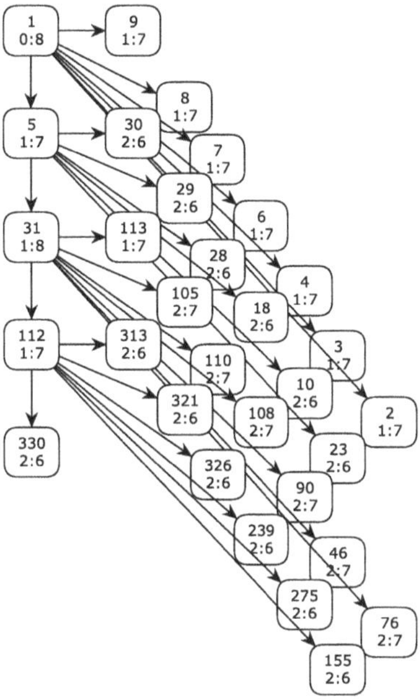

Fig. 8. State Space Graph of 5 selected nodes

Fig. 9. Correct version of dot file processing

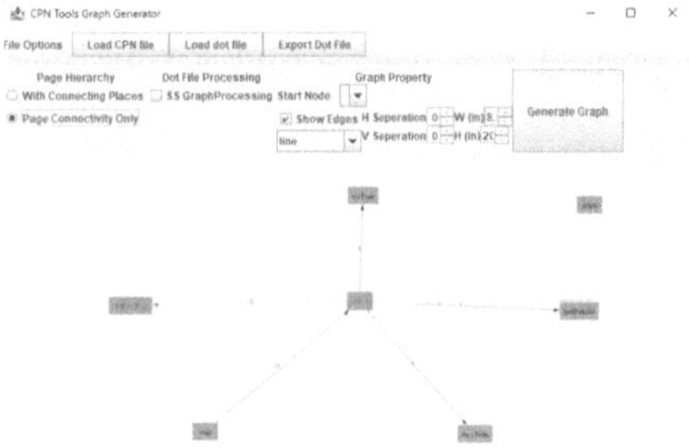

Fig. 10. Page hierarchy graph generated by the tool

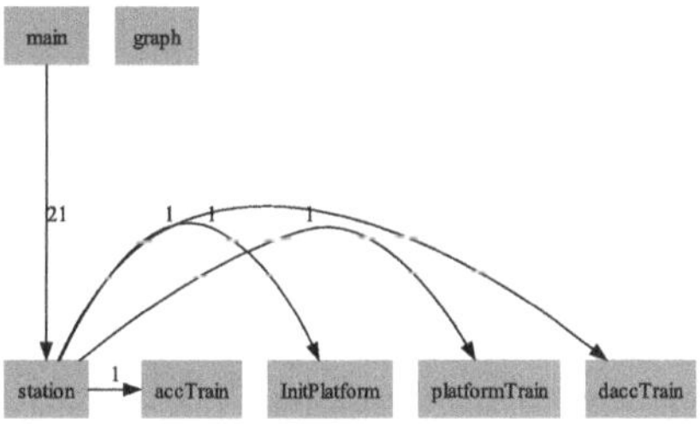

Fig. 11. Page hierarchy with different arc type

for specific visualization needs, such as emphasizing certain connections within the CPN model or fitting the graph within a limited space. The numbers displayed on the arcs represent the number of associated sub-pages. In Fig. 11, the main page has 21 station sub-pages, highlighting the versatility of graph types producible using this tool. This functionality enables users to visualize complex CPN models with hierarchical structures, where each arc can represent a sub-page containing additional details.

7 Conclusion

A satisfactory outcome is achieved based on the project analysis. The program offers significant advantages for analyzing state space graphs. It overcomes a common challenge in CPN analysis by preserving attribute values for the net. This ensures data integrity and eliminates the need for manual re-entry or potential errors during the analysis process. The Java Swing GUI, with its user-friendly interface, prioritizes user experience. It minimizes the learning curve for new users by allowing them to focus solely on the front-end interaction, where they can input their CPN model and desired functionalities. This program caters to both CPN users and novices by enabling them to generate dot files for their specific needs. Furthermore, the program boasts the ability to generate hierarchical colored petri nets (HCPNs) as graphs, with customizable parameters for enhanced net representation. Users can tailor the visual representation of their CPN model to better suit their analysis goals or communication needs. For instance, they can adjust node sizes, or arc styles to emphasize specific aspects of the net. The most user-friendly approach would be to fix this bug directly within CPN Tools itself. This would eliminate the need for an external tool and streamline the workflow for CPN users. Efforts could be directed towards modifying the CPN Tools codebase to handle string values correctly during dot file generation, ensuring compatibility without requiring additional software.

Compliance with Ethical Standards

Funding. We gratefully acknowledge the support of the Science and Engineering Research Board, India under Grant No. EEQ/2022/000844 and the Institutions of Eminence (IoE) Scheme of Banaras Hindu University, Varanasi under Dev. Scheme No. 6031.

Ethical Approval. This article does not contain any studies with human participants or animals performed by any of the authors.

Author Contributions. Vikas Kumar Patel: Conceptualization, Methodology, Software, Validation, Investigation, Writing - Original Draft, Visualization. **Anshul Verma:** Methodology, Resources, Writing - Original Draft, Supervision, Project administration, Funding acquisition. **Pradeepika Verma:** Investigation, Visualization, Writing - Review and Editing.

Data Availability. This manuscript does not involve the use of any new data. All information presented is based on previously published research articles and theoretical considerations. Therefore, there are no data to be made available.

Conflict of Interest. The authors declare that they have no conflict of interest.

References

1. Jensen, K.: Coloured petri nets and the invariant-method. Theoret. Comput. Sci. **14**(3), 317–336 (1981)

2. Alur, R., Holzmann, G., Peled, D.: An analyzer for message sequence charts. Softw.-Concepts Tools **17**(2), 70–77 (1996)
3. Westergaard, M., Slaats, T.: Cpn tools 4: a process modeling tool combining declarative and imperative paradigms. In: BPM Demo Sessions 2013 (Co-located with 11th International Conference on Business Process Management, BPM2013, Beijing, China, August 26-30, 2013), pp. 1–5, CEUR-WS. org, 2013
4. Westergaard, M.: Cpn tools 4: Multi-formalism and extensibility. In: International Conference on Applications and Theory of Petri Nets and Concurrency, pp. 400–409. Springer (2013)
5. Verma, A., Pattanaik, K.: Mobile agent based train control system for mitigating meet conflict at turnout. Procedia Comput. Sci. **32**, 317–324 (2014)
6. Verma, A., Pattanaik, K., Goel, P.: Mobile agent based cbtc system with moving block signalling for Indian railways. In: Proceedings of the 2nd International Conference on Railway Technology: Research, Development and Maintenance, Ajaccio, Corsica, paper, vol. 278, pp. 8–11, 2014
7. Rao, M., Ramakrishnan, S., Dagli, C.: Modeling and simulation of net centric system of systems using systems modeling language and colored petri-nets: A demonstration using the global earth observation system of systems. Syst. Eng. **11**(3), 203–220 (2008)
8. Gansner, E.R.: Drawing graphs with graphviz. AT&T Bell Laboratories, Murray, Technical report (2009)
9. Loy, M., Eckstein, R., Wood, D., Elliott, J., Cole, B.: Java swing. O'Reilly Media, Inc. (2002)
10. Verma, A., Pattanaik, K.: Multi-agent communication based train control system for Indian railways: the structural design. J. Softw. **10**(3), 250–259 (2015)
11. Verma, A., Pattanaik, K.: Multi-agent communication-based train control system for Indian railways: the behavioural analysis. J. Modern Transp. **23**, 272–286 (2015)
12. Ghosh, C., Verma, A., Verma, P.: Real time fault detection in railway tracks using fast fourier transformation and discrete wavelet transformation. Int. J. Inf. Technol. **14**(1), 31–40 (2022)
13. Ellson, J., Gansner, E.R., Koutsofios, E., North, S.C., Woodhull, G.: Graphviz and dynagraph-static and dynamic graph drawing tools. Graph drawing software, pp. 127–148 (2004)
14. Patel, V.K., Verma, A., Verma, P.: A covid-19 vaccine notifier android app "cowin mitra". In: Research Advances in Network Technologies, pp. 223–243, CRC Press (2023)
15. Tsai, Y.: Model integration using sml. DECISION SUPPORT SYSTEMS, vol. 22, pp. 355–377, April 1998
16. Jensen, K., Kristensen, L.M., Wells, L.: Coloured petri nets and cpn tools for modelling and validation of concurrent systems. Int. J. Softw. Tools Technol. Transfer **9**, 213–254 (2007)
17. Patel, V.K., Verma, A., Verma, P.: Comparative study of train operation simulators. J. Sci. Res. **66**(4) (2023)
18. Geoffrion, A.M.: The sml language for structured modeling: Levels 1 and 2. Oper. Res. **40**(1), 38–57 (1992)
19. Jensen, K.: An introduction to the practical use of coloured petri nets. Lectures Petri Nets II: Appl. Adv. Petri Nets **3**, 237–292 (1998)
20. Ragunath, P., Velmourougan, S., Davachelvan, P., Kayalvizhi, S., Ravimohan, R.: Evolving a new model (sdlc model-2010) for software development life cycle (sdlc). Int. J. Comput. Sci. Network Secur. **10**(1), 112–119 (2010)

21. Adenowo, A.A., Adenowo, B.A.: Software engineering methodologies: a review of the waterfall model and object-oriented approach. Int. J. Sci. Eng. Res. **4**(7), 427–434 (2013)
22. Walrath, K.: The JFC Swing tutorial: a guide to constructing GUIs. Addison-Wesley Professional (2004)
23. Zukowski, J.: The definitive guide to Java Swing. Apress (2006)
24. Yadav, A.: Amazon Cloud Computing with Java. Lulu. com (2010)
25. Pill, I., Semprini, S., Cavada, R., Roveri, M., Bloem, R., Cimatti, A.: Formal analysis of hardware requirements. In: Proceedings of the 43rd annual Design Automation Conference, pp. 821–826 (2006)
26. Horstmann, C.S., Cornell, G.: Core Java: Advanced Features, vol. 2. Prentice Hall (2008)

What Does "Remote Driving" Mean for People: A Futurespective Design Approach

Ysanne Yeo[1(✉)] and Federico Trucchia[2]

[1] Graduate School of System Design and Management, Keio University, Tokyo, Japan
ysanneyeo@keio.jp

[2] Institute of Industrial Science, University of Tokyo, Tokyo, Japan

1 Introduction

The mobility industry in recent years has seen tumultuous investment and drawbacks of autonomous vehicle (AV) development, with global automakers including Apple and Samsung Electronics hitting the brakes in research on self-driving cars. The availability of self-driving vehicles promises people the benefits of increased access to transport, the comfort of depending on machines for tasks like driving, and the potential optimisation of logistics and resources [1–3]. However, AV research is stymied for a range of reasons, from the need for immense computing capabilities to make reliable decisions in real-time, to obstacles in gaining public trust, all of which amount to high costs to pursue. The factors underlying technological readiness, social acceptance, and regulatory approval pose speed bumps (or roadblocks) in this direction [4], requiring players in the field to explore alternative pathways to advance vehicle technology.

In this context, remote driving could provide an intermediary solution [5, 6]. Although a comprehensive standard for what constitutes remote driving is yet to be established, SAE International defines remote driving (RD) as: "Real-time performance of part or all of the dynamic driving task (DDT) and/or DDT fallback (including real-time braking, steering, acceleration, and transmission shift) by a remote driver" [7]. Whether fully autonomous vehicle or not, remote driving is generally agreed to involve the operation of a vehicle carried out by a remote location person [8, 9], enabling a 'driverless' passenger experience. In other words, teleoperation is the default mode of remote driving systems; and it is precisely this situation where the operator is remotely located (or beyond line-of-sight) that presents the challenge of ensuring that the operator performs driving reliably and safely.

Furthermore, we expect the drive mission and performance evaluation to become more complex with new driver and passenger roles. As we speculate on the scenario where remote driving is offered as a service in privately owned vehicles, we foresee that the drive mission will extend beyond getting the vehicle from point A to point B. Instead, it would also depend on emerging interactions between the remote driver and the passenger in the car, prompting us to reconsider the fundamental act of driving itself. Based on our review, however, remote driving is often considered in relation to technological feasibility whilst people-centred perspectives are addressed to a limited degree, mainly from the driver's viewpoint. In response to this gap in existing studies,

D. Singh et al. (Eds.): IHCI 2024, LNCS 15557, pp. 241–253, 2025.
https://doi.org/10.1007/978-3-031-88705-5_20

our design-led research aims to bring forth novel human interaction needs for remote driving from both the driver and passenger-ends of the service.

To achieve this objective, we conducted a future respective design workshop where participants embodied the roles of drivers and passengers to create tangible artefacts that articulate the necessary interactions for them to embrace the service. As a result, we managed to glean insights into people's values that would help shape the design of interfaces for remote driving. We distil our findings into three concept directions covering themes of establishing preferences and protocols, conveying intentions, and providing a sense of presence. These directions offer opportunities for further research and development of remotely-driven services, even beyond mobility.

2 Human Needs for Remote Driving

Having and maintaining *situational awareness* is paramount when it comes to driving. Endsley (1988) [10] defined situational awareness as the "perception of the elements in the environment within a volume of time and space, the comprehension of their meaning, and the projection of their status in the near future". Based on this premise, if the driver feels a personal association with the task – as if they were physically embedded within the real-time road conditions – they would presumably achieve more effective driving performance through the control interface [11]. Decrements in driving performance, however, are to be expected, in part due to distractions that will occur in the actual environment of the remote driver [12]. As such, prior researchers have claimed that a lack of *telepresence* is one of the key challenges of remote driving [13]. Sheridan (1992) [16] defined telepresence in the case of remote operation as "the experience of being there". Consequently, enhancing telepresence during remote driving should result in an improvement in the driver's feeling of being present on the roads.

Since remote driving is dependent on an internet communication link, network delay would hinder telepresence and has been identified as a major encumbrance [14]. While network latency continues to improve with advancements in technology such as the impending 6G, however, we argue that technocentric solutions shall not and cannot determine the interactions that will emerge between the remote driver and passenger. The nature of human behaviour is such that it is the synthesis of interactions and subjective meanings, which cannot be fully anticipated within the original physical, perceptual, and cognitive objectives of the design [17]. Based on this knowledge, we hypothesise that there will be more *edge cases* when it comes to remote driving where the driver and passenger are beyond each other's line-of-sight. Edge cases vary significantly but are colloquially seen as 'unknowable unknowns' – unpredictable, novel occurrences which fall outside the parameters defined by classifiers to recognise perceptual information [15]. In the edge case where network connection is lost, for example, how might we be sure that both driver and passenger perform the necessary actions to mitigate safety risks? Even if the connection is restored, how would they (re)establish trust in one another?

This unprecedented remote driver-passenger dynamic raises many questions about what will become of the roles and expectations of drivers and passengers, and by extension the human interface requirements for remote driving. Therefore, amidst a growing body of technical research, we have identified the need to investigate what "remote driving" means for people – not bounded by technological affordances but based on observations of emergent interactions. Since a person's behaviour is subject to perceived information and conveys the synthesis of thinking, knowing and doing [18], observing the organic interactions between drivers and passengers would elicit the perceptions that people implicitly have towards remote driving. This discovery would then help us to identify the human-to-human interaction requirements necessary to be considered in the design of remote driving interfaces.

3 Method

At this juncture of our research, we opted to conduct a futurespective design workshop to develop the concept of remote driving from a value-based perspective as compared to the technocentric, viability-centered approach. Specifically, we sought to evoke the implicit narratives that people have towards the notion of having a stranger drive their car, to explicate their values and concerns. This workshop is an extension of an initial phase of research conducted on the impact of 6G on future mobility in Japan where remote driving surfaced as a plausible scenario [19]. The workshop methodology integrates diverse perspectives and leverages design fiction to discover the unknown unknowns in a speculative project. We invited participants from a range of backgrounds from outside the field of mobility to prevent biasing solutions with feasibility concerns. We had a total of 20 participants of ages 20 to 30 who were mostly graduate students and researchers residing in Tokyo, Japan. Participants were organised into groups of four, with members who do and do not own a driver's license. The reason for the mix is to ensure a basic understanding of the task of driving across the groups.

Our design brief for participants was to identify and represent the features, attributes, characteristics, and behaviours of a remotely driven car and its requisite subsystems by developing the vehicle's Quick Start Guide (QSG). We referenced the idea of creating a QSG from a recent workshop conducted by Near Future Laboratory [20]. For the purpose of our workshop, we explained to participants that a QSG is meant to function like a user manual; it is expected to be used to on board new users to a system and to help them with troubleshooting in a self-explanatory manner. The format of a QSG is a way to surface questions on the human implications of remote driving and to describe ideas in a coherent narrative to be communicated to others. To produce the QSG, our workshop activities geared towards the making of interface prototypes to test the interactions featured in a QSG. The creation of physical artefacts then benefits our retrospective workshop review where we perform deeper synthesis and analysis of our findings and outcomes.

3.1 Activities and Artefacts

To immerse participants in the remote driving future scenario, we curated a sequence of activities that would encourage them to embody the roles of a driver or passenger. By immersion, we mean getting a feel of the tasks they might do first and do often with a remotely driven car, so that their eventual narratives would sound as convincing as possible within the limited amount of time. The activities begin with the identification of the necessary interfaces for remote driving, to illustration of the interaction workflow using those interfaces, and to prototyping and testing of their interface proposals (see Table 1).

Table 1. Table of activities in sequence, their overarching tasks, and expected outputs.

Activity	Task	Output
1) Identify Interfaces	Generate as many ideas as possible to address the touchpoints (i.e., red diamonds) in the user journey map	Multiple ideas of interfaces for drivers or passengers to meet their interaction needs
2) Illustrate Interaction	Translate and add detail to your ideas as an interaction sequence in the provided QSG template	Consolidated interfaces as a cohesive use case scenario in a QSG template
3) Prototyping and Testing	Build physical prototypes to showcase the interfaces in your QSG and how they would work	Rapid interactive prototypes to be used for peer testing of expected interactions

We provided two pieces of contextual information early into the workshop to guide the participants. Each group was given an identical user journey map (printed in A0), which outlines the human-to-human touchpoints required for remote driving (see Fig. 1). The user journey map is used to facilitate ideation by providing initial prompts and parameters to consider. In addition, we pre-assigned each group to represent a specific character of either a remote driver or passenger. A sample character would be a remote driver who is "a professional gamer", or a passenger who is "tech-illiterate". We have observed from our previous endeavours that such character roles invite more creativity to pursue ideas beyond conventional screen-based interactions. A series of photos documenting our workshop activities is shown in Figs. 2, 3, 4 and 5.

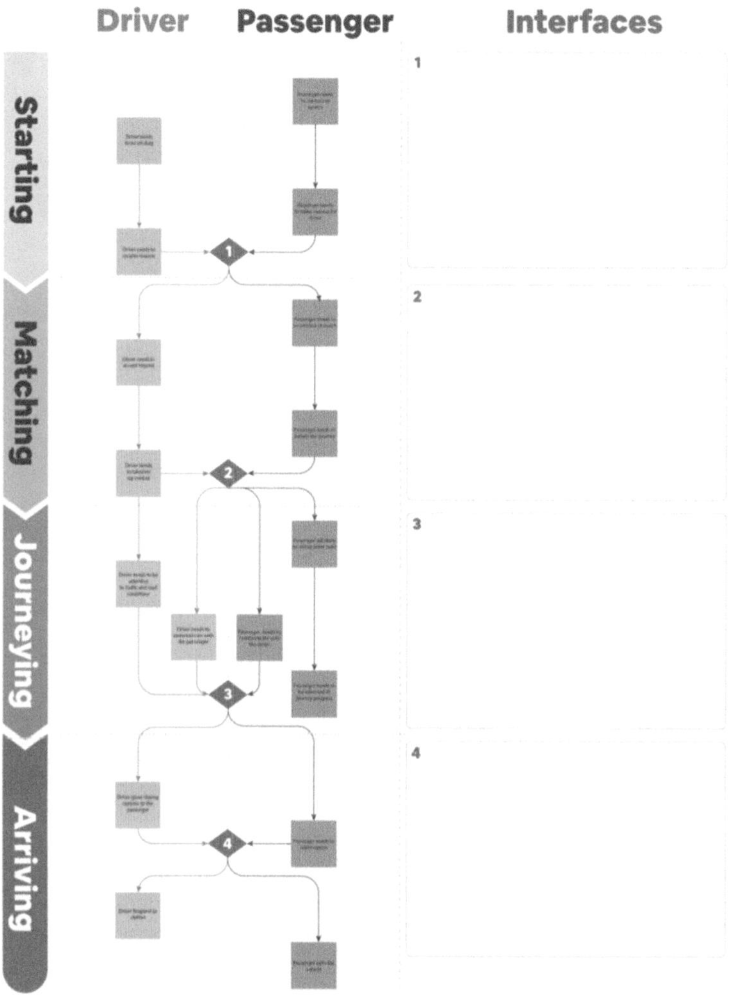

Fig. 1. User Journey Map.

a b

Fig. 2. Teams generating ideas for interfaces in Activity 1) Identify Interfaces.

a b

Fig. 3. Teams making their QSG in Activity 2) Illustrate Interaction.

Each prototype was tested by peer volunteers from other teams and facilitators. Of the five teams, two teams created their interpretation of the driver's interface, two teams focused on the passenger's interface, and one team attempted to tackle both ends of the service. The teams were paired up to roleplay the interaction between remote drivers and passengers using their prototypes. Testers were tasked to read the QSG and interact with the prototypes according to the instructions and/or illustrations as shown in Fig. 6a and b.

a b

Fig. 4. Teams building their prototype interfaces based on their QSG in Activity 3) Prototyping and Testing.

a b

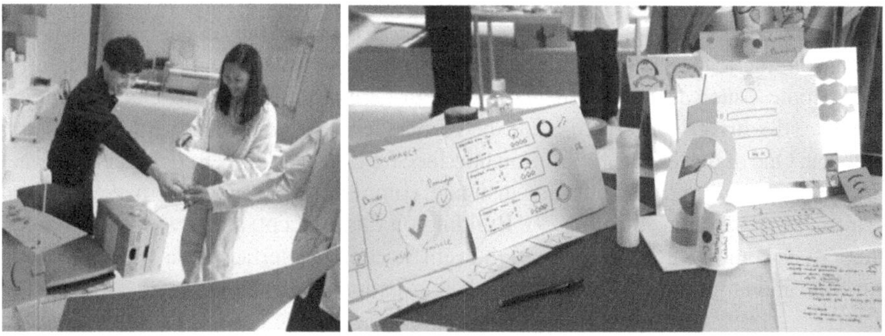

Fig. 5. Teams setting up their prototypes for peer testing.

a b

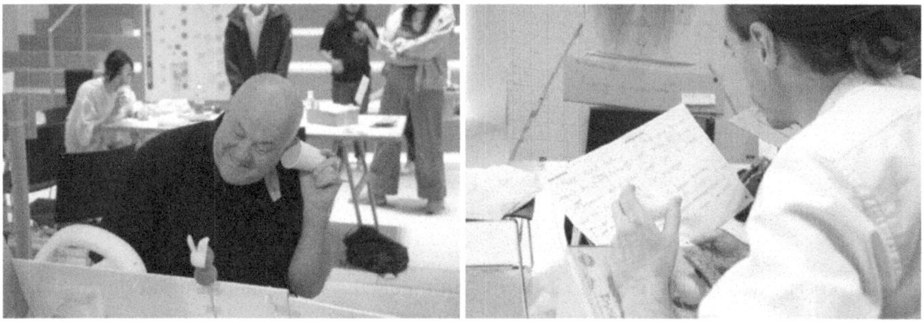

Fig. 6. Testing of interfaces and interactions using the teams' QSG.

4 Results

From the workshop, we derived novel concepts for remote driving vehicle interfaces and surfaced some of people's implicit needs for communication and trust-forming.

Of all the ideas put forth by participants, we found three ideas in particular that stood out in terms of originality and potential meaningfulness to users of the service. The first concept is to augment the steering wheel of remote-driving cars with virtual hands on the wheel that would light up as an indicator that the driver is present and in control (see Fig. 7). The second idea is in response to a need for transparency when it comes to vehicle routing. Due to situational road conditions, participants believe that passengers would want to know why their driver may make a turn or re-route their car along the journey. In the classic scenario where the driver and passenger are physically in the same vehicle, intentions may be understood by sensing or direct communication. In the case of remote driving, these options are unavailable or require an additional interaction layer. This situation sparked the idea of having an interface that would allow passengers to also "drive with their fingers" using a 3D map of real-time traffic conditions (see Fig. 8). In this way, intentions for (re)routing can be conveyed whenever necessary, without verbal communication. Finally, remote driving increases the potential for differently-abled people – such as those with hearing impairments – to partake in the act of driving, therefore, the driver's interface could be reconsidered to provide more multi-sensory feedback. One of the teams proposed an interface for a hearing-impaired driver to interact with various buttons and knobs to instigate communication with the passenger and perform other functions (see Fig. 9). Building on this concept, perhaps different tactile interactions could be used to optimise the vehicle control interface for a wider spectrum of people and at the same time promote their situational awareness.

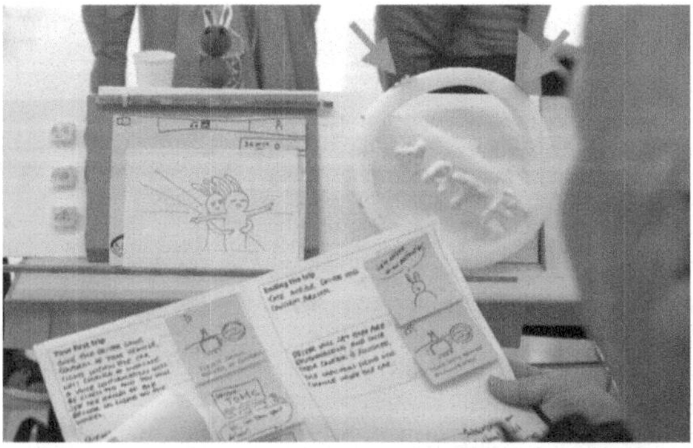

Fig. 7. Prototype describing the idea of having illuminated, virtual hands on the steering wheel (see red arrows) in a passenger's car to provide a sense of the remote driver's presence. (Color figure online)

Overall, the ideas revealed clusters of interests and concerns among participants. We transcribed their post-its of ideas into digital copies on a Miro board and organised them into an affinity map (see Fig. 10). Our synthesis led to a number of key findings listed below, which we then distilled into three directions for further concept development.

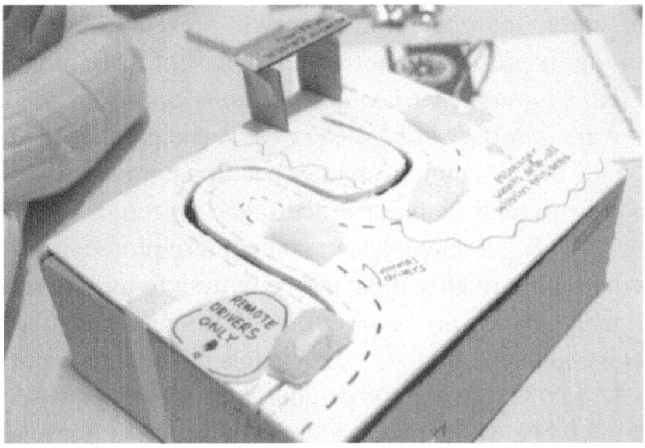

Fig. 8. (left) Showcasing the idea of a map for passengers to manoeuvre their car avatar in real-time to inform the driver of their preferred route, creating an act of "co-driving".

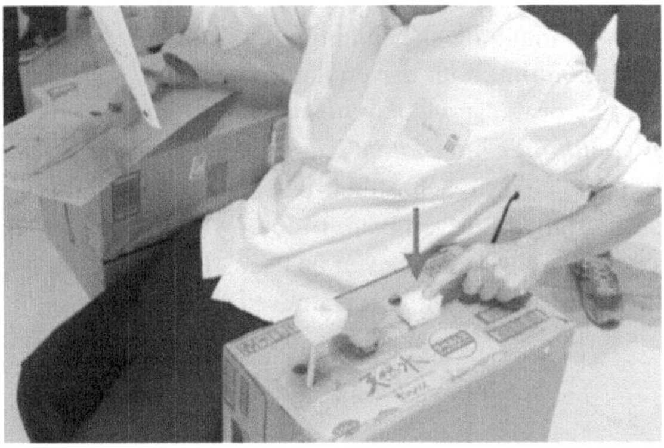

Fig. 9. Proposal of interactive buttons that send various messages to the passenger.

These directions (also expressed as research questions in Table 2) cover themes of establishing preferences and protocols, conveying driver intentions, and providing a sense of telepresence to the passenger. A summary of our findings is as follows:

1. *There appears to be a preference to specify qualitative aspects of the ride, including traits of the driver.* Ideas in this cluster tend to provide passengers with the option to select and customise parameters of their ride, such as the speed, sound (e.g., music, communication etc.), and even a driving modes (e.g., "fast and furious" in the case where the driver is a professional gamer). One of the groups proposed the potential to match passengers with drivers according to the driver's personality traits, such as being a good listener" and based on their Myers Briggs Type Indicator (MBTI).
2. *People's sense of trust in a system is dependent on perceived human connection and choice agency.* As one of our participants wrote, the "clarity at any given point, that both sides [the driver and passenger] are present" is important if such a service were to be available. Ideas in this cluster span the range of technological solutions like an "eye motion tracker" to check the driver's status (e.g., sleeping), to communicate it

to the passenger, and to intervene if necessary to prevent accidents. A way to verify that the remote other is present is an irreducible part of the remote driving system.

3. *It is supposed that thorough verification and onboarding of the driver shall take place before they are qualified to be of service.* Participants listed numerous criteria to be checked on the driver's end, from their health records, driving history and other personal credentials, to their home or office space for remote driving to guarantee a conducive workspace. If passengers can assume these protocols to be done and true, it seems they would more readily trust and be willing to use the service.
4. *Bilateral contingency plans may be required for emergencies.* Considering possible edge cases and emergency situations, participants raised three options as emergency protocols, namely: to revert to manual driving, to call for standby drivers, and for the vehicle itself to intervene. If the network connection is lost, for example, passengers may be requested to take control of their vehicle, or the vehicle could automatically pull to a safe stop. At the very least, it is expected for the vehicle to be equipped with "a backup process for the minimal task", which alludes to built-in protocols and contingency plans that activate either with or without human command. It is worth noting that there were many ideas containing an "emergency control/button", which implies the expectation that both drivers and passengers should have the agency to terminate the journey or the service upon any sign of threat.
5. *Remote driving could offer alternative value-added services.* When the ride comes to an end (i.e., the passenger reaches their destination and alights), the vehicle could be taken to a parking lot or driven back to the owner's home by the remote driver. Some ideas suggest that the vehicle even be used "like a sharing taxi", to ferry other passengers to their destination while it is unused by its owner.

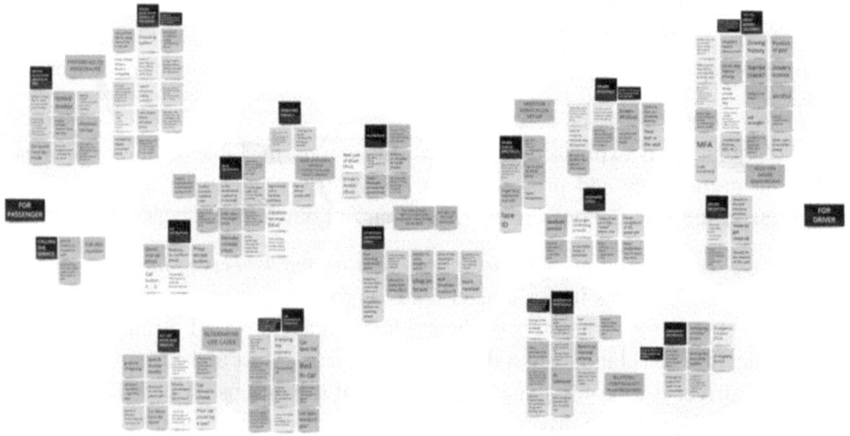

Fig. 10. Synthesis of interface ideas generated by participants to distil their inherent value propositions for drivers and/or passengers. Internally conducted on Miro post-workshop.

Table 2. Concept directions for remote driving research based on workshop insights and preliminary concepts.

Direction No	Theme	Research Question
1	Establishing a system for preferences and protocols	How might personality and preferences change driver-passenger dynamics?
2	Make motivations and intentions transparent in remotely driven services	How might two-way interactive maps influence driver-passenger communication for remote driving services?
3	Have clarity at any given point that both people are present and attentive	What interventions on a steering wheel (for example) could promote a sense of telepresence and reliability?

5 Discussion

The research questions above allow us to frame clearer proposals with which we hope to invite new opportunities for collaboration and engage with the broader community outside of academia. In addition, we have gleaned insights into prospective automotive UI development and design methodology, which we will address in this section.

Firstly, regarding automotive UI, we expect vehicles to offer even more personalised choices and interactions through their interfaces as we discover more of passengers' unspoken needs and expectations of teleoperated services. At the same time, passengers still need a real human presence to be within reach throughout their journey to feel a sense of safety and reliability of the service. Moreover, participants displayed a critical and cautious approach when asked about the security and safety checks required for such a service. Although we suspect that most people would not concern themselves over the identity of their driver or passenger should this service become a social norm (as in the case of ride-hailing), overcoming the current public perception of non-trust in remotely driven services persists to be a challenge. Further testing of interactions and interfaces is required to determine the factors that would ease people's concerns, or that could provide meaning and value significant enough to catalyse a social paradigm shift.

Nonetheless, there are limitations to what we can conclude about the efficacy of ideas that came out from the workshop. Since our workshop essentially created a safe space to contemplate and raise questions on a speculated topic where nobody is an expert or has had firsthand experience, the ideas need to be pulled back to reality to some degree. In this regard, we believe that partnerships with other research labs or industry players are imperative to realise the impact of this design-led research on technology development for remotely driven services.

In terms of design methodology, we have exemplified how technology-enabled innovation should not be driven by technological capabilities. Rather, it is a vital part of the design process to uncover people's implicit needs and values pertaining to the innovation. These tacit sentiments are usually brought forth by organic discussion and debate, which in itself reveals the gaps in existing systems and the opportunities for technology to create social value. With that, we emphasise the act of collaborative thinking through making. In this workshop, we saw how prototyping gave participants a much quicker and more assured sense of which ideas had potential and which ones to drop. By making simple mock-ups of their interfaces with cardboard, they could extract the essence of their ideas and refine their expression. In all, we have also gained a better sense as design facilitators, of how to adapt our workshop activities to other contexts and participants. With design-trained participants, for example, the activities could be rejigged to involve recursive prototyping and interaction refinement. This way, we could achieve more refined artefacts within similar time constraints.

6 Conclusion

By conducting this futurespective design workshop, our team has made progress in answering the question of what "remote driving" means to people. Therein lies the potential for remotely driven services to bring new value to people's lives. With the help of participants, we have shed light on public sentiments and concerns on the topic, especially from the passenger's perspective. With these findings, we can ask the right questions that will reinforce assumptions and concept directions to be tested, thereby supplementing ongoing research on improving the social acceptance of driverless vehicles. At the same time, this paper offers documentation of a design workshop as a tool to explore speculative futures, innovative technologies, and their potential social impact. Design-led workshops are a lightweight way to gain actionable insights on a topic by stress-testing ideas within reasonable resources and time. We recommend that such design workshops be used to the advantage of innovation teams as a means to test and refine early ideas, reframe the problem and solution scope, and find an unorthodox angle to approach a topic from a value-based foundation.

Acknowledgments. These research results were obtained from the commissioned research by National Institute of Information and Communications Technology (NICT), JAPAN.

Disclosure of Interests. The authors declare no competing interests.

References

1. Parekh, D., et al.: A review on autonomous vehicles: progress, methods and challenges. Electronics **11**(14), 2162 (2022)
2. Brell, T., Philipsen, R., Ziefle, M.: SCARy! risk perceptions in autonomous driving: the influence of experience on perceived benefits and barriers. Risk Anal. **39**(2), 342–357 (2018)
3. Zeng, Z., Qu, X.: Optimization of electric bus scheduling for mixed passenger and freight flow in an urban-rural transit system. IEEE Trans. Intell. Transp. Syst. **24**(1), 1288–1298 (2023)
4. Anderson, J.M., Kalra, N., Stanley, K.D., Sorensen, P., Samaras, C., Oluwatola, O.A.: Autonomous vehicle technology: A guide for policymakers. RAND Corporation, Santa Monica, CA (2016)

5. Yu, Y., Lee, S.: A measurement based feasibility study on remote driving system. In: 2021 International Conference on Information Networking (ICOIN), Jeju Island, Korea (South), pp. 17–19 (2021)
6. Amador, O., Aramrattan, M., Vinel, A. A Survey on Remote Operation of Road Vehicles. In: IEEE Access, Vol. 10, pp. 130135–130154 (2022)
7. SAE International (2021a) 3016_202104 Taxonomy and Definitions for Terms Related to Driving Automation Systems for On-Road Motor Vehicles. https://www.sae.org/standards/content/j3016_202104/, last accessed: 2024/7/24
8. Kettwich, C., Schrank, A., Oehl, M.: Teleoperation of highly automated vehicles in Public Transport: user-centered design of a human-machine interface for remote-operation and its Expert usability evaluation. Multimodal Technol. Interaction **5**(5), 26 (2021)
9. Mutzenich, C., Durant, S., Helman, S., et al.: Updating our understanding of situation awareness in relation to remote operators of autonomous vehicles. Cogn. Research **6**, 9 (2021)
10. Endsley, M.R.: Design and Evaluation for Situation Awareness Enhancement. Proceedings of the Human Factors Society Annual Meeting **32**(2), 97–101 (1988)
11. Riley, J.M., Kaber, D.B., Draper, J.V.: Situation awareness and attention allocation measures for quantifying telepresence experiences in teleoperation. Hum. Factors Man. **14**, 51–67 (2004)
12. Chen, J. Y. C., Haas, E. C., Barnes, M. J.: Human Performance Issues and User Interface Design for Teleoperated Robots. In: IEEE Transactions on Systems, Man, and Cybernetics, Part C (Applications and Reviews), vol. 37(6), pp. 1231–1245 (2007)
13. Hosseini, A., Lienkamp, M.: Enhancing telepresence during the teleoperation of road vehicles using hmd-based mixed reality. In: 2016 IEEE Intelligent Vehicles Symposium (IV), pp. 1366–1373 (2016)
14. Sheridan, T. B.: Musings on telepresence and virtual presence. Presence: Teleoper. Virtual Environ., vol. 1, pp. 120–126 (1992)
15. Liu, R., Kwak, D., Devarakonda, S., Bekris, K., Iftode, L.,: Investigating remote driving over the lte network. In: Proceedings of the 9th International Conference on Automotive User Interfaces and Interactive Vehicular Applications, pp. 264–269 (2017)
16. Giacomin, J.: What Is Human Centred Design? Des. J. **17**(4), 606–623 (2014)
17. Koopman, P., Wagner, M.: Autonomous vehicle safety: an interdisciplinary challenge. IEEE Intell. Transp. Syst. Mag. **9**(1), 90–96 (2017)
18. Paul, D.: Where the Action Is: The Foundations of Embodied Interaction, MIT Press (2001)
19. Mackenzie, G., Trucchia, F., Dias, H.: The future archives: a speculative approach for visualising the impacts of 6G-enabled infrastructure in Japan. In: De Sainz Molestina, D., Galluzzo, L., Rizzo, F., Spallazzo, D. (eds.) IASDR 2023: Life-Changing Design, pp. 9–13. Italy, Milan (2023)
20. Bleecker, J.: Quick Start Guide for a Self Driving Car, last accessed: https://nearfuturelaboratory.com/projects/en/qsg-self-driving-car/ 2024/5/24

Controller Adaptive Digital Assistant (CODA): Interim Conclusions on the Development of a New ATC System

Laurie Marsman(✉), Shiva Nischal Lingam, Bart Klein Obbink, Jeanine Vlasblom, Rolf Zon, and Maykel van Miltenburg

Royal NLR, Safety and Human Performance, 1056CM, Amsterdam, The Netherlands
laurie.marsman@nlr.nl

Abstract. This paper aims to outline the development of adaptation strategies and HMI guidelines for a COntroller adaptive Digital Assistant (CODA) to support Air Traffic Controllers in their operational duties by dynamically managing their workload. The strategies and guidelines were constructed using relevant literature and interviews with operational experts. The interim results include areas of specific interest based on operational knowledge, fourteen adaptation strategies and fifteen HMI guidelines. These results will serve as input for subsequent mock-ups that consider technical factors as well, and ensure an effective and usable support to controllers. These interim results provide valuable insights for developing supportive tools in the highly regulated, safety-critical domain of ATC, contributing to the advancement of operational safety and efficiency.

Keywords: Air Traffic Controller · HMI · automation · adaptation strategies · mental state assessment

1 Introduction

Overseeing and controlling air traffic is a highly demanding task that requires a continuous high level of attention and a specific skill set [1]. Air Traffic Controllers (ATCOs) are responsible for acquiring and analyzing various information such as radar data, flight plans, weather conditions, and making critical decisions to ensure safe aircraft separation and operations [1]. Managing ATCO workload is crucial, as high workload can lead to stress and reduced situational awareness, increasing the risk of errors. Dynamic support tailored to the ATCO's mental state is essential for maintaining optimal performance.

Developing automated tools for air traffic management is seen as a key technology to reduce controller workload and enhance airspace capacity [2]. Artificial Intelligence (AI) is increasingly being explored to support ATCOs [1] and may play a crucial role in managing traffic [3], and it has been shown that automation can significantly improve ATCO situation awareness and performance for example, especially when this aids in information acquisition of the ATCO [4]. Hence, such technological developments show great promise in aiding ATCOs in executing their operational tasks.

D. Singh et al. (Eds.): IHCI 2024, LNCS 15557, pp. 254–260, 2025.
https://doi.org/10.1007/978-3-031-88705-5_21

1.1 The Aim of CODA

With the increasing rate of developments in AI and associated benefits, ATCOs are likely to perform their tasks with increasing support of automation. To demonstrate this Human-Machine Teaming (HMT) within the context of ATC, the CODA (COntroller adaptive Digital Assistant) project aims to develop a system in which ATC tasks are dynamically allocated to either the ATCO or AI based on (neuro)physiological measurements of the ATCO and relevant current and future traffic parameters or other relevant aspects. Thus, CODA can then support the ATCO by performing specific tasks or by providing certain information to manage the ATCO's workload.

2 Methods: Development of CODA

The scope of this paper is limited to the development of the HMI of CODA. As a first step, literature was gathered and consulted non-systematically by project members. As there is little to no literature on a system as CODA, papers concerning more general technological developments in ATC were included in the search, as well as papers from domains such as the automotive industry. All information deemed relevant for the development of CODA was subsequently gathered in a document, which was then reviewed and refined by project members based on their expertise. Next, several sessions were organised by the task lead to brainstorm about how these literature findings could be integrated into specific adaptation strategies and HMI guidelines for the CODA system. Once this was finished, an initial draft was written up. Finally, a session was organised with three external ATC experts with operational knowledge to discuss the resulting strategies and guidelines. Key points are discussed in the Results section. The section also details a summary of the resulting adaptation strategies and HMI guidelines that will be used in the development of subsequent HMI mock-ups, a playbook detailing the moment CODA should change one's level of automation based on operational situations, and a resulting CODA system for initial validation.

2.1 Assumptions

To design a functional adaptive assistant in ATC, the ATCO's cognitive state needs to be monitored. However, for the purpose of this paper, it suffices to say that (neuro) physiological parameters are collected for the automation to assess the ATCO's state to be one of the following, other than nominal: *stress, fatigue, mental workload, overload* and *underload*. This classification, in combination with relevant traffic parameters (e.g. number of aircraft in the sector or traffic complexity), will then inform the automation mode. The ATCO's state should always be interpreted in the context of traffic complexity and other external factors, and must be measured robustly and unobtrusively.

In the development of the following strategies and guidelines, it is assumed that the system has current and past knowledge of the ATCO mental state and the current, past and (predicted) future operational situation. Moreover, the ATCO is assumed to have sufficient capabilities to execute all tasks in a nominal state and has the capacity to take over tasks from the automation in case of automation failure. The CODA system will match the flexible nature of ATC operations, yet for the time being, it is assumed that the ATCO is unable to override the CODA system and that the system is not customizable. Importantly however, CODA and the ATCO must share the same operational goals to achieve the objective to effectively and safely manage the air traffic. The adaptation strategies and the HMI guidelines are described below for the CODA system.

3 Results: Adaptation Strategies and HMI Guidelines

3.1 Key Points of ATCO Session

Three male former operational ATCOs were consulted at the premises of a European ATSP regarding their views on the concept of CODA. The session started with an introduction on the project and on the developments thus far. Then, they were asked to explain their views concerning the system and its potential parameters in detail. Some key points are outlined. First, the ATCOs highlighted the flexible nature of ATC operations, and that no operational situation is the same. This is important when planning the actions of CODA, as it needs to ensure it does not interfere with the specific plan an ATCO has in mind for that operational situation. Second, the ATCOs emphasized that experiencing some stress is beneficial to an ATCO and their efficiency, indicating that the detection of stress is not immediately problematic. Third, the CODA HMI should not overflow the ATCO with warnings or notifications once CODA initiates an action, as the ATCO needs to maintain a mental model of the situation. The mental state of the ATCO should be considered when notifying the ATCO. Finally, the notion of criticality should be reconsidered in a CODA setting, as in current terms it suggests a non-nominal situation. It is suggested to define it as 'requiring an action within a limited timeframe' instead, for which the ATCOs think CODA could substantially ease them.

3.2 Resulting Adaptation Strategies and HMI Guidelines

An adaptation strategy can be seen as a prescription of the task allocation between human and machine for different situations. Such strategies are crucial for the functioning of a system like CODA, and for maximizing the positive effects on the ATCO. Research by [1] and [5] highlights the benefits of adaptive automation in reducing mental workload and preventing the Out-Of-The-Loop (OOTL) phenomenon in ATCOs. By incorporating physiological computing, these strategies aim to optimize ATCOs' performance by adjusting automation levels based on their cognitive states and workload. The following adaptation strategies are suggested in summary (Table 1).

Table 1. Adaptation strategies for the CODA system

#	Adaptation strategy
1	CODA and the ATCO can communicate with each other through a HMI, e.g. the CODA dashboard or radar screen, about selected automation modes and other tasks performed by either the ATCO or the CODA system
2	CODA can explain their actions to the ATCO, to help improve trust [6, 7]
3	In case of conflict management, the ATCO's input is always leading
4	CODA will operate in different modes operandi, though the exact order number will have to be backed by empirical studies and operational experts
5	There is a standard automation level, which might be referred to as the baseline and which can be expected to be applied most of the time
6	The implications of each automation mode should be clear and logical to the ATCO
7	The ATCO is informed about the current mode and is notified of any changes
8	CODA selects a mode on the basis of knowledge about current and past mental states of the ATCO and knowledge about the current, past and future operational situation
9	The preferred automation mode counteracts any negative effects of ATCO mental state
10	CODA considers the effectiveness of an ATCO's actions when determining the appropriate automation mode, distinguishing between effective and ineffective stress
11	The automation adaption avoids frequent changes to let the ATCO recognize and adapt to the mode. The maximum frequency of automation changes needs scientific backing
12	Critical tasks are always performed by the ATCO, and not by CODA. The circumstances under which a critical task becomes critical is backed by operational experts
13	CODA shall not take over too many or too demanding tasks, ensuring that if CODA fails, the ATCO can take over all tasks effectively without endangering safety
14	CODA shall take over tasks in such a way that the ATCO will not over-trust or become overly dependent on CODA to ensure ATCOs remain competent and current

Next, the design of HMIs for the CODA system plays a critical role in ensuring efficient and safe operations in ATC, as this is the only means of communication between the ATCO and the system. While the CODA system involves an HMI philosophy, for the purpose of this paper the focus lies on the specific guidelines proposed to enhance the usability and effectiveness of the interfaces (Table 2), again in summary.

Table 2. HMI guideline and their respective categorization for the CODA interface

#	HMI guideline (including categorization)
Visual design	
1	The CODA interface should use symbols, colors and layout conventions that are already familiar or logical to the ATCO, yet not identical or highly similar to existing symbols, colors and codes or that may be used as standards for ATC operations
Layout, navigation and accessibility	
2	The current automation mode should always be directly be accessible and clearly visible
3	Modes and their nomenclature should be clearly distinguishable from one another
4	The (estimated) state of the ATCO should always be accessible to the ATCO with minimal effort. The process of gaining this information is backed by operational experts
5	Updates on the state should be conservative, i.e. not change too often, to minimize distraction. Estimation of state should be across a 'considerable' time period
System feedback	
6	The expected future status or mode of the automation, should be clearly visible to the ATCO when 1) the system is changing modes or functionality imminently, and 2) what the likely next mode is based on the current situation
7	In the event of system failure, the interface must clearly alert the ATCO about the status of the system ("FAIL") and the immediate consequences for their tasks
8	CODA dynamically adjusts the type and modality of alerts (visual, auditory, etc.) based on real-time assessments of the ATCO's cognitive load and situational factors. Taking into account human cognitive limitations, such as change blindness and tunnel vision
9	System feedback from CODA to the ATCO should be integrated as much as possible into the current, familiar format, i.e. into the radar screen and on the flight strips
10	The interface should provide the ATCO with relevant information and suggestions. This should be high level, and should support the ATCO in reaching the goal/task
11	CODA should provide feedback to the ATCO in incremental steps in chronological and critical order, i.e. it shall only take over any tasks if the operational situation calls for it
12	Any notifications from the system to the ATCO should be ordered and colour-coded based on priority and criticality, backed by empirical studies and operational experts
Safety and security	
13	CODA should incorporate best practices in terms of security to protect data
14	CODA should clearly state any error messages when they are encountered
15	There should be the possibility to undo a specific action to prevent any mal-entries and accidental operations from taking effect

4 Discussion and Conclusion

This paper outlines the interim conclusions in the development of the adaptation strategies and HMI guidelines for a COntroller adaptive Digital Assistant (CODA). These strategies and guidelines are subsequently used as input to create a mock-up of the potential CODA system, and can be used as a checklist for the time being. However, any issues arising during their implementation (i.e. due to technical limitations of the current simulation platform) will be discussed and updates to these strategies and guidelines will be made accordingly for the first prototype.

The outlined adaptation strategies enhance key aspects of human-machine teaming, with the aim of improving ATCO's situational awareness, performance and trust. Maintaining predictability of actions and intentions aid in optimizing the performance of CODA, safety, and subsequent trust in the system, which is essential for the implementation of a digital assistant as CODA. Improving adaptation strategies ensures the efficiency of such human-autonomy collaboration.

The HMI guidelines aim to support transparent and effective communication between ATCOs and the CODA system, thereby enhancing user satisfaction and system usability. Information is provided according to priority, criticality, and hierarchy to avoid overloading the ATCO. These guidelines aim to present information in an easily accessible, readable, and intuitive manner, considering the ATCO's state.

Though the proposed adaptation strategies and HMI guidelines were constructed after consulting relevant literature and operational experts, empirical assessment and validation of ensuing mock-ups are required in a high-fidelity setting. Based on findings from this assessment, changes may be made to enhance the operational relevance in one or more iterations. We expect the final iteration of the CODA system to improve situational awareness, user trust, and performance and reduce workload for the ATCOs.

In conclusion, these interim conclusions offer meaningful insights into the development and potential benefits of technical support tools within the highly regulated, safety-critical environment of ATC. While the integration of new digital assistance systems in such domains presents unique challenges due to strict operational standards and the need for ATCO acceptance, the CODA project demonstrates that adaptive support tools hold significant promise for enhancing safety and operational efficiency. By contributing to the scientific literature, this work underscores the importance of further research and iterative development to align technical innovations with the specific needs and preferences of air traffic controllers, paving the way for advancements that support their critical role in maintaining airspace safety.

Acknowledgments. This study is part of the SESAR JU project "CODA", which is a collaborative project by Deep Blue, ENAC, Brainsigns, CRIDA, Universidad Politecnica de Madrid, Eurocontrol, IFATCA, Universidad de Granada, and the NLR.

Disclosure of Interests. The authors have no competing interests to declare that are relevant to the content of this article.

References

1. Di Flumeri, G., et al.: Brain–computer interface-based adaptive automation to prevent out-of-the-loop phenomenon in air traffic controllers dealing with highly automated systems. Front. Hum. Neurosci. **13**, 296 (2019)
2. Paşaoğlu, C., Baspinar, B., Ure, N., Tsourdos, A.: Hybrid systems modeling and automated air traffic control for three-dimensional separation assurance. Proc. Institut. Mech. Eng. Part G J. Aerospace Eng. **230**(9), 1788–1809 (2015)
3. Nikitas, A., Michalakopoulou, K., Njoya, E., Karampatzakis, D.: Artificial intelligence, transport and the smart city: definitions and dimensions of a new mobility era. Sustainability **12**(7), 2789 (2020)
4. Sethumadhavan, A.: Effects of automation types on air traffic controller situation awareness and performance. Proc. Human Factors Ergon. Soc. Annual Meeting **53**(1), 1–5 (2009)
5. Aricò, P., et al.: Adaptive automation triggered by eeg-based mental workload index: a passive brain-computer interface application in realistic air traffic control environment. Front. Hum. Neurosci. **10**, 539 (2016)
6. Smith, P. J., McCoy, C. E., Layton, C.: Brittleness in the design of cooperative problem-solving systems: the effects on user performance. IEEE Trans. Syst. Man Cybernet. – Part A. Syst. Humans **27**(3), 360–371 (1997)
7. Lee, J.D., Moray, N.: Trust, self-confidence, and operators' adaptation to automation. Int. J. Hum. Comput. Stud. **40**(1), 153–184 (1994)

Reimagining the Utility of Dash Cam Data in Autonomous Travel Contexts: A Service Design Proposal

Subi Hahn and Eui-Chul Jung(✉)

Department of Design, Seoul National University, Seoul 08826, South Korea
{subi.hahn,jech}@snu.ac.kr

1 Introduction

On-road travel is often seen as an intrinsically valueless in-between where the passenger looks for ways to kill the time as they wait to arrive at their desired destination [1]. The lack of perceived value in travel time can be easily recognized in the many travel apps, services, and inbuilt GPS systems on the market today that prioritize the fastest routes above all else, with factors such as how scenic a travel route is often being disregarded. With a fully autonomous future well on its way, the role of travelers using on-road vehicles is shifting from a primarily active, to a much more passive travel experience (see Fig. 1). This shift opens the gateway to new avenues of exploration when it comes to designing engaging and enjoyable in-vehicle passenger experiences (PX) that highlight the unique value in the time spent traveling from one place to another.

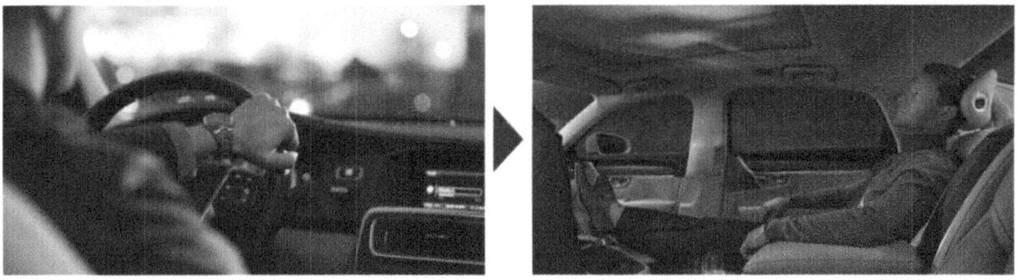

Fig. 1. The shift in travel environment from an active(left) to a passive(right). *(Image Sources: Unsplash.com (left), Volvo S90 Excellence Interior (right)).*

On-road travel is dependent on many factors that either positively or negatively affect a passenger's perceived worthwhileness of travel time. While circumstantial factors, like traffic and weather conditions undoubtedly contribute to one's overall travel experience, there are two inherent elements of travel that are mutually dependent and result in one being transported from one place to another. That is, (1) time and (2) place.

Regardless of the speed or distance traveled, moving from one place to another takes valuable time out of one's day and it is no surprise that it can be difficult to see an innate value in the act of traveling [2]. In response to this, many automotive industry leaders can

D. Singh et al. (Eds.): IHCI 2024, LNCS 15557, pp. 261–266, 2025.
https://doi.org/10.1007/978-3-031-88705-5_22

be seen attempting to battle the negativity surrounding travel time by exploring various engaging PX experiences. It can be noted however, that many of the proposed concepts aim to keep the traveler's attention within the confines of the vehicle, as they incorporate elements of everyday spaces, such as an office, living room, bedroom, or other such spaces of leisure into a vehicle's interior. This approach, while providing a quick fix for the boredom felt while traveling, it can be argued that such simulated spaces mask the element of travel itself, and thus take away from the unique experience and value of on-road travel as they either partially or completely cut off the passenger's connection to and interest in the outside environment.

For the intrinsic value and beauty of on-road travel to truly be appreciated, a shift in travel perception must take place. This shift requires destination-centered travel experiences, where the intrinsic nature of travel is masked by simulated everyday environments, evolves into an engaging, journey-centered experience wherein the traveler can reconnect with the outside environment (see Fig. 2). By manipulating the elements of time and place, this study ultimately aims to bring value to the action of traveling, through re-establishing interest in and connection to the outside environment.

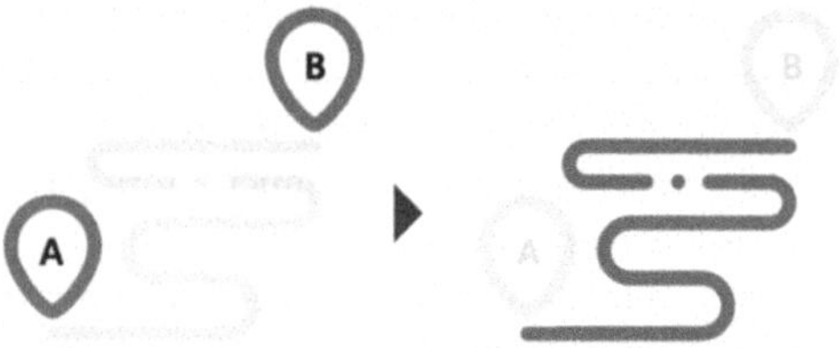

Fig. 2. The desired shift from destination-centered to journey-centered travel.

2 Travel Perception Today

Gaining interest in the 1980s, travel behavior has become a crucial area of research across various industries [3]. Essentially, travel behavior research focuses on studying how and why people move from one place to another. As expected, the automotive industry has shown significant interest in this field of research, and as a result, the 2000s saw the introduction of numerous studies and surveys that developed what is commonly referred to as the 'teleportation question' or 'teleportation test' [4]. Although highly unrealistic in nature, this type of questioning has proven to be useful in revealing how people really feel about their time spent traveling, as many tend to prefer the idea of 'teleporting' or instantly arriving at their desired destination as opposed to spending time traveling there.

Several surveys of interest prove this to be true, with the first, a survey conducted in New Zealand in 2012 that asked 512 commuters if they would prefer to "teleport" to their destination, to which the majority (75%) of the participants indicated that they indeed would prefer to teleport if given the option [5]. Another similar survey also conducted in New Zealand in 2015 posed a very similar question to 1,039 transit passengers, with 66% expressing their preference for teleportation if given the option [6].

A more recent survey conducted in the United States involved 648 commuters [7]. In this survey, 62% of respondents expressed a preference for teleportation over spending time traveling. This survey also collected additional data on 'Mode of Transportation' and 'Travel Perception'. Among participants who travel via automobile, many of the drivers and passengers (73% and 76% respectively) indicated that they would prefer to teleport. As for Travel Perceptions, the findings were as expected, with most participants seeing little to no value in the travel experience itself.

3 Finding the Value in on-Road Travel

Today, advanced in-car GPS systems have been introduced as a convenient tool for drivers to navigate their routes. In the context of a fully autonomous travel landscape, extensive passenger data—covering details like daily schedules, preferred dining spots, and favorite getaway destinations—will enable these vehicles to make more precise route selections tailored to each passenger's specific preferences. The proposed PX concept presented in this paper takes this idea of route customization and explores the use of real-time dash cam footage to blur the physical constrains of place. Essentially, this concept grants passengers full visual control, allowing them to 'hop' to any view point (vehicle view) along a designated route with the help of dash cam footage. This service concept is therefore particularly effective in busy metropolitan areas where there is a constant flow of on-road vehicles throughout the day. For familiar routes, such as commutes or trips to family and friends' homes, passengers may have specific segments of the journey they particularly enjoy. This service entices passengers to take control of the visual timeline of their travel, enabling them to experience the visual journey according to their preferences—in the sequence, order, and at speed they desire (see Fig. 3).

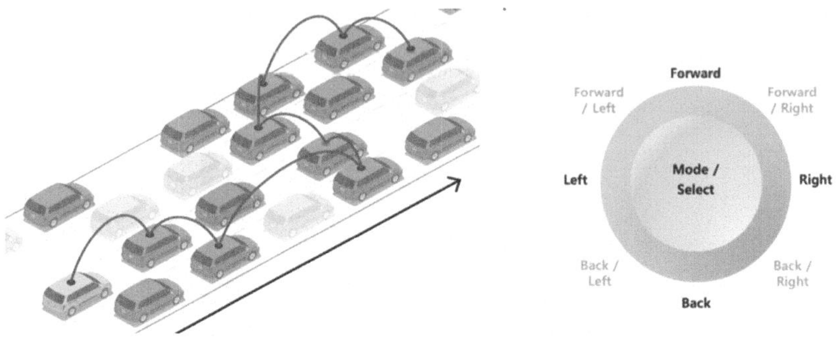

Fig. 3. View hop path perspective diagram (left) and control functions (right).

If at any point a passenger wishes to stop view hopping or has been hopping in one direction for so long that they lose track of their original route, they can simply reset the hopping feature. After resetting, they can resume hopping, starting from their own vehicle and jumping to the view of any other vehicle along their designated travel route. If there's a part of the travel journey that a passenger particularly enjoys, they can continue hopping around in that area for as long as they like, as long as their vehicle is still moving toward the selected destination. Some practical scenarios for this view hopping PX include: (1) visually skipping ahead in traffic, (2) bypassing traffic lights without having to wait for them to change, and (3) enjoying beautiful scenery even if the car isn't physically close enough to offer a good view (see Fig. 4).

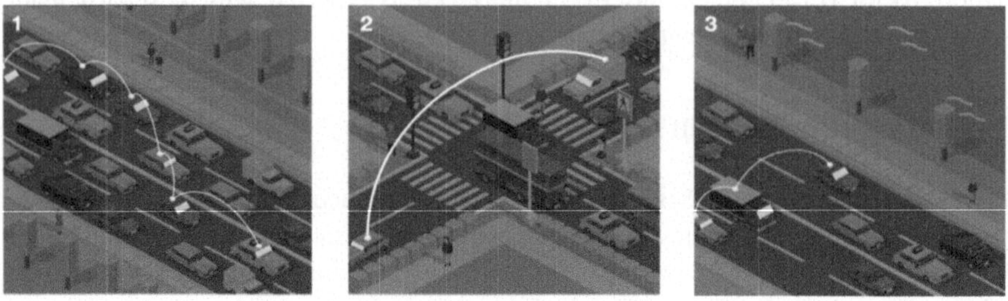

Fig. 4. Three view hop scenarios highlighting relevant advantages of the proposed service.

In developing the visual aspect of this PX concept, special attention was given to creating smooth transitions between views. Various transition types were tested, but it became clear that overly dynamic transitions could not only induce motion sickness in passengers but also make the 'hop' from one vehicle's view to another feel unnatural. Ultimately, a simple blur-out, fade-in effect was chosen, mimicking a camera briefly losing focus before refocusing on a new environment (see Fig. 5).

The visualization of this concept was a 1 min 25-s-long video (see video link at Fig. 6). The video presents a travel scenario that shows how the view hopping service could be used. Each time the user hops to a new view, the music is also seen to change. To the far bottom left-hand corner of the windshield, the number of people currently viewing any vehicle's live view can also be seen.

Fig. 5. Visual transition effects between two views.

Fig. 6. Video edit of a travel scenario utilizing the view hop service (Hahn S., YouTube. Aug 4th 2024. 01:25:00. https://www.youtube.com/watch?v=zaGJc9Bauqs)

By giving passengers full control over the visible external environment they experience with the help of smart window technology, this PX exploration redefines the sense of place while also blurring the element of time. This concept therefore acts as a visual teleportation tool as it enables users to seamlessly hop from one place (view) to another. This allows passengers to savor the most enjoyable parts of their journey for longer while avoiding the unpleasant parts, ultimately leading to a more meaningful travel experience and overall greater travel satisfaction.

4 Conclusion

This study and the PX concept explored throughout challenges the common perception of travel being an unavoidable means to get from one location to another by proposing a service design concept that enables on-road travelers to take full visual control of the journey they experience. This study seizes the opportunity to repurpose dash cam data, offering passengers a dynamic and visually engaging way to experience on-road travel. By removing the constraints of physical location and enabling view hopping along any given route, this proposal seeks to elevate the perceived value of travel time, ultimately contributing to increased travel satisfaction in an autonomous on-road travel context.

Acknowledgments. The proposed concept presented in this paper is included as one of many concept explorations in the author's master thesis titled, 'Re-establishing Passenger Connection to the Outside Environment in Autonomous Vehicles Through Moments in Time,' conducted at the Graduate School of Design at Seoul National University and finalized in August of 2024.

Disclosure of Interests. The author and advising professor of this research have no competing interests to declare that are relevant to the content of this article.

References

1. Dianna, M.: Making the "primary utility of travel" concept operational: a measurement model for the assessment of the intrinsic utility of reported trips. Trans. Res. Part A: Policy Pract. **42**(3), 455–474 (2008)
2. Humagain, P., Singleton, P.A.: Would you rather teleport or spend some time commuting? investigating individuals' teleportation preferences. Travel Res. Part F: Traffic Psychol. Behav. **74**, 458–470 (2020)
3. Davis, A., et al.: Workshop summary and research themes. Mapping the Travel Behav. Genome, 635–673 (2020)
4. Humagain, P., Singleton, P.A.: Exploring satisfaction with travel time profiles towards understanding intrinsic utilities of travel time. Travel Behav. Soc. **24**, 22–33 (2021)
5. O' Fallon, C. E., Wallis, I.: A wider look at how travelers value the quality and quantity of travel time. New Zealand Transport Agency. Res. Rep. **469** (2012)
6. Russell, M., Mokhtarian, P.: How real is a reported desire to travel for its own sake? Exploring the 'teleportation' concept in travel behaviour research. Transportation **42**(2), 333–345 (2014). https://doi.org/10.1007/s11116-014-9546-1
7. Singleton, P.A.: Exploring the Positive Utility of Travel and Mode Choice (doctoral dissertation). Portland State University (2017)

Algorithms and Computer Vision

Keyframe Extraction Based on Large Vision Transformer Model and kNN-DBSCAN Clustering

Huynh Anh Khoa, Huynh Anh Duy, and Phan Duy Hung(✉)

FPT University, Hanoi, Vietnam
{khoa23mse23138,duy23mse23139}@fsb.edu.vn, hungpd2@fe.edu.vn

1 Introduction

Video has become a widespread medium in our daily lives for communication, learning, and entertainment. This is due to increased storage capabilities, improved network technology, and easy access to digital cameras, especially on mobile phones. Popular video platforms like YouTube and TikTok continue to grow rapidly with a vast number of users, increasing the need for video processing techniques. One crucial technique is keyframe extraction, which identifies the most important or representative frames from a video sequence. Keyframe extraction has numerous applications across various areas of video processing, such as video summarization, indexing and retrieval, editing, and compression. Its high application potential has led to the development of various approaches to address this task [1–3]. Additionally, benchmark datasets like Sports-1M [4], YouTube-8M [5], and TVSum [6] have been published to facilitate research efforts in this field. In this paper, we propose a combination of large vision transformer model and deep convolutional neural network for the task of extracting features, then employ a potential adaptive clustering method and post processing based on histogram color to extract Keyframes for videos.

The main contribution of this work is the proposed keyframe extraction algorithm that takes advantage of large vision transformer model for feature extraction and employs the DBSCAN [7] couple with kNN technique [8] as an adaptive clustering method to extract Keyframes efficiently, without the need for complicated parameter tuning. The name of the method is **ViTKDB** (**ViT** represents Vision Transformer, while **KDB** represents the kNN-DBSCAN).

2 Related Works

Methodologies for keyframe extraction can be broadly classified into four distinct categories: uniform sampling based, shot boundary based, clustering based, frame-difference based. While uniform sampling-based methods are straightforward and computationally efficient, they exhibit significant inefficiencies. The number of keyframes generated can be either excessive or insufficient, contingent on the video's length and characteristics, such as its semantic content or static nature [9]. Additionally, there is no assurance that these keyframes will accurately represent the video's content due to the uncontrollable occurrence of duplicate keyframes.

D. Singh et al. (Eds.): IHCI 2024, LNCS 15557, pp. 269–280, 2025.
https://doi.org/10.1007/978-3-031-88705-5_23

An initial approach to keyframe extraction was founded on shot boundary detection [10, 11]. This technique essentially utilizes either the initial or median frame from each identified shot sequence as the representative keyframe following shot boundary determination [12]. While this method ensures a limited quantity of keyframes, with only a single frame selected per shot, thereby preserving the sequential integrity of the keyframes, it fails to guarantee their quality, which constitutes a significant limitation. Moreover, this approach is heavily contingent upon the ability of the detected shot sequences; if the quality of these sequences is suboptimal, the keyframes derived from them will inherently suffer from diminished quality as well.

Clustering-based techniques constitute one of the most prevalent methodologies for keyframe extraction. The fundamental premise involves grouping frames based on their low-level features, such as texture, color, or shape, through the application of clustering algorithms like K-means, with the frames exhibiting the highest similarity or proximity to the cluster centers being selected as keyframes [13]. This clustering-based method was found to be efficient, computationally expedient, and well-suited for online processing. Various techniques, including VSumm [14], SGC [15], and GMC [16], have employed K-means, Minimum Spanning Tree, and Graph Modularity-based Clustering Algorithms, respectively, for keyframe extraction. However, a significant limitation of clustering methods is the inability to preserve the temporal sequence of the keyframes, coupled with the inherent challenge of parameter selection, which remains a pervasive issue for clustering algorithms.

The frame difference-based methodology aims to select keyframes by identifying the frames that exhibit the least dissimilarity from other frames, based on a given similarity metric. VSUKFE [17] and HistDiffKey [18] exemplify this approach, wherein they detect abrupt changes in the distances between consecutive frames and recognize keyframes only when the difference between frames surpasses a predetermined threshold. While this method has the advantage of preserving the temporal order of the keyframes, its major drawbacks lie in the difficulty of determining an optimal threshold value and the potential compromise in keyframe quality.

Deep learning architectures, particularly Convolutional Neural Networks (CNNs), have attained state-of-the-art performance in image processing tasks: object detection [19], object segmentation [20], image classification [21, 22] and object recognition [23]. Numerous studies have successfully applied CNNs to video processing tasks as well. Rukiye et al. integrated CNNs and Recurrent Neural Networks (RNNs) with the objective of leveraging temporal information between video frames to enhance video classification performance [24]. Changhai et al. presented a novel deep learning approach to automatically identify keyframes depicting athletes' salient poses from sports training videos by integrating Fully Convolutional Networks (FCNs) and Convolutional Neural Networks (CNNs) [25]. In 2024, the LMSKE method [26] proposed a potentially promising direction: a large model-based sequential keyframe extraction technique, utilizing large models for shot detection, feature extraction, and a K-means-based clustering method to automatically determine the optimal number of clusters. While the results of this method were promising, there is still scope for improvement in terms of Compression Ratio (CR) and F1 accuracy.

Nevertheless, these existing methods are suboptimal as they fail to fully leverage the advancements in Deep Learning. The results often struggle to strike a balance between precision, recall, and Compression Ratio (CR) [27], with high precision and recall typically corresponding to low Compression Ratio, and vice versa. Additionally, clustering is one of the main methods for identifying Keyframe, but the use of traditional clustering methods poses significant challenges in selecting appropriate parameter values.

The proposed methodology in this study combines shot boundary detection, clustering, and frame difference techniques, with the aim of addressing some significant limitations of the aforementioned approaches. Initially, shot segmentation is performed using the robust TransNetV2 model [28]. Subsequently, features are extracted from each segmented shot using the state-of-the-art CLIP model [29]. These features are then fed into an adaptive clustering model that synergistically combines DBSCAN and kNN algorithms to identify keyframes. Finally, a frame difference mechanism, specifically the histogram criterion, is employed to eliminate frames exhibiting high overlap. By integrating shot boundary detection with adaptive clustering, the temporal order of the keyframes is preserved. Furthermore, the utilization of the kNN algorithm to estimate the epsilon (eps) parameter for DBSCAN enables the algorithm to automatically cluster without the complexity of parameter tuning. The frame difference method serves as the final filtering step, ensuring a concise yet informative representation of the video content through a reduced number of keyframes. Figure 1 generally describes the pipeline of the ViTKDB system.

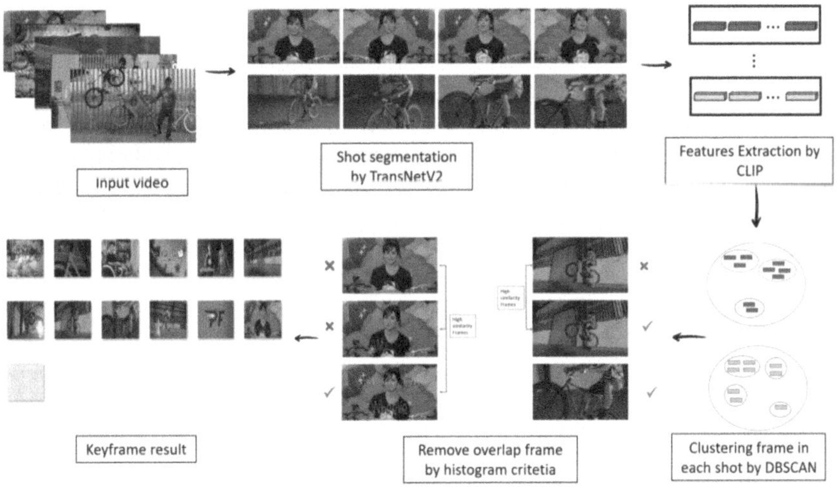

Fig. 1. ViTKDB Keyframe Extraction Pipeline

An empirical evaluation was conducted on the TVSum20 [26] benchmark dataset. The results prove that the proposed methodology not only circumvents the intrinsic limitations of traditional algorithms (Uniform, DiffHist, etc.), but also demonstrates a statistically significant enhancement in F1 score and Compression Ratio compared to algorithms leveraging contemporary deep learning techniques (LMSKE).

3 Method

3.1 Shot Segmentation

Shot boundary-based techniques constitute one of the principal methodologies for keyframe determination [1]. Within the structural composition of a video, a shot represents the smallest temporal unit of visual information, encompassing a sequence of interrelated frames captured uninterruptedly by a single camera. These frames illustrate specific interconnected actions or events occurring over time and in various locations. The incorporation of this technique into the pipeline confers upon the extracted frames the requisite representational capacity to comprehensively encapsulate the video content without omitting any crucial information.

For a given video $V = \{f_0, f_1, \ldots, f_n\}$ which f_i represent $i - th$ frame and n is the number of frame in V, upon passing through the TransNetV2 model, a list of boundaries $B = \{b_0, b_1, \ldots, b_m\}$ is obtained. Each boundary $b_j = \{idx_{start}, idx_{end}\}$ within this list consists of two values representing the start and end frames of a shot while m is the number of shot in V. Subsequently, the model proceeds to segment the video into a list of shots $S = \{s_0, s_1, \ldots, s_m\}$ based on the aforementioned boundary list B. It is noteworthy that the shots within the resulting list S maintain the original sequential order of the input video.

3.2 Feature Extraction

Subsequent to the extraction of the shot list S, each individual shot s_i within S undergoes the feature extraction process utilizing the large vision transformer model CLIP [29]. With each shot s_i, we obtain a list of features $F = \{x_0, x_1, \ldots, x_l\}$, $x_i \in R^{768}$ denote $i - th$ frame's feature vector and l is the number of frames of shot s_i. The quality of the extracted features holds significant bearing on the quality of the resultant keyframes. Consequently, the employment of CLIP, a state-of-the-art model endowed with Attention mechanism [30] and robust image understanding capabilities, ensures a rich semantic comprehension and yields features of optimal quality.

3.3 Adaptive Clustering

To group visually similar frames and facilitate keyframe selection, we employ an adaptive density-based clustering approach using the DBSCAN algorithm. DBSCAN is a powerful unsupervised clustering technique that is well suited for the keyframe extraction problem. DBSCAN can effectively identify clusters of arbitrary shape and size, unlike algorithms like K-means which assume spherical cluster shapes. This flexibility is advantageous for keyframe extraction, as the distribution of frames within a video can be complex and non-uniform. Besides that, DBSCAN has the ability to automatically identify and label noise points or outliers, which are data points that do not belong to any cluster. This property is valuable for keyframe extraction, as some frames within a video may not be representative or informative, and can be effectively filtered out as noise. Crucially, DBSCAN does not require prior knowledge of the number of clusters like other clustering algorithms, making it well-suited for the context of keyframe extraction, where the optimal number of keyframes (clusters) can vary significantly across different videos and is generally unknown a Priori.

The concept of density-reachable is the main problem that DBSCAN targets [7]. The following quantities are initially computed for this purpose:

i. **minSamples** – It represents the minimum number of neighboring points required for a point to be considered a core point. This parameter essentially sets the minimum cluster size. Typically, for 2-D data, a value of 3 or 4 is sufficient. In our proposal, the value is not critically impactful on performance, hence it is set to 4.
ii. **eps (ε)** – A point p is considered a neighbor of point q if it lies within the eps distance. Originally, DBSCAN employed a fixed value of ε defined by the user, and the output heavily depends on it. In other words, determining an appropriate value for the ε parameter is the key challenge when using DBSCAN, as an excessively small value may lead to excessive fragmentation, while an excessively large value may cause distinct clusters to merge incorrectly. Furthermore, choosing a fixed ε value for Keyframe Extraction is completely inconsistent with the diverse and dynamic nature of video content. The optimal value of ε is highly dependent on the distribution and density of the data points within the dataset. A fixed ε value may work well for certain density regions but fail to capture the appropriate clusters in other regions with different densities. This can lead to either over-clustering (splitting dense clusters) or under-clustering (merging distinct clusters). Besides that, real-world datasets, including video frames, often exhibit varying densities and complex structures. A fixed ε value cannot effectively adapt to these variations, resulting in suboptimal clustering performance. Different regions of the dataset may require different ε values to accurately identify the underlying clusters.

To adaptively estimate an appropriate ε value tailored to the specific characteristics of the input data, we employ a kNN (k-nearest neighbors) based approach. It is computed as the mean of all $k - th$ nearest neighbor distances of all points. Regarding the k value of kNN, we have plotted the extracted features to visually inspect the separability and structure of the data. Based on the visualizations of the feature space, the data exhibits a relatively uniform or continuous distribution without distinct, well-separated clusters. In such cases, the kNN density estimation method is less sensitive to the choice of k, as the underlying density structure of the data does not vary significantly within a certain range of k values. For example, in two cases in Fig. 2, the data points appear to be distributed in a relatively uniform or evenly spaced manner across the 2D embedding space. There are no clear distinct clusters or dense regions visible. This uniformity in the data distribution could lead to similar nearest neighbor distances, regardless of the value of k chosen, resulting in similar ε estimates. In the other cases in Fig. 3, when the points are more densely distributed, the distances to the nearest neighbors within the same cluster is relatively small and consistent. As long as k is small enough to capture the local density within each cluster, the ε value will not change significantly when varying k.

The number of frames for clustering is diverse, ranging from less than 10 to hundreds per shot. To maintain generality, we experimented with different k values, ranging from 4 to 50, in different shots to verify this insensitivity. The results show that the predicted

ε value is almost unchanged. Therefore, despite the insensitivity, we choose value of $k = 10$. This value is neither too large nor too small to produce the optimal results while still providing a reasonable density estimation and clustering results, striking a balance between capturing the local density structure of the data and being robust to potential noise or outliers. The clustering results in Fig. 2 and Fig. 3 are obtained with the value $k = 10$. The value ε is calculated based on kNN with the formula:

$$\varepsilon = mean_{p \in F} \|p, p_{10}\|_2$$

where F is the list of extracted features from one shot s, p_{10} is the $10-th$ nearest neighbor of p. In case that F itself has fewer data points than 10 (number of features in the shot is less than 10), the length of features k will be used instead of 10, and ε is computed as:

$$\varepsilon = mean_{p \in F} \|p, p_k\|_2$$

This approach ensures that the ε radius encompasses at least k points in the local neighborhood of each frame, while accounting for variations in local density.

To summarize, we first use kNN to find the appropriate number ε for that set of features, then use that number ε along with the predefined minSamples parameter (set to 4) for the DBSCAN algorithm to find clusters as well as noise points. After the clusters have been formed, the frame features closest to the centroid clusters will be add to the set of potential keyframes P.

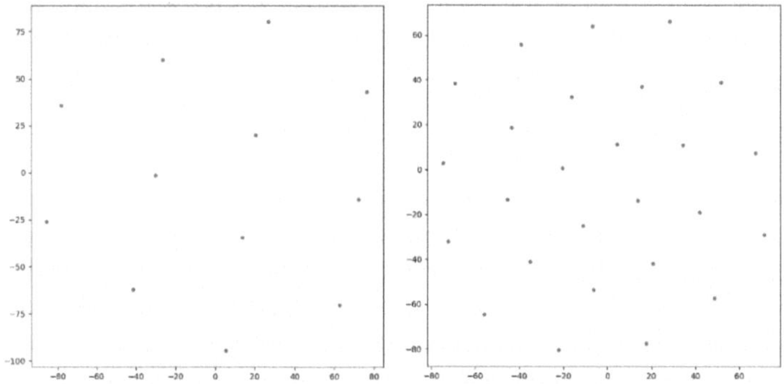

Fig. 2. Feature Space Visualization (uniform data point)

3.4 Keyframe De-duplication

Subsequent to the clustering process, we obtain a list of potential keyframes. These potential keyframes represent each shot, with the possibility of one or more keyframes per shot. However, depending on the specific nature of the shot, the quality of these potential keyframes is not guaranteed, as they may exhibit redundancy or lack of information. Therefore, in this step, we perform post-processing operations to filter out redundant keyframes.

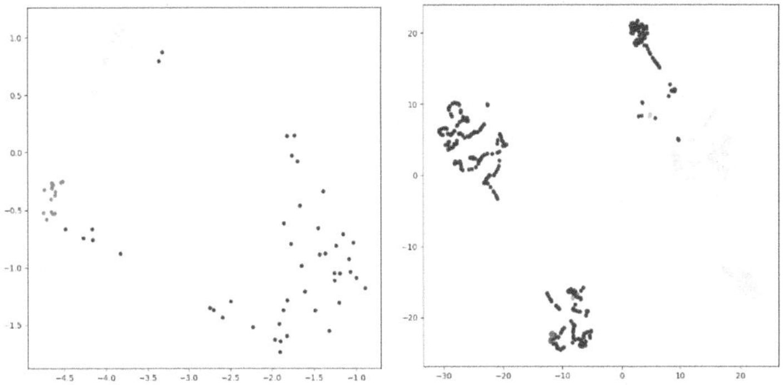

Fig. 3. Feature Space Visualization (dense data point)

Certain shots consist solely of transitions and are predominantly displayed in black. Consequently, the potential keyframes filtered from such shots convey almost no information representative of the video content. Removing these keyframes is necessary to optimize the final result. Utilizing the histogram criterion, we assess which frames possess a sufficient number of non-zero bin counts above a certain threshold. In greater detail, if the threshold is set too low, the removal process will be ineffective, whereas an excessively high threshold may lead to the exclusion of frames containing important information (some frames may have a black background). For example, Fig. 4 and Fig. 5 respectively represent a frame with no information (the number of zero bin-counts is too high) and a frame with a black background (the number of zero bin-counts is high but still contains information). After conducting several experiments, a threshold value of 8 has been determined as appropriate for eliminating frames with insufficient information.

Fig. 4. Frame with no information

Fig. 5. Frame with black background

Furthermore, for extended shots, the number of potential keyframes can be substantial, and a high degree of similarity between them is likely. Therefore, still relying on the histogram criterion, we compare the similarity between potential keyframes. If the similarity between two frames exceeds a certain threshold, one of them will be eliminated. A matrix Q representing the similarity between potential keyframes is established.

$$i, j = argmax_{i,j} UpTri(Q[i, j])$$

$UpTri()$ is utilized to obtain the upper triangular portion of the similarity matrix Q. For example, couple of frame i and j has a similarity index higher than 0.7. Subsequently, the $j-th$ frame, considered redundant, is eliminated from the set P, and Q is updated by removing the $j - th$ column. This iterative process continues until no keyframe pair has a similarity index exceeding 0.7. Note that, since the frames are removed from the end during each iteration, the relative order of the remaining frames is preserved. This method ensures that the original sequence of frames is maintained as closely as possible.

4 Experiments and Analyze

This section provides a description of the datasets utilized, the experimental configuration, and the evaluation approach.

4.1 Dataset and Setting

The study employs the TVSum20 Dataset, a benchmark dataset derived from the TVSum Dataset. The TVSum20 comprises 20 videos curated from YouTube, spanning 10 distinct categories, with each category encompassing 2 videos, wherein every video possesses a sequence of keyframes. All clips within the TVSum20 dataset exhibit a fixed resolution of 640x360 pixels. The durations of the clips range from a minimum of 1.38 min to a maximum of 4.03 min.

4.2 Evaluation Method

In this experiment, we use two metrics to evaluate results: F1 score and Compression Ratio (CR). These two popular metrics demonstrate high efficiency in assessing the quality of keyframe extraction models.

The F1 Score is a useful metric for keyframe extraction, balancing precision (fraction of truly representative extracted keyframes) and recall (proportion of important frames correctly identified). It accounts for false positives and false negatives, with higher values indicating better video summarization accuracy.

The Compression Ratio (CR) is a critical metric in the evaluation of keyframe extraction techniques, as it quantifies the degree of video summarization achieved. It is defined as the ratio of the total number of frames in the original video to the number of extracted keyframes. The formula is:

$$CR = 1 - \frac{N_k}{N_f}$$

where, N_k is the number of keyframe models extracted, N_f is the number of frames in the video.

A higher CR value is desirable, as it indicates that fewer keyframes are required to effectively summarize the video content. This translates to a more concise and compact video summary, reducing redundancy and storage requirements compared to the original video. Conversely, a lower CR implies a less condensed representation, potentially including redundant or less informative frames. An effective keyframe extraction method should aim to maximize the CR while preserving the salient visual information and maintaining a coherent narrative structure within the extracted keyframes. A high CR highlights the method's ability to extract a small set of highly representative keyframes that capture the essence of the video, enabling efficient storage and transmission of the summarized content.

4.3 Result

We conducted experiments on the TVSum20 benchmark dataset, employing several popular methods described earlier. The methods evaluated include: Uniform (employing a predetermined frame interval of 30), HistDiffKey, GMC, VSUMM, K-means, LBP-Shot and LMSKE.

Table 1. Result

Method	F1	CR
Uniform [9]	0.206	0.966
HistDiffKey [18]	0.338	0.983
GMC [16]	0.483	0.988
VSUMM [14]	0.489	0.990
K-means [13]	0.504	0.990
LBP-SHOT [31]	0.505	0.991
LMSKE [26]	0.531	0.992
ViTKDB	**0.573**	**0.993**

Table 1 presents the comparative results across the evaluated methods. It is evident that the proposed architecture, ViTKDB, which integrates TransNetV2, CLIP, Adaptive Clustering with DBSCAN combined with kNN for keyframe extraction, surpasses the previously existing methods, attaining F1 score of 0.573 and CR of 0.993, respectively.

Figures 6 and 7 illustrate examples of applying the proposed ViTKDB method to two videos from the TVSum20 dataset. It can be observed that the difference between the number of predicted keyframes and the ground truth labels is relatively small, with 13 and 13, as well as 16 and 12 keyframes, respectively. Similarly, the number of correctly predicted keyframes closely aligns with the labeled data, exhibiting 11 and 13, as well as 10 and 12 correct predictions. Furthermore, it is noteworthy that the order of the extracted keyframes is preserved, maintaining the same sequence as the original video.

5 Conclusion and Perspectives

This paper presents an approach that effectively utilizes the power of Deep learning, especially the large vision transformer model with adaptive clustering techniques and post-processing strategies for the task of keyframe extraction from video data. The proposed ViTKDB method leverages TransNetV2 and CLIP for high-quality shot segmentation and feature extraction tasks, respectively. Subsequently, an adaptive clustering approach utilizing DBSCAN combined with kNN is employed to identify keyframes. Lastly, a post-processing step is performed to eliminate highly similar keyframes and keyframes with minimal informative content based on histogram criteria, while preserving the order of frames relative to the original video. Experimental results demonstrate that ViTKDB achieves the highest F1 score and compression ratio among the evaluated methods.

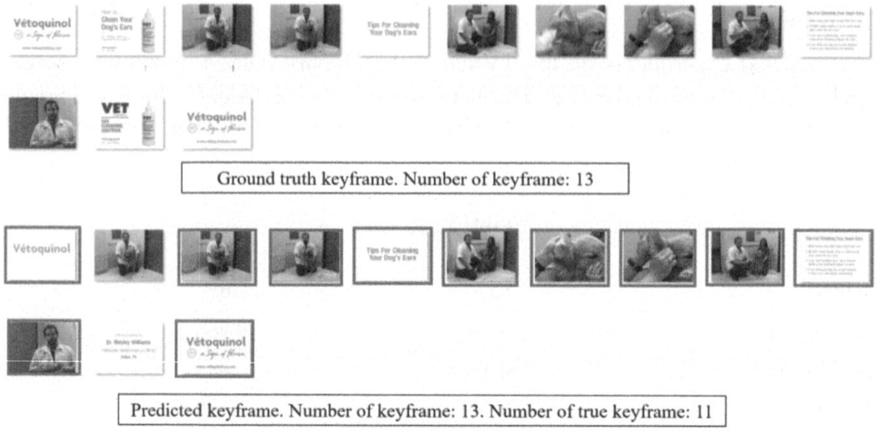

Fig. 6. Case study 1

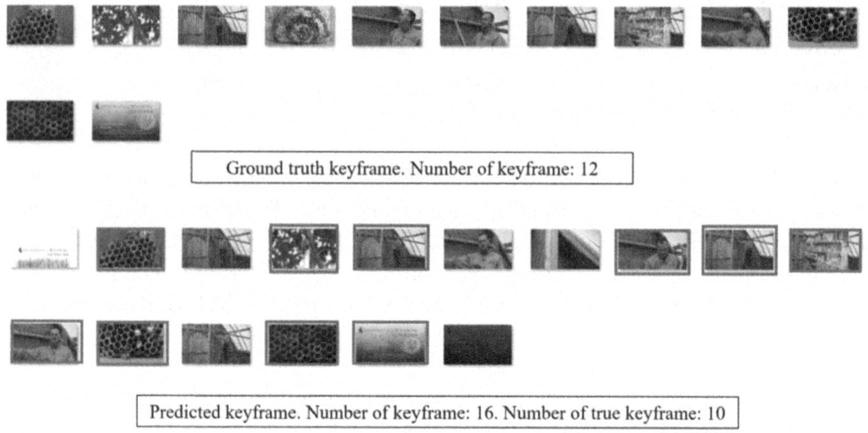

Fig. 7. Case study 2

Despite the proposed method outperforming commonly used keyframe extraction techniques, this study is subject to limitations regarding technical infrastructure. Specifically, the experiments were conducted on a single Tesla T4 GPU with limited time availability through Google Colab, batch size for vision transformer models was constrained, preventing the full utilization of these powerful model's capabilities and exploration of alternative architectures. Future work could focus on applying the proposed algorithm using more advanced architectures to address real-world video summarization problems.

References

1. Sadiq, B.O., Muhammad, B., Abdullahi, M.N., Onuh, G., Muhammed, A.A., Babatunde, A.E.: Keyframe extraction techniques: a review. ELEKTRIKA- J. Elect. Eng. **19**(3), 54–60 (2020)

2. Paul, M.K.A., Kavitha, J., Rani, P.A.J.: Key-frame extraction techniques: a review. Recent Patents Comput. Sci. **11**(1), 3–16 (2018)
3. Meena, P., Kumar, H., Yadav, S.K.: A review on video summarization techniques. Eng. Appli. Artifi. Intell. **118**, 105667 (2023)
4. Karpathy, A., Toderici, G., Shetty, S., Leung, T., Sukthankar, R., Fei-fei, L.: Large-scale video classification with convolutional neural networks. In: Proceedings of the IEEE Conference on Computer Vision and Pattern Recognition, CVPR 2014, Columbus, OH, USA (2014)
5. Abu-El-Haija, S., Kothari, N., Lee, J., Natsev, P., Toderici, G., Varadarajan, B., Vijayanarasimhan, S.: Youtube-8m: a large scale video classification benchmark. arXiv: 1609.08675 (2016)
6. Song, Y., Vallmitjana, J., Stent, A., Jaimes, A.: Tvsum: summarizing web videos using titles. In: Proceedings of the IEEE Conference on Computer Vision and Pattern Recognition, CVPR 2015, Boston, MA, USA (2015)
7. Ester, M., Kriegel, H.S., Sander, J., Xu, X.: A density-based algorithm for discovering clusters in large spatial databases with noise. In: KDD 1996 Proceedings (1996)
8. Zhou, H.: K-nearest neighbors. In: Learn Data Mining Through Excel, pp. 93–108. Apress, Berkeley, CA (2020). https://doi.org/10.1007/978-1-4842-5982-5_7
9. Truong, B.T., Venkatesh, S.: Video abstraction: a systematic review and classification. ACM Trans. Multimedia Comput. Commun. Appli. **3**(3) (2007)
10. Boreczky, J.S., Rowe, L.A.: Comparison of video shot boundary detection techniques. J. Electron. Imaging **5**, 122–129 (1996)
11. Nagasaka, A., Tanaka, Y.: Automatic video indexing and full-video search for object appearances. J. Inform. Process. **15**, 113–127 (1992)
12. Pal, G., Rudrapaul, D., Acharjee, S., Ray, R., Chakraborty, S., Dey, N.: Video shot boundary detection: a review. AISC, pp 119–127. Springer (2015). https://doi.org/10.1007/978-3-319-13731-5_14
13. Zhuang, Y., Rui, Y., Huang, T.S., Mehrotra, S.: Adaptive key frame extraction using unsupervised clustering. In: International Conference on Image Processing, Chicago, IL, USA (1998)
14. de Avila, S.E.F., Lopes, A.P.B., da Luz, A., de A. Araujo, A.: VSUMM: a mechanism designed to produce static video summaries and a novel evaluation method. Pattern Recognition Lett. **32**, 56–68 (2011)
15. Sima, M.: Key frame extraction for human action videos in dynamic spatio-temporal slice clustering. In: Proceedings of the International Conference on Computer Information Science and Application Technology, Lanzhou, China (2021)
16. Gharbi, H., Bahroun, S., Massaoudi, M., Zagrouba, E.: Key frames extraction using graph modularity clustering for efficient video summarization. In: Proceedings of the International Conference on Acoustics, Speech and Signal Processing, New Orleans, LA, USA (2017)
17. Ding, Y., Hou, S., Yang, X.: Key frame extraction based on frame difference and cluster for person re-identification. In: Symposia and Workshops on Ubiquitous, Autonomic and Trusted Computing, Atlanta, GA, USA (2021)
18. Rodriguez, J.M.D., Yao, P., Wan, W.: Selection of key frames through the analysis and calculation of the absolute difference of histograms. In: International Conference on Audio, Language and Image Processing, Shanghai, China (2018)
19. Tra, N.D., Tri, N.C., Hung, P.D.: Improving warped planar object detection network for automatic license plate recognition. In: Dang, T.K., Küng, J., Chung, T.M. (eds.) Future Data and Security Engineering. Big Data, Security and Privacy, Smart City and Industry 4.0 Applications. FDSE 2023. CCIS, vol 1925. Springer, Singapore (2023). https://doi.org/10.1007/978-3-319-13731-5_14

20. Manh, D.Q., Khoi, T.M., Hieu, D.M., Hung, P.D.: TextFocus: efficient Multi-scale Detection for Arbitrary Scene Text. In: Hà, M.H., Zhu, X., Thai, M.T. (eds.) Computational Data and Social Networks. CSoNet 2023. LNCS, vol 14479. Springer, Singapore (2024). https://doi.org/10.1007/978-981-97-0669-3_4
21. Krizhevsky, A., Sutskever, I., Hinton, G.E.: Imagenet classification with deep convolutional neural networks. In: Advances in Neural Information Processing Systems, pp 1097–1105 (2012)
22. Mai, C.T.H., Dung, D.T.P., Le Anh Duc, P., Hung, P.D.: Skeleton-Based fall detection using computer vision. In: Luo, Y. (eds.) Cooperative Design, Visualization, and Engineering. CDVE 2023. LNCS, vol 14166. Springer, Cham (2023). https://doi.org/10.1007/978-3-031-43815-8_15
23. Hieu, P.N.D., Hung, P.D.: A convolutional autoencoder approach for weakly supervised anomaly video detection. In: Nguyen, N.T., et al. Computational Collective Intelligence. ICCCI 2023. LNCS, vol 14162. Springer, Cham (2023). https://doi.org/10.1007/978-3-031-41456-5_11
24. Savran Kızıltepe, R., Gan, J.Q., Escobar, J.J.: A novel keyframe extraction method for video classification using deep neural networks. Neural Comput. Appl., 1–12 (2021). https://doi.org/10.1007/s00521-021-06322-x
25. Lv, C., Li, J., Tian, J.: Key frame extraction for sports training based on improved deep learning. Sci. Program. **2021**, 1–8 (2021)
26. Tan, K., Zhou, Y., Xia, Q., Liu, R., Chen, Y.: Large model based sequential keyframe extraction for video summarization. In: Proceedings of the International Conference on Computing, Machine Learning and Data Science (CMLDS 2024). Association for Computing Machinery, New York, USA, Article 52, pp. 1–5 (2024)
27. Gharbi, H., Bahroun, S., Zagrouba, E.: A novel key frame extraction approach for video summarization. In: Proceedings of the 11th Joint Conference on Computer Vision, Imaging and Computer Graphics Theory and Applications (VISIGRAPP 2016), vol. 3, pp. 148–155 (2016)
28. Soucek, T., Lokoc, J.: Transnet V2: an effective deep network architecture for fast shot transition detection. arXiv: 2008.04838 (2020)
29. Radford, A., et al.: Learning transferable visual models from natural language supervision. In: ICML 2021, pp. 8748–8763 (2021)
30. Vaswani, A., et al.: Attention Is All You Need. arXiv:1706.03762 (2017)
31. Nandini, H.M., Chethan, H.K., Rashmi, B.S.: Shot based keyframe extraction using edge-lbp approach. J. King Saud Univ.-Comput. Inform. Sci. **34**(7), 4537–4545 (2022)

Improving Polyps Segmentation in Colonoscopy Images Using Modified UNet3 + Network

Huynh Anh Duy, Huynh Anh Khoa, and Phan Duy Hung(✉)

FPT University, Hanoi, Vietnam
{duy23mse23139,khoa23mse23138}@fsb.edu.vn, hungpd2@fe.edu.vn

1 Introduction

Cancer affecting the colon and rectum is a major health issue worldwide. For women, it's the second most common cancer they face, while for men, it ranks third. Most of these cancers—about 80% to 95%—start as small growths called polyps in the intestines. Doctors use a procedure called colonoscopy to look for these polyps and diagnose cancer. It's considered the best method available, but how well it works depends a lot on the doctor's skill and ability to control the scope accurately. Because of this, it's really important to find ways to spot polyps early and correctly in the images from colonoscopies. Many researchers have been working on this problem, trying to come up with better ways to segment polyps [1–6]. In recent years, the rapid advancement of deep learning has led to deep learning-based solutions being prioritized for this problem. Specifically, segmentation models leveraging convolutional neural networks have demonstrated remarkable proficiency. Notably, the UNet [7] architecture, employing an encoder-decoder paradigm and skip connections to preserve spatial information, has exhibited impressive performance on medical image segmentation tasks. However, the utilization of skip connections can lead to imbalance in information within the connection of low-level features from the encoder and high-level features from the decoder. To address this challenge, many enhanced variants of the U-net model have been proposed, such as ResUNet [8], ResUNet ++ [9], MultiResUNet [10], which incorporate ResPath, or UNet ++ [11], which utilizes nested and dense connections to calibrate the semantic information exchange between the feature maps of encoder and decoder. The UNet3 + model [12] is an upgrade version of UNet + + [11] architecture which redesign the connections between the encoder and decoder, as well as the connections between the decoder components, to effectively capture information across the whole scale. However, merely utilizing traditional backbones such as VGG19 [13] and ResNet101 [14], as well as not fully exploiting features from powerful pretrained models, makes the model suboptimal in terms of both efficiency and size.

Transfer learning is a subfield of machine learning and artificial intelligence that was first introduced in 1992 [15]. It is designed to apply knowledge gained from one task to another but similar task. In a nutshell, transfer learning can be simply understood as enhancing learning in a new task by transferring knowledge from a related task that has already been learned. Nowadays, transfer learning is a powerful resource for humans to build new AI solutions and continue pushing the boundaries of machine learning.

D. Singh et al. (Eds.): IHCI 2024, LNCS 15557, pp. 281–292, 2025.
https://doi.org/10.1007/978-3-031-88705-5_24

In this work, we aim to improve polyps segmentation results on colonoscopy images through upgrading and applying transfer learning to the UNet3 + model. Experimental results on a benchmark dataset show that the proposed method improves the original model in both performance and efficiency.

This paper is structured into five main sections. Following the introduction, we provide a background of existing research in Sect. 2. Section 3 details our research approach and methods. The findings of our study and their interpretation are presented in Sect. 4. To conclude, Sect. 5 summarizes our key discoveries and proposes areas for further investigation in this field.

2 Related Works

The UNet3 + model has been widely used in many practical applications, including areas outside of medical imaging. Along with this, many solutions have been proposed in recent years to improve the performance of the model. Some notable works are described below:

In paper [16], the authors introduce an innovative approach to tackle the cloud and snow detection problem by leveraging an improved version of the UNet3 + model. The proposed methodology begins with a color space conversion applied to remote sensing images, wherein both RGB and HIS images serve as input to the UNet3 + network. To facilitate a more profound extraction of remote sensing image features, the feature extraction component of the original UNet3 + is replaced with ResNet50 [14]. Additionally, the Convolutional Block Attention Module (CBAM) is incorporated into ResNet50, enhancing the network's ability to focus on cloud and snow regions. Lastly, a weighted cross-entropy loss function is constructed to address the class imbalance issue stemming from the disproportionately large background areas present in the images. The empirical evaluation demonstrates that the proposed methodology can successfully remove irrelevant information and accurately identify cloud and snow regions from remote sensing imagery of different land types.

The study presented in [17] proposes SwinE-UNet3 +, a novel end-to-end framework that seamlessly integrates Transformer architectures with convolutional neural networks (CNNs), built upon the foundation of the UNet3 + model for medical image segmentation tasks. The traditional convolutional network in the UNet3 + encoder is replaced by the Swin Transformer structure [18], while the decoder component retains the use of convolutional blocks. The SwinE-UNet3 + architecture employs a hybrid approach, Comparative evaluations against the UNet series models and the Attention-UNet model [19], conducted on the TipDM Cup rectal cancer dataset [20] and the ISIC-2017 dataset [21], demonstrate the superior accuracy and enhanced segmentation performance of the proposed methodology in delineating cancerous regions within tumors.

Another innovative strategy aimed at enhancing the UNet3 + model is presented in [22]. The authors propose a full-scale deeply supervised network architecture based on the U-Net3 + framework, which achieves more effective feature fusion by modifying the fusion strategy within the deeply supervised modules. Furthermore, an attention mechanism is incorporated into the model to preserve spatial information, thereby enabling improved suppression of background regions. The experimental evaluation, conducted on a set of clinical SPECT skeletal scans, assesses the performance of the proposed model using the Dice Similarity Coefficient (DSC), Class Pixel Accuracy (CPA), and Recall metrics, achieving respective scores of 0.6433, 0.6120, and 0.6833.

The author in paper [23] introduces another transformer-based approach for improving UNet3+ model named TA-UNet3+. This approach employs a visual transformer to replace the encoder component of the network architecture. Notably, the researchers incorporate a novel attention mechanism, termed the encoder-decoder transformer (EDformer), within the skip connection, enabling the effective learning of local features. Additionally, the study proposes a unique skip connection strategy that facilitates the seamless integration of low-level semantic features with high-level semantic features. The evaluation of the proposed methodology is conducted on the KisTS19 dataset [24], and the results demonstrate its superiority over three advanced segmentation methods with the mean dice coefficients achieved for kidney and kidney tumor segmentation tasks are 0.9885 and 0.8638, respectively.

A new segmentation network for remote sensing image named MA-UNet3+ is proposed in paper [25]. This approach aims to enhance the segmentation accuracy by incorporating feature maps of varying scales from the input image during the coding stage. Moreover, the authors integrate the ECA Block [26] into the UNet3+ network architecture, thereby improving the feature extraction capabilities for the input images. Additionally, the study implements a pruning strategy on the UNet3+ model, designed to reduce computational complexity by removing certain skip connections and optimizing the network structure. The experimental evaluation, conducted on the Vaihingen dataset, demonstrates the superior performance of the proposed segmentation algorithm compared to other benchmarked methods.

Some other studies with the same purpose of improving the UNet3+ model can be mentioned such as: integrating the channel and spatial attention mechanism, commonly referred to as CBAM, in conjunction with the cross-entropy loss function, into the network architecture of the UNet3+ model for the specific task of walnut image segmentation [27], combining full-scale skip connection with the spatial attention gate mechanism into UNet3+ model to solve the problem of mapping water surface [28], substituting the connections between feature maps of the encoder and decoder components within the UNet3+ model, which exhibit the same resolution, with a lightweight attention mechanism [29].

In general, the mentioned studies share a common trend of incorporating the attention mechanism from transformer models, which has emerged as a significant and powerful technique in recent years. However, focusing too much on this transformer-based direction unintentionally causes traditional CNN-based methods to be neglected and not fully exploited. In fact, appropriately combining traditional techniques like CNN-based methods and transfer learning still has the potential to improve the model's performance and reduce its size.

This work proposes applying transfer learning in a flexible way to improve the UNet3+ model for the Polyp Segmentation task. Specifically, intermediate layers with compatible shapes will be used to construct a new Encoder component for the UNet3+ model, transfer learning is applied by preserving the pretrained weights of these layers. Five well-known pretrained models are selected to evaluate this method, including:

VGG16 [13], ResNet50 [14], DenseNet121 [30], MobileNet [31], and MobileNetV2 [32]. Experimental results on the Kvasir-SEG dataset [33] demonstrate that our approach works well with most of pretrained models tested. Specifically, four out of five models achieve better segmentation performance and has fewer parameters compared to the original model.

3 Method

3.1 UNet3+ Model

In image processing, it is a well-established notion that feature representations at diverse scales capture varying levels of granular details and overarching structural patterns. The hierarchical architecture of the UNet framework offers a promising strategy for leveraging multiscale information effectively. A comparative analysis of UNet variants is depicted in Fig. 1.

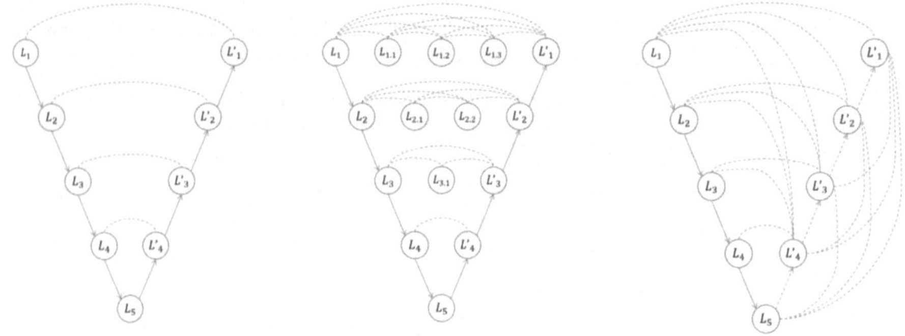

(a) UNet (b) UNet++ (c) UNet3+

Fig. 1. Comparison of UNet, UNet++ and UNet3+ Architecture

The original UNet [7] employs skip connections to synergistically combine low-level semantic cues from the encoder with their corresponding high-level semantic counterparts from the decoder. This mechanism facilitates the efficient extraction and retention of both fine-grained and coarse-grained information. In 2018, the UNet++ [11] variant introduced nested and dense skip connections, thereby augmenting the connectivity across different representational levels. The goal was to capture features spanning multiple scales and synthesize them through feature amalgamation. Thanks to the varying sensitivities of different receptive fields to objects of diverse shapes and sizes, the incorporation of multiple receptive field sizes in UNet++ resulted in enhanced information extraction and mitigated the loss of edge information during upsampling and downsampling operations. However, despite its improved performance, UNet++ still exhibited limitations in adequately capturing information from multiscale data. The UNet3+ [12] architecture introduced a significant advancement by eschewing short connections and capturing both fine-grained and coarse-grained semantics at full scale through comprehensive skip connections coupled with a loss function. Drawing inspiration from this, feature maps at different scales can extract diverse semantic information in image denoising tasks. Low-level detail maps encode rich spatial information and accentuate the overall structure of the image, while higher-level semantic maps capture nuanced information such as image contours. Specifically, the UNet3+ 's full-scale skip connection is performed by modifying the interconnectivity between the encoder and decoder, as well as the internal connections between decoder subnets, small-scale, same-scale, and large-scale feature maps were fully fused in each layer of the decoder to comprehensively exploit multiscale features, thereby enhancing the model's performance.

3.2 Proposed Method

In this section, we propose a new approach for the UNet3+ model to maximize the potential of available pretrained models. We will replace all the layers in the Encoder part of the original UNet3+ model with suitable layers extracted from the pretrained model. Specifically, layers from the pretrained model are considered suitable when their shape matches the shape of the layer they replace in the original UNet3+ model. All weights of these layers will be kept intact to serve the purpose of transfer learning. In the case where the utilized pretrained model has multiple layers of suitable shape, the layer at the deepest position will be chosen to ensure the quality of the selected features. Figure 2 illustrates the proposed method with the utilized pretrained model being VGG16 [13] and the input shape being 256 × 256. All other pretrained models will be handled similarly.

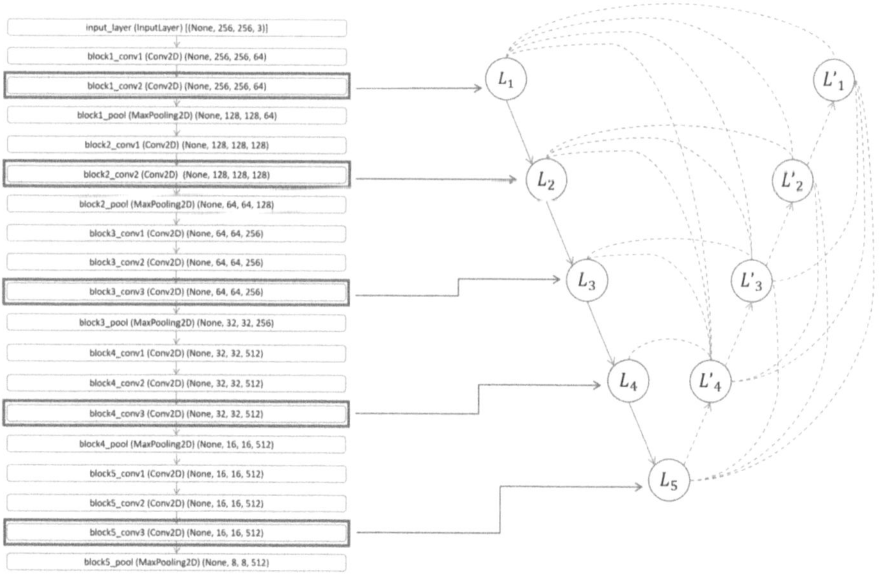

Fig. 2. Proposed Architecture with Pretrained VGG16

For this case, it can be observed that with the presented rule, the selected layers from the pretrained model are, respectively, "*block1_conv2*" with shape 256 ×

256, "*block2_conv2*" with shape 128 × 128, "*block3_conv3*" with shape 64 × 64, "*block4_conv3*" with shape 32 × 32, and "*block5_conv3*" with shape 16 × 16.

4 Experiments and Analyze

4.1 Dataset and Setting

We validate the effectiveness of the proposed method on the practical task of polyp segmentation, a crucial problem that has garnered significant attention from the medical community in recent years. For the segmentation task involving polyp images, it is a prerequisite that each pixel in the training image data is categorized as either belonging to the polyp class or the non-polyp class. To evaluate the proposed methodology, we employ the Kvasir-SEG dataset [33], comprising 1,000 polyp images and their corresponding ground truth masks meticulously annotated by expert endoscopists from the Oslo University Hospital (Norway). In our experiment, the dataset is divided into 80% training, 10% validation, 10% testing and all images are resized to 256 × 256. Figure 3 illustrates sample images from the Kvasir-SEG dataset on the left, along with their respective ground truth masks on the right.

Keras is used as the main framework for building and training deep learning model. All experiments were conducted on the cloud environment provided by Kaggle, equipped with an Intel(R) Xeon(R) CPU @ 2.20 GHz, 32 GB of RAM, and an Nvidia Tesla P100 GPU with 16 GB of memory.

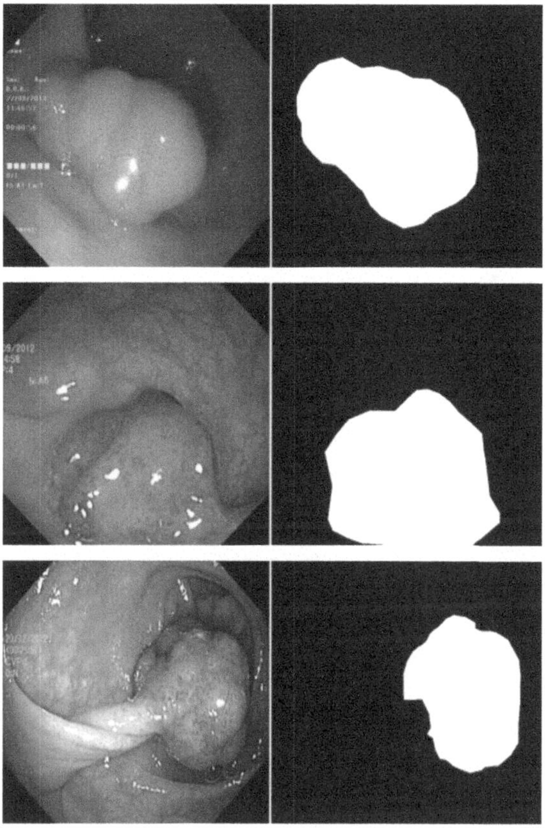

Fig. 3. Sample for Kvasir-SEG dataset

4.2 Evaluation Metrics

During the experimental evaluation, we employ two distinct evaluation metrics to quantify the performance of the proposed model – the Dice coefficient and the Intersection over Union (IoU) score. These metrics serve as measures of similarity, enabling a comparative analysis between the ground truth annotations, and the segmentation results predicted by the model. Values approaching unity for the Dice coefficient and IoU score indicate a high level of resemblance, implying that the model's predictions closely mirror the ground truth annotations. Conversely, values nearing zero signify a substantial divergence between the model's output and the ground truth, thereby indicating a poor performance in accurately predicting the segmentation maps. Below are details about the formulas of these two metrics.

$$IoU = \frac{TP}{TP + FP + FN + \epsilon}$$

$$Dice = \frac{2 \times TP}{2 \times TP + FP + FN + \epsilon}$$

In the above formula, TP, FP, and FN represent True Positive, False Positive, and False Negative, respectively. Meanwhile, ϵ is a constant added to the denominator to avoid the case where the denominator is equal to zero.

4.3 Comparison with Original Model

In this section, we compare the proposed method with the original UNet3+ model. The selected pretrained models include: VGG16 [13], ResNet50 [14], DenseNet121 [30], MobileNet [31], and MobileNetV2 [32]. These are powerful pretrained models that have achieved impressive performance on the ImageNet benchmark dataset [33].

4.3.1 Quantitative Comparison

The performance of the proposed method is compared against the original UNet3+ model using criteria such as the number of parameters and the values of two employed metrics, Dice coefficient and mean IoU (mIoU). The detailed results on test set are presented in Table 1.

Table 1. Performance comparison with original model

Model	Params	*Dice*	*mIOU*
Based model	26.97M	0.8114	0.7211
Pretrained VGG16	21.66M	0.7928	0.7051
Pretrained ResNet50	16.43M	0.8727	0.7993
Pretrained DenseNet121	12.14M	0.8680	0.7904
Pretrained MobileNet	8M	0.8586	0.7422
Pretrained MobileNetV2	8.48M	0.8782	0.7792

From Table 1, it can be observed that all models employing the proposed method have significantly fewer parameters compared to the original model. Among them, only the model utilizing the VGG16 pretrained model exhibits lower Dice and mIoU scores than the original model, while the remaining four pretrained models achieve superior Dice and mIoU values. Notably, the models employing the MobileNet and MobileNetV2 pretrained models, with significantly fewer parameters (8M and 8.48M, respectively) compared to the original model (26.97M), demonstrate remarkably superior performance. This evidence substantiates that our proposed method not only enhances the segmentation results but also reduces the model's size, a crucial factor for optimizing training and inference times.

4.3.2 Qualitative Comparison

Figure 4 illustrates the segmentation results of the original model and the model employing the proposed method, specifically the model utilizing the ResNet50 pretrained model. It can be visually observed that the results of the model applying the proposed method are closer to the ground truth compared to the initial original model. Specifically, the cases of over-segmentation, incomplete segmentation, or fragmented segmentation present in the original model have been significantly improved in the new model.

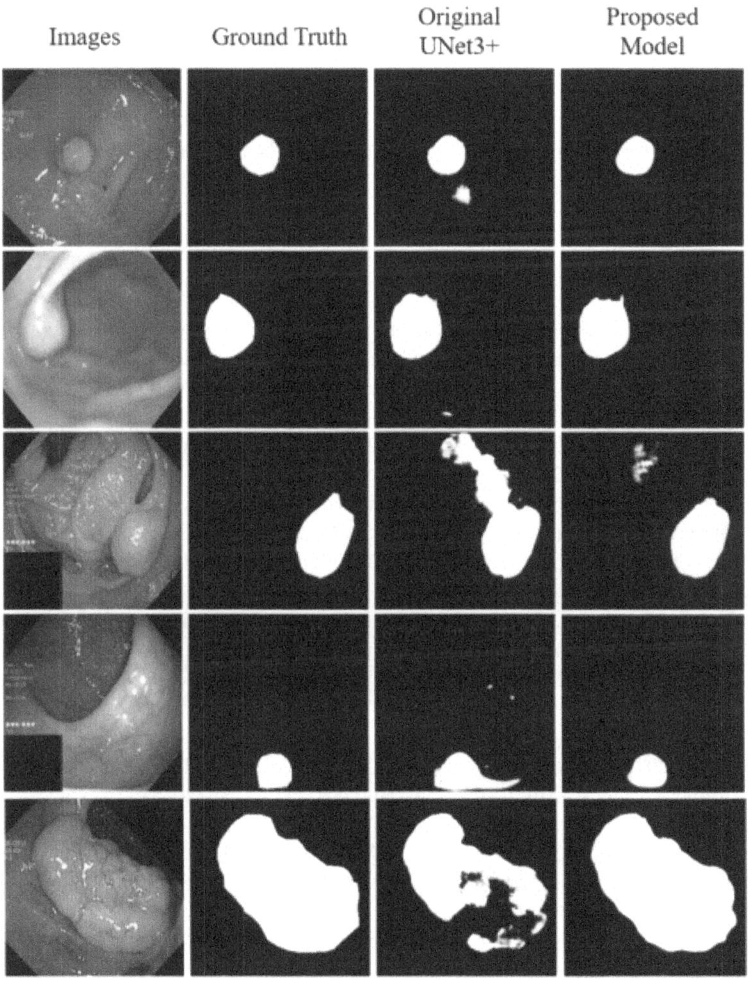

Fig. 4. Qualitative comparisons of original model and proposed model

5 Conclusion and Perspectives

This paper proposes a modification approach for the UNet3+ model to maximize the potential of available pretrained models through transfer learning. Specifically, the layers of the Encoder part are replaced by suitable layers from the pretrained model, while preserving the values of the trained parameters. The pretrained models utilized for evaluating the proposed method include: VGG16, ResNet50, DenseNet121, MobileNet, and MobileNetV2. These are all famous pretrained models, which have achieved very impressive results on benchmark datasets. Experimental results on the Kvasir-SEG dataset demonstrate that the proposed method not only improves the segmentation performance but also significantly reduces the number of parameters in the model. In future study, investigating the way to apply this proposed method to other CNN-based deep learning models is an interesting direction. This approach has great potential to enhance the effectiveness of these models, not just for medical images but also in many other real-world problems.

References

1. Ganz, M., Yang, X., Slabaugh, G.: Automatic segmentation of polyps in colonoscopic narrow-band imaging data. IEEE Trans. Biomed. Eng. **59**(8), 2144–2151 (2012)
2. Zhao, X., Zhang, L., Lu, H.: Automatic polyp segmentation via multi-scale subtraction network. In: de Bruijne, M., et al. (ed.) Medical Image Computing and Computer Assisted Intervention, MICCAI 2021. LNCS, vol. 12901. Springer, Cham (2021). https://doi.org/10.1007/978-3-030-87193-2_12
3. Fan, D.P. et al.: PraNet: parallel reverse attention network for polyp segmentation. In: Martel, A.L., et al. (ed.) Medical Image Computing and Computer Assisted Intervention, MICCAI 2020. LNCS, vol. 12266. Springer, Cham (2020).https://doi.org/10.1007/978-3-030-59725-2_26
4. Huynh, L.D., Doan, T.C., Hung, P.D.: A Lightweight Deep Semi-supervised Student Model for Medical Image Segmentation. In: Luo, Y. (eds) Cooperative Design, Visualization, and Engineering. CDVE 2024. Lecture Notes in Computer Science, vol 15158. Springer, Cham. https://doi.org/10.1007/978-3-031-71315-6_25 (2024)
5. Akbari, M., et al.: Polyp segmentation in colonoscopy images using fully convolutional network. In: 2018 40th Annual International Conference of the IEEE Engineering in Medicine and Biology Society (EMBC), Honolulu, HI, USA, pp. 69–72 (2018)
6. Ronneberger, O., Fischer, P., Brox, T.: U-Net: convolutional networks for biomedical image segmentation. In: Navab, N., Hornegger, J., Wells, W., Frangi, A. (eds.) Medical Image Computing and Computer-Assisted Intervention, MICCAI 2015. LNCS, vol. 9351. Springer, Cham (2015). https://doi.org/10.1007/978-3-319-24574-4_28
7. Zhang, Z., Liu, Q., Wang, Y.: Road extraction by deep residual U-Net. IEEE Geosci. Remote Sensing Lett. **15**(5), 749–753 (2018)
8. Jha, D., et al.: ResUNet++: an advanced architecture for medical image segmentation. In: 2019 IEEE International Symposium on Multimedia (ISM), San Diego, CA, USA, pp. 225–2255 (2019)
9. Ibtehaz, N., Sohel Rahman, M.: MultiResUNet : Rethinking the u-net architecture for multimodal biomedical image segmentation. Neural Netw. **121**, 74–87 (2020) ISSN 0893–6080

10. Zhou, Z., Rahman Siddiquee, M.M., Tajbakhsh, N., Liang, J.. UNet++: a nested U-Net architecture for medical image segmentation. In: Stoyanov, D., et al. (eds.) Deep Learning in Medical Image Analysis and Multimodal Learning for Clinical Decision Support. DLMIA ML-CDS 2018 2018. LNCS, vol. 11045. Springer, Cham (2018). https://doi.org/10.1007/978-3-030-00889-5_1
11. Huang, H., Lin, L., Tong, R., et al.: Unet 3+: A full-scale connected UNet for medical image segmentation. In: Proceedings of the IEEE international conference on acoustics, speech and signal processing (ICASSP), pp 1055–1059. IEEE (2020)
12. Simonyan, K., Zisserman, A.: Very deep convolutional networks for large-scale image recognition. In: Proceedings of the 3rd International Conference on Learning Representations (ICLR 2015) (2015)
13. He, K., Zhang, X., Ren, S., Sun, J.: Deep residual learning for image recognition. In: Proceedings of the IEEE Conference on Computer Vision and Pattern Recognition (CVPR), Las Vegas, NV, USA, pp. 770–778 (2016)
14. Pratt, L.Y.: Discriminability-based transfer between neural networks. In: Proceedings of the 5th International Conference on Neural Information Processing Systems (NIPS 1992). Morgan Kaufmann Publishers Inc., San Francisco, CA, USA, pp. 204–211 (1992)
15. Yin, M., Wang, P., Ni, C.: Cloud and snow detection of remote sensing images based on improved Unet3+. Sci. Rep. **12**, 14415 (2022)
16. Zou, P., Wu, J.S.: SwinE-UNet3+: swin transformer encoder network for medical image segmentation. Prog Artif Intell **12**, 99–105 (2023)
17. Liu, Z., et al.: Swin Transformer: Hierarchical Vision Transformer using Shifted Windows. arXiv:2103.14030 (2021)
18. Oktay, O., et al.: Attention U-Net: Learning Where to Look for the Pancreas. arXiv:1804.03999 (2018)
19. Bipat, S., Glas, A.S., Slors, F.J.M., Zwinderman, A.H., Bossuyt, P.M.M., Stoker, J.: Rectal cancer: local staging and assessment of lymphnodeinvolvement with endoluminal US, CT, and MR imaging—a meta-analysis. Radiology **232**(3), 773–783 (2004)
20. Codella, N.C., et al.: Skin lesion analysis toward melanoma detection: a challenge at the 2017 international symposium on biomedical imaging (isbi), hosted by the international skin imaging collaboration (isic). In: Proceedings of the IEEE 15th International Symposium on Biomedical Imaging (ISBI 2018), pp. 168–172. IEEE (2018)
21. Liu, C., Cao, Y., Lin, Q., Man, Z., He, Y., Peng, L.: Segmentation of metastatic lesions on bone scan images based on improved UNet3+ network. In: Proceedings of the 4th International Conference on Computer Engineering and Application (ICCEA), Hangzhou, China, pp. 916–920 (2023)
22. Hu, X.: TA-UNet3+: a transformer-based method for kidney tumor segmentation. In: Proc. SPIE 12634, International Conference on Optics and Machine Vision (ICOMV 2023), 126340D (2023)
23. Heller, N., et al.: The kits19 challenge data: 300 kidney tumor cases with clinical context, ct semantic segmentations, and surgical outcomes. arXiv:1904.00445 (2019)
24. Gao, L., Wang, J., Ye, X.: MA-Unet3+ segmentation network of remote sensing image based on ECA block. In: Proceedings of the 6th International Conference on Computer Network, Electronic and Automation (ICCNEA), Xi'an, China, pp. 5–9 (2023)
25. Wang, Q., Wu, B., Zhu, P., et al.: ECA-Net: Efficient channel attention for deep convolutional neural networks. In: Proceedings of the IEEE/CVF Conference on Computer Vision and Pattern Recognition, pp. 11534–11542 (2020)
26. Tie, J., Wu, W., Zheng, L., Wu, L., Chen, T.: Improving walnut images segmentation using modified UNet3+ algorithm. Agriculture **14**(1), 149 (2024)

27. Jiang, C., Zhang, H., Wang, C., Ge, J., Wu, F.: Water surface mapping from sentinel-1 imagery based on attention-UNet3+: a case study of Poyang Lake Region. Remote Sensing **14**(19), 4708 (2022)
28. Tran, N.D., Nguyen, D.Q.D., Nguyen, N.L.C., Pham, V.T., Tran, T.T.: A Multi Context Decoder-based Network with Applications for Polyp Segmentation in Colonoscopy Images. In: Nguyen, N.T., Le-Minh, H., Huynh, C.P., Nguyen, Q.V. (eds.) The 12th Conference on Information Technology and Its Applications. CITA 2023. LNNS, vol 734. Springer, Cham (2023). https://doi.org/10.1007/978-3-031-36886-8_13
29. Huang, G., Liu, Z., Van Der Maaten, L., Weinberger, K.Q.: Densely Connected Convolutional Networks. In: Proceedings of the IEEE Conference on Computer Vision and Pattern Recognition (CVPR), Honolulu, HI, USA, 2017, pp. 2261–2269 (2017)
30. Howard, A., et al.: MobileNets: Efficient Convolutional Neural Networks for Mobile Vision Applications. arXiv:1704.0486 (2015)
31. Sandler, M., Howard, A.G., Zhu, M., Zhmoginov, A., Chen, L.: MobileNetV2: inverted residuals and linear bottlenecks. In: Proceedings of the IEEE/CVF Conference on Computer Vision and Pattern Recognition, pp. 4510–4520 (2018)
32. Jha, D. et al.: Kvasir-SEG: a segmented polyp dataset. In: Ro, Y., et al. (ed.) MultiMedia Modeling. MMM 2020. Lecture Notes in Computer Science(), vol 11962. Springer, Cham (2020).https://doi.org/10.1007/978-3-030-37734-2_37
33. Deng, J., Dong, W., Socher, R., Li, L.-J., Li, K., Li, F.-F.: ImageNet: a large scale hierarchical image database. In: Proceedings of the IEEE Conference on Computer Vision and Pattern Recognition, pp. 248–255 (2009)

Scheduling for Lecturers Using Genetic Algorithm with Fuzzy Constraints

Vu Thanh Lam, Nguyen Duc Minh, Nguyen Ngoc Lan, and Phan Duy Hung(✉)

FPT University, Hanoi, Vietnam
{lamvthe160382,minhndhe160265,lannnhe150254}@fpt.edu.vn, hungpd2@fe.edu.vn

Abstract. University timetabling is one of the classic schedule optimization problems that has attracted the attention of researchers for many years. Case studies in each school have unique characteristics, so this is a challenging task without a general solution. Through a long period of in-depth research, it can be divided into parts, and teaching assignment is one of them. Assigning tasks to lecturers needs to consider many complex constraints to ensure that classes are taught by people with good expertise as well as satisfy the lecturers' wishes. In this study, we build a model from hard constraints that cannot be violated to ensure compliance with institutional regulations, quality standards and especially soft constraints that can be violated, and even add fuzzy elements that represent lecturers' expectations. Then, the genetic algorithm will be used to optimize the model on the actual data set at FPT University and achieve reasonable and logical results. Moreover, the experiment was also tried on data sets with changes in the number of lecturers and their positions.

Keywords: University scheduling · Teaching assignment · Genetic algorithm · Fuzzy logic

1 Introduction

1.1 Problem

The Teaching Assignment Problem (TAP) is critical in university course scheduling, influencing educational quality. TAP entails allocating instructors to courses while fulfilling a variety of limitations. Finding a universal answer is difficult due to each university's differences in operations, infrastructure, and academic resources. The FPT University Training Department. will arrange classes based on student enrollment and available resources. Department managers of each major will assign speakers to courses. This study focuses on the final stage, professor assignment. The timetable, prepared before the semester, applies weekly and is unaffected by post-semester difficulties. Absences due to illness or emergency will be paid for later. If a professor is unable to continue teaching, the situation will be handled on an individual basis, with departmental assistance.

The schedule, shown in Table 1, groups classes by name and topic code, resulting in subject classes. Each lecturer is recognized by a unique code in the Lecturer-task aim

D. Singh et al. (Eds.): IHCI 2024, LNCS 15557, pp. 293–305, 2025.
https://doi.org/10.1007/978-3-031-88705-5_25

column, which is classified as management, full-time, part-time, or researcher depending on position and teaching time. The slot column on the schedule indicates exact timings for subject classes. At FPT, each workday is divided into two morning and two afternoon periods. To make it easier to arrange lecturers for subject classes, the two positions have been consolidated and designated in Table 2.

In this case study, we encounter various constraints. For starters, classes must have a single appointed lecturer with no overlap in hours. Lecturers lecture within their areas of competence and within their allotted hours. Second, circumstances promote lecturer satisfaction by connecting employment with skills and providing flexible times. This involves accommodating hour preferences, emphasizing teaching talents, scheduling full-time sessions, limiting tasks, and avoiding undesirable timeslots. These requirements, while demanding and possibly contradicting, assure student pleasure and instructional quality. The more their satisfaction, the better the timetable.

Table 1. Demo schedule

Class	Subject	Slot	Room	Room2	Lecturer
SE1878	CEA201	A63	BE-118		duongnv44
SE1898	CEA201	P63	BE-115		duongnv44
SE1882	CEA201	A24	BE-101	BE-118	ducdm39
SE1880	CEA201	P42	BE-115	BE-116	minhnt79
SE1879	CEA201	A53	BE-106		hieudv2
SE1899	CEA201	P53	BE-118		hieudv2
SE1876	CEA201	A46	BE-108		anhkd
SE1883	CEA201	P64	BE-101		ducdm39
SE1890	CEA201	A25	BE-109		minhnt79
SE1881	CEA201	P25	BE-102		sonhx

Table 2. Slot in FPT University

	Morning			Afternoon	
Slot	Slot 1	Slot 2	Slot	Slot 3	Slot 4
A24	Mon	Wed	P24	Mon	Wed
A42	Wed	Mon	P42	Wed	Mon
A35	Tue	Thu	P35	Tue	Thu
A53	Thu	Tue	P53	Thu	Tue
A46	Wed	Fri	P46	Wed	Fri
A64	Fri	Wed	P64	Fri	Wed
A52	Thu	Mon	P52	Thu	Mon
A25	Mon	Thu	P25	Mon	Thu
A63	Fri	Tue	P63	Fri	Tue
A36	Tue	Fri	P36	Tue	Fri

1.2 Related Works

Timetabling is a very well-known problem with multiple solutions. Chen et al. [1] did a survey that included 35 approaches to solving timetabling problems. Some major approaches that are widely used are: Operational research (OR) methods, Single solution-based meta-heuristics, Population-based meta-heuristics, etc. In reality, no method is completely effective, different datasets will have varying state-of-the-art, for instant: best result in Socha dataset is using Single solution-based meta-heuristics (Tabu Search) [2], best result in ITC2007-Track2 dataset is using population-based meta-heuristics (Genetic Algorithm) [3]. With real-world data, R. Tkaczyk et al. uses Agent-based timetabling system to solve this problem and it outperformed the state-of-the-art timetabling software based on the genetic algorithms [4], etc. When applied to real-world dataset, the state-of-the-art techniques become even more varied. The methods that are trending in use are primarily meta-heuristic, especially genetic algorithms.

Similar to the evolution of biological species, genetic algorithm [5] rely on mechanisms such as crossover, selection, mutation, etc., to evolve. It will create a population in which individuals reproduce to create the next generation. Some recent studies propose using genetic and combined variants. Junrie B et al. using A Hybrid Genetic Algorithm to solve teaching management and course scheduling problem [6]. In [7], Wong et al. combined genetic algorithm and tabu search to solve course timetabling and showed results that were about 54% superior to manual. A. Hussain et al. use Genetic Algorithm to provide a good initialization to the Hill Climbing algorithm in university timetabling problem [8]. Using NASH equilibrium and genetic algorithm [9], Chi et al. proposed an approach that balances stakeholders' interests but does not deal with complex, ambiguous constraints.

Above studies have achieved certain results and proven that optimal algorithm such as GA is applicable. However, the fuzzy and uncertain factor "Ensuring high expectations and satisfaction of lecturers or students about the quality of teaching or learning" has not been mentioned. This requirement in university teaching is even more critical because it affects both teachers and learners. For scheduling problems with unclear constraints, people often use fuzzy logic to build the model. Malek et al. built fuzzy modeling and combined genetic algorithm to solve resource leveling in scheduling Helimaintenance project [10]. C. Di Stefano using evoluationary algorithm and fuzzy logic to solve timetabling problem [11]. H Asmuni et al. using fuzzy method to combine heuristic orderings for exam schedule and reach good result [12]. Kohshori et al. applied fuzzy logic and genetic algorithm to University Course Timetabling [13] but has not focused on optimizing benefits for lecturers. Son et al. used the genetic algorithm and fuzzy logic to reduce the meantime when lecturers are assigned non-consecutive slots each day, such as 1 morning slot and 1 afternoon slot [14].

1.3 Contribution

This research aims to enhance teaching quality by proposing an improved scheduling system tailored to lecturers' capacities. We use Genetic Algorithms and Fuzzy logic to address conflicting expectations and constraints, ensuring balanced class loads and maximizing satisfaction. Our study resolves scheduling challenges at FPT University and demonstrates the effectiveness of our approach, contributing to broader task assignment problems.

2 Methodology

2.1 Problems Formulation

To solve this problem, we define some denotations:

- L, C, S is the number of lecturers, courses, pair slots.
- A is the total number of classes that need to be assigned a lecturer.
- l_a, c_a, j_a is lecturer, course, pair slot of ath class.
- T is a specific slot in a week (from Monday to Friday, each day has 4 slots, 2 in the morning, 2 in the afternoon)
- Matrix $B = \{b_{l,t} \in N, 1 \leq l \leq L, 1 \leq t \leq T\}$, which $b_{l,t}$ is the total number of classes the lecturer l had to teach at slot t.
- Matrix $E = \{e_{l,c} \in [1,4], 1 \leq l \leq L, 1 \leq c \leq C\}$, which $b_{l,c}$ represents the teaching quality of the lecturer l for course c. If the lecturer does not have any skill in teaching course c, $c_{l,s} = 1$.
- Matrix $D = \{d_{l,a} \in [0,1], 1 \leq l \leq L, 1 \leq a \leq A\}$, which $d_{l,a} = 1$ if the lecturer l has assigned to class a, $d_{l,a} = 0$ otherwise.
- Matrix$K = \{k_{l,s} \in [0,1], 1 \leq l \leq L, 1 \leq s \leq S\}$, which $k_{l,s} = 1$ if the lecturer l does not want to teach at pair slot s. $k_{l,s} = 0$ otherwise.
- Vector $M = \{m_l \in [0,10], 1 \leq l \leq L\}$, which m_l is the minimum class that the lecturer l^{th} needs to teach.
- Vector $N = \{n_l = (a, b) 1 \leq l \leq L, 0 \leq a \leq b \leq 10\}$, which n_l is the desired range of classes that the lecturer l^{th} wants to teach.

2.2 Constraint

The schedule must follow certain rules and constraints. Hamed Babaei has defined 16 constraints for the problem university course timetabling problem [15]. In the case of FPT University, we propose 5 hard constraints and 1 soft constraint.

Hard constraints:

- Every class must be assigned one and only one lecturer.

$$l_a = 1 \ \forall a \in [1, A] \tag{H1}$$

- Lecturers can not simultaneously be in more than 1 class.

$$b_{l,t} \leq 1 \forall l \in [1, L], t \in [1, T] \tag{H2}$$

- Lecturers can not teach a course they don't have skill at.

$$d_{l,a} * e_{l,c_a} \geq 1 | d_{l,a} = 1, \forall l \in [1, L], a \in [1, A] \tag{H3}$$

- Managers can not teach more than 4 classes in a week.

$$\sum_{x=1}^{T}\left(d_{manager,x}\right) \leq 4 \tag{H4}$$

- Every lecturer must satisfy the minimum number of classes.

$$\sum_{x=1}^{A}\left(d_{l,x}\right) \geq m_l \forall l \in [1, L] \tag{H5}$$

Soft constraints:

- Lecturers did not teach at the pair slot they did not desire.

$$\sum_{x=1}^{T}\left(b_{l,x} * k_{l,x}\right) = 0 \forall l \in [1, L] \tag{S1}$$

In reality, the above constraints are not enough. Additional constraints are required to satisfy lecturers' expectations. The following is a description of the expectation functions:

- The total number of classes of a lecturer is within the desired range of that lecturer.

$$min\left(\left(\sum_{x=1}^{A} d_{l,x}\right) - n_l\right) \forall l \in [1, L] \tag{E1}$$

- Maximize the lecturers' expectations for the course they are good at teaching.

$$max\left(\sum_{x=1}^{A} d_{l,x} * e_{l,c_x}\right) \forall l \in [1, L] \tag{E2}$$

- Maximize lecturers' expectations about the part of the day they need to teach. For example, assigning them to a full morning on a couple of days, rather than spreading their classes throughout the week.

$$max(epd(l)) \forall l \in [1, L] \tag{E3}$$

To achieve the best results, the schedule needs to satisfy all hard conflicts and most of soft conflicts and maximize the expectations of lecturers.

2.3 Fuzzy Constraint

Instead of rigidly defining constraints, they can be represented as fuzzy sets, allowing greater extensibility and flexibility in applying them to the scheduling process. Using fuzzy will help handle complex logic from some very simple rules.

This paper will use fuzzy logic to handle 2 problems (E1) (E2):

- The total number classes of a lecturer are within the desired range of that lecturer: Normally, depending on the role of each lecturer, there will be a different number of desired teaching sessions in the range (a, b) | $0 < a < b < 10$. Fuzzy set: low, average, high. Average is the best, low or high result in reduced expectation of lecturer. Based on (a, b) we will have a membership function as follows (Fig. 1). Each lecturer will have a different membership function.

$$num(l) = \left(\sum_{x=1}^{A} d_{l,x}\right) \forall l \in [1, L]$$

- Lecturers are assigned courses based on their proficiency: When assigning courses, we categorize a lecturer's ability to teach into four levels: good, normal, forced, or unable to teach. We calculate an average score across all courses to gauge their overall performance, using the function roc(). Fuzzy set: good, normal, force, bad. The membership function is defined as shown in Fig. 1.

$$roc(l) = \frac{\sum_{x=1}^{A} point(l,x)}{\sum_{x=1}^{A} d_{l,x}} \qquad point(l,x) = \begin{cases} 1, & e_{l,c_x} = 4 \\ 2/3, & e_{l,c_x} = 3 \\ 1/3, & e_{l,c_x} = 2 \\ 0, & e_{l,c_x} = 1 \end{cases}$$

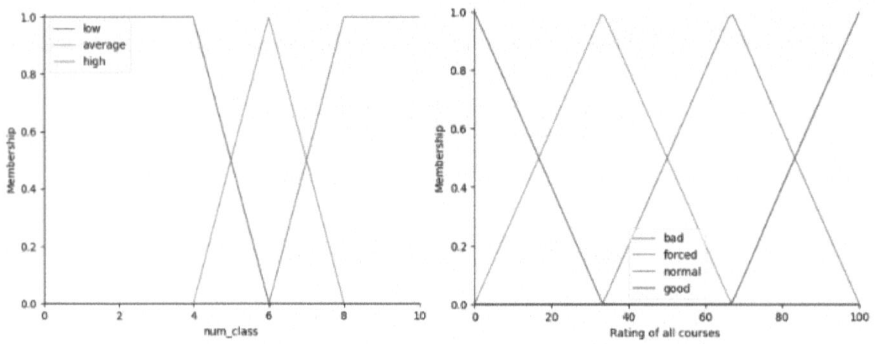

Fig. 1. Membership function of Number of classes and Rating of all courses

The expected result is expectation of lecturer in range [0,1] represents how high of lecturer's expectation is. Fuzzy set: good, average, decent, poor. The membership function is defined as shown in Fig. 2.

After acquiring 3 membership functions, we will set up rules: Fuzzy rules are IF-THEN rules that represent the relationship between input variables and output variables in a fuzzy way. They are used for reasoning in Fuzzy Logic systems. Based on the fuzzy sets set out above, 12 rules (Table 3) are generated to help calculate results.

From here, we can calculate the expectation of each lecturer ($fcs(l)$) based on the number of classes and $roc(l)$ using fuzzy control system on scikit-fuzzy library.

$$fcs(l) = fuzzyControlSytem(num(l), roc(l))$$

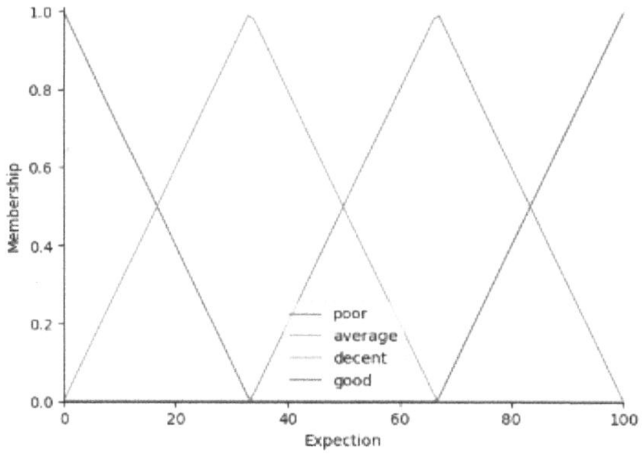

Fig. 2. Membership function of Expectation

Table 3. Fuzzy rule of Expectation

Number of classes	Rating of all courses	Expectation
Low	Bad	Poor
Low	Forced	Poor
Low	Normal	Average
Low	Good	Decent
Average	Bad	Poor
Average	Forced	Average
Average	Normal	Decent
Average	Good	Good
High	Bad	Poor
High	Forced	Average
High	Normal	Average
High	Good	Decent

The expectation of lecturers about part of day that they need to teach (E3) also can be solved by using fuzzy membership function as follows:

Function $epd(l)$.

1	$n = \sum_{x=1}^{A} d_{l,x}$	33	If w1s = 0 and k = n + 1 and 5 ≤ n ≤ 6 and w = (k/2 + 2) return 70
2	k: total number of part days that lecturer had to teach	34	If w1s = 0 and k = n + 1 and 5 ≤ n ≤ 6 and w = (k/2 + 1) return 80
3	w: total number of days that lecturer had to teach	35	If w1s = 0 and k = n + 1 and 5 ≤ n ≤ 6 and w = k/2 return 90
4	w1s: total number of days that lecturer teaches only 1 slot	36	If w1s = 0 and k = n + 1 and n ≥ 7 and w = k return 70
5	w1s ≥ 3 return 0	37	If w1s = 0 and k = n + 1 and n ≥ 7 and w ≥ (k/2 + 3) return 75
6	If w1s = 2 and k = n and n ≤ 1 return 0	38	If w1s = 0 and k = n + 1 and n ≥ 7 and w = (k/2 + 2) return 80
7	If w1s = 2 and k = n and n ≥ 2 return 35	39	If w1s = 0 and k = n + 1 and n ≥ 7 and w = (k/2 + 1) return 85
8	If w1s = 2 and k = n + 1 and n ≤ 1 return 0	40	If w1s = 0 and k = n + 1 and n ≥ 7 and w = k/2 return 90
9	If w1s = 2 and k = n + 1 and n ≥ 2 return 20	41	If w1s = 0 and k = n + 2 and n ≤ 1 return 0
10	If w1s = 2 and k = n and n ≤ 1 return 0	42	If w1s = 0 and k = n + 2 and 2 ≤ n ≤ 4 and w = k return 30
11	If w1s = 2 and k = n + 2 and n ≥ 2 return 10	43	If w1s = 0 and k = n + 2 and 2 ≤ n ≤ 4 and w ≥ (k/2 + 3) return 40
12	If w1s = 2 and k ≥ n + 2 and n ≥ 2 return 0	44	If w1s = 0 and k = n + 2 and 2 ≤ n ≤ 4 and w = (k/2 + 2) return 50
13	If w1s = 1 and k = n and n ≤ 1 return 0	45	If w1s = 0 and k = n + 2 and 2 ≤ n ≤ 4 and w = (k/2 + 1) return 70
14	If w1s = 1 and k = n and n ≥ 2 return 35	46	If w1s = 0 and k = n + 2 and 2 ≤ n ≤ 4 and w = k/2 return 85
15	If w1s = 1 and k = n + 1 and n ≤ 1 return 0	47	If w1s = 0 and k = n + 2 and 5 ≤ n ≤ 6 and w = k return 40
16	If w1s = 1 and k = n + 1 and n ≥ 2 return 30	48	If w1s = 0 and k = n + 2 and 5 ≤ n ≤ 6 and w ≥ (k/2 + 3) return 50
17	If w1s = 1 and k = n and n ≤ 1 return 0	49	If w1s = 0 and k = n + 2 and 5 ≤ n ≤ 6 and w = (k/2 + 2) return 60
18	If w1s = 1 and k = n + 2 and n ≥ 2 return 25	50	If w1s = 0 and k = n + 2 and 5 ≤ n ≤ 6 and w = (k/2 + 1) return 70
19	If w1s = 1 and k ≥ n + 2 and n ≥ 2 return 20	51	If w1s = 0 and k = n + 2 and 5 ≤ n ≤ 6 and w = k/2 return 85
20	If w1s = 0 and k = n and w = k return 80	52	If w1s = 0 and k = n + 2 and n ≥ 7 and w = k return 65
21	If w1s = 0 and k = n and w ≥ (k/2 + 3) return 85	53	If w1s = 0 and k = n + 2 and n ≥ 7 and w ≥ (k/2 + 3) return 70
22	If w1s = 0 and k = n and w = (k/2 + 2) return 90	54	If w1s = 0 and k = n + 2 and n ≥ 7 and w = (k/2 + 2) return 75
23	If w1s = 0 and k = n and w = (k/2 + 1) return 95	55	If w1s = 0 and k = n + 2 and n ≥ 7 and w = (k/2 + 1) return 80
24	If w1s = 0 and k = n and w = k/2 return 100	56	If w1s = 0 and k = n + 2 and n ≥ 7 and w = k/2 return 85
25	If w1s = 0 and k = n + 1 and n ≤ 1 return 0	57	If w1s = 0 and k ≥ n + 3 and n ≤ 1 return 0
26	If w1s = 0 and k = n + 1 and 2 ≤ n ≤ 4 and w = k return 35	58	If w1s = 0 and k ≥ n + 3 and 2 ≤ n ≤ 4 return 30
27	If w1s = 0 and k = n + 1 and 2 ≤ n ≤ 4 and w ≥ (k/2 + 3) return 45	59	If w1s = 0 and k ≥ n + 3 and 5 ≤ n ≤ 6 return 30
28	If w1s = 0 and k = n + 1 and 2 ≤ n ≤ 4 and w = (k/2 + 2) return 60	60	If w1s = 0 and k ≥ n + 1 and n ≥ 7 return and w = k return 40
29	If w1s = 0 and k = n + 1 and 2 ≤ n ≤ 4 and w = (k/2 + 1) return 80	61	If w1s = 0 and k ≥ n + 1 and n ≥ 7 return and w ≥ (k/2 + 3) return 50
30	If w1s = 0 and k = n + 1 and 2 ≤ n ≤ 4 and w = k/2 return 90	62	If w1s = 0 and k ≥ n + 1 and n ≥ 7 return and w = (k/2 + 2) return 65
31	If w1s = 0 and k = n + 1 and 5 ≤ n ≤ 6 and w = k return 50	63	If w1s = 0 and k ≥ n + 1 and n ≥ 7 return and w = (k/2 + 1) return 70
32	If w1s = 0 and k = n + 1 and 5 ≤ n ≤ 6 and w ≥ (k/2 + 3) return 60	64	If w1s = 0 and k ≥ n + 1 and n ≥ 7 return and w = k/2 return 80

2.4 Genetic Algorithm

2.4.1 Individual Representation

The individuals (i) in genetics will be the weekly schedules of all lecturers. An individual is a matrix consisting of L rows and S columns. Row l^{th} is the schedule of lecturer l^{th}. $cell(l, s)$. This class that the lecturer l^{th} teaches in the pair slot s. If lecturer l^{th} does not have class at pair slot s then $cell(l, s) = 0$.

2.4.2 Fitness Function

The fitness function is used to rate an individual after generated. The fitness function will return that individual's point in the range [0,1]. The larger of fitness function, the better of individual. It evaluates the schedule as follows:

$$fitness = \mu * Con(i) + (1 - \mu) * Expec(i)$$

$Con(i)$ is used to deal with (H2)(H3)(H4)(H5)(S1) by counting the total number of hard and soft conflicts of every lecturer in the schedule. $Con(i)$ max when total number of conflicts min. The initialization process will handle constraint (H1) will be described in Sect. 2.4.3. $Expec(i)$ is used to handle with (E1)(E2). $Expec(i)$ will calculate the average expectation of every lecturer in the schedule in the range [0,1].

$$Con(i) = \frac{1}{1 + (hard * \theta + (1 - \theta) * soft)} \qquad Expec(i) = \frac{\sum_{x=1}^{L} (\lambda * fcs(l) + (1 - \lambda) * epd(l))}{L}$$

2.4.3 Algorithm Flow

Here is our proposed genetic algorithm:

Genetic algorithm
1. Create population P have α Schedules.
1.1. Loop α time.
1.1.1. Create an empty individual.
1.1.1.1. Loop A time.
1.1.1.1.1. Randomly fill a^{th} class in any cell in column $t_{a^{th}}$ has value = 0 (to slove (H1))
1.1.2. Add individual to P.
2. Loop until convert:
2.1. Calculate fitness of all schedules in P.
2.2. Sorted schedule in order fitness decrease.
2.3. Creat population P'. Add β the best schedule of P to P'.
2.4. Loop $(\alpha - \beta)$ div 2 time.
2.4.1. Selection: Choice pair of schedules p_{father}, p_{mother} from P (based on Tournament Selection [16]).
2.4.2. Crossover: Apply partially mapped crossover method [17] for p_{father}, p_{mother} to create p_{child1}, p_{child2}.
2.4.3. Add p_{child1}, p_{child2} to P'.
2.5. Mutation: Apply scramble mutation method.
P = P'

3 Experiments and Results

3.1 Data Collection

Table 4. Sample of course data

Class	Subject	Slot	Room
SE1878	CEA201	A63	['BE-118']
SE1898	CEA201	P63	['BE-115']
SE1882	CEA201	A24	['BE-101', 'BE-118']
SE1880	CEA201	P42	['BE-115', 'BE-116']
SE1879	CEA201	A53	['BE-106']

Experiments were conducted using real data from FPT University. Course data contains 76 class-courses across 15 subjects comprises class names, subject codes, session slots, and classrooms (Table 4). The ITS department - a department in charge of computer science subjects will provide lecturer data (Table 5) including lecturer ID, roles, sex, times when instructors do not want to teach. Furthermore, each lecturer for each subject will have a score scale from 1 to 4 based on the lecturer's suitability in that subject.

Table 5. Sample of lecturer data

lecturer_id	sex	role	except_time	CEA201	IOT102	...
lec_mng_m	man	manager	['P36', 'P46']	3	3	...
lec_rs1_m	man	researcher	0	2	4	...
lec_ft9_wm	woman	fulltime	0	3	3	...
lec_ft10_wm	woman	fulltime	['A24', 'A42', 'A52', 'A25']	4	2	...
lec_pt1_m	man	part-time	['P24', 'P42', 'P52', 'P25']	3	3	...

In addition, we experiment with two other datasets based on changing the ratio of lecturers' positions, the number of slots in except_time and the teacher's grading scale for the subject while ensuring real situation. So, the research conducted experiments on 3 datasets with different lecturers: Data0 (real data): 13 lecturers including 1 manager, 1 researcher, 1 part-time, 10 full-time. Data1 (have more lecturers): 15 lecturers including 1 manager, 1 researcher, 1 part-time, 12 full-time. Data2 (have less lecturers): 11 lecturers including 1 manager, 1 researcher, 2 part-time, 7 full-time.

3.2 Experiments and Analytics

The experiments were conducted on a Windows personal computer with the following parameters: Core i5 10300H 2.5GHz, 16 GB RAM and Python 3.9. The experiments focus on the effectiveness of fuzzy constraints and their applicability in real life.

Firstly, in the fitness function, we set θ in Con(l) is 0.9, λ in Expec(l) is 0.5 and z in Con(l) is 5. With μ is 0.8 we optimize all constraints (include fuzzy constraints), so that function is called fuzzy fitness. On other hand, θ is 1, function is called non-fuzzy fitness because function does not optimize fuzzy constraints. Hyperparameters of the genetic algorithm are referred stable range values in similar research [17] and adjusted many times with data0. Therefore, we chose the following set of parameters: population size is 150, tournament size is 7, and mutation rate is 0.15.

Secondly, each semester, there may be slight changes in faculty policies and expectations. General policy data for each role can be configured easily, including the minimum teaching time, number of expectation teaching class-courses. The list of teachers participating in guiding the capstone project will have the number of teaching classes reduced. Besides, we will add priority according to the lecturer's role by multiplying the number of conflicts and their satisfaction with the corresponding role weight. Currently, the lecturer parameters mentioned above are specified in Table 6.

Finally, we conducted experiments on three datasets in two cases: have fuzzy constraints or not in optimization fitness function. We ran 10000 generation or until optimization fitness reach 1 each case and Fig. 3 shows the process of optimization fitness. The results of three datasets that have fuzzy constraints or not are shown in Table 7 and 8. We can see that fitness, expectation in cases have fuzzy constraints are always. Data 1 still has hard conflict lead to the worst fitness because two lecturers have number class

Table 6. Config for Lecturers following role

Role	Min teaching (class-course)	Default expectation class (class-course)	Priority
Manager	3	3 - 4	8
Researcher	3	3- 5	2
Part-time	0	3 - 5	2
Full-time	6	6 – 8	1

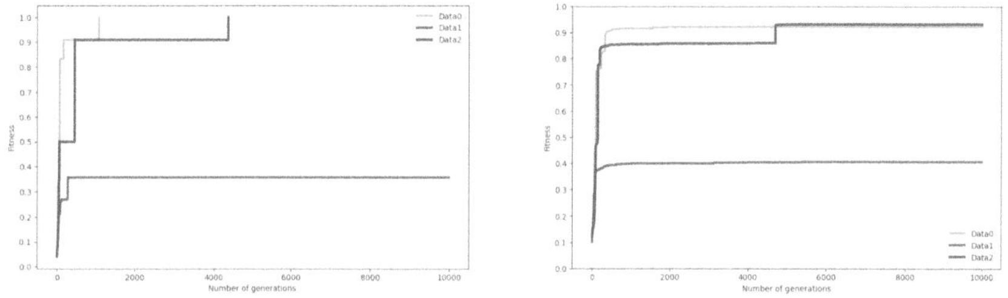

Without fuzzy constraints   With fuzzy constraints

Fig. 3. Process of fitness of three datasets when have fuzzy constraints or not

smaller than min. Easy to understand in data1, number of lecturers is too many so that do not have enough class for lecturers. Detailed schedules of each role in two cases are shown in Table 9 and results are better than manual schedules (Table 8).

Table 7. Final result in case without fuzzy constraints

Dataset	Fitness fuzzy	Number hard conflict	Number soft conflict	Average expectation preference	Average expectation part day
Data0	0.870	0	0	36.592	33.462
Data1	0.347	2	0	37.732	23.667
Data2	0.889	0	0	47.739	40.909

Table 8. Final result in case has fuzzy constraints

Dataset	Fitness fuzzy	Number hard conflict	Number soft conflict	Average expectation preference	Average expectation part day
Data0	0.922	0	0	55.373	66.538
Data1	0.405	2	0	47.944	71.667
Data2	0.931	0	0	58.957	72.273

Table 9. Result scheduling of data0 following role

Without fuzzy constraints

schedule of lec_mng_m:

	Monday	Tuesday	Wednesday	Thursday	Friday
1	NaN	NaN	IOT102(3)	NaN	CEA201(3)
2	IOT102(3)	CEA201(3)	NaN	NaN	NaN
3	NaN	NaN	NaN	CEA201(3)	CEA201(3)
4	NaN	CEA201(3)	CEA201(3)	NaN	NaN

With fuzzy constraints

schedule of lec_mng_m:

	Monday	Tuesday	Wednesday	Thursday	Friday
1	NaN	CEA201(3)	NaN	CEA201(3)	NaN
2	NaN	CEA201(3)	NaN	CEA201(3)	NaN
3	NaN	IOT102(3)	NaN	AIL303M(2)	NaN
4	NaN	AIL303M(2)	NaN	IOT102(3)	NaN

4 Conclusion and Future Works

This paper presents an approach to the problem of teaching assignments in university that prioritizes faculty satisfaction by optimizing optional and fuzzy constraints. Using fuzzy logic, we created a comprehensive function to evaluate faculty satisfaction based on schedule convenience and balance between faculty abilities and workload. Our planning model delivers better results than manual methods, with a user-friendly interface for easy monitoring, analysis, and adjustment. Future research aims to conduct experiments with diverse datasets and resource conditions, enhancing faculty benefits by addressing additional constraints. This paper is also a good reference for applied research directions of artificial intelligence in education [18, 19].

References

1. Chen, M.C., Sze, S.N., Goh, S.L., Sabar, N.R., Kendall, G.: A survey of university course timetabling problem: perspectives trends opportunities. IEEE Access **9**, 106515–106529 (2021)
2. Nagata, Y.: Random partial neighborhood search for the post-enrollment course timetabling problem. Comput. Oper. Res. **90**, 84–96 (2018)
3. Gozali, A.A., Kurniawan, B., Weng, W., Fujimura, S.: Solving university course timetabling problem using localized island model genetic algorithm with dual dynamic migration policy. IEEJ Trans. Electr. Electron. Eng. **15**(3), 389–400 (2019)
4. Tkaczyk, R., Ganzha, M., Paprzycki, M.: AgentPlanner - agent-based timetabling system. Informatica (Slovenia) **40**(1), 3–17 (2016)
5. Scott, M.T.: An introduction to genetic algorithms J. . Comput. Sci. Coll. **20**(1), 115–123 (2004)

6. Matias, J.B., Fajardo, A.C., Medina, R.P.: A hybrid genetic algorithm for course scheduling and teaching workload management. In: Proceedings of the IEEE 10th International Conference on Humanoid, Nanotechnology, Information Technology, Communication and Control, Environment and Management (HNICEM), Baguio City, Philippines, pp. 1–6 (2018)
7. Wong, C.H., Goh, S.L., Likoh, J.: A genetic algorithm for the real-world university course timetabling problem. In: Proceedings of the IEEE 18th International Colloquium on Signal Processing & Applications (CSPA), Selangor, Malaysia (2022)
8. Hussain, A., Ashas, H., Shahid, A., Qureshi, S., Karrila, S.: Hybrid approach involving genetic algorithm and hill climbing to resolve the timetable scheduling for a university. In: Arai, K. (eds.) Advances in Information and Communication. FICC 2024. LNCS, vol. 919. Springer, Cham (2024). https://doi.org/10.1007/978-3-031-53960-2_6
9. Chi, L.P, Thang, H.Q., Huong Giang, V.T., Son, N.T., Anh, B.N.: Teaching assignment based on NASH equilibrium and genetic algorithm. In: Proceedings of the IEEE Symposium on Industrial Electronics & Applications (ISIEA), Kuala Lumpur, Malaysia (2023)
10. Masmoudi, M., HaïT, A.: Project scheduling under uncertainty using fuzzy modeling and solving techniques. Eng. Appl. Artif. Intell. **26**(1), 135–149 (2013)
11. Di Stefano, C., Tettamanzi, A.G.B.: An evolutionary algorithm for solving the school timetabling problem. In: Boers, E.J.W. (eds.) Applications of Evolutionary Computing. EvoWorkshops 2001. LNCS, vol. 2037. Springer, Heidelberg (2001). https://doi.org/10.1007/3-540-45365-2_47
12. Asmuni, H., Burke, E.K., Garibaldi, J.M., McCollum, B.: Fuzzy multiple heuristic orderings for examination timetabling. In: Burke, E., Trick, M. (eds.) Practice and Theory of Automated Timetabling V. PATAT 2004. Lecture Notes in Computer Science, vol. 3616. Springer, Heidelberg (2005).
13. Kohshori, M.S., Liri, M.S.: Multi population hybrid genetic algorithms for university course timetabling. economics and applied informatics, "Dunarea de Jos" University of Galati, Faculty of Economics and Business Administration, issue 2, pp. 5–16 (2012)
14. Tung, N.S., Jaafar, J., Aziz, I.A., Anh, B.N.: A compromise programming for multi-objective task assignment problem. Computers **10**(2), 15 (2021)
15. Babaei, H., Karimpour, J., Hadidi, A.: A survey of approaches for university course timetabling problem. Comput. Ind. Eng. **86**, 43–59 (2015)
16. Miller, B.L., Goldberg, D.E.: Genetic algorithms, tournament selection, and the effects of noise. Complex Systems, pp. 193–212 (1995)
17. Abdoun, O., Abouchabaka, J.: A comparative study of adaptive crossover operators for genetic algorithms to resolve the traveling salesman problem. arXiv:1203.3097
18. Hung, P.D., Minh, N.C.: Application of fuzzy logic in university suggestion system for Vietnamese high school students. In: Dang, T., Küng, J., Takizawa, M., Bui, S. (eds.) Future Data and Security Engineering. FDSE 2019. LNCS, vol. 11814. Springer, Cham (2019). https://doi.org/10.1007/978-3-030-35653-8_44
19. Luan, N.K., Ha, P.T.T., Hung, P.D.: An automated proctor assistant in online exams using computer vision. In: Luo, Y. (eds.) Cooperative Design, Visualization, and Engineering. CDVE 2022. LNCS, vol. 13492. Springer, Cham (2022). https://doi.org/10.1007/978-3-031-16538-2_12

No Code Computer Vision

Phan Nguyen Duc Hieu and Phan Duy Hung(✉)

FPT University, Hanoi, Vietnam
hieupndhe153303@fpt.edu.vn, hungpd2@fe.edu.vn

1 Introduction

IP Cameras are widely used nowadays. It takes a cheap cost to be installed. Along with that, many AI models are developed for analyzing images from cameras, mainly focusing on object detection, abnormal detection, segmentation, or object tracking. However, almost all IP Cameras are just used for viewing live videos. Despite the development of Computer Vision and General AI, the work that integrates AI techniques with AI cameras is still very time-consuming and non-cost-effective. YOLO [1] started an era of fast and accurate object detection architecture. YOLO frames object detection as a regression problem. Bounding boxes and class probabilities are separately predicted on different regression heads. YOLOv2 [2] made a number of improvements based on YOLO, including BatchNorm, higher resolution, and anchor boxes. The YOLOv3 [3] model continues developing by proposing a score for evaluating the bounding box prediction and adding connections to backbone network layers. From the presentation of YOLOv4 [4], a bag of freebies for data augmentation was officially announced. The newest YOLO and other object detection models are using those augmentation strategies. The bag of freebies consists of the crop, rotation, flip, hue, saturation, exposure, aspect, MixUp, CutMix, Blur, and Mosaic techniques, which change an origin entire image to have multiple different images. We called the augmentation strategies which based on the entire image are image-level augmentation. Those strategies proved their efficiency in improving accuracy, but there remain some challenges: 1. The image-level augmentation techniques cannot increase the number of different objects. 2. Those techniques also cannot increase the number of different backgrounds. So it limits the ability to make more various objects for various different backgrounds. The state-of-the-art model, which is called DAMO-YOLO [5] proposes a robust architecture for real-time object detection but this model still uses only image-level augmentation techniques for increasing the number of data. That's a crucial point for us working on data-centric development, this study proposes a new process for data augmentation, which is named object-level augmentation. We extract objects from images using an accurate general segmentation model, and change their features: color, shape, size, etc. but also keep their main characteristics to recognize. Then the augmented objects are added on various backgrounds in different locations. In this way, we can generate thousands of quality different images from a few amount of data.

With fewer necessary amounts of data for training, the effort to develop a computer vision model will be decreased, but it's not enough. Computer vision projects consume

D. Singh et al. (Eds.): IHCI 2024, LNCS 15557, pp. 306–315, 2025.
https://doi.org/10.1007/978-3-031-88705-5_26

a huge amount of effort in coding and deploying AI models. To reduce the maximum development time but also keep the good accuracy and inference time, we developed CVall. CVall is a web-based system, that allows to development of object detection models, segmentation problems, and object tracking projects with no line of code. Following a standard machine learning life cycle, CVall supports all processes of AI development: data preparation, data analysis, model development, model evaluation, and deployment. CVall integrates object-level augmentation for data, DAMO-YOLO [5] for object detection, FastSAM [6] for segmentation, and tracking algorithms, which can connect directly with the IP Cameras. Our system proved its efficiency in both accuracy and development time by using state-of-the-art architectures and our improvements.

2 Business Context

Table 1. Sample effort estimation for an open-source project

Role	Number	Effort (MM)
Project Manager	1	4
AI Scientist	1	4
AI Engineer	2	9
QA Officer	1	4

Table 1 shows a sample of the effort needed for computer vision projects. A team that consists of 1 project manager, 1 AI scientist, 2 AI engineers, and 1 quality assurance officer and 4 months of work is needed for a complete product. The total effort will be around 25 man-months. So the cost of developing computer vision models is huge. It's impossible for non-tech companies to have their own AI models, especially small and medium companies. Because of high cost, instead of self-developing their own computer vision model, many companies decide to use third-party APIs. In this way, only the data preparation process requires effort. However, the data must be pushed to the cloud, which may violate data privacy. Even the cost of using APIs is higher than self-developing an AI model (Table 2).

Table 2. Computer Vision APIs Pricing of providers

Provider	Area	Cost / 1000 requests	Yearly fee estimation
Azure AI Vision	Central US	0.53–0.75 USD	12960–45360 USD
Google Cloud Vision	Any	0.60–1.50 USD	14400–57000 USD
Amazon Rekognition	US (Ohio)	0.25–1.00 USD	6000–24000 USD

CVall solved all cost and data privacy problems. By developing as an open source, CVall allows everyone to run a local web server on their own computers. The web system provides full control of AI development processes with no line of code and also simplifies each process to be more accurate and effective. The CVall visions are increasing the applicability of AI in real life, and building a strong community through open source to explore more efficient applications of computer vision.

3 Market Research

Microsoft, Google, and Amazon offered their services for developing computer vision projects with small effort through API calls. However, the development processes on their clouds are complicated and require strong engineering skills. The cost of using APIs also remains huge (shown in Table 2), and data privacy is still a concern.

Some open source projects integrate state-of-the-art AI models and provide an environment for training AI models, typical of open source projects of MMLab: MMDetection [7], MMTracking [8], MMSegmentation [9], Although these projects have nice work on integrating various architectures and models in 1 repository, the process of developing an AI model is still very complicated, even more complicated than using the origin models, which haven't merged to MMlab's repositories.

Snorkel AI - a unicorn start-up that provided efficient and time-saving solutions for developing AI models, mainly focusing on data-centric development. By focusing on data, customers may fine-tune, pre-train, and align AI models many times faster and better. Their services support multiple applications in various industry use cases: banking, government, healthcare, insurance, and telecom. Almost all use cases are NLP tasks, Snorkel AI lacks thousands of applications for computer vision. Their customers are big companies or giant corporations, so cost savings are not always a priority in their plans. But does it mean no cost, no AI?

CVall focuses on specific tasks for computer vision, which do not require huge computing infrastructures. We are publishing CVall as an open-source project, so it requires no cost for developing AI models. The cost may be reduced as much as possible, but the accuracy of the overall system is impressive by using our data-centric strategies and state-of-the-art model architectures.

4 Overall Requirements

The machine learning lifecycle includes 5 main steps: 1. Data Preparation, 2. Data Analysis, 3. Modeling, 4. Evaluation, 5. Error Analysis (Fig. 1).

To achieve the best efficiency, CVall must follow the machine learning life cycle steps (Fig. 1). Users are going to do it step-by-step using an easy-to-understand web UI.

4.1 Data Preparation

The data preparation consists of 3 steps: upload data, annotate data, and augment data. The upload data step receives image input in the format of jpg, png, or jpeg. The system also allows users to upload videos and input fps, CVall extracts frames based on video and fps parameters that users provided.

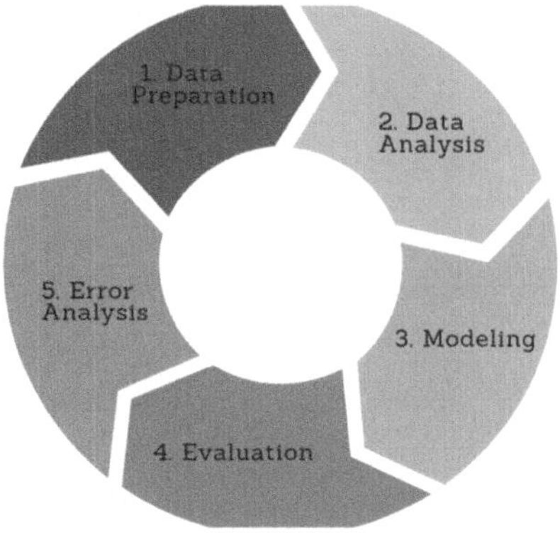

Fig. 1. The machine learning life cycle

The annotate data step allows users to use a trained model for auto-annotating data. If there is no trained model, CVall provides a pre-trained DAMO-YOLO on the COCO dataset [10], which can predict up to 80 classes. Then users may review the annotation and modify it if necessary.

The augment data step uses object-level augment. Users need to run the segmentation using FastSAM pre-trained model and review the segmentation output to make sure that there is no error. Then the objects are extracted from images, and randomly put on backgrounds, which are uploaded by the user.

4.2 Data Analysis

CVall visualizes the number of objects per class (Fig. 2, left) and the object size/image size distribution (Fig. 2, right). The diagram of the number of objects per class shows class inequality and statistically the objects on images. The diagram of object size/image size distribution shows statistics about small, medium, or large objects in the dataset, which are calculated by:

$$r = \sqrt{\left({}^{So}/_{Si}\right)} \tag{1}$$

where r is the ratio of object size/ image size, Si is the area of the entire image, and So is the area of the object. The square root is used for normalizing the value of r.

4.3 Modeling

After finishing the preparation for train datasets and test datasets. Users choose 1 train dataset and 1 test dataset accordingly to train their own detection model using DAMO-YOLO architecture. By providing the number of epochs, users can control when the training should finish, or they can terminate them if there is no more improvement in accuracy. The accuracy of the test dataset and training loss are updated continuously during the training process (Fig. 3).

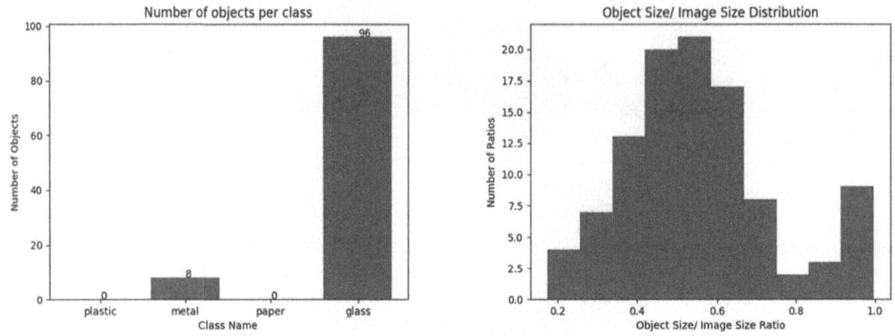

Fig. 2. Samples of data analysis diagrams.

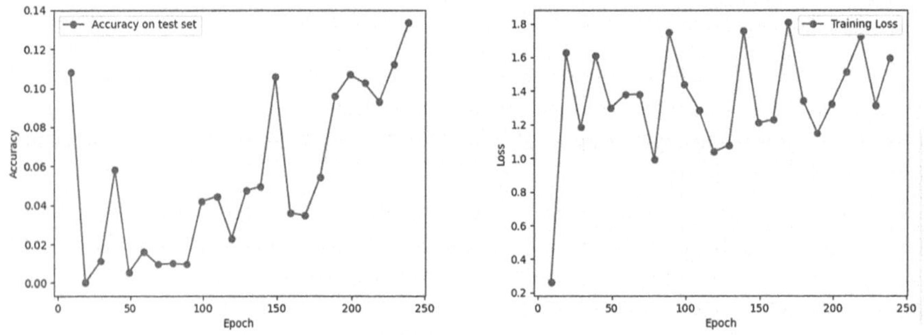

Fig. 3. Sample diagrams of accuracy on the test dataset (left) and training loss (right)

4.4 Evaluation and Error Analysis

Users may evaluate their model with an arbitrary test dataset. The results show in detail the accuracy of each class in the test dataset (Fig. 4). Number of true indicates the number of true predictions in that class, where excessive, missing, and wrong indicate the excessive bounding boxes, the missing bounding boxes and the false prediction of the class accordingly. Based on those parameters, the user can see which class has the most wrong cases, or what types of wrong cases lead to poor accuracy.

4.5 Deployment

We don't put the deployment process life cycle, because the machine learning life cycle can loop multiple times before the official is deployed. CVall supports users to immediately connect to IP Cameras through IP address, username, and password. The detection model can run as soon as the camera is connected. The CVall also supports segmentation using FastSAM and tracking using IOU-based algorithms directly by UI checkboxes without any further training or preparation.

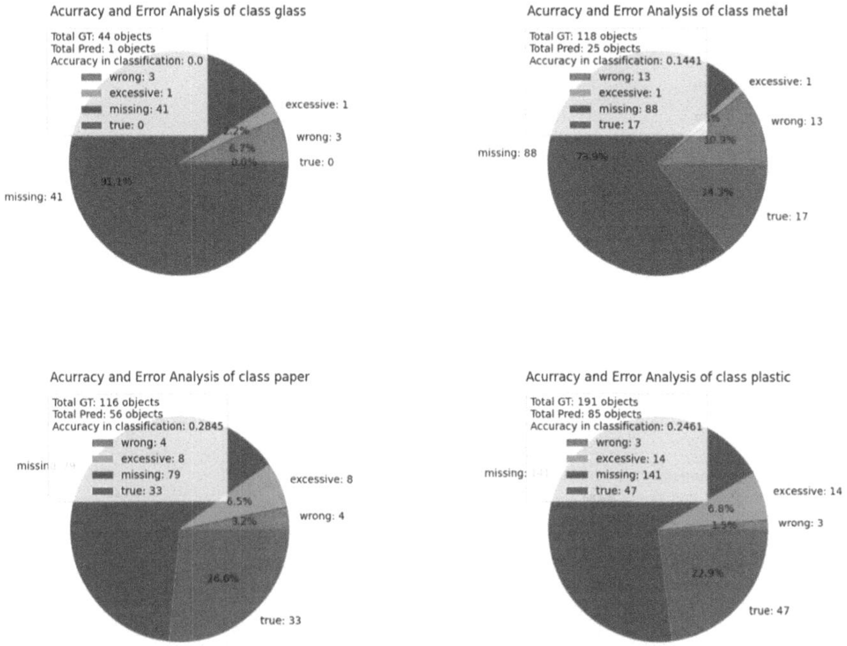

Fig. 4. Each class error analysis of the test dataset.

5 Proposed Techniques

5.1 Overall Architecture

The efficient design of the CVall system (Fig. 5) will reduce the development time of a computer vision project. Using a trained model in data preparation for annotation reduces up to 95% annotation effort. The use of a pre-trained FastSAM segmentation model directly outputs the polygons or masks of objects without any training for the segmentation model. However, the integration of those models and processes just makes CVall time-saving. So CVall proposed a data-centric strategy on data augmentation, called object-level augmentation. This is the most important point for the impressive overall accuracy of CVall.

5.2 Object-Level Augmentation

The object-level augmentation is separated into 2 parts: 1. Extract objects from images (Fig. 6). 2. Augment data by objects and backgrounds (Fig. 7).

CVall extract objects from images by using the FastSAM pre-trained model (Fig. 6). The FastSAM model use the origin image and the bounding boxes from the annotation process as input, and output the masks of objects. Then CVall extracts objects from the origin image by masks.

In part 2 (Fig. 7), The extracted objects are augmented in various ways: random resize, color, rotate, transform perspective, brightness, shear, and cutout. From just one

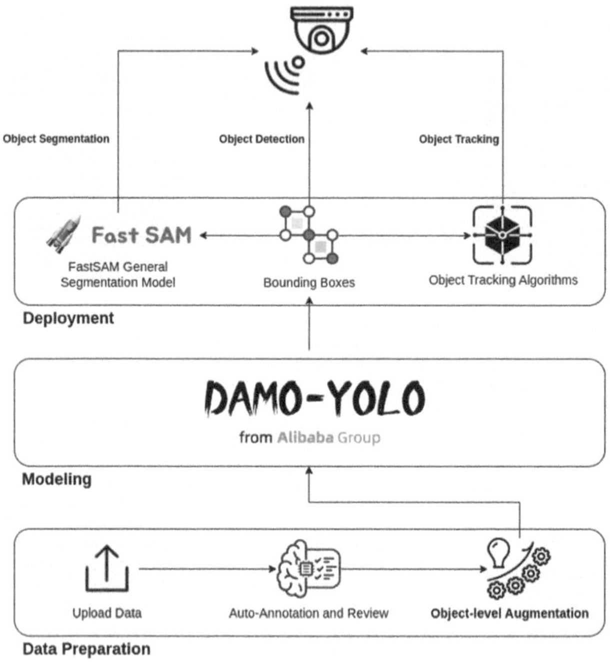

Fig. 5. Overall architecture of the CVall system

object in the beginning, CVall can generate hundreds of different things. After that, by using the augmented objects and various different backgrounds, thousands of images will be generated.

Fig. 6. Part 1 of object-level augmentation: Extract objects from images

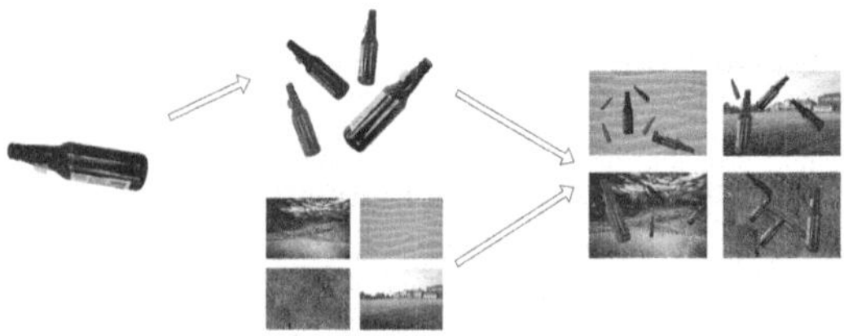

Fig. 7. Part 2 of object-level augmentation: Augment data by objects and backgrounds

6 Experiments

6.1 Datasets

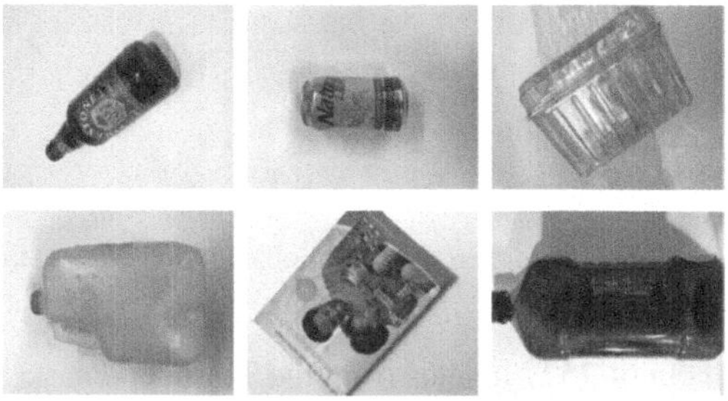

Fig. 8. Some images in the Trashnet dataset

Trashnet [11] is a dataset of 2527 waste images (glass, paper, cardboard, plastic, metal, trash) taken on white poster board with sunlight/room lighting, resized to 512 x 384 pixels, captured with Apple iPhone 7 Plus/5S/SE. Because of the setup to take images in the laboratory, the images in this dataset don't represent the objects in real life (Fig. 8). To evaluate the CVall system's efficiency on a small dataset and model-drift prevention, Trashnet is used as a training dataset.

Fig. 9. Some images in the Aquatrash dataset

Aquatrash [12] contains 369 images of trash with a total of 470 bounding boxes for 4 classes: metal, plastic, glass, and paper (Fig. 9). The images from this dataset were taken in reality, and contain various different objects and background scenes. CVall uses this dataset for evaluating performance.

6.2 Results

We used CVall to train an object detection model, the training dataset is Trashnet and the evaluated dataset is Aquatrash. First, we have to augment data from the Trashnet dataset by CVall's object-level augmentation (Fig. 10).

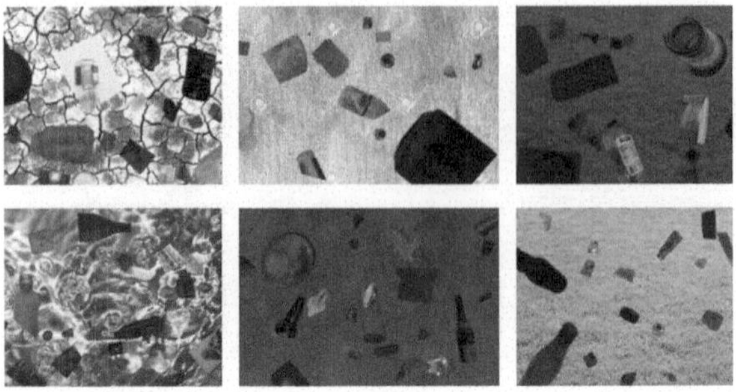

Fig. 10. Some generated images of object-level augmentation

We just used 400 out of 2527 images from the Trashnet dataset for the experiment. These 400 images were used to generate a dataset that includes 3000 images by CVall's object-level augmentation. We train the DAMO-YOLO object detection model on the generated dataset. The hardware used for training and testing is a GTX 3090 Ti GPU. The parameters of the training process consist of batch size 16, learning rate 1.5625e-05, number of epochs 300, weight decay 0.0005, max image range 640. The model is evaluated on the Aquatrash dataset and achieves impressive results. Compared to AquaVision [12], we used 5 times less on the amount of training images but achieved an 86.8 mAP score, which is 5.3% higher than AquaVision (Fig. 11).

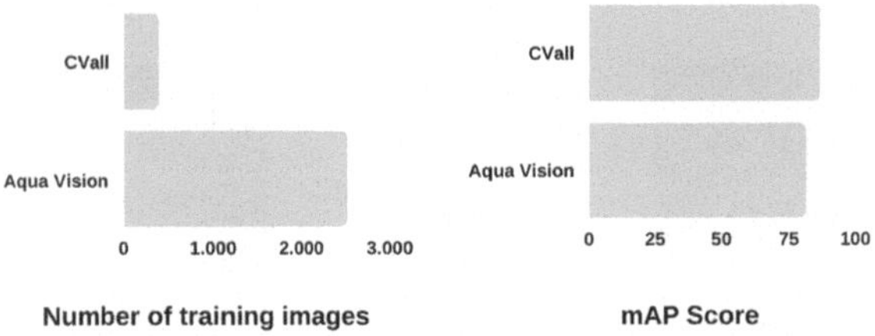

Fig. 11. Number of data and mAP score of CVall's work compared to AquaVision

The inference time of DAMO-YOLO [5] and FastSAM [6] was calculated. By using CVall on GTX 3090ti GPU, we achieve 34 FPS when running both DAMO-YOLO and FastSAM simultaneously on 1024x1024 images (Table 3).

Table 3. Inference time of object detection and segmentation on CVall

Task	Image size	Inference time (s)	FPS
Object Detection	1024	0.0144	69
Segmentation	1024	0.0150	66
Both	1024	0.0294	34

7 Conclusion and Future Works

The current version of CVall is open source and we commit that the new updated version of CVall is also public for everyone. This commitment is made to achieve our vision that we are building a strong community for applying computer vision to reality, with a cheap cost and time-saving processes. The source code is published at: https://github.com/duchieuphan2k1/cvall.

The CVall system needs to improve on various aspects: 1. More controlable parameters for model training including backbone architecture, activate function, and loss function, etc. 2. More model architecture options, 3. Image augment studio for reviewing and choosing the best suitable parameters of data augmentation. 4. Security and encryption of sensitive data, 5. Working with group features needs to be added for efficiency development.

References

1. Joseph, R., Santosh, D., Ross, G., Ali, F.: You only look once: unified, real-time object detection. arXiv:1506.02640 (2015)
2. Joseph, R., Ali, F.: YOLO9000: better, faster, stronger. arXiv:1612.08242 (2016)
3. Joseph, R., Ali, F.: YOLOv3: an incremental improvement. arXiv:1804.02767 (2018)
4. Alexey, B., Chien-Yao, W., Hong-Yuan, M.L.: YOLOv4: optimal speed and accuracy of object detection. arXiv:2004.10934 (2020)
5. Xianzhe, X., Yiqi, J., Weihua, C., Yilun, H., Yuan, Z., Xiuyu, S.: DAMO-YOLO: a report on real-time object detection design. arXiv:2211.15444v2 (2022)
6. Xu, Z., et al.: Fast segment anything. arXiv:2306.12156 (2023)
7. MMDetection contributors: MMDetection: open MMLab detection toolbox and benchmark. arXiv:1906.07155 (2019)
8. MMTracking Contributors: video perception toolbox and benchmark (2020). https://github.com/open-mmlab/mmtracking
9. MMSegmentation contributors: OpenMMLab semantic segmentation toolbox and benchmark (2020). https://github.com/open-mmlab/mmsegmentation
10. Tsung-Yi, L., et al.: Microsoft COCO: common objects in context. arXiv:1405.0312 (2014)
11. Trashnet Contributors: Trashnet: dataset of images of trash (2017). https://github.com/garythung/trashnet
12. Harsh, P., et al.: AquaVision: automating the detection of waste in water bodies using deep transfer learning. Case Stud. Chem. Environ. Eng. **2**, 100026 (2020)

Transfer Learning to Identify Multilingual Machine Genenrated Text

Samyak Jain, Anshul Bhardwaj, Mohit Somani, Alka Kumari, Disha Goyal, and Pritish Kumar Varadwaj(✉)

Indian Institute of Information Technology-Allahabad, Prayagraj 211012, India
{iit2021104,iit2021057,iit2021105,iit2021103, iit2021102,pritish}@iiita.ac.in

Abstract. In recent years, the rise of AI-driven tools has threatened academic integrity, with students increasingly relying on AI-generated content and compromising their scholarly efforts. To address the challenge of distinguishing machine-generated text from human-written content, this study utilizes the dataset provided by SemEval-2024. We explore various methodologies, comparing traditional machine learning (ML) techniques enhanced with natural language preprocessing (NLP) against the fine-tuning of Large Language Models (LLMs) like XLM-RoBERTa for text classification. Through systematic evaluation across SemEval-2024's training, development, and testing sets, we uncover key insights. While traditional ML methods with NLP refinements offer a pragmatic approach, transformer models, particularly XLM-RoBERTa, demonstrate superior efficacy. Our findings underscore the transformative potential of transformer models in detecting machine-generated texts, especially in multilingual environments. Additionally, we highlight the effectiveness of majority voting mechanisms in enhancing precision in multilingual contexts.

Keywords: LLM detection · Multilingual · Transformer · NLP

1 Introduction

In contemporary society, AI-driven tools have eroded academic integrity, with students often compromising their scholarly efforts by using AI-generated content. Our research seeks to counteract this trend by providing a deterrent against plagiarism, safeguarding academic credibility. Furthermore, the unbridled dissemination of false narratives through ChatGPT poses a threat to democratic principles, as automated bots using LLM-generated text can manipulate public opinion and distort political discourse. Identifying and mitigating such misinformation is imperative to uphold the integrity of democratic processes. Large Language Models (LLMs) mark a significant milestone in artificial intelligence, featuring neural networks with parameters often in the hundreds of billions or trillions. Trained on extensive textual corpora, these models have revolutionized natural language processing by emulating human speech patterns and tackling

D. Singh et al. (Eds.): IHCI 2024, LNCS 15557, pp. 316–324, 2025.
https://doi.org/10.1007/978-3-031-88705-5_27

intricate text generation tasks. ChatGPT, for instance, has gained widespread acclaim and is the fastest-growing application in history. The predominant focus in LLM detection has been on identifying AI-generated text within monolingual frameworks. However, the widespread adoption of LLM applications globally necessitates expanding detection capabilities to multilingual contexts. Accurately delineating boundaries between mixed LLM and human-generated texts is crucial for understanding the origins and influences shaping textual compositions. The unique challenge in this regard is to identify instances where human and LLM-generated content coexist within the exact text. This mixed composition poses significant obstacles to traditional detection methods, requiring nuanced approaches to discern subtle differences in writing styles and linguistic nuances. Addressing these challenges, our research aims to deepen our understanding of LLM-generated content and advance detection methodologies to unprecedented accuracy. By delving into the intricacies of mixed text compositions, we seek to unravel the complex interplay between human and AI contributions, shedding light on the evolving landscape of digital discourse. Ultimately, our objective is to equip academic and broader communities with robust detection mechanisms to navigate the complexities of diverse linguistic landscapes. By fortifying defences against LLM misuse in the digital age, we endeavour to safeguard the integrity of online communication and uphold the principles of trust and authenticity in information dissemination.

2 Related Study

Numerous studies have been conducted on LLM-generated text detection, and a curated list of these resources is available in the review [1], wherein several notable works were discussed for their contributions to the field. The "GPT Pat" [2] proposed a novel approach for detecting AI-generated text using the the genetic inheritance of GPT models, through generating question and answers using ChatGPT, and comparing their similarity to classify the text. The method achieved high accuracy (94.57%) and robustness against obfuscation techniques, demonstrating its applicability in projects requiring reliable AI-generated text detection. Bakhtin et al. introduced a study [3], to investigate the effectiveness of Energy-based Models (EBMs) in detecting machine-generated text. The findings highlighted the adaptability of EBMs to various methods of generating fake text but also underscored the influence of specific training data on model performance, leading to nuanced accuracy rates across different scenarios. Similarly, CheckGPT [4] proposed a two-module system for detecting LLM-generated academic writing. The approach has achieved high accuracy (98–99%) in identifying LLM-generated content across diverse academic disciplines while emphasizing the need for further investigation into out-of-distribution scenarios. Another study [5] focused on deepfake text detection in real-world settings with unknown text sources and domains. The comprehensive dataset encompassed diverse human-written and LLM-generated content, revealing surprising similarities in linguistic features between the two sources. Despite achieving high average recall rates (over 90%) with fine-tuned pre-trained language models (PLMs), the performance significantly declined in out-of-distribution scenarios, highlighting the challenge of domain adaptation. OUTFOX [6] aimed to improve the robustness of LLM-generated-text detectors against adversarial attacks. The framework, OUTFOX, incorporated considerations from both the detector and the attacker, enhancing the overall robustness of LLM-generated-text detectors.

Our research addresses the dual facets of LLMs, particularly ChatGPT, as both an innovation beacon and a potential source of deception. We aim to develop a robust deep-learning model capable of discerning between human-authored and AI-generated text in monolingual and multilingual contexts, with significant implications for real-world applications beyond academic inquiry.

3 Methods

We use two different methods to address these tasks. Firstly, we utilize classical machine learning techniques, which are specifically adapted for natural language preprocessing (NLP). Secondly, we employ a strategy that involves fine-tuning transformer-based Language Models (LLMs).

3.1 Classical Machine Learning

In this study, we adopted a two-pronged approach to address the task of multilingual text classification. Firstly, employing classical machine learning techniques, we initiated our methodology with comprehensive preprocessing of the textual data. This involved the removal of extraneous elements such as URLs, stopwords, and whitespaces. Subsequently, TF-IDF vectorization [14] was executed using a vectorizer configured to accommodate a maximum of 8000 features. To enrich the feature set and capture linguistic intricacies, we incorporated metrics including the Gunning Fog Index and Flesch Reading Ease Score.

The Gunning Fog Index is a readability metric developed by Robert Gunning in 1952 [7].The Gunning Fog Index is a readability metric that estimates the years of formal education needed to understand a text on the first reading. It estimates the readability of a piece of text by considering the average number of words per sentence and the percentage of complex words (words with three or more syllables). A higher Fog Index indicates more complex text, while a lower index suggests simpler text. Incorporating this metric into our feature set allowed us to capture the complexity of the text, which could be indicative of certain linguistic patterns or content characteristics relevant to the classification task. The Flesch Reading Ease Score, developed by Rudolf Flesch in 1948 [8], measures the ease of readability of a text based on the average sentence length and the average number of syllables per word Fig. 1. The score ranges from 0 to 100, with higher scores indicating easier readability. By including the Flesch Reading Ease Score as a feature in our model, we aimed to capture the overall readability level of the text, which could be relevant for understanding language nuances and aiding classification.

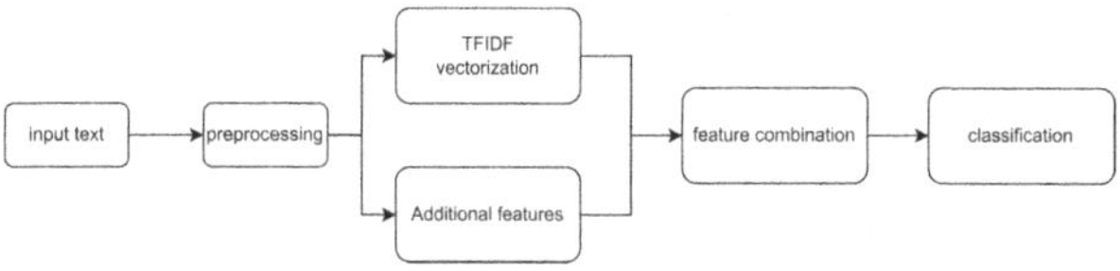

Fig. 1. machine learning methods

Following feature engineering, we proceeded to model selection, employing classifiers such as Logistic Regression [9] and XGBoost [10] for classification [5]. Logistic Regression [9] is a widely used linear classifier that models the probability of a binary outcome. It's particularly suitable for binary classification tasks like sentiment analysis or spam detection. On the other hand, XGBoost [10] is an ensemble learning method that builds a series of decision trees to make predictions. It's known for its efficiency and effectiveness in handling large datasets and complex features.

3.2 Fine Tuning LLMs

In our endeavor to explore advanced language models for multilingual text classification, we leveraged two distinct yet impactful architectures: XLM-RoBERTa [11] and DistilBERT [13].

XLM-RoBERTa. As established by Conneau et al. (2019), XLM-RoBERTa [11] stands as a cornerstone in transformer-based architectures, representing a multilingual adaptation of the original RoBERTa model [12] Fig. 2.

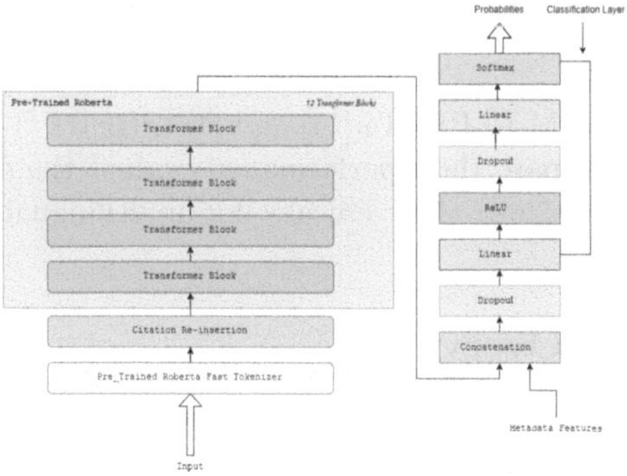

Fig. 2. Prediction accuracies of all models for Subtask A

Specifically tailored to comprehend and process a diverse array of languages, XLM-RoBERTa [11] benefits from extensive pre-training on a substantial dataset comprising 2.5TB of filtered CommonCrawl data (Zhang et al., 2020). This corpus spans text in 100 different languages, enabling the model to capture nuanced language features and patterns across a broad linguistic spectrum. The utilization of such a diverse training dataset facilitates the acquisition of a robust understanding of various linguistic structures and vocabularies, rendering XLM-RoBERTa [11] highly effective for tasks involving multiple languages.

DistillmBERT. While RoBERTa and XLM-RoBERTa models offer exceptional performance, they are computationally expensive, necessitating the exploration of more computationally efficient alternatives. In this context, we investigated DistilBERT base multilingual cased [13], a distilled version of the BERT base multilingual model. Pre-trained on the concatenation of Wikipedia articles in 104 different languages, it features 6 layers, each comprising 768 dimensions and 12 attention heads, totaling 134 million parameters Fig. 3. This configuration strikes a balance between model efficiency and significant representational power, making DistilBERT [13] an attractive option for resource-constrained environments.

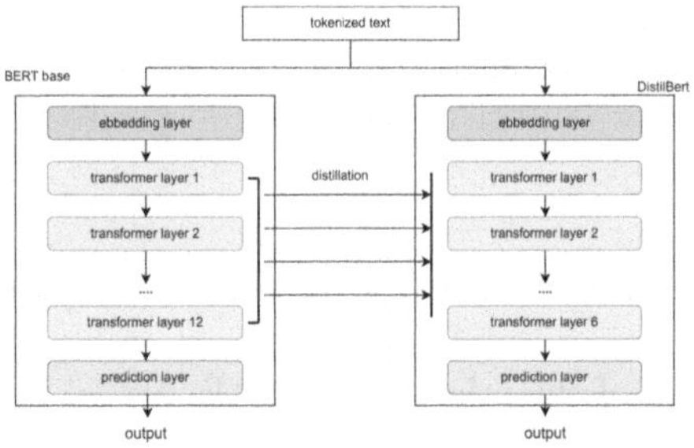

Fig. 3. Model Architecture for DistilBERT

By incorporating XLM-RoBERTa [11] and DistilBERT [13] into our methodology, we aimed to compare their performance against the established models, thereby gaining insights into their efficacy for multilingual text classification tasks.

4 Experimental Overview

In our experiment, we have developed two modules while concentrating on two specific subtasks from the SemEval 2024 Task 8 dataset: Subtask A and Subtask B. Subtask A involved classifying text as either human-written or machine-generated, while Subtask B required identifying boundaries within mixed text to distinguish between human-written and machine-generated segments. The dataset sizes are detailed in the table. The dataset used for Subtask A has 172,417 training data and 4,000 test data, whereas Subtask B was implemented with 3,649 training and 505 training instances.

4.1 Subtask A: Classification of Text as Human-Written vs. Machine-Generated

For Subtask A, we conducted experiments using both classical machine learning models with default parameters and fine-tuning methods with advanced language models. Classical Machine Learning Models: We experimented with a variety of machine learning models, including Logistic Regression [9] and XGBoost [10], utilizing default parameters. These models were trained on the labeled dataset to classify text into human-written or machine-generated categories.

Fine-tuning XLM-RoBERTa: In the fine-tuning approach using XLM-RoBERTa [11], we employed a learning rate of 2.0e-5 and a batch size of 16. During training, all parameters were initialized and trained for the first epoch. Subsequently, only the classifier layer parameters were fine-tuned for the next two epochs, while the initial transformer layers remained frozen. Fine-tuning DistilBERT: Similarly, for the fine-tuning of DistilBERT [13], we utilized a batch size of 16 and a learning rate of 1.0e-3. The model was trained for a total of 5 epochs, with the objective of optimizing its performance on the classification task.

4.2 Subtask B: Boundary Detetction

For Subtask B, where the goal is to distinguish between human-written and machine-generated segments within mixed text, we devised a novel approach combining fine-tuning techniques and regression methodology. Initially, we mapped the given partition index in the text to its corresponding index in the tokenized representation Fig. 4. This alignment ensured seamless processing within our model by maintaining consistency between the original text and its tokenized form.

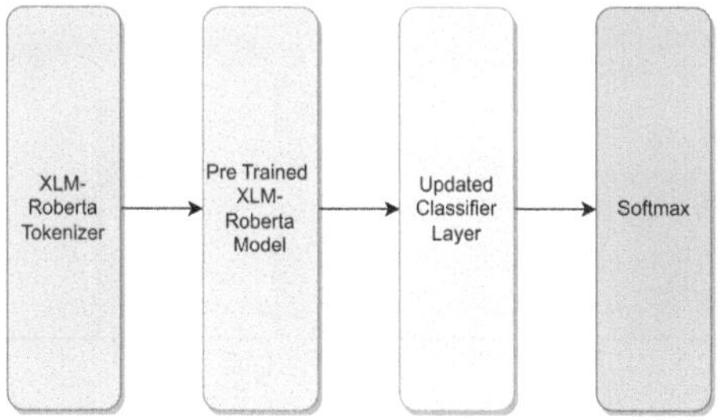

Fig. 4. using trained model with softmax

To tackle the task, we repurposed a model initially trained for Subtask A, which was originally designed for classification. Our adaptation involved modifying the model's classifier layer to a regression layer. By replacing the classifier with regression functionality and incorporating a softmax activation function, we could compute the probability of each index being the partition index within the tokenized text representation.

During training, the modified model underwent 20 epochs using a learning rate set at 1.0e-3. Given that the task involved regression rather than classification, we employed the Mean Squared Error (MSE) loss function to optimize the model parameters. This loss function allowed us to minimize the discrepancy between predicted partition indices and the ground truth labels.

Following prediction, we reverse-mapped the predicted partition index from the tokenized representation to its corresponding position in the original text. This reverse mapping facilitated the identification of the actual index within the original text, enabling accurate delineation of boundaries between human-written and machine-generated segments.

5 Result and Analysis

5.1 Subtask A

In Subtask A, which involved the classification of text into human-written or machine-generated categories, our experiments yielded promising results. Among the methodologies explored, the fine-tuning of the XLM-RoBERTa [11] model emerged as the most effective approach, showcasing notable performance metrics. The XLM-RoBERTa [11] model achieved an accuracy of 0.734, with an F1 score of 0.659, precision of 0.394, recall of 0.509, and specificity of 0.964. These results underscore the efficacy of advanced language models in accurately classifying multilingual text, demonstrating the superiority of XLM-RoBERTa [11] in this classification task Fig. 5.

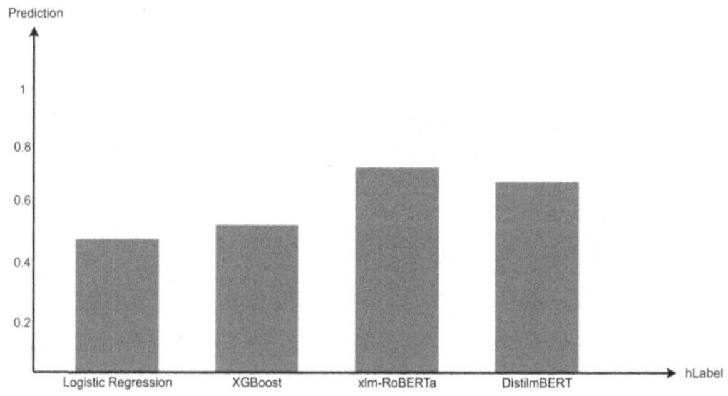

Fig. 5. prediction accuracies of all models for subtask A

5.2 Subtask B

For Subtask B, which focused on identifying boundaries within mixed text to distinguish between human-written and machine-generated segments, our approach centered on regression methodology. The Mean Squared Error (MSE) loss achieved for this task was 75.4671, indicating a certain level of discrepancy between predicted and ground truth partition indices. Additionally, the Mean Absolute Error (MAE) loss was 49.60, providing a measure of the average distance between predicted and actual boundaries within the text. A lower MAE signifies a closer match between predicted and actual boundaries, suggesting better performance in identifying segment boundaries within mixed text.

6 Conclusion and Future Scope

Our study investigates Subtasks A and B of SemEval 2024 Task 8, addressing the challenges of multilingual text analysis. In Subtask A, which involves classifying text as either human-written or machine-generated, we highlight the superior performance of the fine-tuned XLM-RoBERTa model. With impressive accuracy and F1 scores, XLM-RoBERTa excels in capturing nuanced linguistic features across diverse languages, establishing itself as a leading model in multilingual text classification. For Subtask B, which focuses on identifying boundaries within mixed text to distinguish human-written from machine-generated segments, we employ a regression-based approach. Our model achieves lower Mean Absolute Error (MAE) scores, indicating its effectiveness in accurately delineating segment boundaries. While some discrepancies between predicted and actual boundaries remain, our study emphasizes the importance of MAE as a metric for assessing accuracy and guiding future improvements. Our findings demonstrate the efficacy of advanced language models like XLM-RoBERTa in classification tasks and underscore the significance of MAE in boundary detection. These insights advance multilingual text analysis and provide valuable considerations for future research in natural language processing, enabling robust analysis across diverse domains and applications.

References

1. Wu, J., et al.: A survey on LLM-generated text detection: necessity, methods, and future directions (2024). arXiv:2310.14724
2. Yu, X., et al.: LLM paternity test: generated text detection with LLM genetic inheritance (2024). arXiv:2305.12519
3. Bakhtin, A., Gross, S., Ott, M., Deng, Y., Ranzato, M.A., Szlam, A.: Real or fake? Learning to discriminate machine from human generated text (2019). arXiv:1906.03351
4. Liu, Z., Yao, Z., Li, F., Luo, B.: On the detectability of ChatGPT content: benchmarking, methodology, and evaluation through the lens of academic writing (2024). arXiv:2306.05524
5. Li, Y., et al.: Deepfake text detection in the wild (2023). arXiv:2305.13242
6. Koike, R., Kaneko, M., Okazaki, N.: OUTFOX: LLM generated essay detection through in-context learning with adversarially generated examples (2024). arXiv:2307.11729
7. BRIAN SCOTT. The Gunning's Fog Index (or FOG) Readability Formula (2023)

8. Rogers, R.L., Kincaid, J.P., Fishburne Jr, R.P., Rogers, R.L., Chissom, B.S.: Derivation of new readability formulas (automated readability index, fog count and Flesch reading ease formula) for navy enlisted personnel (1975)
9. Cox, D.R.: The regression analysis of binary sequences. J. Royal Stat. Soc. Ser. B (Methodol.) **20**(2), 215–232 (1958)
10. Chen, T., Guestrin, C.: XGBoost: a scalable tree boosting system. In: Proceedings of the 22nd ACM SIGKDD International Conference on Knowledge Discovery and Data Mining, KDD '16, New York, NY, USA. ACM, pp. 785–794 (2016)
11. Conneau, A., et al.: Unsupervised cross-lingual representation learning at scale. CoRR, abs/1911.02116 (2019)
12. Liu, Y., et al.: and Veselin Stoyanov. A robustly optimized bert pretraining approach, Roberta (2019)
13. Sanh, V., Debut, L., Chaumond, J., Wolf, T.: DistilBERT, a distilled version of BERT: smaller, faster, cheaper and lighter. ArXiv, abs/1910.01108 (2019)
14. Jing, L.P., Huang, H.K., Shi, H.B.: Improved feature selection approach TFIDF in text mining. In: Proceedings International Conference on Machine Learning and Cybernetics, vol. 2, pp. 944–946 (2002)

A Novel Preprocessing Method for Transforming Federal Sentencing Data to Ensure Unbiased AI Adjudication Research Using Large Language Models

Nafiul Alam, Hannah Cho, M. I. R. Shuvo, and Jong-Hoon Kim(✉)

Advanced Telerobotic Research Lab, Computer Science, Kent State University, Kent, OH, USA
jkim72@kent.edu
https://www.atr.cs.kent.edu/

Abstract. In an effort to create objective AI for judicial decision-making, this study offers a unique preprocessing method for the JUST-FAIR data-set of federal criminal sentencing records, which will be used in Large Language Models (LLMs). While LLMs struggle with numerical and categorical data, traditional machine learning methods struggle with the complex, linked nature of judicial data. Our approach tackles these problems in a multi-phase manner: (1) data cleansing with MongoDB; (2) textual categorization of numerical data using USSC and JUSTFAIR codebooks; (3) enlarging legal acronyms; and (4) narrative story creation from discrete data points with Claude 3.5 Sonnet. LLMs can now more fully understand the complex relationships involved in sentencing decisions because to this shift. Our method establishes the groundwork for LLMs to be trained on less biased court data, which could transform AI applications in the legal space. The work offers ramifications outside of the legal system to other industries with sensitive, complicated data, contributing to AI ethics and bias reduction. In this paper, we present a detailed exposition of our methodological framework and offer a comprehensive evaluation of outcomes, thereby establishing the credibility and significance of our findings.

Keywords: AI Judge · Machine Learning · Biased Judgement · Large Language Model · Generative AI

1 Introduction

Artificial intelligence and the law bring extraordinary prospects and ethical challenges. As AI systems become more prevalent in decision-making, their use in criminal sentencing raises concerns about fairness, transparency, and bias [1]. This research asks: Can an AI system make unbiased court decisions, especially when educated on historically biased data? Traditional machine learning

D. Singh et al. (Eds.): IHCI 2024, LNCS 15557, pp. 325–337, 2025.
https://doi.org/10.1007/978-3-031-88705-5_28

algorithms, including deep learning, are prone to biases in training data [2]. This bias is especially troublesome in the US court system, since historical data shows long-standing socioeconomic inequalities. This dataset, JUSTFAIR (Judicial System Transparency through Federal Archive Inferred Records), contains almost 600,000 federal criminal sentence records from 2001–2018, providing a thorough but potentially biased foundation for AI research in this area. We propose a unique strategy using Large Language Models (LLMs) [4] to handle this difficulty while minimizing biases. LLMs can grasp and reason about complicated, interdependent elements in judicial decisions, unlike typical machine learning models that may amplify numerical and category biases. LLMs face challenges in handling and understanding numerical and categorical data [5].

Our research focuses on creating a novel preprocessing mechanism for the JUSTFAIR dataset. The process consists of several essential steps: cleansing and organizing the data, transforming numerical and categorical information into explanatory text, expanding legal acronyms, and ultimately producing narrative stories that effectively depict the intricate connections among different factors in sentencing determinations. Our objective is to convert individual data points into cohesive narratives in order to generate a dataset that enables LLMs to comprehend the complete context and intricacy of each case. This initial preprocessing stage is essential but only marks the beginning of our comprehensive study plan. Our primary objective is to create an artificial intelligence (AI) system called FAIRS (Fair Artificial Intelligent Response System) that can provide impartial suggestions in legal matters. After completing the data preparation, our next step is to meticulously choose and optimize a suitable LLM (Language Model) using the preprocessed dataset. In the upcoming stages of our research, we will concentrate on appraising the model's effectiveness, evaluating its capacity to make impartial judgments, and progressively enhancing our methodology to reduce any identified biases.

This research has far-reaching consequences that go well beyond the immediate judicial decision-making situation. Our study addresses the problem of building unbiased AI systems trained on historically biased data, and makes a contribution to machine learning fairness, AI ethics, and responsible AI development. The research's methods and discoveries may help develop strategies for reducing bias in AI systems in a variety of fields where complicated, delicate, and possibly biased historical data are present. We continue to be acutely aware of the ethical issues and possible societal effects of artificial intelligence in the legal system as we advance this study. The essential rights to due process, fairness, and transparency serve as the guiding principles of our work. Our goal in developing an unbiased AI adjudication system is to offer a tool that may support and possibly even enhance the fairness of judicial decision-making processes, rather than to completely replace human judgment.

We will describe our preprocessing strategy, address LLM adaptation for court data difficulties and solutions, and explain our model development and evaluation strategies in the following sections. We hope this research contributes to the AI in society debate and promotes more just and equitable AI systems.

2 Literature Review

Racial bias in the US criminal justice system is a long-standing issue. Critics argue that institutional discrimination is behind minority arrest and imprisonment rates, notably African American ones. According to early estimates, African Americans make up only 12% of the national population yet are responsible for 48% of arrests and imprisonments [6]. This gap has led to charges of racial bias in areas where criminal justice officials have discretion. System defenders say these figures show minority community crime, not institutional discrimination. According to street crime data, 51% of African American males in large cities are arrested for an index offense, compared to 14% of white males.

Recent study has tried to reconcile these opposing views. A study of 52 multivariate statistical studies completed since 1970 on presentencing supports the discrimination thesis (DT) [7]. This research showed that minority offenders are still treated differently, notably in presentencing judgments, showing that racial bias continues to impact criminal justice decisions. The potential biases in datasets used for analysis and decision-making make tackling racial injustice in the criminal justice system harder. The JUSTFAIR database contains nearly 600,000 federal criminal sentencing records from 2001–2018, while the USSC [8] maintains individual offender files (IOFs) with detailed sentencing information for federal courts from 2002–2018. While useful for study, these datasets may perpetuate prejudices if not properly examined and contextualized. Extracting valuable information without amplifying historical biases is challenging.

The use of artificial intelligence (AI) in legal systems has become more popular worldwide, with several nations investigating how it might improve justice and efficiency. However, there are important concerns regarding bias, transparency, and ethical considerations when applying AI in this delicate field. The integration of AI in China's legal system demonstrates a step-by-step evolution, moving from auxiliary instruments to decision-making assistance, ultimately leading to the creation of self-governing "robot judges" [9]. Although this method has increased efficiency, the public's doubts about fairness and transparency have arisen because of worries about the "black box" nature of AI algorithms.

In India, AI is being considered to reduce judicial backlogs and speed up case processing [10]. AI could be used for legal research, reasoning, and prediction. To ensure fair and effective AI integration in the judicial process, researchers emphasize overcoming biases, honoring legal precedents, and evaluating implementation costs. AI is being used in the Crown Courts of England and Wales to reduce administrative costs and enhance case management efficiency [11]. AI standardizes outcomes and may reduce bias and judicial discretion without adjudicating cases.

AI's use in legal processes, especially sentencing, raises ethical concerns. AI systems may reinforce biases in their training data, which is a major worry. This is especially problematic when using historical data on long-standing social inequalities. New research suggests a bias impact assessment paradigm similar to pharmaceutical trials to address these challenges [12]. This paradigm promotes a holistic approach to data science that integrates philosophical and sociological viewpoints to comprehend and account for social preconceptions. AI in court systems has potential benefits but also ethical and practical drawbacks, according to the literature. AI systems that interpret complex legal information and minimize historical biases must be developed in the future. This requires interdisciplinary legal, ethics, computer science, and social science skills. We must strike a balance between using AI to improve court efficiency and upholding core concepts of fairness, transparency, and due process [13]. Creating unbiased AI systems for judicial decision-making is a technical difficulty but a necessary step toward a more egalitarian justice system.

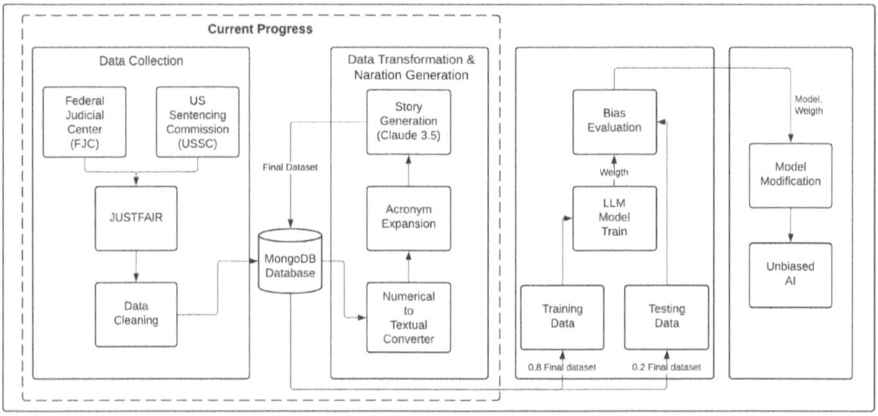

Fig. 1. Basic Architecture of the Proposed FAIRSystem

3 Proposed Methodology

Our research approach focuses mostly on novel data preparation methods to tackle the difficult problem of developing an impartial AI system for judicial decision-making shown in Fig. 1. The procedure consists of multiple critical steps, all of which are essential to converting unprocessed, perhaps biased data into a format that can be used to train Large Language Models (LLMs) in an objective manner. The detailed preprocessing is shown in Fig. 2

3.1 Data Collection and Initial Processing

Dataset Acquisition: The basis of our research relies on the JUSTFAIR (Judicial System Transparency via Federal Archive Inferred Records) dataset. This extensive compilation consists of about 600,000 records of federal criminal sentence decisions spanning from 2001 to 2018. It offers a substantial and potentially partial source of information regarding the process of judicial decision-making in the United States.

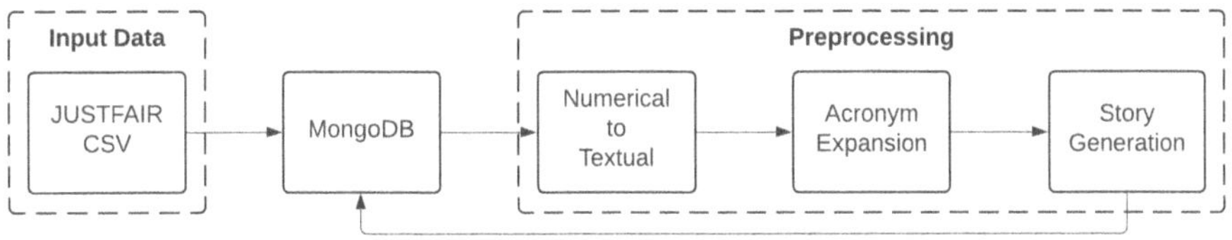

Fig. 2. Preprocessing of the Data

Data Cleaning and Storage: Our first step was to import the raw data, which included numerical values, classification categories, and acronyms, into a MongoDB database [14]. We selected MongoDB due to its versatility and efficacy in managing extensive and intricate datasets. Through a rigorous methodology, we detected and addressed missing values and irregularities in the dataset, establishing a well-structured database as the foundation for further preprocessing.

3.2 Data Transformation

Conversion of Numerical and Categorical Data: The initial round of data translation involved employing a blend of Pandas and MongoDB queries, in conjunction with codebooks and dictionaries supplied by the United States Sentencing Commission (USSC) and JUSTFAIR [15]. We have developed an extensive JSON file that links numerical codes to their accompanying textual descriptions. Subsequently, this mapping was employed to transform the numerical categorical data into descriptive text throughout the entire dataset, so enhancing the contextual information for each data point.

Acronym Expansion: The second phase of the project aimed to broaden the scope of the many legal and dataset-specific abbreviations found in the data [16]. We generated a bespoke JSON file by utilizing USSC and JUSTFAIR codebooks. This file has a comprehensive list of acronyms and their corresponding complete forms. By utilizing this resource, we created and implemented scripts to fully expand all acronyms present in the dataset, thereby augmenting the contextual depth of the data.

3.3 Narrative Generation

Story Creation: Having transformed the data into a more detailed and explanatory textual structure, we proceeded to the cutting-edge essence of our approach: generating narratives. We utilized Claude 3.5 Sonnet [17], an advanced Large Language Model, for this undertaking. We created meticulously designed cues that directed the LLM to produce logical and connected stories based on the individual data points of each record.

Every updated record was inputted into Claude 3.5 Sonnet, which then produced a narrative that encapsulated the connections among many factors influencing sentence. This technique successfully converted the individual, possibly prejudiced data points into narratives that are rich in context, enabling a more detailed comprehension of each instance.

Quality Assurance: In order to guarantee the excellence and precision of these produced narratives, we have established a meticulous quality assurance procedure. This entailed a manual examination of a subset of the produced narratives, verifying their precision, logical consistency, and possible prejudices. Using feedback from these reviews, we continuously improved our prompts and the creation process to enhance the quality and dependability of the narratives.

3.4 Final Dataset Preparation

The produced narratives were organized into a systematic structure, with metadata connecting each story to its associated original record identifier. The final dataset was meticulously curated for the specific goal of training and assessing Large Language Models for impartial judicial decision-making.

3.5 Future Steps (Planned)

LLM Selection and Training: Our next actions involve the meticulous selection of suitable LLMs for the purpose of judicial decision-making. Our objective is to develop a fine-tuning procedure that employs our preprocessed narrative dataset, with the goal of creating a model that can engage in impartial reasoning inside legal contexts.

Model Evaluation: We will create reliable standards for evaluating the effectiveness and prejudice of models, employing thorough testing procedures to detect any lingering biases in the system's results.

Iterative Refinement: We expect a process of continuous improvement, where we will study the outputs of the model to find any biases or inconsistencies. The investigation will provide valuable insights for enhancing our preprocessing methodologies and model architecture, hence ensuring a consistent enhancement in the system's impartial performance.

3.6 Ethical Considerations

At every level of our technique, we prioritize ethical considerations.

1. We guarantee the protection of data privacy and the process of anonymization in accordance with both legal and ethical guidelines.
2. We systematically evaluate the possibility of adding additional biases during the preprocessing and modeling steps.
3. We collaborate with legal and ethical experts to enhance our decision-making, ensuring it upholds principles of fairness and justice.

Ultimately, our methodology offers a fresh and innovative way to prepare intricate judicial material for integration into AI systems, aiming to create an impartial AI judge. Our objective is to enhance the understanding of Large Language Models by converting unprocessed data, which may contain biases, into detailed narratives that provide contextual information. This approach helps to minimize the possibility of reinforcing existing prejudices when making sentencing judgments. This technique not only tackles the special difficulties of judicial decision-making but also provides potential uses in other areas characterized by intricate, delicate data.

4 Implementation and Results

Our methodology was implemented using several technical components and tools, each of which had a vital part in the stages of data preparation and narrative development. This section provides a comprehensive overview of the precise methodologies and advanced technology utilized in our research.

4.1 Data Processing with MongoDB and Pandas

MongoDB Setup and Data Import: We employed MongoDB version 7.0 for our database requirements. The JUSTFAIR dataset was loaded into MongoDB via a sequence of procedures:

Converting data from its original format to JSON. This is shown in Table 1. Performing a mass insertion of JSON data using MongoDB's import tool. Creating indexes on frequently requested data to enhance performance.

Data Cleaning and Transformation from Numerical to Textual Form: We utilized Pandas version 1.3.0 to do data manipulation and cleansing tasks. Primary tasks encompassed: Retrieving data from MongoDB and storing it in Pandas DataFrames. Handling missing values through appropriate filling or imputation techniques. Transforming categorical codes into descriptive text by utilizing pre-established mappings. This transformation is shown in Table 1.

4.2 Acronym Expansion

We created a Python script that utilizes our unique JSON mapping to enlarge acronyms. The result is shown in Table 1. This process entailed: Retrieving the acronym dictionary from a JSON file. Utilizing the expansion function on pertinent text fields inside the dataset. Conducting spot-checks to ensure the accuracy of expansions.

4.3 Narrative Generation with Claude 3.5 Sonnet

For narrative generation, we utilized the Claude 3.5 Sonnet API. The process involved the following steps:

Table 1. Merged JUSTFAIR Data: Original MongoDB, Descriptive, and Acronym-Expanded Formats

Original MongoDB Format	Descriptive MongoDB Format	Acronym Expanded Data
"CIRCDIST": 1	"CIRCDIST": "Dist of Columbia"	"Circuits and Districts": "Dist of Columbia"
"AGE": 26	"AGE": 26	"Age of Defendant (years)": 26
"CITIZEN": 3	"CITIZEN": "Illegal Alien"	"Citizenship Status": "Illegal Alien"
"SENTTOT": 24	"SENTTOT": 24	"Total Sentence (months)": 24
"NEWRACE": 3	"NEWRACE": "Hispanic"	"Race of Defendant (Recoded)": "Hispanic"
"BOOKERCD": 7	"BOOKERCD": "Government Sponsored - Below Range"	"Post-Booker Reporting Category (Detailed)": "Government Sponsored - Below Range"
"judge": "Amy Berman Jackson"	"judge": "Amy Berman Jackson"	"Judge's Name": "Amy Berman Jackson"

Table 2. Comparison of Preprocessed JUSTFAIR Data and Corresponding Narrative

Preprocessed Data	Narrative
"Circuits and Districts": "Dist of Columbia"	In the District of Columbia in late 2011,
"Age of Defendant (years)": 26	a 26-year-old male defendant,
"Citizenship Status": "Illegal Alien" "Country of Citizenship": "El Salvador"	an illegal alien from El Salvador.
"Judge's Name": "Amy Berman Jackson"	a case came before Judge Amy Berman Jackson
"Statute of Conviction (First)": "81326A" "Statute of Conviction (Second)": "81326B2"	The defendant was charged with illegal re-entry after deportation in violation of 8 U.S.C. §1326(a) and (b)(2).
"Criminal History Points": 2 "Final Criminal History Category": 2	The defendant had a limited criminal history, scoring 2 criminal history points which placed him in Criminal History Category II under the U.S. Sentencing Guidelines.
"Base Offense Level": 8	His base offense level for the illegal re-entry crime was 8 under U.S.S.G. §2L1.2.
"Specific Offense Characteristic B": 16	However, a 16-level enhancement was applied under §2L1.2(b)(1)(A)
"Adjusted Offense Level": 24	resulting in an adjusted offense level of 24.

API Setup: We utilized the official Claude API client for Python, guaranteeing the secure management of API keys and appropriate authentication.

Prompt Engineering: We created a template prompt to assist Claude in producing cohesive narratives. The purpose of this prompt is to offer background information regarding the sentencing statistics. Direct Claude to concentrate on crucial variables and interconnections in every instance. Highlight the importance of impartial and accurate storytelling. Refrain from making extra assumptions that go beyond the given data.

Narrative Generation: We implemented the narrative generating method on every entry in our dataset, guaranteeing that each narrative was exclusively derived from the provided case data. The resulting text had a persistent length and organization. The procedure was streamlined to enhance efficiency in handling extensive datasets. Some part of the dataset is shown in Table 2

4.4 Final Dataset Preparation

The ultimate dataset was produced via Pandas and saved in both CSV and JSON formats to allow for versatility. This phase entailed the process of carefully choosing the appropriate columns to be included in the final dataset. Ensuring uniform formatting throughout all records. Conducting last checks to ensure the integrity of the data.

This application utilizes MongoDB for quick data storage and retrieval, Pandas for versatile data processing, and the Claude 3.5 Sonnet API for advanced story synthesis. By utilizing these methods, we were able to convert the unprocessed JUSTFAIR data into a comprehensive dataset that is well-suited for training impartial AI models for judicial decision-making.

To further validate the effectiveness of our methodology, we compared a narrative generated by FAIRS to a real court case report from The New York Times (NYT) [18], as shown in Fig. 3. The AI-generated narrative (Fig. 3b) was based on a similar case from the JUSTFAIR dataset, while the NYT story (Fig. 3a) detailed the specifics of the Jesse Jackson Jr. fraud case. Despite the limited information available in the JUSTFAIR dataset, FAIRS was able to generate a narrative that closely resembled the structure and content of the real news report. We used the following prompt to guide the AI in generating the narrative:

"Please use the law references provided to express the story more elaborately. Include relevant legal principles, precedents, and any significant details from the case. Structure the story in a logical manner, covering the background, legal issues, arguments, decision, and reasoning in a narrative form."

Table 3 presents a comparison of the similarities between the two stories. Using Claude 3.5 Sonnet, we determined that the similarity between the AI-generated narrative and the New York Times report was approximately 80–85%. This high level of similarity is evident in the table, which compares various

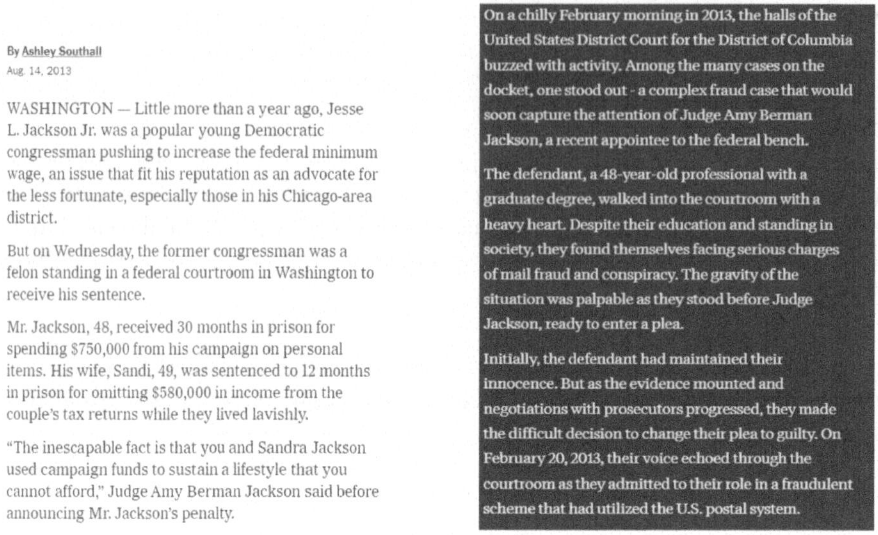

By Ashley Southall
Aug. 14, 2013

WASHINGTON — Little more than a year ago, Jesse L. Jackson Jr. was a popular young Democratic congressman pushing to increase the federal minimum wage, an issue that fit his reputation as an advocate for the less fortunate, especially those in his Chicago-area district.

But on Wednesday, the former congressman was a felon standing in a federal courtroom in Washington to receive his sentence.

Mr. Jackson, 48, received 30 months in prison for spending $750,000 from his campaign on personal items. His wife, Sandi, 49, was sentenced to 12 months in prison for omitting $580,000 in income from the couple's tax returns while they lived lavishly.

"The inescapable fact is that you and Sandra Jackson used campaign funds to sustain a lifestyle that you cannot afford," Judge Amy Berman Jackson said before announcing Mr. Jackson's penalty.

(a) The New York Times Story

On a chilly February morning in 2013, the halls of the United States District Court for the District of Columbia buzzed with activity. Among the many cases on the docket, one stood out - a complex fraud case that would soon capture the attention of Judge Amy Berman Jackson, a recent appointee to the federal bench.

The defendant, a 48-year-old professional with a graduate degree, walked into the courtroom with a heavy heart. Despite their education and standing in society, they found themselves facing serious charges of mail fraud and conspiracy. The gravity of the situation was palpable as they stood before Judge Jackson, ready to enter a plea.

Initially, the defendant had maintained their innocence. But as the evidence mounted and negotiations with prosecutors progressed, they made the difficult decision to change their plea to guilty. On February 20, 2013, their voice echoed through the courtroom as they admitted to their role in a fraudulent scheme that had utilized the U.S. postal system.

(b) AI(Claude 3.5) generated Story

Fig. 3. Original and AI generated Story

Table 3. Revised Comparison of Two Real Fraud Cases from Different Sources

Element	Story 1 (Jesse Jackson Jr.)	Story 2 (AI Generated Dataset)	Similarity
Data Source	The NewYork Times	JustFair dataset and Claude	Partial
Subject	Named: Jesse Jackson Jr. fraud case	Unnamed defendant fraud case	Partial
Defendant	Jesse Jackson Jr., 47	48-year-old professional with graduate degree	High
Charges	Conspiracy to defraud campaign funds	Mail fraud and conspiracy	High
Amount	$750,000	Not specified	Low
Plea	Guilty	Guilty	High
Timeline	Events in 2013, release in 2015	Events in 2013	High
Specific Details	Use of funds for personal items	Courtroom proceedings	Partial
Judge	Not specifically named	Named as "Judge Amy Berman Jackson"	Low
Overall Narrative Style	Factual news report	Detailed legal proceedings	Partial

elements of the two stories. We conducted this comparison for a total of 20 stories and observed consistent results, further demonstrating the robustness of our approach in generating realistic and accurate legal narratives from limited datasets.

5 Discussion

The utilization of artificial intelligence to aid in judicial decision-making was simultaneously promising and perilous. Our objective was to create an AI system capable of producing unbiased judicial decisions, even when trained on historically prejudiced data. Our investigation into Large Language Models (LLMs) and innovative data preprocessing extended beyond conventional machine learning. The use of our approach of transforming raw data into compelling stories yielded positive results. With a mixture of admiration and trepidation, we saw as FAIRS comprehended the intricate connections among the factors used to determine sentences, as if granting the machine the ability to perceive the profound human narratives that lie beneath each individual case, transcending just numerical data.Our next objective is to utilize the created stories to train a Language Model (LLM) and evaluate the extent to which the LLM's conclusions correspond with the actual judicial decisions. With stories that have an accuracy rate of 80–85%, there is a great possibility for making wise decisions. However, this also highlights the crucial necessity to evaluate the impartiality of the LLM, as guaranteeing equity is fundamental in establishing a morally upright AI.

However, like any journey, our route was not without obstacles. We encountered the risk that our biases might infiltrate our data-driven narratives. We endeavored to enhance the transparency of FAIRS's decision-making process. Throughout the judicial proceedings, we were consistently reminded of the intricate equilibrium between artificial intelligence and human intellect. The purpose of FAIRS is to enhance human understanding, rather than supplant it. Upon contemplation of our trajectory, we acknowledge that our role extends beyond the realm of criminal justice. By employing a novel strategy to address prejudice in a critical domain, we have made a valuable contribution to the ongoing discourse on the ethical development of artificial intelligence. The methodology we have created for judicial decision-making could be valuable in other industries that have issues with historical data biases.

6 Conclusion and Future Works

At this juncture, we are cautiously optimistic as we contemplate the path towards a legal system that is both fair and efficient, while also leveraging the capabilities of artificial intelligence. We have made substantial progress, even though our journey is still far from finished. As we contemplate the future, there are both new possibilities and difficulties that await us in our endeavor to extend the application of this technology from criminal sentencing to other domains of law, and maybe even across international borders to aid judicial systems globally. Nevertheless, successfully pursuing this direction necessitates meticulous maneuvering inside strong ethical and legal structures to guarantee that as FAIRS grows in capability, it continues to function as a tool for promoting fairness rather than assuming the role of a judge.

FAIRS is continuously improving its ability to identify and rectify its own biases, becoming more rational and flexible through rigorous testing in actual courtrooms, under the careful observation of juries and attorneys. The upcoming stage necessitates cooperation among computer scientists, legal scholars, ethicists, and legislators. Collectively, we have the ability to shape a forthcoming era in which artificial intelligence augments human discernment, so improving the fairness and efficiency of our justice systems. As we progress, we extend an invitation to everyone to join us in this captivaA Novel Preprocessing Method for Transforming Federal Sentencing Data to Ensure Unbiased AI Adjudication Research Using Large Language Modelsting story where technology and justice collaborate together to construct a more equitable society. The journey ahead is arduous and meandering, even though with each stride, breakthrough, and revelation, we draw nearer to achieving a fully impartial, AI-enhanced legal framework.

We strongly believe that this research will extend beyond merely gathering and analysing judicial data. The potential of this technology to transform the creation of artificial data in different domains, such as marketing, anonymized data management, and the medical sector where data is frequently inadequate or unavailable, is significant. By utilizing more intelligent prompts and sophisticated language models such as ChatGPT-4o, which have the capability to search the internet, we can enhance the precision and comprehensiveness of the generated data. This undertaking lays the foundation for wider applications and a fairer world that is guided by data.

References

1. Van Cleve, N.G.: Crook County: Racism and Injustice in America's Largest Criminal Court, Redwood City: Stanford University Press (2016). https://doi.org/10.1515/9780804799201
2. Mehrabi, N., Morstatter, F., Saxena, N., Lerman, K., Galstyan, A.: A survey on bias and fairness in machine learning. ACM Comput. Surv. Aluvalu, Rajanikanth, and M. A. Jabbar. Handling data analytics on unstructured data using MongoDB (2018) **54**(6), 1–35, Article 115 (2022). https://doi.org/10.1145/3457607
3. Ciocanel, M.V., Topaz, C.M., Santorella, R., Sen, S., Smith, C.M., Hufstetler, A., et al.: JUSTFAIR: judicial system transparency through federal archive inferred records. PLoS ONE **15**(10), e0241381 (2020). https://doi.org/10.1371/journal.pone.0241381
4. Chang, Y., et al.: A survey on evaluation of large language models. ACM Trans. Intell. Syst. Technol. **15**(3), 1–45 (2024)
5. Dong, H., Wang, Z.: Large language models for tabular data: progresses and future directions. In: Proceedings of the 47th International ACM SIGIR Conference on Research and Development in Information Retrieval (2024)
6. Petersilia, J.: Racial Disparities in the Criminal Justice System, vol. 2947. Rand Corporation, Santa Monica, CA (1983)
7. Free Jr, M.D.: Racial bias and the American criminal justice system: race and presentencing revisited. Crit. Criminol. **10**(3), 195–223 (2001)
8. The United States Sentencing Commission - https://www.ussc.gov/
9. Wang, N.: Black Box Justice: robot Judges and AI-based Judgment Processes in China's court system. In: 2020 IEEE International Symposium on Technology and Society (ISTAS). IEEE (2020)

10. Gorlamudiveti, L.P., Sethu, S.G.: Role of artificial intelligence in the Indian judicial system. In: 2023 International Conference on Computational Intelligence and Knowledge Economy (ICCIKE). IEEE (2023)
11. Buckland, R.: AI, Judges and Judgment: Setting the Scene. M-RCBG Associate Working Paper Series (2023)
12. Belenguer, L.: AI bias: exploring discriminatory algorithmic decision-making models and the application of possible machine-centric solutions adapted from the pharmaceutical industry. AI Ethics **2**(4), 771–787 (2022)
13. Kandeel, M.E., Elrefae, G.: The impact of artificial intelligence on achieving the efficiency of justice AI and Speedy Justice. In: 2023 24th International Arab Conference on Information Technology (ACIT). IEEE (2023)
14. Aluvalu, R., Jabbar, M.A.: Handling data analytics on unstructured data using MongoDB. pp. 5–41 (2018)
15. USSC CodeBook - https://www.ussc.gov/research/datafiles/commission-datafiles
16. Abidin, T.F., et al.: Recognizing Indonesian acronym and expansion pairs with supervised learning and mapreduce. Information **11**(4), 210 (2020)
17. Claude/Anthropic - https://www.anthropic.com/news/claude-3-5-sonnet
18. The NewYork Times - https://www.nytimes.com/2013/08/15/us/politics/jesse-jackson-jr-sentenced-to-30-months.html

Human Factors in Cybersecurity

Computational Analysis of User Experience of Password-Based Authentication Systems

Migle Abromaityte[1], Elise Dubero[1], Quintan Kruger[1], Benne Lucassen[1], Tessa Vos[1], Niek Jan van den Hout[1], Debby Bouma[1,2], Jan Treur[1,2(✉)], and Peter H. M. P. Roelofsma[1]

[1] Centre of Expertise Cybersecurity, Risk Management and Cybersecurity Group, The Hague University of Applied Sciences, The Hague, Netherlands
{N.J.vandenHout,D.Bouma,P.H.M.P.Roelofsma}@hhs.nl, j.treur@vu.nl

[2] Department of Computer Science, Vrije Universiteit Amsterdam, Social AI Group, Amsterdam, Netherlands

Abstract. Using a network modelling approach to model adaptive dynamical systems based on self-modeling networks, this paper present a computational analysis of password-based authentication systems. It reveals several key insights into the interplay between user experience (UX) and security. An important finding is the tension between security and usability in password systems. While these systems are ubiquitous due to their relative ease of implementation and user familiarity, they come with significant challenges. As for users often convenience is more important than security, they use weak, easily guessable passwords and password reuse across multiple accounts. This behaviour, also driven by the cognitive burden of managing numerous passwords, has a bad impact on the risk of unauthorized access and security breaches. Computational analysis also showed that with adequate training, the usability of 2FA systems improved, leading to a reduction in user stress over time.

Keywords: user experience · computational analysis · authentication systems

1 Introduction

The growing number of cyber threats in our increasingly connected digital landscape requires a thorough understanding of the vulnerabilities in password-based authentication systems. As the average online user has to manage an ever-growing number of accounts, the cognitive burden of remembering numerous complex passwords has given rise to 'password fatigue'. This fatigue often leads to risky behaviors such as password reuse or the recycling of password patterns, creating significant security risks (NordPass, https://nordpass.com/). To address this issue, the research proposes the innovative application of computational analysis based on network modelling, a methodology traditionally employed in cyber risk management, to analyze and optimize the user experience (UX) of password-based authentication systems.

D. Singh et al. (Eds.): IHCI 2024, LNCS 15557, pp. 341–352, 2025.
https://doi.org/10.1007/978-3-031-88705-5_29

Network modeling, with its origins in graph theory and complex systems analysis, offers a framework for representing the complex relationships between various components of a system as nodes and links (Treur, 2016; 2020). By adopting this approach to password UX, we can visualize users actions as nodes and their interactions with the authentication process as links, gaining insights into the user and potential areas for improvement. This computational analysis approach allows for a deeper understanding of how user behavior impacts security outcomes, enabling the development of more user-friendly authentication mechanisms.

This work is part of a research programme on computational analysis of safety and security in and through cyberspace. In (Akers et al., 2025; Börcsök, Bos, Buter, Van Der Hoeven, Velev, Van den Hout, Bouma, Treur, Roelofsma, 2025; Den Oudsten, Naffakh, Waliji, Treur, Roelofsma, 2025; Caneva, Oliver, Rietze, Šileris, Barelds, Hoffmans, Treur, Roelofsma, 2025; Jeffery, Malsagov, Tijssen, Zeisig, Hoffmans, Treur, Roelofsma, 2025; Keijzer, Lochtenbergh, Marsman, Voorhoeve, Zwarts, Bouma, Treur, Roelofsma, 2025) various other themes within cybersecurity were addressed. Moreover, in (Roelofsma, Jabeen, Taal, Treur, 2025) it was analysed how an AI coach can support safety and security for medical teams in hospitals.

2 Background Knowledge and Scenario Description

In the early days of digital systems, passwords were an effective means of authentication due to the limited number of users and systems. Today, however, the landscape has drastically changed. Users now manage numerous passwords across various platforms, from email and banking to social media and workplace systems. This increase in the number of passwords has led to significant usability challenges. Users often rely on memory aids, such as writing passwords down or using similar patterns across different accounts, to cope with this burden. While these aids can help users remember their passwords, they introduce substantial security risks if compromised (Conklin, Dietrich, Walz, Conklin et al., 2004).

The dual nature of password-based systems—balancing security and usability—poses a significant challenge. On one hand, passwords are relatively easy to implement, and users appreciate that they don't require carrying additional physical objects or exerting physical effort beyond pressing keys (Bonneau Van Oorschot, Stajano, 2024 n.d.). They are also easy to learn and recover, making them a convenient choice for many users. However, passwords are fraught with security problems. Users frequently express frustration such as stress, it takes more time and that all leads to annoyance with password requirements, and the common practice of reusing passwords across multiple accounts exacerbates these security vulnerabilities.

The multifaceted nature of security and usability further complicates the evaluation and deployment of password systems. Users often fail to adhere to best practices, such as creating strong, unique passwords for each account. This behavior increases the risk of theft and unauthorized access, especially if passwords are poorly implemented or easily guessable (Zimmermann and Gerber, 2020). Additionally, there is a notable lack of progress in finding universally accepted alternatives to passwords, partly due to their familiarity and perceived ease of use, despite their acknowledged drawbacks (Bonneau et al., n.d.). To address the security limitations of passwords, two-factor authentication (2FA) has become an essential component of web security. By requiring a secondary verification step—such as a code sent to a phone—2FA significantly reduces the risk of unauthorized access even if a password is compromised (2-Factor Authentication Frequently Asked Questions, 2024). This extra layer of security actively involves users in maintaining their digital safety, making them more aware and vigilant. However, while 2FA enhances security, it can negatively impact usability compared to single-factor authentication. Users often find 2FA more cumbersome and time-consuming, which can lead to frustration (Gunson et al., 2011).

Research into password creation strategies has shown that providing users with guidelines and persuasive messages can lead to stronger and more memorable passwords. For instance, participants who received instructions along with motivational messages created more secure and easier-to-remember passwords than those who followed strict policy rules without additional guidance (Yıldırım and Mackie, 2019). This approach highlights the importance of user education in enhancing password security without overburdening users' memories.

A comprehensive understanding of user behavior is crucial for improving password security. Studies have identified common strategies users follow when choosing passwords, such as reusing passwords and adhering to the constraints imposed by websites during password creation (Wash and Rader, 2021). By examining these behaviors, researchers can identify potential vulnerabilities and develop strategies to mitigate them. Furthermore, empirical analyses of password usability and security provide valuable insights into the strengths and weaknesses of text-based passwords, informing better practices and policies (Walia, Shenoy, Cheng, 2020).

To address these challenges, it is essential to design authentication systems that consider both security and user experience. This involves not only implementing robust security measures like 2FA but also providing clear, user-friendly instructions and support for recovering access. Additionally, educating users about best practices for password management and encouraging the use of secure, memorable passwords can help reduce the frequency of lockouts and improve overall user satisfaction.

In conclusion, while password-based authentication systems are deeply entrenched in our digital lives, their usability and security challenges must be addressed through a multifaceted approach. By understanding user behaviors, enhancing password creation methods, and balancing security with usability, we can develop more effective and user-friendly authentication systems that mitigate the frustrations associated with user lockouts.

3 Design of the Adaptive Network Model

In this section the the introduced adaptive network model is described. In temporal-causal networks, nodes Y in a network are also called states which have activation levels $Y(t)$ that vary over time. Connections $X \rightarrow Y$ between states are interpreted as causal relations that indicate how a state X can affect a state Y over time. A temporal-causal network is defined by three types of characteristics: connectivity characteristics, timing characteristics, aggregation characteristics. For the connections $X \rightarrow Y$ their *weight* (or *strength)* $\boldsymbol{\omega}_{X,Y}$ indicate their strengths. For timing, speed factors $\boldsymbol{\eta}_Y$ that represent the rate at which nodes Y within the network change over time based on the impacts they receive. The initial values for all states determine the activation levels at the start. Multiple causal relations can affect a given state Y. To aggregate multiple impacts, a combination function $\mathbf{c}_Y(..)$ is indicated. Different states and networks may use different combination functions. In the network model introduced here, two combination functions are used: the *advanced logistic sum function*: $\mathbf{alogistic}_{\sigma,\tau}(..)$ with parameters $\boldsymbol{\sigma}$ for steepness and $\boldsymbol{\tau}$ for threshold, as well as the $\mathbf{stepmod}_{\rho,\delta}(..)$ function is used. The **stepmod** combination function we used to describe input from the environment is a function that has periodic steps but with a changeable duty cycle. The intervals between when the function is 'on' or 'off' are determined by the tipping point $\boldsymbol{\delta}$ and the total time of the entire cycle is determined by the repeated time period $\boldsymbol{\rho}$. In the context of our temporal-causal network, we chose the **stepmod** function to model the external events that influence the network at regular intervals, which are the Password Rotation (X10) and the Second Authentication System (X7).

$$\mathbf{stepmod}_{\rho,\delta(..)} = 0 \text{ if } \mathrm{mod}(t,\ \rho) < \delta,\ \text{ else } 1 \text{ where } t \text{ indicates time}$$

All other states apply the in neural networks often used combination function **alogistic** which is defined as follows with $\boldsymbol{\sigma},\boldsymbol{\tau}$ representing steepness and excitability threshold parameters respectively (Treur, 2020):

$$\mathbf{alogistic}_{\sigma,\tau}(V_1, \ldots, V_k) = \left[\frac{1}{1 + \mathrm{e}^{-\sigma(V_1 + \cdots + V_k - \tau)}} - \frac{1}{1 + \mathrm{e}^{\sigma\tau}}\right]\left(1 + \mathrm{e}^{-\sigma\tau}\right)$$

In adaptive network models, some of the network's characteristics change over time. This can be modeled by adding self-model states representing these characteristics. For example, the value of a network characteristic $\boldsymbol{\tau}_Y$ that indicates the excitability threshold of the combination function **alogistic** for state Y can be represented by a self-model state $\mathbf{T}_Y$ that is added to the network, also called a **T**-state. For more details of the adaptive network modeling approach, see (Treur, 2020).

The addressed scenario officially starts at Password complexity (X1). See Fig. 1 for the overall network picture and Table 1 for explanations of all states. The complexity of passwords has different influences on the user experience such as added stress (connection from X1 to X3) and it leads to reconstruction difficulties for the user (connection from X1 to X2) since the more complex the password is, the more difficult it is for users to remember them. Then, the reconstruction difficulty leads to users inputting the incorrect password (connection from X2 to X4) which then leads to additional stress (connection from X4 to X3) but also to the state of Account Recovery (X5). The reason for this state and the connection with X4 is that users are more likely to recover their password and thus account before they have used all their remaining password attempts and subsequently get locked out of their accounts automatically. Furthermore, the state of incorrect password attempts also leads to the state of Allowed Password Attempts Remaining (X13) in the self-model level. This state is the focus of the research since the main research objective is to see how this number of passwords remaining can influence the user experience, in this case, demonstrated by the downward causal connections from the X13 to the state for Stress (X3) and Annoyance (X6). The additional connection is to the state of Account Recovery (X5) in case the user does use all the remaining

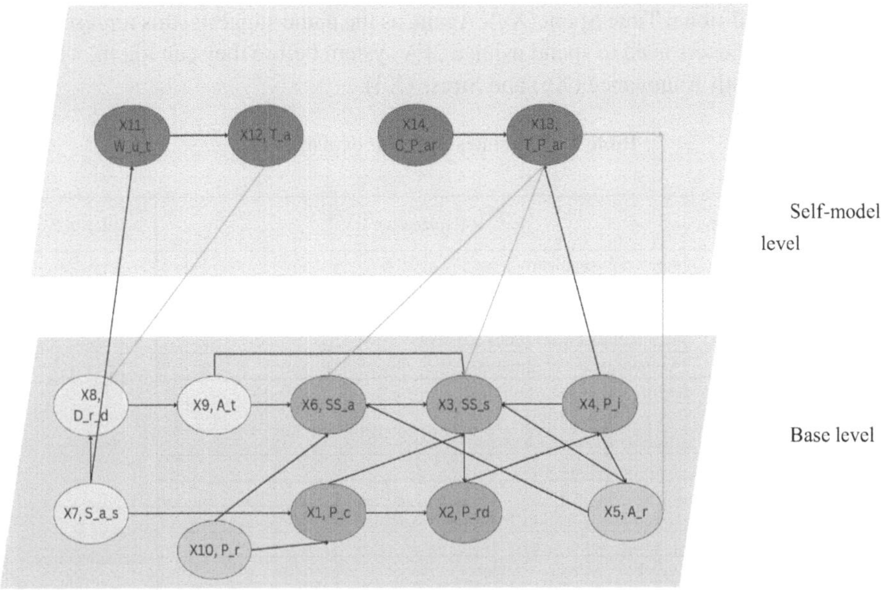

Fig. 1. Graphical conceptual representation of the adaptive network model

password attempts and is automatically locked out of the account. State X14 is added as a context state for X13 to be able to balance the activation of X13. This balancing is determined by the weights of the connection to X13, which will be shown in the role matrices explanation.

An added state outside the 'main' loop is the state for Password Rotation. This state represents when users are required to change their password every set period of time to enhance security. Although this measure enhances the security because when old passwords are leaked, they are invalid and useless, the user experience is impacted negatively since it requires people to come up with new and complex passwords, increasing Password Complexity (connection from X10 to X1), and causing annoyance to the user (X6, connection from X10 to X6).

The last loop in this model is the loop for the second authentication system, which starts at the Second Authentication System (X7). This leads to both more Password Complexity, which in this case for this particular state is more so considered log-in difficulties, and to Difficulties Reaching the Second Device (X8). This is a common user-experience problem with using a second authentication system since almost every 2FA system nowadays requires an authentication application on a phone, which causes the issue of difficulties using a second device. Furthermore, the state of the Second Authentication System (X7) also leads to a state representing User Training On How To Deal With A Second Authentication System (X11). As the name suggests, this state represents training and information sessions for users about the usage of 2FA systems which should in turn cause a decrease in the Difficulties Reaching Second Device (X8), depicted by the downward causal connection from X12 to X8. State X11 is a weighted state and X12 is the **T**-state used for the changing of the sensitivity of the state over time. Additionally, the Difficulties Reaching the Second Device (X8) are then connected to the state of Additional Time Spent (X9). Again, as the name suggests, this represents the additional time users need to spend using a 2FA system before they can log in, which in turn leads to both Annoyance (X6) and Stress (X3).

Table 1. The states and their explanations

State nr.	State name	Explanation	Level
X1	P_c	Password complexity	Base level
X2	P_rd	Password reconstruction difficulty	
X3	SS_s	Sensory state for *stress*	
X4	P_i	Incorrect password (false negative)	
X5	A_r	Account recovery initiated	
X6	SS_a	Sensory state for *annoyance*	
X7	S_a_s	Second authentication system	
X8	D_r_d	Difficulties reaching second device	
X9	A_t	Additional time spent	
X10	P_r	Password rotation	
X11	W_u_t	User training on how to deal with a second authentication system	First-order self-modeling level
X12	T_a	T state	
X13	T_P_ar	Allowed password attempts remaining (E.g. max 10)	
X14	C_P_ar	Control state	

The state with the most dependencies is the Stress state (X3). Since it has the most incoming connections, 5 in total (X1, X4, X5, X6, X9), the weights for these connections have been adjusted to the importance and the amount of stress gained from that particular state. For example, the connection incoming from Annoyance (X6) has a weight of 0.1 indicating that that connection is less influential on the stress levels than the incoming connection from Incorrect Password (X4) which carries a connection weight of 0.3. With these differences in the connection weights, the aim is to accurately describe and isolate the differences in influences that the other states have on the Stress state (X3). Furthermore, the speed factors of the states in the first-order self-model levels are reduced by a factor of 10 in comparison to the other states to ensure a clearer result in the simulations from the downward causal connections. More details of the introduced model can be found in the Appendix available as Linked Data at URL https://www.researchgate.net/publication/382975936.

4 Simulation Experiments

A number of simulation experiments have been conducted. One example of them will be discussed in this section. More variations are analysed in the above mentioned Appendix. We used the dedicated software environment implemented in MATLAB as described in (Treur, 2020), Ch. 9 for all our simulations, including the simulations for the What-If analysis described in the subsequent sections. Figure 2 shows the full simulation of the example simulation with curves for all the states displayed.

Due to the number of states, we have made several plots with only the relevant states that impact each other to gain a better understanding of the influences in each 'loop'. The simulation curves in Fig. 3 show the Second Authentication loop, including the Training

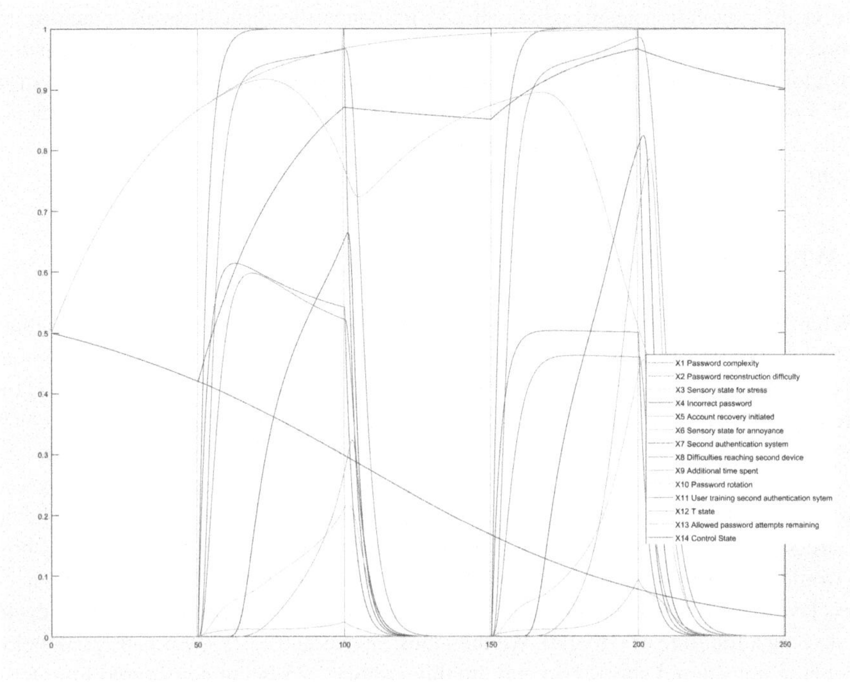

Fig. 2. Simulation 1 with all states depicted

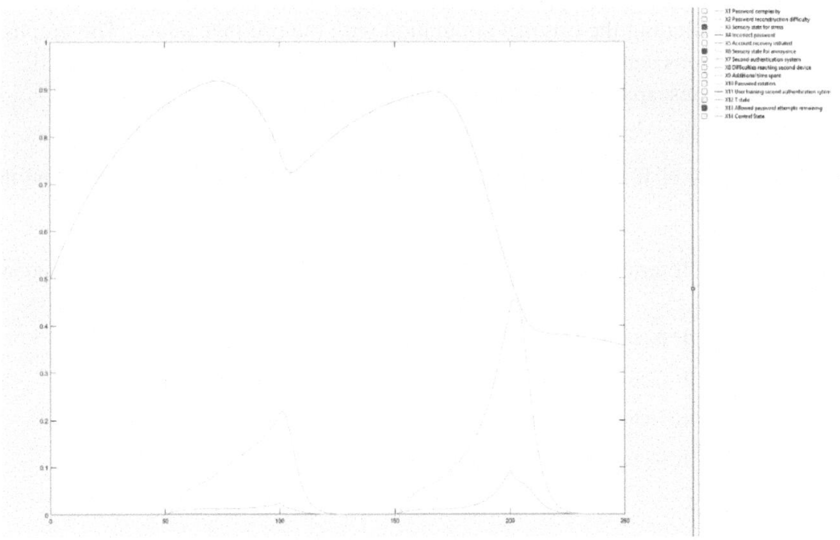

Fig. 3. Simulation 1 with focus on the second Authentication loop

state in the adaptation level, which should influence the difficulties for reaching the second device for users. As is shown in the second loop of the simulation, over time with additional training, the difficulties of reaching the second device for users seems to be decreasing every loop, which in turn increases the user experience. More simulation details can be found in the Appendix available as Linked Data at URL https://www.researchgate.net/publication/382975936.

5 What-If Analysis

A What-If analysis can be used to explore different scenarios in a systematic manner. Adding various different values to a simulation will provide different possible outcomes. This What-If analysis includes 2 different If-factors, each with two options, that lead to 4 different What factor combinations. If factor 1 is the threshold of state X13 which can be low or high. If factor 2 is the weight of the connection from X14 to X13 which also can be low or high. The If factors eventually lead to the following What factors: stress level low and high on short term and stress level low and high on long term. To decide whether the stress level is high or low, this analysis used a middle point of 0.5. Any results above 0.5 would be ruled as a high stress level and any results below 0.5 would be ruled as a low stress level. To establish whether it is short term or long term we decided to run 4 cycles (peaks), on a timeline of 0 to 450. We then took an average of 225, with any peak below 225 being considered short term and anything above 225 being considered long term. Table 2 shows in bold the different values we put to create the different simulations: just the middle point of the value intervals for the options. The images below show an overview of how the sensory state for stress changes when different values are put into the If factors, including the original simulation with the original values. The graphs of the four simulations can be found in the Appendix available as Linked Data at URL https://www.researchgate.net/publication/382975936.

Table 2. Overview of If factors options, the values chosen for simulations in bold, and their assumed probabilities

If factors	If factor options	If factor option labels	If factor option value intervals	If factor option probabilities
If factor 1: Treshold of X13	If factor 1 option 1	low	0–0.5 (**0.25**)	0.2
	If factor 1 option 2	high	0.5–1 (**0.75**)	0.8
If factor 2: Weight of X14 → X13	If factor 2 option 1	low	0–0.5 (**0.25**)	0.8
	If factor 2 option 2	high	0.5–1 (**0.75**)	0.2

Next, we expand on the conceptual What-If analysis by estimating probabilities for the different If factor options. Table 2, last column shows the probabilities we assumed for the different If factor options. The first three columns in Table 3 show how we assumed independence and used the product rule p(A and B) = p(A)*p(B) to estimate the probabilities for the If factor option combinations. Moreover, Table 3 includes indications of the impact on user annoyance, ranging from 1 to 10, and calculates the overall risk by multiplying the probability of each What factor by its impact on annoyance. The analysis reveals that the highest risk is associated with a high threshold for X13 and a high weight from X14 to X13. This scenario leads to a high probability of long-term stress and a significant impact on user annoyance, resulting in the highest calculated risk. Conversely, the lowest risk is associated with a low threshold for X13 and a low weight from X14 to X13, indicating a lower probability of stress and a reduced impact on user experience. A significant number of companies maintain a policy that limits the number of allowed password attempts. Therefore, the figures in the table above may be considered plausible.

Table 3. Overview of the What-If analysis: If factor option combination probabilities and risk assessment

If factor 1 option probabilities	If factor 2 option probabilities	If factor option combination probabilities	What factor 1 Stress low Short term	What factor 2 Stress high Short term	What factor 3 Stress low Long term	What factor 4 Stress high Long term
If factor 1 option 1: Low 0.2	If factor 2 option 1: Low 0.8	0.2 × 0.8 = 0.16	0.16			0.16
	If factor 2 option 2: High 0.2	0.2 × 0.2 = 0.04	0.04		0.04	
If factor 1 option 2: High 0.8	If factor 2 option 1 Low: 0.8	0.8 × 0.8 = 0.64		0.64		0.64
	If factor 2 option 2: High 0.2	0.8 × 0.2 = 0.16		0.16		0.16
Sum			0.20	0.80	0.04	0.96
Impact	Annoyance 1–10		2	8	3	7
Risk			0.20 × 2 = 0.4	0.80 × 8 = 6.4	0.04 × 3 = 0.12	0.96 × 7 = 6.72

6 Discussion

Using a network modelling approach from (Treur, 2016; 2020; 2021) to model adaptive dynamical systems based on self-modeling networks, our study on password-based authentication systems reveals several key insights into the interplay between user experience (UX) and security. As far as the authors know, no other approaches of computational analysis from such an adaptive dynamical system perspective can be found in the literature. A primary finding is the tension between security and usability in password systems. While these systems are ubiquitous due to their relative ease of implementation and user familiarity, they are burdened by significant challenges. Users often prioritize convenience over security, creating weak, easily guessable passwords and the practice of password reuse across multiple accounts. This behaviour, driven by the cognitive burden of managing numerous passwords, significantly increases the risk of unauthorized access and security breaches (Zimmerman and Gerber, 2020).

To mitigate these risks, two-factor authentication (2FA) has emerged as a crucial security measure. By requiring a secondary verification step, 2FA significantly reduces the risk of unauthorized access even if a password is compromised (2-Factor Authentication Frequently Asked Questions, n.d.). However, while 2FA enhances security, it can negatively impact usability due to its perceived complexity and time-consuming nature, leading to user frustration and potentially lower adoption rates (Gunson et al., 2011).

Our research also underscores the importance of understanding user behaviour in designing effective authentication systems. Simulations using our temporal-causal network model demonstrated that as the number of allowed password attempts decreases, user stress and annoyance escalate rapidly. This finding aligns with previous research highlighting the negative impact of password-related frustrations on user experience (Wash and Rader, 2021).

Furthermore, our study emphasizes the critical role of user education and training in promoting secure password practices and facilitating the adoption of more robust authentication methods like 2FA. Simulations showed that with adequate training, the usability of 2FA systems improved, leading to a reduction in user stress over time. This highlights the importance of not only implementing strong security measures but also providing clear, user-friendly instructions and support to enhance the overall effectiveness and user satisfaction of authentication systems (Yıldırım and Mackie, 2019).

While our study provides valuable insights into the UX of password-based authentication systems, several limitations must be acknowledged. Firstly, the data used in the study, primarily derived from academic sources, may not fully represent the diverse range of user behaviours and practices across different demographics and regions. Secondly, the rapid pace of technological evolution in the field of authentication means that new methods and security threats are constantly emerging. Consequently, the study's findings may have limited temporal validity and may need to fully account for emerging technologies and strategies (Umejiaku, Dhakal, Sheng, 2023).

By addressing these limitations in future research, we can achieve a more comprehensive and nuanced understanding of the user experience in password-based authentication systems. This will enable the development of more effective, user-friendly, and secure authentication solutions that cater to the diverse needs and behaviours of online users.

Any smooth dynamical system has a canonical representation as a network and any smooth (multi-order) adaptive dynamical system has a canonical representation as a (multi-order) self-modeling network, as is shown in (Treur, 2021; Hendrikse et al, 2023). Therefore, compared to adaptive dynamical systems in general, the network-oriented modeling approach used does not introduce any fundamental limitations concerning what it can model. This has also been confirmed by applications to complex adaptive processes in different disciplines, such as mental models in cognitive science, organisational learning in management science, synchronisation and bonding in social psychology, and teamwork in health science; e.g., (Treur and Van Ments, Treur et al., 2022; 2023; Hendrikse, et al, 2025; Roelofsma et al., 2025).

References

Two-Factor Authentication Frequently Asked Questions|Two-factor authentication (2FA)|Binghamton University. (n.d.). Two-factor Authentication (2FA) - Binghamton University. https://www.binghamton.edu/its/two-fa/multi-factor-faqs.html (2024)

Bonneau, J., Van Oorschot, P.C., Stajano, F.,. (n.d.). The Quest to Replace Passwords: A Framework for Comparative Evaluation of Web Authentication Schemes. https://www.cl.cam.ac.uk/~fms27/papers/2012-BonneauHerOorSta-password.pdf (2024)

Börcsök, T., et al.: Adaptive defence mechanisms: an adaptive network model to protect HiX systems against cyber attacks. In: Proceedings of the 8th International Conference on Computational Methods in Systems and Software, CoMeSySo'24. Lecture Notes in Networks and Systems, Springer Nature (2025)

Bouma, D., Den Oudsten, T., Naffakh, A, Waliji, M., Treur, J., Roelofsma, P.H.M.P.: Cyber risk management and organisational alignment using Endley's risk management model: computational analysis from an adaptive dynamical system perspective. In: Proceedings of the 8th International Conference on Computational Methods in Systems and Software, CoMeSySo'24. Lecture Notes in Networks and Systems, Springer Nature (2025)

Bouma, D., et al.: Cyber security in hospitals: a network-oriented model for behavioural learning of employees during phishing simulations. In: Proceedings of the 17th International Conference on Computational Intelligence in Security for Information Systems, CISIS'24. Lecture Notes in Networks and Systems, vol. 957, pp 98–111, Springer Nature (2024)

Canbaloğlu, G., Treur, J., Wiewiora, A. (eds.): Computational modeling of multilevel organisational learning and its control using self-modeling network models. Springer Nature (2023). https://doi.org/10.1007/978-3-031-28735-0

Caneva, J., et al.: The unforeseen blackout and cyber risk management for business continuity: network-oriented computational analysis. In: Proceedings of the 8th International Conference on Computational Methods in Systems and Software, CoMeSySo'24. Lecture Notes in Networks and Systems, Springer Nature (2025)

Conklin, A., Dietrich, G., Walz, D.: Password-based authentication: a system perspective. In: Proceedings of the 37th Annual Hawaii International Conference on System Sciences, (p. 10) (2004). IEEE Xplore (2004)

Gunson, N., Marshall, D., Morton, H., Jack, M.: User perceptions of security and usability of single-factor and two-factor authentication in automated telephone banking. Comput. Secur. **30**(4), 208–220 (2011). https://doi.org/10.1016/j.cose.2010.12.001

Hendrikse, S.C.F., Treur, J., Koole, S.L.: Modeling emerging interpersonal synchrony and its related adaptive short-term affiliation and long-term bonding: a second-order multi-adaptive neural agent model. Int. J. Neural Syst. **33**(7), 2350038 (2023)

Hendrikse, S.C.F., Treur, J., Koole, S.L. (eds.): New Analysis and Modeling Directions in Social Interaction Science: Emergent Multimodal Interpersonal Synchrony, Affiliation, and Bonding. Springer Nature (2025)

Jeffery, E., et al.: Computational analysis of the effectiveness of the intelligence cycle for organizational cyber risk management. In: Proceedings of the 8th International Conference on Computational Methods in Systems and Software, CoMeSySo'24. Lecture Notes in Networks and Systems, Springer Nature (2025)

Keijzer, S., et al.: Managing classified information by a third-party contractor: a computational cybersecurity analysis. Proceeidngs of the 8th International Conference on Computer-Human Interaction Research and Applications, CHIRA'24. Communications in Computer and Information Science, vol. 2371, pp 299–315, Springer Nature (2024)

Roelofsma, P.H.M.P., Jabeen, F., Taal, H.R., Treur, J., (eds.): Using Shared Mental Models and Organisational Learning to Support Safety and Security through Cyberspace: a Computational Analysis Approach. Springer Nature (2025)

Treur, J.: Network-Oriented Modeling: Addressing Complexity of Cognitive, Affective and Social Interactions. Springer Nature (2016)

Treur, J.: Network-Oriented Modeling for Adaptive Networks: Design Higher-Order Adaptive Biological, Mental and Social Network. Springer Nature (2020)

Treur, J.: On the dynamics and adaptivity of mental processes: relating adaptive dynamical systems and self-modeling network models by mathematical analysis. Cogn. Syst. Res. **70**, 93–100 (2021)

Treur, J., Van Ments, L. (eds.): Mental models and their dynamics, adaptation, and control. Springer Nature (2022). https://doi.org/10.1007/978-3-030-85821-6

Umejiaku, A. P., Dhakal, P., Sheng, V. S.: Balancing password security and user convenience: exploring the potential of prompt models for password generation. Electronics **12**(10), 2159 (2023). https://doi.org/10.3390/electronics12102159

Walia, K.S., Shenoy, S., Cheng, Y.: An empirical analysis on the usability and security of passwords. In: 2020 IEEE 21st International Conference on Information Reuse and Integration for Data Science (IRI), pp. 1–8. IEEE (2020)

Wash, R., Rader, E.: Prioritizing security over usability: strategies for how people choose passwords. J. Cybersecur. **7**(1), tyab012 (2021)

Yıldırım, M., Mackie, I.: Encouraging users to improve password security and memorability. Int. J. Inf. Secur. **18**(6), 741–759 (2019)

Zimmermann, V., Gerber, N.: The password is dead, long live the password – A laboratory study on user perceptions of authentication schemes. Int. J. Human-Comput. Stud. **133**, 26–44 (2020). https://doi.org/10.1016/j.ijhcs.2019.08.006

Computational Analysis of Human Factors in Spear-Phishing Attacks: An Adaptive Network Model

Ana Maria Briana Capra[1], Emil Dekker[1], Lielie Hoffman[1], Charlotte Hoffmans[1,2], Tristan Mueller[1], Niek Jan van den Hout[1], Jan Treur[1,2](✉), and Peter H. M. P. Roelofsma[1]

[1] Centre of Expertise Cybersecurity, Risk Management and Cybersecurity Group, The Hague University of Applied Sciences, Hague, Netherlands
{N.J.vandenHout,P.H.M.P.Roelofsma}@hhs.nl, j.treur@vu.nl

[2] Department of Computer Science, Vrije Universiteit Amsterdam, Social AI Group, Amsterdam, Netherlands

Abstract. This paper presents an adaptive network model for spear-phishing and analyzed how a refined and formalised version of the the COM-B model could provide a computational analysis for possible solutions in mitigating spear-phishing threats. This is mainly achieved through improving behavior, such as reflectiveness traits and traits of awareness. In our case study, through continuous learning and adaptation, the employee begins their work with certain knowledge and experience and thereafter begin to adapt and improve throughout training and simulation. Here reflection and behaviors change based primarily on the feedback the employee receives since, as they progress, they learn which mistakes they have made and how to avoid them in the future.

1 Introduction

Spear phishing is a sophisticated form of cyber-attack that targets specific individuals or organizations to extract sensitive information, conduct financial theft, or install malicious software Burns et al. (Illustrative Media Items, 2024). Unlike general phishing, which targets a broad audience, spear phishing involves meticulously crafted messages that appear to come from a trustworthy source tailored to a specific recipient to increase the likelihood of deception, Burns et al., (Illustrative Media Items, 2024). Spear phishing poses a substantial cybersecurity threat due to its targeted nature and the sophistication of the tactics employed. It occurs with a high frequency, with major organizations receiving several phishing emails daily and 50% of large organizations being targeted by spear phishing by Smith in 2022, see (Illustrative Media Items, 2024). According to a report, spear phishing accounts for 65% of all phishing attacks, showcasing its prevalence in the cyber threat landscape Symantec, see (Illustrative Media Items, 2024). Phishing emails, in general, increased by 58% between 2022 and 2023, Desai and Hegde, (Illustrative Media Items, 2024). Spear phishing only comprises a small amount of all the phishing emails sent but accounts for 66% of breaches of all phishing attacks: Barracuda (Illustrative Media Items, 2024). The potential financial damage is also significant, with the

D. Singh et al. (Eds.): IHCI 2024, LNCS 15557, pp. 353–366, 2025.
https://doi.org/10.1007/978-3-031-88705-5_30

average spear phishing attack costing companies 1.6 million dollars, Makowski (Illustrative Media Items, 2024). An example of the potential financial damage is the 2016 incident involving Facebook and Google, where $100 million was stolen through spear phishing emails disguised as a company that Google and Facebook were doing business with (Huddleston, 2019) see (Illustrative Media Items, 2024).

The objective of this project is to examine the issue of spear phishing by developing a deeper understanding of the interplay of different variables. Network-oriented modeling will be used to model and simulate the decision-making processes that underlie responses to spear phishing attacks. This will involve the construction and analysis of a mental and social process model to identify key behavioral factors. The project will use simulation experiments based on network-oriented modeling to dynamically visualize and analyze how various factors influence susceptibility to spear phishing attacks. This model allows for testing of the hypothetical scenario and aids in understanding the complex interplay of factors that contribute to the success or failure of spear phishing attacks.

The COM-B framework will be the point of departure used to analyse and model the dynamics of the behaviour displayed in a spear phishing attack. The COM-B framework considers Capability, Opportunity, and Motivation as factors influencing behavior, see Table 1. For this abstract global framework, (1) a more detailed, refined version will be introduced, by adding more detailed factors and adaptivity, which (2) will be extended to obtain an adaptive dynamical system describing the process in more detail, which in turn (3) will be formalized as an adaptive network model based on the self-modeling network concept. This will guide our computational analysis of why targeted individuals might respond to spear phishing emails and how their capabilities, opportunities, and motivations affect the likelihood of falling victim to a spear phishing attack. For instance, increasing an employee's capability through training can enhance their ability to recognize phishing attempts, adjusting their work environment can reduce their exposure to attacks, and changing motivational incentives can encourage safe behaviors. Our project will use the developed adaptive network model to structure our simulation experiments and derive actionable insights that can be implemented within organizational cybersecurity strategies.

This work is part of a research programme on computational analysis of safety and security in and through cyberspace. In (Börcsök, Bos, Buter, Van Der Hoeven, Velev, Van den Hout, Bouma, Treur, Roelofsma, 2025; Bouma, Den Oudsten, Naffakh, Waliji, Treur, Roelofsma, 2024; Caneva, Oliver, Rietze, Šileris, Barelds, Hoffmans, Treur, Roelofsma, 2025; Jeffery, Malsagov, Tijssen, Zeisig, Hoffmans, Treur, Roelofsma, 2025) various other themes within cybersecurity were addressed. Moreover, in (Roelofsma, Jabeen, Taal, Treur, 2025) it was analysed how an AI coach can support safety and security for medical teams in hospitals.

Table 1. The main elements in the standard COM-B Framework for Behavioural Change

COM-B element	Description
Capability	The individual's psychological and physical capacity to engage in the activity. It includes having the necessary knowledge and skills
Opportunity	All the factors that lie outside the individual that make the behavior possible or prompt it
Motivation	All brain processes that energize and direct behavior. It includes habitual processes, emotional responding, as well as analytical decision-making

2 Background Knowledge

In 2023, the healthcare industry witnessed a major increase in sophisticated phishing attacks, leading to a sharp increase in cybersecurity threats. Spear phishing, a targeted type of phishing attack, is especially worrying for healthcare. Last year, 50% of healthcare organizations reported being experiencing spear phishing attacks. This targeted attack caused 66% of all recent healthcare data breaches, emphasizing its severe impact (Security & Contributor, 2024) see (Illustrative Media Items, 2024).

The consequences of these attacks are far-reaching. Nearly every healthcare organization that fell victim to spear phishing faced significant repercussions, including malware infections, data breaches, account takeovers, financial losses, and reputational damage. Since 2022 there has been a 92% increase in patient record breeches, with over 100 million victims (Security & Contributor, 2024) see (Illustrative Media Items, 2024).

Hospitals have become more vulnerable to cyber risks due to the increased dependence on information technology. Reliance on electronic health records, telehealth platforms, and communication systems creates numerous entry points for phishing attacks, Security (Illustrative Media Items, 2024). Since spear phishing typically requires the recipient to perform an action, it relies heavily on social engineering techniques; thus, posing as a trusted person makes it more likely for the victim to do what the email requires them to. Healthcare professionals often experience high-stress and fast-paced environments. This is a vulnerability exploited by the attackers, since staff may miss the signs of a phishing email due to fatigue, sense of urgency crafted in the email or lack of thoroughness in administrative tasks (Rizzoni et al., 2022). Since spear phishing attacks are highly targeted, attackers often focus on individuals with access to sensitive information or authority to approve financial transactions. An example of such an attack is whaling.

Whaling, a subset of spear phishing, targets high-profile individuals such as executives, administrators, and other senior leaders within an organization. This is a meticulously crafted attack, which often involves extensive research on the target to personalize the attack and increase the likelihood of deception. An example of a successful whaling attack is back in 2015, when a networking technology company, Ubiquiti Networks Inc., lost $46.7 million. The attackers managed to spoof communications from executives to initiate unauthorized international wire transfers through the company's finance department (Krebson Security, 2015) see (Illustrative Media Items, 2024). The company's investigation found no evidence of system penetration or any employee involvement, indicating that the attack relied solely on social engineering tactics. Falling for a scam can be embarrassing for executives and organizations, particularly when the CEO or other high-rank employee leads to the incident. This often leads to a big, unreported rate of such cyberattacks. Moreover, whaling attacks can prioritize data over cash, information such as employees' payrolls, and tax records.

This case underscores the effectiveness and danger of whaling attacks, which can also be highly relevant and damaging in the healthcare sector. Just like in the Ubiquiti case, healthcare executives and senior officials have access to vast amounts of sensitive information and financial resources, thus making them prime targets for whaling attacks. Additionally, in the healthcare sector, a breach could expose not only financial information but also patient data. This sector is particularly susceptible to whaling attacks due to the variety of interconnected systems they use for patient records, billing, communication, and more. These systems need to be accessible to a wide range of personnel, and their complexity creates security gaps that attackers can easily exploit. With so many potential entry points, it's easier for attackers to find ways into the system.

In order to increase secure employee behavior, organizations can opt for various strategies and interventions. An essential part is to take proactive measures by educating employees to identify phishing emails and report them (Jampen et al., 2020). One effective intervention is conducting phishing training simulations, where the organization's IT security department creates fake phishing scenarios by sending out mock phishing emails to employees. When an employee clicks on a link in these emails, they are redirected to a notification page informing them that they have participated in a phishing simulation. Recent studies show that these simulations have been proven rather effective in the healthcare sector, especially when the training has been carried out regularly, Gordon et al. (Illustrative Media Items, 2024). One important aspect to consider is the employees' motivation to engage in these simulations. The most effective way is to vary the training content, email scenarios, and challenge levels to address the capability, opportunity, and motivational factors influencing behavior.

As mentioned earlier, through phishing simulation training, employee's behaviour can be enhanced, thus helping them identify whaling intentions. The training operates by notifying employees when they click on a link in a simulated phishing email, indicating that they have fallen for a whaling attempt. Alternatively, if the employee reports the simulated phishing email, they receive a success notification. This process aims to educate employees, influencing their capability, motivation, and opportunity to identify and respond to whaling attempts. To showcase the impact of this training and the effectiveness of various interventions on overall cybersecurity, a hospital environment scenario is proposed and used as an example.

Consider an individual working as a hospital executive, receiving numerous emails daily. He recently learned of a whaling case happening at another hospital in the same city. In an effort to improve the hospital's cybersecurity, the phishing simulation training is implemented immediately. The individual is working on the computer when he receives an email. Due to the amount of emails he must go through, he often clicks on emails without paying too much attention. When he receives a phishing email, he must click or report the email. This decision is influenced by his COM-B characteristics and the quality of the email. Whichever decision is made, it influences his feedback in the form of learning. This feedback lowers the threshold levels, making him more sensitive and aware the next time he receives a phishing email.

3 Introduction to the Network Modelling Approach Used

The network-oriented modeling approach used from (Treur, 2020) defines temporal-causal network models by three types of network characteristics:

Connectivity Characteristics. These concern nodes (also called states) X and Y in a network and their connections $X \rightarrow Y$ indicating that X causally affects Y with connection weight $\boldsymbol{\omega}_{X,Y}$. Such states have activation values $X(t)$ and $Y(t)$ over time t and the (causal) impact of X on Y at time t is defined by: $\mathbf{impact}_{X,Y}(t) = \boldsymbol{\omega}_{X,Y}\ X(t)$.

Aggregation Characteristics. To aggregate the (possibly multiple) impacts on Y from different states $X_1, \ldots, X_k$ a combination function $\mathbf{c}_Y$ is chosen to obtain the aggregated impact on Y: $\mathbf{aggimpact}_Y(t) = c_Y(\mathbf{impact}_{X_1}, Y(t), \ldots, \mathbf{impact}_{X_k}, Y(t))$.

One of the most often used combination functions (also in neural networks) is the function **alogistic** defined as follows:

$$\mathbf{alogistic}_{\sigma,\tau}(V_1, \ldots, V_k) = \left[\frac{1}{1 + \mathrm{e}^{-\sigma(V_1 + \cdots + V_k - \tau)}} - \frac{1}{1 + \mathrm{e}^{\sigma\tau}}\right]\left(1 + \mathrm{e}^{-\sigma\tau}\right)$$

where $\boldsymbol{\sigma}$, $\boldsymbol{\tau}$ are parameters representing steepness and excitability threshold, respectively (Treur, 2016).

Timing Characteristics. The aggregated impact on Y makes that $Y(t)$ changes over time thereby taking into account a specified speed factor $\boldsymbol{\eta}_Y$ using the following differential or difference equation for each state Y:

$$dY(t)/dt = \boldsymbol{\eta}_Y\left[\mathbf{c}_Y\left(\boldsymbol{\omega}_{X_1,Y}X_1(t), \ldots, \boldsymbol{\omega}_{X_k,Y}X_k(t)\right) - Y(t)\right]$$

$$Y(t + \Delta t) = Y(t) + \boldsymbol{\eta}_Y\left[\mathbf{c}_Y\left(\boldsymbol{\omega}_{X_1,Y}X_1(t), \ldots, \boldsymbol{\omega}_{X_k,Y}X_k(t)\right) - Y(t)\right]\Delta t$$

In this way a network model with n states defines a dynamical system that can also be described by a system of n coupled differential equations. In (Treur, 2021) and (Hendrikse, Treur, Koole, 2023) it is shown in detail how any (smooth) dynamical system can be represented as a temporal-causal network.

To model adaptive dynamical systems in this way, some of the network characteristics can be made adaptive, so that their values can change over time too. By using the self-modeling principle for networks, such adaptive characteristics can be modeled by adding self-model states for them to the network. An example for aggregation characteristics for the combination function **alogistic** is a self-model state $\mathbf{T}_Y$ (called a **T**-state) that can be added representing excitability threshold $\boldsymbol{\tau}_Y$. In the adaptive network model introduced here, **T**-states will be used to model adaptation.

What is evident in particular for the cybersecurity domain from other cybersecurity network models or frameworks is that these models share similarities with each other, specifically in relevance to which attributes to measure, such as with the NIST cybersecurity framework (National Institute of Science and Technology). Specifically, with the NIST framework, the model identifies the key roles of responding to and identifying cybersecurity threats, for example. (The NIST Cybersecurity Framework (CSF) 2.0," 2024). This relates to the COM-B model, where behaviour of individuals within the network determines one's overall awareness of impending spear-phishing threats, such as carefully crafted e-mails that seek to gain unlawful access to critical IT systems (The NIST Cybersecurity Framework (CSF) 2.0, 2024). By improving an individual's awareness of these spear phishing threats, one simultaneously affects the COM-B model, whereby awareness is improved through continuous learning and improvement. As mistakes are identified and overcome, so does one's awareness improve. As one's awareness improves in identifying spear phishing e-mails, the rate of one's mistakes decreases.

4 Design of the Adaptive Network Model

As a refinement of the standard COM-B model used as point of departure, here instead of three, six categories were used as main factors affecting the behaviour. These six categories can be adjusted depending on the type of person which will result in different outcomes pertaining to the Clicking and Reporting of the emails. Based on the different outcomes the feedback will influence the learning over time. Firstly, Capability is refined into categories of Psychological and Physical Capability. Psychological Capability (X2) shows a person's knowledge and skills, while Physical Capability (X3) shows strength and stamina. Second, the categories refining Opportunity are Physical and Social Opportunity. Physical Opportunity (X4) pertains to the opportunities in the environment while the Social Opportunity (X5) concerns the behavior of the other employees. Lastly, the categories refining Motivation are Reflective and Automatic Motivation. Reflective Motivation (X6) involves planned behavior while Automatic Motivation (X7) involves impulsive behavior see (Fig. 1).

The employee in this scenario was given higher connection weight values for connections to X2, X5, and X7. There were mediate levels for connections to X3 and X4, while there were low levels for connections to X6. This was done to reflect the employee's highly impulsive personality and show the influence of the environment around him. These six subcategories are affected by the threshold parameters for these states, which are represented by self-model **T**-states X21 to X26. The latter states model threshold sensitivity in an adaptive manner from feedback on the behaviour. To model this form of learning the **T**-states are influenced by the feedback given from the Clicking (X13) and Reporting (X14) email states for an overview of all states, see (Table 2).

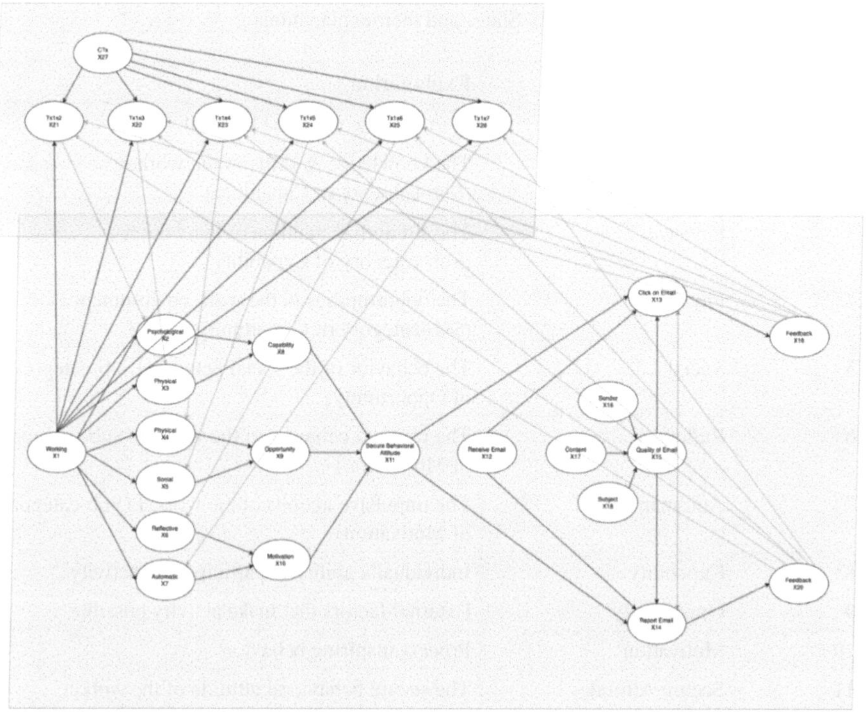

Fig. 1. Conceptual Design of the Adaptive Network Model

If the threshold is lower, then there is more sensitivity. This will cause the employee to learn from their behavior and, over time, make a favorable decision. In order for this to occur, mutual negative connections between X13 and X14 are included. There is a strong negative connection from X11 to X13 and a positive one from X11 to X14. This is to simulate the learning to positively influence the Reporting rather than the Clicking. The connection from X13 to X14 and vice versa are both negative due to one action influencing the outcome of the other. From X15 to X13, there is a strong positive connection, while there is a strong negative connection to X14. This is due to the influence of the Quality of the Email X15.

The constant states are X1, X16, X17, X18. The initial values of all states are 0 except for X16, X17, and X18 because these represent contextual factors. Additionally, the threshold states have an initial value of 0.2, and the context state for them has an initial value of 1. The speeds indicated in the role matrices vary. States X2 – X7 are slower, while X1, X16, X17, and X18 have a very high speed. The speeds for the threshold and context state for them are the lowest. The remaining states have mediate speeds.

Table 2. States and their explanations

Base States		**Explanation**
X1	Working	Starting to work in the hospital
X2	Psychological	The knowledge & skills of the worker (Sub-category of Capability)
X3	Physical	The strength & stamina of the worker (Sub-category of Capability)
X4	Physical	The opportunities in the work environment (Sub-category of Opportunity)
X5	Social	The behavior of the social network (Sub-category of Opportunity)
X6	Reflective	The planned behavior of the worker (Sub-category of Motivation)
X7	Automatic	The impulsive actions of the worker (Sub-category of Motivation)
X8	Capability	Individual's ability to participate in activity
X9	Opportunity	External factors that make activity possible
X10	Motivation	Process inspiring behavior
X11	Secure Attitude	The secure behavioral attitude of the worker
X12	Receive Email	The act of receiving a phishing email
X13	Click on Email	Clicking on (links) the phishing email
X14	Report Email	Report the email as suspicious
X15	Quality of Email	Quality of the email influenced by the sender, content, and subject states
X16	Sender	The posed sender of the email
X17	Content	The content of the email
X18	Subject	The topic of the email
X19	Feedback	Feedback from clicking on the phishing email
X20	Feedback	Feedback from reporting the phishing email
1st order self-model states		**Explanation**
X21	$\mathbf{T}_{X2}$	Excitability threshold **T**-state feeding the Feedback to X2
X22	$\mathbf{T}_{X3}$	Excitability threshold **T**-state feeding the Feedback to X3
X23	$\mathbf{T}_{X4}$	Excitability threshold **T**-state feeding the Feedback to X4

(*continued*)

Table 2. *(continued)*

Base States		**Explanation**
X24	$\mathbf{T}_{X5}$	Excitability threshold **T**-state feeding the Feedback to X5
X25	$\mathbf{T}_{X6}$	Excitability threshold **T**-state feeding the Feedback to X6
X26	$\mathbf{T}_{X7}$	Excitability threshold **T**-state feeding the Feedback to X7
X27	C_{Tx}	Context State for the **T**-states

In this scenario, the employee ideally learns over time and eventually reports the email every time rather than clicking on it. The combination functions used in this simulation include **alogistic** and **stepmod**. The **stepmod** is used for states X1, X12, X16, X17, X18. All other states use **alogistic** (see Sect. 4). Function **stepmod** is used to simulate an ongoing process where the context repeats and the behaviour improves over time. These states include values for the **ρ** for repetition period and **δ** for the tipping point (change of the value) based on time *t*:

$$\mathbf{stepmod}_{\rho,\delta}(\cdot) = 1 \text{ if } t \text{ modulo } \boldsymbol{\rho} > \boldsymbol{\delta} \text{ and } 0 \text{ else}$$

For the **alogistic** function all states are given a 5 for the steepness parameter **σ**, whereas the excitability threshold parameters **τ** vary. States X8, X9, X10, X13, X14, X15, X16, X17, X18, X19, X20, and X27 are given mediate threshold values. State X11 is given the highest threshold value at 1.5. States X2 to X7 have adaptive threshold values represented by the **T**-states. Lastly, the **T**-states are all given a threshold of 1. For all **stepmod** states, the repetition period **ρ** was set to 100. The **δ** values are 50 for X1 and 60 for X12, X16, X17, and X18. For more details, see the Appendix available as Linked Data at https://www.researchgate.net/publication/382975864.

5 Simulation Results

The simulation shown in Fig. 2 indicates two cycles of the scenario where the effects of the learning can be seen. The simulation starts with the employee working, through which his behavior is determined by the subcategories on the COM-B model.

The X2 – X7 states are seen to increase slightly over time since they are influenced by the **T**-States. In the simulation, X11 begins slightly after X2 – X10, where it rises to the top and falls after the Clicking or reporting action has been performed. Because receiving the email is a constant, it rises and falls the same in each round of the simulation. The biggest changes can be seen in X13 and X14. The simulation shows that Clicking on the Email (X13) gets lower in the next round of the scenario while Reporting the Email (X14) gets higher in the next round. This indicates that the feedback given by the actions successfully influences the employee's behavior in choosing the favorable action in the next round of the simulation. Although the quality of the email does not change over

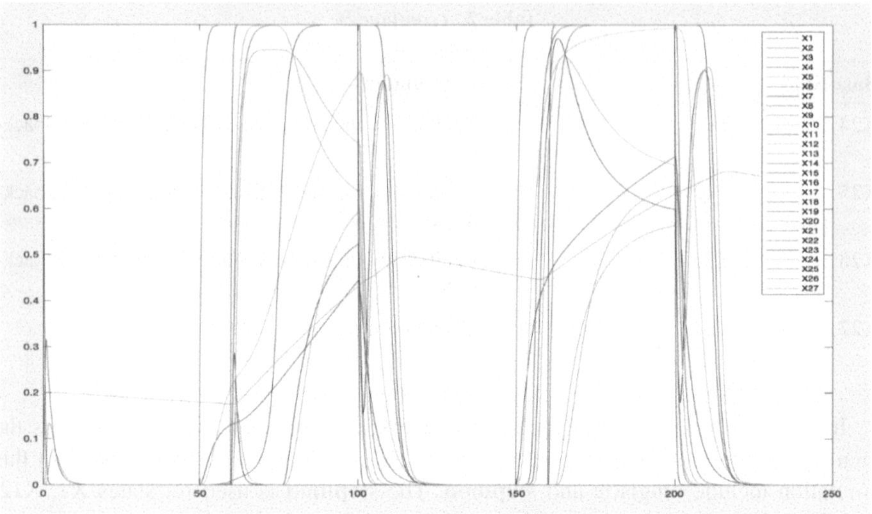

X1: Working	X7: Automatic	X13: Click on Email	X19: Feedback	X25: $\mathbf{V}_{X6}$
X2: Psychological	X8: Capability	X14: Report Email	X20: Feedback	X26: $\mathbf{V}_{X7}$
X3: Physical	X9: Opportunity	X15: Quality of Email	X21: $\mathbf{V}_{X2}$	X27: C_{V_x}
X4: Physical	X10: Motivation	X16: Sender	X22: $\mathbf{V}_{X3}$	
X5: Social	X11: Secure Behavioural Attitude	X17: Content	X23: $\mathbf{V}_{X4}$	
X6: Reflective	X12: Receive Email	X18: Subject	X24: $\mathbf{V}_{X5}$	

Fig. 2. Simulation Results

time, it has an effect on X13 and X14. X16, X17, and X18 are constants, so they start and stop with each round.

What can be seen in the simulation is the Feedback (X19) from Clicking on the Email (X13) rises over time. The Feedback (X20) from Reporting the Email (X14) also rises over time. This indicates that the employee's feedback and learning are increased each round. Lastly, the threshold states X21-X26 are seen to decrease slightly. However, once the simulation starts, they all increase slightly. It can be seen that they slightly decrease after the simulation has ended, before they start rising after the next round starts. This drop and rise can be explained by human learning. The employee likely learns from the actions taken (X13 or X14); however, it is likely that they will forget some of the learning before the next time they receive an email. Although there is a slight drop after each episode, there is a steady increase seen in the **T**-States over time.

6 What-If Analysis

What-If Analyses are critical components in the context of risk assessment and decision-making, particularly in scenarios where the outcomes are influenced by multiple factors with varying degrees of uncertainty (ISACA, 2021), see Illustrative Media Items (2024). This report uses the analysis to quantify the probabilities and risks associated with specific actions, such as clicking on or reporting a phishing email. The analysis focuses on two key factors: the three Connections to Quality of Email and the Threshold of it, each divided into three categories: low, mediate, and high, with low ranging from 0 to 0.35, mediate from 0.35 to 0.65, and high from 0.65 to 1, see Table 3. By evaluating the probabilities of various combinations of these factors, normalizing them, and assigning impact levels, the likelihood and risk of each outcome are determined. Thus, organization can enhance their resilience and adaptability, ensuring they are better equipped to handle unexpected events, SafetyCulture (Illustrative Media Items, 2024).

Table 3. If factor options and their labels and value intervals

If factors	If factor options	If factor option labels	If factor option value intervals	If factor option probabilities
If factor 1: Connections to Quality	If factor 1 option 1	low	0–0.35	0.2
	If factor 1 option 2	mediate	0.35–0.65	0.3
	If factor 1 option 3	high	0.65–1	0.5
If factor 2: Threshold	If factor 2 option 1	low	0–0.35	0.25
	If factor 2 option 2	mediate	0.35–0.65	0.35
	If factor 2 option 3	high	0.65–1	0.4

To add a probabilistic What-If analysis, Table 3 includes estimated probabilities for each option of the two If factors in the last column. Table 4 shows in its first three columns the probabilities for each combination of the options of the two If factors. It lists the individual probabilities for each If factor option level and the combined probabilities for each combination of options.

For example, the probability of having a low connection to quality and a low threshold is 0.05. This probability is calculated (assuming independence so that the rule p(A and B) = p(A)*p(B) can be applied) by multiplying the probability 0.2 for If factor 1 option 1 with probability 0.25 of If factor 2 option 1. These combined probabilities are important for the risk assessment in which the sum of the relevant probabilities is multiplied by the associated impact (1 or 9) of either clicking on the email or reporting the email. By focusing only on the relevant probabilities for each What factor, the risk for What Factor 1 and What Factor 2 could be calculated accurately.

Table 4. If factor option combination probabilities and risk assessment

If factor 1 options Connections to Quality	If factor 2 options Threshold of Quality	If factor option combination probabilities	What factor 1: Clicking on Email probabilities	What factor 2: Reporting Email probabilities
If factor 1 option 1: Low Connections to Quality (0.2)	If factor 2 option 1: Low Threshold (0.25)	0.2*0.25 = 0.05		0.05
	If factor 2 option 2: Mediate Threshold (0.35)	0.2*0.35 = 0.07		0.07
	If factor 2 option 3: High Threshold (0.4)	0.2*0.4 = 0.08	0.08	
If factor 1 option 2: Mediate Connections to Quality (0.3)	If factor 2 option 1: Low Threshold (0.25)	0.3*0.25 = 0.075		0.075
	If factor 2 option 2: Mediate Threshold (0.35)	0.3*0.35 = 0.105	0.105	
	If factor 2 option 3: High Threshold (0.4)	0.3*0.35 = 0.120	0.120	
If factor 1 option 3: High Connections to Quality (0.5)	If factor 2 option 1: Low Threshold (0.25)	0.5*0.25 = 0.125	0.125	
	If factor 2 option 2: Mediate Threshold (0.35)	0.5*0.35 = 0.175	0.175	
	If factor 2 option 3: High Threshold (0.40)	0.5*0.4 = 0.2		0.2
	Sum of probabilities		0.605	0.395
	Impact Indicator		9	1
	Risk assessment		0.605*9 = 5.445	0.395*1 = 0.395

7 Discussion

To conclude, our network model approached spear-phishing and analyzed how the COM-B model could benefit and provide solutions in mitigating spear-phishing threats. This is mainly conducted through improving behavior, such as reflectiveness traits and traits of awareness. Through continuous learning and adaptation, our chosen employee begins their work with certain knowledge and experience and thereafter begin to adapt and improve throughout the training and simulation. This is because reflection and behaviors change based primarily on the feedback the employee receives since, as they progress, they learn which mistakes they have made and how to avoid them in the future. What is equally interesting to note is that as the trait of reflectiveness increases in the employee, so does the trait of impulsiveness decrease, since the worker first contemplates any e-mail's significance, relevance, context, and who sent the e-mail for which purpose before opening it. The mere act of awareness of spear phishing e-mails, for example, increases the employees' reflectiveness. However, awareness of such spear phishing e-mails coupled with continuous learning with a continuous feedback loop on reporting these e-mails and learning which ones to avoid is what constitutes a lasting change in the employee's behavior regarding spear phishing e-mails.

In (Treur, 2021) and (Hendrikse, Treur, Koole, 2023) it is shown in detail how any (smooth) adaptive dynamical system (of any order of adaptivity) can be represented as a self-modeling network, so the network format does not introduce any fundamental limitation.

References

Börcsök, T., et al.: Adaptive defence mechanisms: an adaptive network model to protect HiX systems against cyber attacks. In: Proceedings of the 8th International Conference on Computational Methods in Systems and Software, CoMeSySo'24. Lecture Notes in Networks and Systems, Springer Nature (2025)

Bouma, D., Den Oudsten, T., Naffakh, A., Waliji, M., Treur, J., Roelofsma, P.H.M.P.: Cyber risk management and organisational alignment using Endley's risk management model: computational analysis from an adaptive dynamical system perspective. In: Proceedings of the 8th International Conference on Computational Methods in Systems and Software, CoMeSySo'24. Lecture Notes in Networks and Systems, Springer Nature (2025)

Bouma, D., Hoffmans, C., Van den Hout, N.J., Zwarts, N., Treur, J., Roelofsma, P.H.M.P.: Cyber security in hospitals: a network-oriented model for behavioural learning of employees during phishing simulations. In: Proceedings of the 17th International Conference on Computational Intelligence in Security for Information Systems, CISIS'24. Lecture Notes in Networks and Systems, Springer Nature, in press (2024)

Caneva, J., et al.: The unforeseen blackout and cyber risk management for business continuity: network-oriented computational analysis. In: Proceedings of the 8th International Conference on Computational Methods in Systems and Software, CoMeSySo'24. Lecture Notes in Networks and Systems, vol. 957, pp. 98–111. Springer Nature (2025)

Hendrikse, S.C.F., Treur, J., Koole, S.L.: Modeling emerging interpersonal synchrony and its related adaptive short-term affiliation and long-term bonding: a second-order multi-adaptive neural agent model. Int. J. Neural Syst. **33**(7), 2350038 (2023). https://doi.org/10.1142/S0129065723500387

Illustrative Media Items. Available as Linked Data at URL. https://www.researchgate.net/publication/382975864 (2024)

Jampen, D., Gür, G., Sutter, T., Tellenbach, B.: Don't click: towards an effective anti-phishing training. A comparative literature review. Human-centric Comput. Inf. Sci. **10**(1) (2020). https://doi.org/10.1186/s13673-020-00237-7

Jeffery, E., et al.: Computational analysis of the effectiveness of the intelligence cycle for organizational cyber risk management. In: Proceedings of the 8th International Conference on Computational Methods in Systems and Software, CoMeSySo'24. Lecture Notes in Networks and Systems, Springer Nature (2025)

Michie, S., Van Stralen, M.M., West, R.: The behaviour change wheel: a new method for characterising and designing behaviour change interventions. Implementation Sci. **6**(1) (2011). https://doi.org/10.1186/1748-5908-6-42

Rizzoni, F., Magalini, S., Casaroli, A., Mari, P., Dixon, M., Coventry, L.: Phishing simulation exercise in a large hospital: a case study. Digital Health **8**, 205520762210817 (2022). https://doi.org/10.1177/20552076221081716

Roelofsma, P.H.M.P., Jabeen, F., Taal, H.R., Treur, J., (eds) Using Shared Mental Models and Organisational Learning to Support Safety and Security through Cyberspace: a Computational Analysis Approach. Springer Nature (2024). https://doi.org/10.1007/978-3-031-72075-8

The NIST Cybersecurity Framework (CSF) 2.0. NIST.gov. (2024). https://nvlpubs.nist.gov/nistpubs/CSWP/NIST.CSWP.29.pdf

Treur, J.: Network-Oriented Modeling: Addressing Complexity of Cognitive, Affective and Social Interactions. Springer Nature (2016). https://doi.org/10.1007/978-3-319-45213-5

Treur, J.: Network-Oriented Modeling for Adaptive Networks: Design Higher-Order Adaptive Biological, Mental and Social Networks. Springer Nature (2020). https://doi.org/10.1007/978-3-030-31445-3

Treur, J.: On the dynamics and adaptivity of mental processes: relating adaptive dynamical systems and self-modeling network models by mathematical analysis. Cogn. Syst. Res. **70**, 93–100 (2021). https://doi.org/10.1016/j.cogsys.2021.07.003

Sha, Z., et al.: A network-based approach to modeling and predicting product coconsideration relations. Complexity **2018**, 2753638 (2018). https://doi.org/10.1155/2018/2753638

Analysing the Rise of Biohacking in a Tech-Driven City: Assessing Risks Through Adaptive Network Modeling

Charlotte C. A. Hoffmans[1,2], Lisa van der Does[1], Merel van der Kolk[1], Timo van Rooyen[1], Nick Barelds[1], Jan Treur[1,2(✉)], and Peter H. M. P. Roelofsma[1]

[1] Centre of Expertise Cybersecurity, Risk Management and Cybersecurity Group, The Hague University of Applied Sciences, The Hague, Netherlands
{N.F.Barelds,P.H.M.P.Roelofsma}@hhs.nl, j.treur@vu.nl

[2] Department of Computer Science, Vrije Universiteit Amsterdam, Social AI Group, Amsterdam, Netherlands

Abstract. In this paper we analyse the world and network connections of biohacking within a medium-sized city, focusing on the interactions between the stakeholders and the potential risks and benefits. Through the design of a network model and simulation experiments with it, we get insight into the dynamics of biohacking activities and the impact of potential protective measures in a medium-sized city. One the findings from the simulation scenario was the identification of the risks associated with unregulated biohacking practices, in particular those conducted by DIY biohacking enthusiasts. The conducted simulations showed how with the implementation of structured protective measures, such as the public forum and task forces, the risk levels can be reduced. The simulations also showed that the benefits from biohacking such as enhanced physical and cognitive abilities can outweigh the risks of complications over time, providing that protective measures have been provided and are effective.

1 Introduction

With the advancements of technology emerges new practices that challenge existing norms, ethics and regulations. One such emerging trend is biohacking, which is the altering of the body through methods like genetic engineering, cybernetic implants and chemical enhancements, see, for example, (Ikemoto, 2017; Meyer and Vergnaud, 2020; Seiden, Baggili, Ali-Gombe, 2023; Šupa and Kruopštaitė, 2022; Yetisen, 2018; Zettler, Guerrini, Sherkow, 2019).

This paper explores a scenario set in a medium-sized city known for its diversity in population and growing technology industry, where biohacking has gained popularity. With the rise of biohacking it brings both opportunities and challenges with it. Biohacking offers potential increase in physical and cognitive abilities, which could lead to medical advancements and improve the quality of life. However, biohacking also raises safety and ethical concerns, especially with unregulated practices by DIY enthusiasts. These opportunities and challenges create a complex landscape for stakeholders, including biohackers, medical professionals, ethicists, law enforcement, and policymakers.

D. Singh et al. (Eds.): IHCI 2024, LNCS 15557, pp. 367–379, 2025.
https://doi.org/10.1007/978-3-031-88705-5_31

This paper aims to provide an analysis of the biohacking trend in this city, focusing on the development and application of a network model to understand the connections and their dynamics involved. The paper will outline the scenario in more detail, present a conceptual network model, and conduct analyses to explore the implications of biohacking practices.

This work is part of a research programme on computational analysis of safety and security in and through cyberspace. In (Börcsök, Bos, Buter, Van Der Hoeven, Velev, Van den Hout, Bouma, Treur, Roelofsma, 2025; Bouma, Den Oudsten, Naffakh, Waliji, Treur, Roelofsma, 2025; Caneva, Oliver, Rietze, Šileris, Barelds, Hoffmans, Treur, Roelofsma, 2025; Jeffery, Malsagov, Tijssen, Zeisig, Hoffmans, Treur, Roelofsma, 2025) various other themes within cybersecurity were addressed from a computational network based adaptive dynamical systems perspective, which as far as the authors know was not done before. Moreover, in (Roelofsma, Jabeen, Taal, Treur, 2024) it was analysed how an AI coach can support safety and security for medical teams in hospitals.

2 Background Information and Scenario Description

In the scenario, the concept of biohacking is investigated within a small city distinguished by its diverse population and developing technology industry. Biohacking involves making changes to the body through the use of genetic engineering, cybernetic implants, and chemical enhancements.

The rise of biohacking has started a debate among stakeholders, including bioethicists, members of the DIY biohacking community, and law enforcement officials. Bioethicists express their concern about the absence of regulations surrounding biohacking practices, drawing attention to the fear of potential health risks and ethical implications. On the other side, members within the DIY biohacking community advocate for the personal freedom and the potential benefits of biohacking for individuals, for example enhanced physical and cognitive abilities. However, the efforts made by a journalist to investigate has uncovered incidents where experiments have been conducted with DIY biohacking practices that resulted in health complications and raised safety concerns for the public. The incidents have displayed challenges for law enforcement officials in enforcing safety standards and legal uncertainty surrounding biohacking practices.

In response to the concerns, the municipality organizes a public form to address the growing challenges concerning biohacking. Stakeholder from various backgrounds, including biohackers, medical professionals, ethicists and concerned citizens, come to the forum to discuss potential solutions and strategies. Following the forum, the city takes steps to address issues raised by the forum. A task force is established to develop guidelines for safe biohacking practices, including expertise's from fields such as medicine, ethics, and law enforcement. Additionally, the city explores the possibility of using an AI-coach to detect biohacking activities and provide guidance and insights on how to address the activities. Public awareness campaigns are also launched to educate and inform citizens of the risks and benefits of biohacking, aiming to promote decision-making that is informed and will encourage a safer biohacking environments within the city. For more information, see (Ikemoto, 2017; Meyer and Vergnaud, 2020; Seiden,

Baggili, Ali-Gombe, 2023; Šupa and Kruopštaitė, 2022; Yetisen, 2018; Zettler, Guerrini, Sherkow, 2019).

3 The Network Modeling Approach Used

Network modelling is a computational approach used to represent and analyse the structure and behaviour of complex systems. This method is commonly applied in fields such as neuroscience, artificial intelligence, and systems biology. Key concepts in network modelling approach (Treur, 2020) used here include states and network characteristics such as connections and their connection weights (connectivity characteristics), combination functions and their parameters (aggregation characteristics), speed factors and initial values (timing characteristics). Moreover, self-model states are used to model adaptivity.

States refer to nodes Y of the network with the activation values $Y(t)$ associated with them over time t. States are dynamic, changing over time based on inputs and the network's characteristics. Connections, or edges, link the nodes in the network and define how information flows from one node to another. Each connection $X \rightarrow Y$ has an associated weight $\boldsymbol{\omega}_{X,Y}$ that determines the strength and direction (positive or negative) of this influence. They determine the importance of the input from one node to another. During learning or adaptation, weights can be adjusted to improve performance. Positive weights enhance the activation level of nodes, while negative weights inhibit it.

Combination functions $\mathbf{c}_Y$ determine how inputs to a node Y are combined (or aggregated) to produce an updated activation level as output. Parameters of combination functions can include excitability thresholds $\boldsymbol{\tau}_Y$ or scaling factors $\boldsymbol{\lambda}_Y$ and other coefficients that modulate the output.

Speed factors $\boldsymbol{\eta}_Y$ specify the rate at which states change their activation level over time, controlling how quickly a node responds to changes in its inputs in dynamic systems. Initial values are the starting states of the nodes at $t = 0$ before any input is processed.

Self-model states refer to a network's ability to model and adapt its own structure and parameters, allowing the network to adjust its behaviour based on experience or changes in the environment. Self-modelling can be achieved through mechanisms like learning rules, feedback loops, and adaptive weights. For example, if the excitability threshold $\boldsymbol{\tau}_Y$ of state Y is adaptive, then to model this a self-model state $\mathbf{T}_Y$ can be added to the network which represents this threshold, and if the speed factor of a state is adaptive, to model this a self-model state $\mathbf{H}_Y$ can be added. This can also be done in combination to obtain multiple orders af adaptivity. For example a second-order self-model state $\mathbf{H}_{\mathbf{T}_Y}$ can be added to represent the speed factor of first-order self-model state $\mathbf{T}_Y$.

In summary, network modelling involves nodes representing states and connections between nodes with weights that define the strength of their interaction. Combination functions aggregate inputs using these weights, while parameters such as speed factors and initial values influence the system's dynamics. Adaptive networks use self-model states to continuously adjust and improve their performance based on feedback. This approach is foundational in various computational and biological systems, enabling the simulation and understanding of complex behaviours and interactions.

Based on these network characteristics dynamics of a network are defined by the following difference equation:

$$Y(t + \Delta t) = Y(t) + \eta_Y\left[\mathbf{aggimpact}_Y(t) - Y(t)\right]\Delta t$$

where

$$\mathbf{aggimpact}_Y(t) = \mathbf{c}_Y\left(\boldsymbol{\omega}_{X_1,Y}X_1(t), \ldots, \boldsymbol{\omega}_{X_k,Y}X_k(t)\right)$$

with X_1, ..., X_k the states from which Y has incoming connections. In Table 1 the used combination functions are shown.

Table 1. Descriptions of the used combination functions

Function	Notation	Formula	Parameters
Advanced logistic sum	$\mathbf{alogistic}_{\boldsymbol{\sigma},\boldsymbol{\tau}}(V_1, \ldots, V_k)$	$[\frac{1}{1+e^{-\sigma(V_1+\cdots+V_k-\tau)}} - \frac{1}{1+e^{\sigma\tau}})](1 + e^{-\sigma\tau})$	Steepness $\boldsymbol{\sigma}$ Ecitability threshold $\boldsymbol{\tau}$
Stepmod	$\mathbf{stepmod}_{\boldsymbol{\rho},\boldsymbol{\delta}}(.)$	0 if $\boldsymbol{\delta} > t \bmod \boldsymbol{\rho}$ 1 if $\boldsymbol{\delta} \leq t \bmod \boldsymbol{\rho}$	Duration to start $\boldsymbol{\delta}$ Duration of repetition $\boldsymbol{\rho}$

In this way adaptive dynamical systems can be modeled in a network-oriented manner. As a modeling format, this does not introduce any fundamental limitation, as in (Treur, 2021) and (Hendrikse, Treur, Koole, 2023) it has been shown in detail by mathematical analysis how any (smooth) adaptive dynamical system can be represented as a temporal-causal self-modeling network model.

4 The Designed Adaptive Network Model Explained

The network model created in this paper is a social (temporal-causal) network model. This model is used to provide a graphical overview of the interactions between different processes and stakeholders concerning biohacking and the influence of protection barriers established to contain the biohacking boom and reduce its risks. In addition, the model shows that the benefits ultimately outweigh the risks due to the implementation of protection measures. The model starts with biohacking X1, which is connected to its three specific components, cybernetic implants X2, genetic engineering X3, and chemical enhancements X4. These three each are linked to states for risks X5 and benefits X6. The benefits involve enhanced physical abilities X7 and medical advancements X8. The risks involve health risks X9, lack of regulations X10, and difficulty enforcing safety standards X11. The stakeholders all will shed light on some of the aforementioned benefits and risks, and therefore these are connected to the stakeholders according to who-said-what. Stakeholders consist of a DYI Biohacker X12, a BioEthicist X13, and the Police X14. All stakeholders are linked to the municipal public forum X15, as all stakeholders discuss these things with each other. This is the first protection barrier and it therefore impacts the risk level (by having the role of a **T**-state for Risks). The public forum is in turn linked to the task force X16, which is another Protection Measure X17, which takes the opinions of stakeholders about risks and benefits into account in order to create guidelines. This task force is linked to the AI agent X21, because the agent is sort of a virtual assistant to this process and it also helps create more protection barriers. Protection barriers is then linked as **T**-state to X2, X2, and X4 for the specific types of biohacking. Initially, biohacking occurs, followed by a second biohacking, and this process repeats. The model is designed such that the more biohackings occur, the lower the risk becomes. This is due to the fact that the aforementioned protection barriers are adaptive states, together X17 and X18, with the speed of the effect of the protection barriers X20 and the speed of the impact on the risk X19.

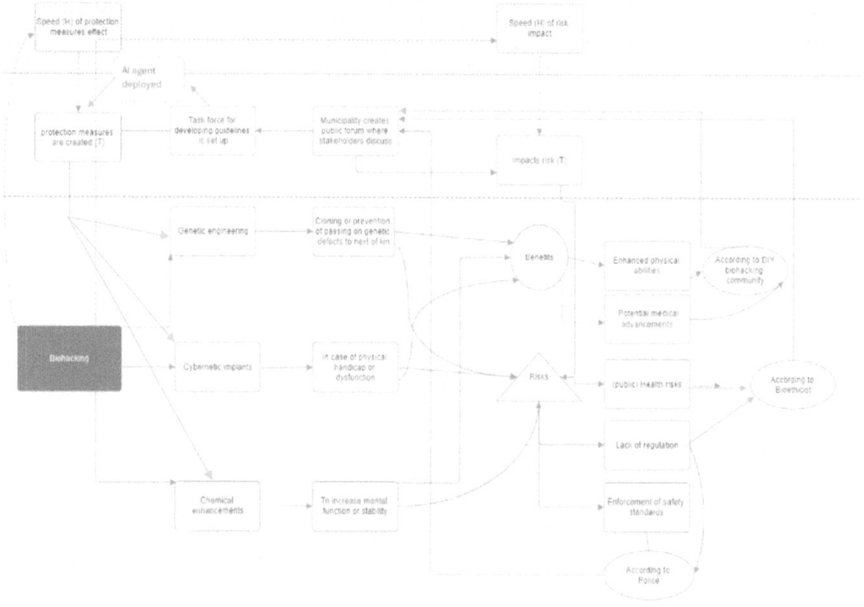

Fig. 1. Picture of the second-order adaptive network model of the biohacking trend in a medium-sized city

In this paper, we designed a network model to analyze the dynamics of biohacking within a medium-sized city, see Fig. 1. The states are explained in Table 2, the connections in Table 3, and the role matrices are available in the Appendix in Table 4 to Table 8. The model is designed to capture interactions between the different factors influencing biohacking activities. The elements of the model include states, connections and their weights, speed factors and combination functions. The role matrices are representations of the network characteristics such as connections and other influences of different states in our network model. Each state is influenced by other states to different degrees, and the influences are measured in the role matrices. In the scenario the role matrices help to understand how factors such as motivations, emotions, decision options, social influences and adaptivity interact with each other. The connection and their weights represent the ways through which states influence each other. Each of the connections has their associated weights, which tells the strength of the influences.

Table 2. States and their explanation

X1	Biohacking	Biohacking entails the practice of altering the human body through genetic engineering, cybernetic implants, and chemical enhancements
X2	Cybernetic Implants	A cybernetic implant could replace lost function for someone with disability, for example the bionic ear.
X3	Genetic Engineering	Genetic engineering (also called genetic modification) is a process that uses laboratory-based technologies to alter the DNA makeup of an organism. This may involve changing a single base pair (A-T or C-G), deleting a region of DNA or adding a new segment of DNA.
X4	Chemical Enhancements	Chemical enhancement is defined as the enhancement of the effectiveness of X-ray irradiation by catalytic nanomaterials dissolved or embedded in media. Similar to physical enhancement, chemical enhancement can be quantified by the ratio of signal with nanomaterials to without nanomaterials when there are little or no physical or other enhancements. The detected signals are considered equivalent to the amount of chemicals of interest.
X5	Risks	Biohacking has possible risks concerning bad or severe health complications
X6	Benefits	The use of biohacking can lead to multiple health benefits
X7	Enhanced Abilities	A benefit of biohacking concerning enhanced physical abilities
X8	Medical Advancements	A benefit of biohacking concerning potential medical advancements
X9	Health Risks	A possible risk concerning the public health of those who undergo biohacking procedures
X10	Lack of Regulation	Lack of regulations cause risks for those who undergo biohacking procedures
X11	Enforcement Safety Standards	The act of compelling observance of compliance with the law concerning Biohacking
X12	DIY Biohacking	Unprofessional biohackers who do not work to the same regulations as professional biohacking labs
X13	Bio Ethicist	Professionals who are concerned about the absence of regulations and fear health risks

(*continued*)

Table 2. (*continued*)

X1	Biohacking	Biohacking entails the practice of altering the human body through genetic engineering, cybernetic implants, and chemical enhancements
X14	Police	Civil force of state concerned with the prevention and detection of crime and the maintenance of public order. In this particular case facing the incidents, challenges with lack of regulations and difficulties enforcing safety standards regarding biohacking
X15	Municipality Forum	A public forum organized by the municipality of the city to address the growing concerns
X16	Task Force	A set-up group to develop guidelines for safe biohacking practices and to investigate the possibility to implement an AI-coach to detect biohacking activities and provide advice on how to deal with them
X17	Protection Measures	Protection measures regarding the safe use of biohacking in order to minimalize the risks
X18	Impacts Risk	How the protection measures, AI-Coach, task force, the municipality forum impact the risk level of biohacking
X19	Speed Risk Impact	The speed of the impact risk
X20	Speed Protection Measure	The speed of the effect of the protection measures
X21	AI-Agent	An AI-coach made to detect biohacking activities and provide advice on how to deal with them

Weights are values that determine how much of a state will affect another state within the model. Higher weights show stronger influences while lower weights show weaker influences. Speed factors control how quick a state reacts to changes. For example, if the speed factor from state 1 is high, it means that state 1 will quickly react to changes. Combination functions show how multiple influences combine to affect one single state. The functions take into account their own respective weights and influences from other states to make an output that is the new state value.

The role matrices provide a framework for understanding the dynamics within our scenario of biohacking within a medium-sized city. They allow for a quantitative analysis to asses the potential risks and benefits of biohacking practices within the city. They also allow running simulations based on specific scenarios to asses changes to the risks and benefits with the influence of factors, including the public forum, task forces and AI-coaching. For more details of the model, see the Appendix available as Linked Data at URL https://www.researchgate.net/publication/382975862.

5 Simulation Results

An example simulation is shown in Fig. 2 (overall), Fig. 3 (base level states), Fig. 4 (first-order adaptation states) and Fig. 5 (second-order adaptation states).

Table 3. Connections and their explanation

Connections	Explanation
X1 → X2-X4, X17, X19, X20	This demonstrates that biohacking (X1) drives the use of and improvements in cybernetic implants (X2), genetic engineering (X3), and chemical enhancements (X4). This influences the speed at which protection measures (X20) and the risk impacts (X19) develop, and it necessitates the implementation of protection measures (X17)
X2-X4 → X1, X5, X6	This demonstrates that X2-X4 are important components of biohacking (X1). Each of the technologies contributes the field of biohacking through human modification and enhancement. These technologies/methods create various risks (X5) and benefits (X6) to the human well-being
X5→ X2-X4, X9, X10, X11, X18	This demonstrates that the risks associated with technologies/methods for biohacking (X2-X4) lead to specific health risks (X9), and show the need for regulations (X10), necessitate safety standards enforcement (X11), and impact the overall risk profile (X18)
X6→ X2-X4, X7, X8	This demonstrates that the benefits (X6) associated with technologies/methods for biohacking (X2-X4) contribute to enhanced abilities (X7) and potential medical advancements (X8)
X7-X8→ X6, X12	This demonstrates that enhanced abilities (X7) and medical advancements (X8) are results of biohacking benefits (X6) and is the encouragement for DIY biohacking community (X12)
X9, X11→ X5, X13	This demonstrates that the health risk (X9) associated with the risks (X5) of biohacking are an important concern for the bioethicist (X13) and that the enforcements of safety standards (X11) to minimize the risks (X5) aligns with their concern
X10→ X5, X13. X14	This demonstrates that the risks (X5) associated with biohacking is increased by the lack of regulation (X10), and is a concern for the bioethicists (X13) and the police (X14)
X12→ X7, X8	This demonstrates that DIY biohacking (X12) is driven by the potential enhanced abilities (X7) to the human body and medical advancements (X8)
X13→ X9, X10	This demonstrates that bioethicists (X13) show concerns with the health risk (X9) and the lack of regulation (X10) of biohacking
X14→ X10, X11	This demonstrates that the police (X14) show concerns with the lack of regulation (X10) and the need for enforcement of safety standards (X11) of biohacking

(*continued*)

Table 3. (*continued*)

Connections	Explanation
X15→ X12, X13, X14, X16, X18	This demonstrates that the municipality forum (X15) brings the opinions together from the stakeholders; DIY biohacking (X12), bioethics (X13), and law enforcement (X14) to form a task force (X16) and address the impact risk (X18)
X16→ X15, X21	This demonstrates that the task force (X16), established by the municipality forum (X15), aims to implement an AI-agent (X21) to monitor and advise on biohacking activities
X17→ X1, X21	This demonstrates that the protection measures (X17) are essential for safe biohacking (X1) and are supported by the AI-agent (X21)
X18→ X5, X15, X18, X19	This demonstrates that the impact risk (X18) assesses the risks (X5) discussed in the municipality forum (X15) and its implications on the overall risk profile (X18) and speed of risk impact (X19)
X21→ X16, X17	This demonstrates that the AI-agent (X21), developed by the task force (X16), supports protection measures (X17) to ensure safe biohacking practices
X19→ X1, X18	This demonstrates that the speed of risk impact (X19) is influenced by the prevalence of biohacking (X1) and the assessed impact risk (X18)
X20→ X1, X17	This demonstrates that the speed at which protection measures (X20) are implemented affects biohacking (X1) and the effectiveness of protection measures (X17)

In the base level, X1 shows when hacking is being done. This state is activated according to a repetitive pattern of period 100. This affects, among others, X2, X3, and X4. It is shown that X2, X3, and X4 decrease over time since the thresholds (represented by X17) increase. Partly due to this, X5 Risk also decreases but also because X5's own threshold (represented by X18) increases as well. State X6 Benefits also decrease, but relative to X5, the decrease is minimal. Thus, the positives here outweigh the negative effects. States X7 to X14 start at 0 but quickly increase and maintain high values.

When examining the simulations of the first-order self-adaptive level, it is evident that X17 Protection Measures and X18 Impacts Risk increase. State X17 represents the excitability thresholds of X2 Cybernetic Implants, X3 Genetic Engineering, and X4 Chemical Enhancements, while X18 represents the excitability threshold of X5 Risk. So what was observed at the base level, namely that X2 to X4 and X5 decreased is explained by the patterns for X17 and X18 which show increasing trends over longer time periods for both. State X15 Municipality Forum, influenced by the base level, also increases. This influences X18 and X16 Task Force, which also increase, and this, in turn, influences X21 AI-Agent, which also increases. State X21 also influences X18.

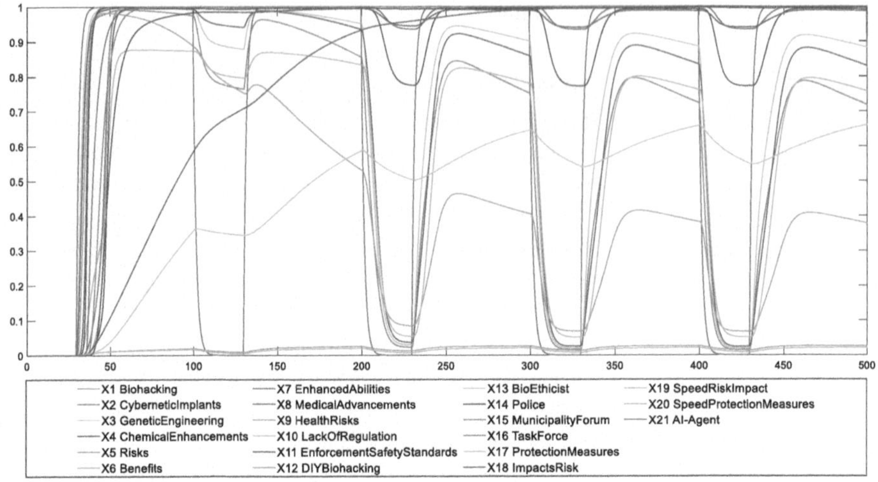

Fig. 2. Example simulation: all states

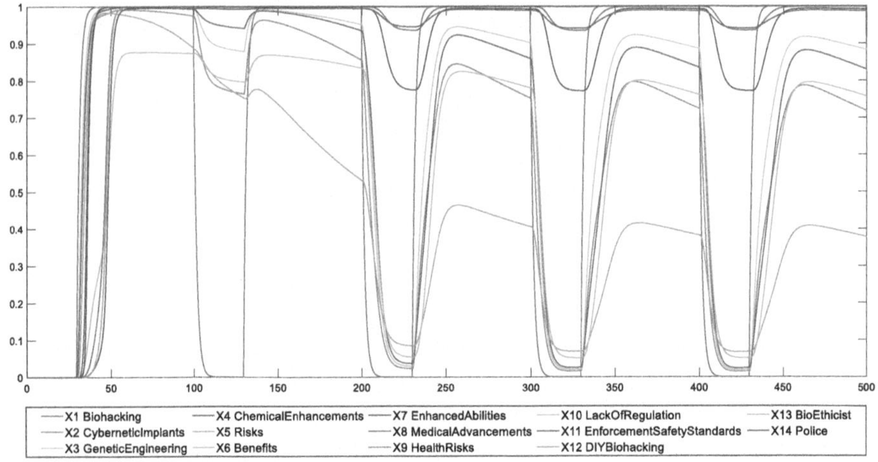

Fig. 3. Example simulation: base level states

The speed of the **T**-states X17 and X18 is determined by two states in the second-order adaptive level. State X19 (an $\mathbf{H_T}$-state) represents the speed of X18 and X20 (another $\mathbf{H_T}$-state) represents the speed of X17. Both increase. Although the effects seem minimal (note that the scale at the vertical axis is this time the [0, 0.05] interval), an increase can be observed both short term during the active episodes and long-term over multiple episodes with dips in between the activation episodes of X1 Biohacking.

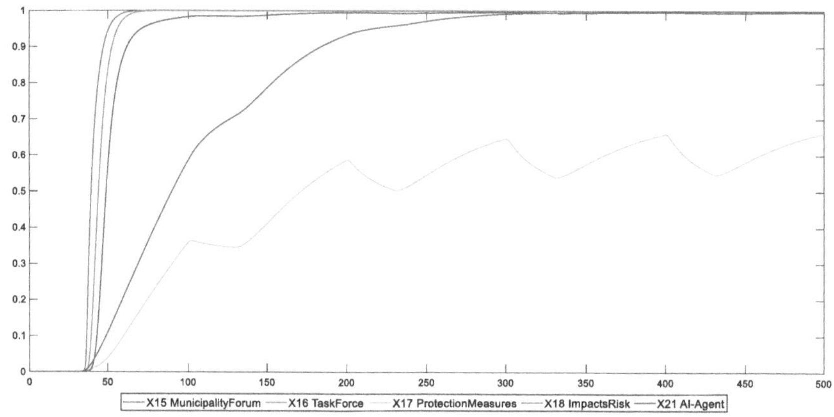

Fig. 4. Example simulation: first-order adaptaption states

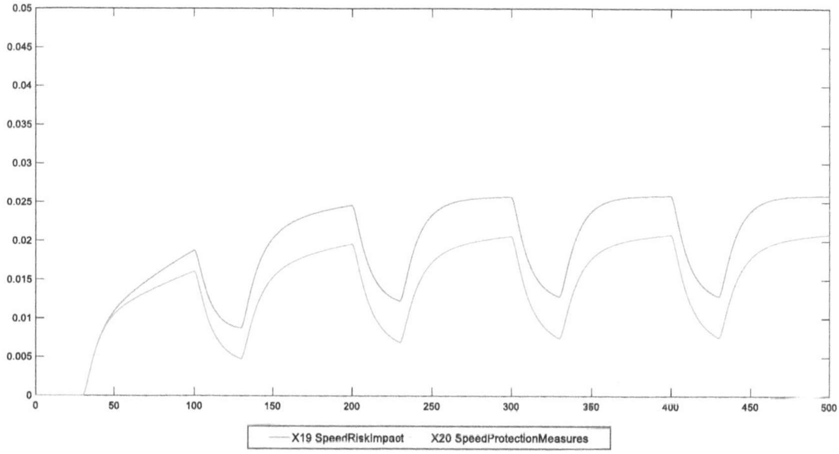

Fig. 5. Example simulation: second-order adaptation states

6 Discussion

In this paper we have explored the world and connections of biohacking within a medium-sized city, focusing on the interactions between the stakeholders and the potential risks and benefits. Through the design and implementation of a network model, we have gained insight into the dynamics of biohacking activities and the impact of potential protective measures in a medium-sized city.

One of our key findings from the scenario was the identification of the risks associated with unregulated biohacking practices, in particular those conducted by DIY biohacking enthusiasts. The risks associated with unregulated biohacking practices include potential health risks and complications, ethical concerns and the challenges for law enforcement in maintaining safety standards. The simulation conducted showed that with the implementation of structured protective measures, such as the public forum, task forces, and the AI-coaching, the risk levels could be reduced. The simulation also showed that the benefits from biohacking for example, enhanced physical and cognitive abilities could outweigh the risks of complications over time, providing that the protective measure have been provided and are effective.

Looking forward to the future there are implications for cyber security models, frameworks, and assessment methods. To begin, the need for flexible models is important, given the evolution of biohacking technologies and practices. Next the involvement of stakeholders in the development and implementation of protective measures is important to address the ethical, legal, and social aspects of biohacking. Finally, monitoring and bringing up to date of guidelines and regulations will be important to keep up with technological advancements and emerging risks.

The future research could develop on this by exploring biohacking in different scenarios with other demographics and technological aspects. This can also assist in investigating long-term impacts of biohacking on the health and social norms of the public. As well as the ability to be able to integrate other AI and machine learning methods and techniques into the network model to enhance the accuracy and adaptability.

To finish, the relationship between technological innovation and regulatory measures in biohacking presents both challenges and opportunities. By making use of network modelling, it can help us understand this scenario of biohacking trend in a medium-sized city, ultimately helping use to potentially increase the safety and ethical environment of biohacking.

References

1. Bennett, G., Gilman, N., Stavrianakis, A., Rabinow, P.: From synthetic biology to biohacking: are we prepared? Nat. Biotechnol. **27**(12), 1109–1111 (2009)
2. Börcsök, T., et al.: Adaptive defence mechanisms: an adaptive network model to protect HiX systems against cyber attacks. In: Proceedings of the 8th International Conference on Computational Methods in Systems and Software, CoMeSySo'24. Lecture Notes in Networks and Systems, Springer Nature (2025)
3. Bouma, D., Den Oudsten, T., Naffakh, A, Waliji, M., Treur, J., Roelofsma, P.H.M.P.: Cyber risk management and organisational alignment using Endley's risk management model: computational analysis from an adaptive dynamical system perspective. In: Proceedings of the 8th International Conference on Computational Methods in Systems and Software, CoMeSySo'24. Lecture Notes in Networks and Systems, Springer Nature (2025)
4. Bouma, D., Hoffmans, C., Van den Hout, N.J., Zwarts-Wojtowicz, N., Treur, J., Roelofsma, P.H.M.P.: Cyber security in hospitals: a network-oriented model for behavioural learning of employees during phishing simulations. In: Proceedings of the 17th International Conference on Computational Intelligence in Security for Information Systems, CISIS'24. Lecture Notes in Networks and Systems, Springer Nature, vol. 957, pp. 98–111 (2024)
5. Caneva, J., et al.: The unforeseen blackout and cyber risk management for business continuity: network-oriented computational analysis. In: Proceedings of the 8th International Conference on Computational Methods in Systems and Software, CoMeSySo'24. Lecture Notes in Networks and Systems, Springer Nature (2025)
6. Hendrikse, S.C.F., Treur, J., Koole. S.L.: Modeling emerging interpersonal synchrony and its related adaptive short-term affiliation and long-term bonding: a second-order multi-adaptive neural agent model. Int. J. Neural Syst. **33**(7), 1–41 (2023). https://doi.org/10.1142/S0129065723500387

7. Ikemoto, L.C.: DIY bio: hacking life in biotech's backyard. UCDL Rev. **51**, 539 (2017)
8. Jeffery, E., et al.: Computational analysis of the effectiveness of the intelligence cycle for organizational cyber risk management. In: Proceedings of the 8th International Conference on Computational Methods in Systems and Software, CoMeSySo'24. Lecture Notes in Networks and Systems, Springer Nature (2025)
9. Keijzer, S., et al.: Managing classified information by a third-party contractor: a computational cybersecurity analysis. In: Proc. of the 8th International Conference on Computer-Human Interaction Research and Applications, CHIRA'24Volume: Communications in Computer and Information Science, vol. 2371, pp. 299–315. Springer Nature (2025)
10. Meyer, M., Vergnaud, F.: The rise of biohacking: tracing the emergence and evolution of DIY biology through online discussions. Technol. Forecast. Soc. Chang. **160**, 120206 (2020)
11. Roelofsma, P.H.M.P., Jabeen, F., Taal, H.R., Treur, J., (eds): Using Shared Mental Models and Organisational Learning to Support Safety and Security through Cyberspace: a Computational Analysis Approach. Springer Nature (2025). https://doi.org/10.1007/978-3-031-72075-8
12. Seiden, S., Baggili, I., Ali-Gombe, A.: I've Got You, Under my skin: biohacking augmentation implant forensics. In: Goel, S., Nunes de Souza, P.R. (eds.) Digital Forensics and Cyber Crime. ICDF2C 2023. Lecture Notes of the Institute for Computer Sciences, Social Informatics and Telecommunications Engineering, vol. 571, pp. 315–332. Springer Nature, Cham (2023)
13. Šupa, M., Kruopštaitė, I.: Deviance and ethical considerations in online biohacking communities. Deviant Behav. **43**(10), 1245–1266 (2022)
14. Treur, J.: Network-Oriented Modeling for Adaptive Networks: Designing Higher-Order Adaptive Biological, Mental and Social Network Models. Springer Nature (2020)
15. Treur, J.: On the dynamics and adaptivity of mental processes: relating adaptive dynamical systems and self-modeling network models by mathematical analysis. Cogn. Syst. Res. **70**, 93–100 (2021)
16. Yetisen, A.K.: Biohacking. Trends Biotechnol. **36**(8), 744–747 (2018)
17. Zettler, P.J., Guerrini, C.J., Sherkow, J.S.: Regulating genetic biohacking. Science **365**(6448), 34–36 (2019)

An Adaptive Network Model to Analyse Cyber Security: Understanding Cyber Threats in Financial Institutions

Wissal Mestour[1,2], Niek Jan van den Hout[2], Jan Treur[1,2](✉), and Peter H. M. P. Roelofsma[2]

[1] Department of Computer Science, Vrije Universiteit Amsterdam, Social AI Group, Amsterdam, Netherlands
wiwimestour@gmail.com, j.treur@vu.nl
[2] Centre of Expertise Cyber Security, Risk Management and Cyber Security Group, The Hague University of Applied Sciences, The Hague, Netherlands
{N.J.vandenHout,P.H.M.P.Roelofsma}@hhs.nl

Abstract. This study investigates cybersecurity threats in financial organizations by developing an adaptive network model to mitigate attacks. Using the cyber kill chain framework, the performed simulations reveal the adaptive dynamics between attackers and defenders over time. In the first scenario, the attacker's rapid learning leads to successful attacks and data breaches. In the second scenario, the organization's superior learning results in effective defenses, with no data stolen. The findings underscore the importance of continuous learning and adaptive defense mechanisms in cybersecurity.

Keywords: Adaptive network · cyber security · cyber threats · financial institutions · What-If analysis · risk assessment

1 Introduction

The last few years have witnessed a significant escalation in both the scale and complexity of cyber-attacks targeting organizations. According to a Symantec report, over one million cyber-attacks occur daily, impacting private companies, governments, and individuals (Aslan and Samet, 2017). Another global report by IBM highlighted that the average cost of a data breach is now \$3.86 million (Ahmed, Asyhari, Rahman, 2021). These alarming frequency underscores the critical need for robust and effective cybersecurity strategies to mitigate the potentially devastating consequences of these attacks. Cybercriminals frequently gain unauthorized access to remote systems or deploy malicious software by exploiting vulnerabilities and bugs in hardware, software, and networks (Yadav and Rao, 2015). Despite significant investments in security measures, the number of cybersecurity incidents continues to rise (Refsdal, Solhaug, Stølen, 2015). Traditional security approaches have shown limited success against the increasingly sophisticated cyber-attacks. This limited success is primarily due to the complexity of modern attacks and the availability of powerful tools that cybercriminals can leverage (Sebastian, 2022).

D. Singh et al. (Eds.): IHCI 2024, LNCS 15557, pp. 380–395, 2025.
https://doi.org/10.1007/978-3-031-88705-5_32

In response to these evolving threats, cybersecurity professionals have developed frameworks to better understand and combat cyber-attacks. One such framework is the cyber kill chain, originally developed by Lockheed Martin (Sebastian, 2022; Cho, Han, Jeong, Kim, Koo, Oh, Park, 2018).

To further enhance our understanding and prevention of cyber-attacks, this paper proposes the development of an adaptive network model that simulates cyber security attacks. By creating a realistic simulation of these attacks, we can gain deeper insights into how they occur, evolve, and impact systems. This model aims to replicate the dynamic nature of cyber threats, allowing for the analysis of different attack vectors and the effectiveness of various defensive measures. Through this simulation, organizations can gain a deeper understanding of the factors, techniques, and procedures that influence cyber-attacks. This approach allows for the development of more effective strategies to prevent such attacks.

In addition to developing new models and frameworks, a crucial aspect of enhancing cybersecurity involves implementing robust risk assessment and risk management practices (Scala, Reilly, Goethals, Cukierm, 2019). This paper delves into a comprehensive What-If analysis to evaluate various scenarios and potential vulnerabilities. By simulating different attack vectors and their impacts, the analysis provides valuable insights into how various factors can affect an organization's security posture. This approach not only helps in identifying potential weaknesses but also enables the development of tailored risk management strategies. Through this detailed examination, organizations can better prepare for and mitigate the effects of cyber threats, ultimately strengthening their overall cybersecurity defenses (Scala et al, 2019).

2 Related Work

The cyber kill chain breaks down the stages of a cyber-attack into identifiable phases, providing a systematic approach to detecting, responding to, and mitigating attacks (Sebastian, 2022). By dissecting the sequence of actions an adversary undertakes, from reconnaissance via weaponize, delivery, exploitation, installation, comand and control to exfiltration, organizations can better anticipate and prevent potential attacks (Cho, et al, 2018). Integrating the cyber kill chain into cybersecurity strategies enhances an organization's ability to disrupt the lifecycle of cyber threats, thereby strengthening the overall security (Mirza, Brown, Halling, Shand, Alam, 2021).

To further enhance understanding and operational effectiveness, Network-Oriented Modeling was applied as a useful method to represent complex real-world processes concerning humans and their interaction (Treur, 2018b; 2020). This approach involves creating detailed network models that simulate various interactions and processes within a system based on principles of network adaptation such as Hebbian learning (Treur, 2020). In educational and organizational settings, network-oriented modeling helps visualize and analyze the flow of information, decision-making pathways, and potential outcomes of different scenarios. It has been applied to model networks for a wide range of processes, especially biological, mental and social processes. The adaptive Network-Oriented Modeling approach described in (Treur, 2020) can be used to model such processes.

In the context of cybersecurity, Network-Oriented Modeling can be particularly useful. It enables the simulation of how individuals and organizations respond to cyber threats by modeling the interactions and adaptations within their mental and operational states. For example, this approach can be used to simulate the progression of a cyber-attack and the subsequent defensive responses, thereby providing valuable insights into the effectiveness of different security measures. By incorporating principles of network adaptation, organizations can better understand and improve their resilience against cyber threats.

3 The Self-modeling Network Modeling Approach Used

In this paper the modeling approach used is network-oriented and addresses adaptation of networks by self-modeling networks (Treur, 2020), as it has been demonstrated to be effective for modeling higher-order adaptive dynamical systems (Hendrikse, Treur, Koole, 2023). Temporal-causal network models are characterized by three main types of network characteristics. Here X and Y denote nodes of the network, also called states and $X(t)$ and $Y(t)$ denote their activation values at time t (Treur, 2018a):

- **Connectivity characteristics** Connections from a state X to a state Y and their weights $\boldsymbol{\omega}_{X,Y}$.
- **Aggregation characteristics** For any state Y, some combination function $\mathbf{c}_Y(..)$ defines the aggregation that is applied to the impacts $\boldsymbol{\omega}_{X,Y}X(t)$ on Y from its incoming connections from states X.
- **Timing characteristics** Each state Y has a speed factor $\boldsymbol{\eta}_Y$ defining how fast it changes for a given aggregated causal impact.

The following difference equations for temporal-causal network models, used for the simulation and analysis of such network models, incorporate these network characteristics $\boldsymbol{\omega}_{X,Y}$, $\mathbf{c}_Y(..)$ and $\boldsymbol{\eta}_Y$ in a standard numerical format:

$$Y(t + \Delta t) = Y(t) + \boldsymbol{\eta}_Y\left[\mathbf{c}_Y\left(\boldsymbol{\omega}_{X_1,Y}X_1(t), \ldots, \boldsymbol{\omega}_{X_k,Y}X_k(t)\right) - Y(t)\right]\Delta t \quad (1)$$

where X_1 to X_k represent states with incoming connections to Y. Table 1 shows the combination functions used in this paper.

Table 1. The combination functions used in this paper.

$\mathbf{alogistic}_{\boldsymbol{\sigma},\boldsymbol{\tau}}(V_1, \ldots, V_k)$ advanced logistic sum	$[\frac{1}{1+e^{-\sigma(V_1+\cdots+V_k-\tau)}} - \frac{1}{1+e^{\sigma\tau}}]\,(1 + e^{-\sigma\tau})$	steepness $\boldsymbol{\sigma}$ threshold $\boldsymbol{\tau}$
$\mathbf{stepmod}_{\boldsymbol{\rho},\boldsymbol{\delta}}(V_1, \ldots, V_k)$ step-modulo	time t 0 if t mod($\boldsymbol{\rho}$) < $\boldsymbol{\delta}$, else 1	repeated duration $\boldsymbol{\rho}$ tipping point $\boldsymbol{\delta}$

The three types of network characteristics $\boldsymbol{\omega}_{X,Y}$, $\mathbf{c}_Y(..)$ and $\boldsymbol{\eta}_Y$ that define a temporal-causal network are typically constant, thus modeling a non-adaptive network. However, in realistic scenarios, these characteristics change over time, making the network adaptive. The self-modeling principle, or reification principle, for networks introduced in Treur, 2018b) and further outlined in (Treur, 2020), states that networks can address adaptivity by incorporating self-model states that represent the adaptive network characteristics. For instance, a self-model state $\mathbf{W}_{X,Y}$ can represent an adaptive connection weight $\boldsymbol{\omega}_{X,Y}$, and a self-model state $\mathbf{H}_Y$ can represent a speed factor $\boldsymbol{\eta}_Y$. These types of self-model states are referred to as **W**-states and **H**-states, respectively. These self-model states are part of the temporal-causal network but operate at a higher level. They have incoming connections for causal impacts and outgoing connections for effectuating their impacts on other (lower-level) states; if these characteristics are adaptive, then in Eq. (1) the values $\mathbf{H}_Y(t)$ are used for η_Y and $\mathbf{W}_{X,Y}(t)$ for $\boldsymbol{\omega}_{X,Y}$.

In this way adaptive dynamical systems can be modeled in a network-oriented manner. As a modeling format, this does not introduce any fundamental limitation, as in (Hendrikse et al, 2023) it has been shown in detail by mathematical analysis how any (smooth) adaptive dynamical system can be represented as a temporal-causal self-modeling network model.

4 Scenario Description

All organizations must be well-prepared for security incidents and crises to ensure resilience and continuity. This preparation is typically managed through Business Continuity Management (BCM) and Disaster Recovery Planning (DRP) (Hoong and Marthandan, 2014). These practices involve defining and implementing controls that enhance the organization's ability to withstand and recover from disruptions. Controls may include formalizing a Business Continuity Plan or a Disaster Recovery Plan, which outline procedures for maintaining operations and recovering systems after an incident (Jorrigala, 2017).

One of the main challenges in BCM and DRP is the inherent uncertainty surrounding the exact nature of potential disasters (Hoong and Marthandan, 2014). Because the specific details of a crisis cannot be predicted with certainty, organizations often resort to crisis simulations. During these simulations, stakeholders engage in artificial training scenarios designed to test the organization's response mechanisms. These exercises help to evaluate the effectiveness of the response and recovery plans in a controlled environment (Jorrigala, 2017).

To further improve the accuracy and effectiveness of these simulations, network-oriented modeling can be employed. This paper describes an approach that involves developing a detailed network model that simulates a crisis scenario. By applying the cyber kill chain framework, organizations can model and understand the stages of a cyber-attack, from reconnaissance to exfiltration (Sebastian, 2022). For example, a simulation might include a scenario where an attacker goes through the cyber kill chain phases, allowing stakeholders to test their response at each stage, such as identifying vulnerabilities, responding to exploitation attempts, and mitigating potential data breaches.

Through these simulations, companies may evaluate the effectiveness of current control systems, watch how various stakeholders respond to different threat scenarios, and find out if disaster recovery plans are met.

5 Design of the Adaptive Network Model

In order to visualize the processes and interactions demonstrated in our scenario, a graphical representation of the temporal-causal network has been made, see Fig. 1. In Fig. 1, ovals represent nodes, and arrows indicate the connections between them, with the arrowheads showing the direction of causal impact. The downward causal connections (pink arrows) illustrate the specific causal impact each self-modeling state has. The upward (or leveled) causal connections (blue arrows) to the self-modeling states give them the dynamics as desired. These are used, along with the chosen combination function and the downward connections, to specify the particular adaptation principle being addressed. For an overview of all states and their explanations see Table 2.

As depicted in Fig. 1, our network model incorporates elements from the cyber kill chain to simulate the lifecycle of cyber-attacks. By integrating stages such as reconnaissance, weaponization, delivery, exploitation and command and control, our model provides a detailed visualization of how cyber-attacks unfold. This approach allows us to analyze each phase in depth, understanding the tactics, techniques, and procedures used by attackers. The benefit of incorporating the cyber kill chain lies in its systematic framework, which helps in identifying potential vulnerabilities and attack vectors at each stage (Sebastian, 2022). By doing so, we can develop targeted defensive measures to disrupt the attack progression effectively. For more details including the role matrices for the scenarios see the Appendix available as linked data at URL

https://www.researchgate.net/publication/382975671.

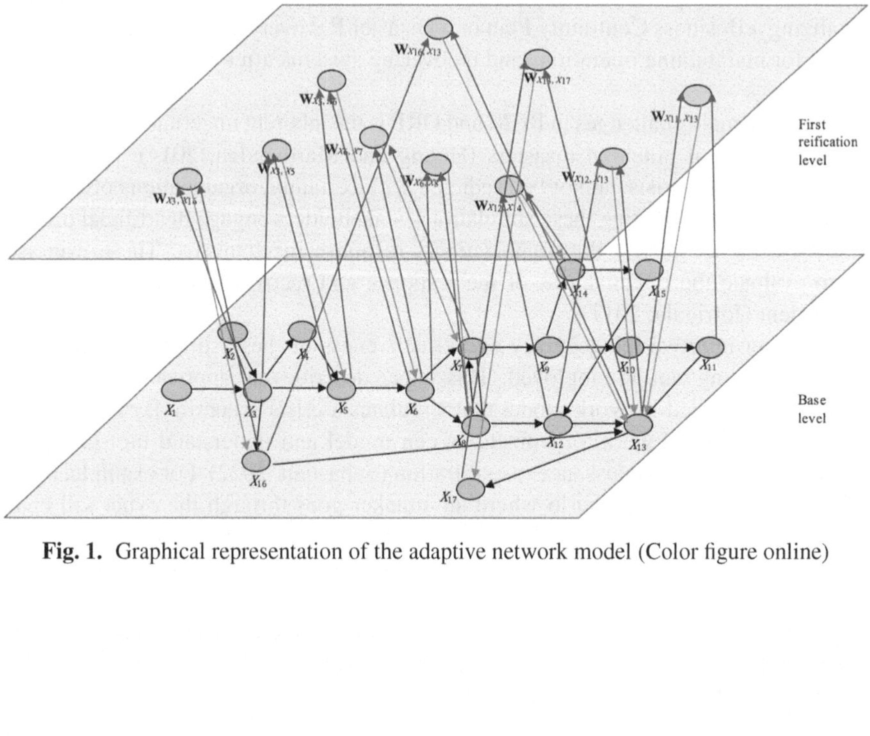

Fig. 1. Graphical representation of the adaptive network model (Color figure online)

Table 2. State names and their explanation: base level and first-order self-model level

State nr	State name	Explanation	Level
X_1	Initial State	Initial State for Organization	Base level
X_2	Reconnaissance	Identification, Selection and Profiling of Target	
X_3	Vulnerability	Extent of Vulnerability	
X_4	Weaponization	Attacker creates the attack vector that will be used in the attack	
X_5	Delivery	Transmission of the cyber weapon to the targeted environment	
X_6	Exploitation	Malicious code is executed within the victim's system	
X_7	Success	Successfulness of the attack	
X_8	Fail	Failure of the attack	
X_9	Persistence	Attacker tries to maintain access	
X_{10}	Command and Control	Attacker uses the attack vector to control devices remotely within the target's network	
X_{11}	Data Stolen	Data Exfiltration and System Disruption	
X_{12}	Suspicious activity detected	Team detects a suspicious activity	
X_{13}	Team Takes Action	Team Takes Action against the attacker	
X_{14}	Fail Persistence	Failure of Persistence	
X_{15}	No Data Stolen	Failure of Data Exfiltration	
X_{16}	Detection	Vulnerability detected by team	
X_{17}	Recovery	Vulnerability fixed by team and Recovery State	
X_{18}	$\mathbf{W}_{X3,X16}$	Self-model state representing connection weight $\boldsymbol{\omega}_{X3,\ X16}$	First self-model level
X_{19}	$\mathbf{W}_{X3,X5}$	Self-model state representing connection weight $\boldsymbol{\omega}_{X3,\ X5}$	
X_{20}	$\mathbf{W}_{X5,X6}$	Self-model state representing connection weight $\boldsymbol{\omega}_{X5,\ X6}$	
X_{21}	$\mathbf{W}_{X6,X7}$	Self-model state representing connection weight $\boldsymbol{\omega}_{X6,\ X7}$	

(continued)

Table 2. (*continued*)

State nr	State name	Explanation	Level
X_{22}	$\mathbf{W}_{X_6,X_8}$	Self-model state representing connection weight ω_{X_6,X_8}	
X_{23}	$\mathbf{W}_{X_{16},X_{13}}$	Self-model state representing connection weight $\omega_{X_{16},X_{13}}$	
X_{24}	$\mathbf{W}_{X_{12},X_{13}}$	Self-model state representing connection weight $\omega_{X_{12},X_{13}}$	
X_{25}	$\mathbf{W}_{X_{12},X_{14}}$	Self-model state representing connection weight $\omega_{X_{12},X_{14}}$	
X_{26}	$\mathbf{W}_{X_{13},X_{17}}$	Self-model state representing connection weight $\omega_{X_{13},X_{17}}$	
X_{27}	$\mathbf{W}_{X_{11},X_{13}}$	Self-model state representing connection weight $\omega_{X_{11},X_{13}}$	

6 Simulation Results

In this paper, we will examine two distinct scenarios to understand cyber threats in organizations, focusing on the adaptive dynamics between attackers and defenders. By analyzing these scenarios, we aim to shed light on how different learning patterns between the two parties influence the effectiveness of cyber defenses. In order to obtain the simulation processes of these scenarios, the dedicated software described in (Treur, 2020, Ch 9) was used. The first scenario involves a strong attacker who learns and adapts more effectively than the organization. In this case, the attacker quickly identifies vulnerabilities within the organization's network, often using advanced techniques. As the attacker penetrates the network, they adapt to defensive measures almost in real-time, leveraging their superior learning and adaptability. This results in successful exploitation of the network, allowing the attacker to install malicious software or exfiltrate sensitive data. The organization's response is typically reactive, struggling to keep up with the attacker's rapidly evolving methods.

Looking at the simulation results of our first scenario, a detailed illustration of the dynamics between the attacker and the organization is shown. Note that both the attacker and the organization engage in continuous learning and adaptation over time. In this scenario, the attacker's learning curve is significantly faster, allowing the attacker to learn more quickly and efficiently than the organization. The graph shows multiple attacks over time, with each attack period lasting 40 s. Initially, there are alternations between the red (succesful attack) and green (failing attacks) curves (see Fig. 2), suggesting a competitive struggle where sometimes the organization manages to fend off attacks. However, as the simulation progresses, around the fourth attack, at $t = 160$, the attacker's learning becomes remarkably faster and more effective. As shown in Fig. 3, the blue curve which represents the learning curve of the attacker, surpasses the value 0.9 at $t = 160$. This shift is evident as the red curve starts to dominate, indicating repeated successful attacks over time. By the fifth attack, the red curve remains consistently above the green curve, which also results in the outcome of the data being stolen. To achieve this scenario, specific settings were used for the connection weights of the base-level states as well as to the persistence factors of the **W**-States, X_{21} and X_{22}, influencing the learning within the model. In this first scenario, the persistence factor of the state influencing the learning of the success of the cyber attack which is state X_{21} ($\mathbf{W}_{X_6,X_7}$), was given a value of 0.98, whereas the persistence factor of the state influencing the fail of the attack which is the state X_{22} ($\mathbf{W}_{X_6,X_8}$) was given the value of 0.8.

The persistence factor of 0.98 for state X_{22} implies that the attacker's successful actions have a long-lasting impact on their learning process. This high persistence value means that once the attacker learns a successful strategy, they retain this knowledge over a prolonged period, allowing them to refine and repeat successful attacks with increasing efficiency. This is reflected by the red curve dominating over time. On the other hand, the lower persistence factor of 0.8 for state X_{23} indicates that the organization's defensive measures do not have as long-lasting an effect.

The organization's learning from failed attacks does not persist as effectively, leading to a slower improvement in their defensive capabilities compared to the attacker. This lower persistence factor results in the organization needing to repeatedly learn from each attack, thus slowing their overall learning process. The simulation results show that as the attacker's learning becomes more efficient and persistent, their success rate increases, leading to more frequent and successful attacks.

The second scenario, see Fig. 4, explores an environment where the organization learns and adapts more effectively than the attacker. Here, the organization's cyber defense team utilizes advanced threat intelligence and continuous monitoring to anticipate and respond to potential threats.

During the exploitation stage of our model, the organization detects the attacker's attempt to exploit a vulnerability and quickly implements countermeasures. The attacker's inability to adapt quickly to these defensive actions leads to the failure of the exploitation attempt. Similarly to the first scenario, note that both the attacker and the organization engage in continuous learning and adaptation over time. Looking at the simulation results of our second scenario in more detail, we notice that this time the organization's learning curve is significantly faster, allowing the organization to learn and adapt more quickly and efficiently than the attacker, see Fig. 5.

However, this does not lead to immediate results as between time 200 and 250 the fourth and fifth attack are still successful. But as the simulation progresses, around the seventh attack, at $t = 280$, the organization's learning becomes remarkably more effective. This shift is evident as the green curve starts to dominate, indicating successful defense against repeated attacks over time. The continuous learning of both the attacker and the organization is a result of the alternations between the green and red curves, representing the success and failure of attacks. Initially, these alternations suggest a competitive struggle, as we can see between $t = 0$ and $t = 280$ in Fig. 4, but the organization's superior learning and adaptation lead to a more consistent defensive performance, which is shown over time by the repeated failed attacks, depicted by the green curve in Fig. 4.

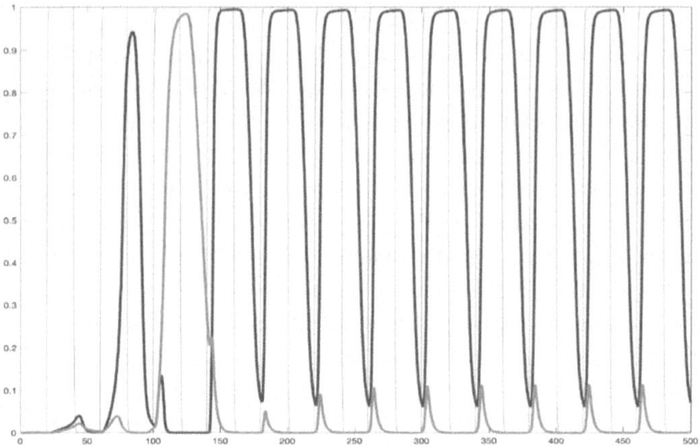

Fig. 2. Simulation results Scenario 1. The red curve represents the successful attacks. The green curve represents the failed attacks. (Color figure online)

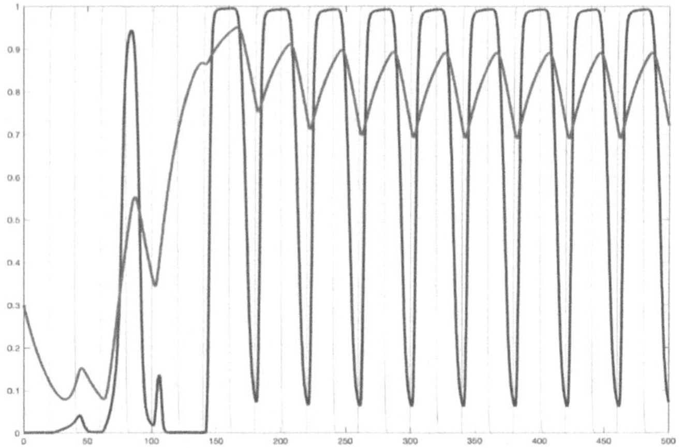

Fig. 3. Simulation results Scenario 1. The red curve represents the successful attacks. The blue curve represents the learning of the attacker. (Color figure online)

Moreover, the continuous efforts of the cyber defense team to 'take action' and prevent attacks are illustrated by the light blue and yellow curve, which steadily rise (Fig. 5). The light blue curve represents the implementation of countermeasures against the attacker, which is represented by the state X_{13}, see Table 2 for the overview of the states. The organization's defense mechanisms are further depicted by the yellow curve, highlighting the ongoing learning and efforts to improve their defensive strategies (state X_{24}). As we can see in Fig. 5, between $t = 0$ and $t = 130$, the light blue curve fluctuates between the values of 0 and 0.9, before it peaks and reaches the value of 1 at $t = 160$. After that the learning (yellow curve in Fig. 7) reaches its peak as well, at value 1, and stabilizes which explains the repeated pattern over time.

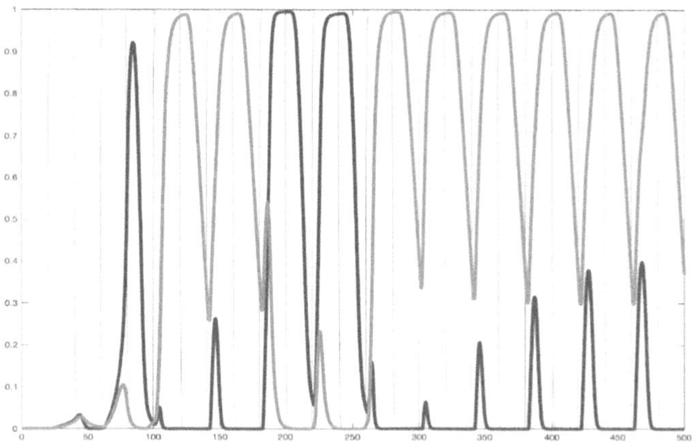

Fig. 4. Simulation results Scenario 2. The red curve represents the successful attacks. The green curve represents the failed attacks.

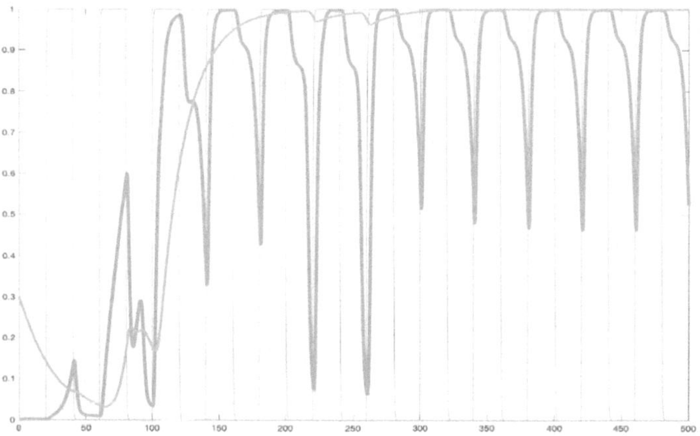

Fig. 5. Simulation results for Scenario 2. The light blue curve represents the state X_{13}, which is Team Takes Action. The yellow curve represents the learning curve for state X_{13}. (Color figure online)

To achieve this scenario, specific variations were made of the connection weights of the base-level states as well as to the persistence factors of the **W**-states, X_{21} and X_{22}, influencing the learning within the model. In this scenario, the persistence factor of the state influencing the learning of the success of the cyber-attack which is the state X_{21} ($\mathbf{W}_{X_6,X_7}$), was given a value of 0.8, whereas the persistence factor of the state influencing the failure of the attack which is the state X_{22} ($\mathbf{W}_{X_6,X_8}$) was given a value of 0.98. The persistence factor of 0.8 for state X_{22} implies that the attacker's successful actions have a shorter-lasting impact on their learning process. This lower persistence value means that even when the attacker learns a successful strategy, they do not retain this knowledge for long, making it harder for them to refine and repeat successful attacks consistently. This is reflected by the red curve in Fig. 6 struggling to dominate over time. In contrast, the higher persistence factor of 0.98 for state X_{22} indicates that the organization's defensive measures have a long-lasting effect.

The organization's learning from successful attacks persists more effectively, leading to a more rapid improvement in their defensive capabilities compared to the attacker. This higher persistence factor results in the organization being able to retain learned defensive strategies over extended periods, enhancing their overall learning process and making their defenses progressively stronger. Overall, the simulation results show that as the organization's learning becomes more efficient and persistent, their success rate in defending against attacks increases, leading to failed cyber-attacks.

7 What-If Analysis and Risk Assessment

What-If analysis and risk assessment are fundamental methodologies for evaluating the potential outcomes of complex scenarios, particularly in the realm of cybersecurity (Wexler, Pushkarna, Bolukbasi, Wattenberg, Viégas, Wilson, 2019). What-If analysis involves a systematic examination of the effects of various 'If factors', critical variables that can significantly influence outcomes. By varying these factors across specified value ranges (e.g., low, medium, high), it is possible to simulate different attack vectors and assess the likelihood of their success or failure. The results of these simulations, known as 'What factors', provide valuable insights into the possible success or failure of an attack, the persistence of a threat, and the potential for data exfiltration or detection of suspicious activities. Complementing this approach, risk assessment involves assigning probabilities to each combination of options for If factors, thereby offering a quantitative understanding of the risks associated with each potential outcome. For the What-If analysis presented in this paper, the following If factors were chosen:

- **Speed of Vulnerability Detection**: This factor refers to the time it takes for the security team to identify a vulnerability within the system.
- **Threshold of Exploitation**: This is the critical threshold at which an identified vulnerability can be exploited by an attacker.
- **Speed of Learning**: This refers to the rate at which the system or security team can learn and adapt to new threats, particularly in relation to the specific state X_{22}, $\mathbf{W}_{X6,X8}$ (see Table 2 for an overview of all states).

The What factors we are examining are the Data Theft at different points in time.

1. **Data Stolen at $t = 40$**: This outcome evaluates whether data theft occurs at time = 40.
2. **Data Stolen at $t = 80$**: This outcome assesses whether data theft occurs at time = 80.
3. **Data Stolen at $t = 320$**: This outcome assesses whether data theft occurs at time = 320.

The severity of this factor is further classified as:

- **High:** If the likelihood of data theft exceeds 0.3.
- **Low:** If the likelihood of data theft is below 0.3.

By analyzing these specific If factors and What factors, we can assess the potential impact on data security and the effectiveness of the organization's response mechanisms during a cybersecurity incident. Table 3 shows a detailed overview of the If factor with their options, labels, value intervals, and assumed probabilities.

Table 3. Overview of If factor options and their labels and value intervals.

If factors	If factor options	If factor option labels	If factor option value intervals	If factor option probabilities
If factor 1: Speed of Vulnerability detection	If factor 1 option 1	low	0–0.35	0.4
	If factor 1 option 2	mediate	0.35–0.65	0.3
	If factor 1 option 3	high	0.65–1	0.3
If factor 2: Threshold of Exploitation	If factor 2 option 1	low	0–0.5	0.6
	If factor 2 option 2	high	0.5–1	0.4
If factor 3: Speed of learning	If factor 3 option 1	low	0–0.05	0.3
	If factor 3 option 2	high	0.05–0.1	0.7

Table 4 presents a structured analysis of the potential outcomes of the scenario presented in this paper by combining the three If factors and their associated probabilities. For combining these probabilities, it is assumed they are independent so that the product rule p(A and B) = p(A)*p(B) can be used. Based on this, the table calculates the combined probabilities for each scenario by multiplying the probabilities of the three If factors, such as 0.072 = 0.4*0.6*0.3 for a low speed of detection (0.4), a low threshold of exploitation (0.6), and a low speed of learning (0.3). These combined probabilities are then used to assess the likelihood of data theft at different time points 40, 80, and 320. In conducting simulations based on these combinations, careful attention was given to monitoring whether the curve representing the likelihood of data theft surpasses the critical threshold of 0.3. This threshold has been established as the point at which the likelihood of data theft becomes sufficiently significant to raise concern. If it exceeds this 0.3 threshold at the given time intervals, it is classified as a case of data theft. This classification suggests that the system's security measures have potentially failed to prevent the unauthorized exfiltration of sensitive information.

Table 4. Structured analysis of the potential outcomes of the scenario presented in this paper.

If factor 1 options	If factor 2 options	If factor 3 options	If factor option combination probabilities		What factor 1 Data Theft $t = 40$ probabilities	What factor 2 Data Theft $t = 80$ probabilities	What factor 2 Data Theft $t = 320$ probabilities
If factor 1 option 1: Low 0.4	If factor 2 option 1: low 0.6	If factor 3 option 1: low 0.3	0.4*0.6*0.3	= 0.072		X 0.072	X 0.072
		If factor 3 option 2: high 0.7	0.4*0.6*0.7	= 0.168		X 0.168	X 0.168
	If factor 2 option 2: high 0.4	If factor 3 option 1: low 0.3	0.4*0.4*0.3	= 0.048			
		If factor 3 option 2: high 0.7	0.4*0.4*0.7	= 0.112			
If factor 1 option 2: Mediate 0.3	If factor 2 option 1: low 0.6	If factor 3 option 1: low 0.3	0.3*0.6*0.3	= 0.054		X 0.054	X 0.054
		If factor 3 option 2: high 0.7	0.3*0.6*0.7	= 0.126		X 0.126	
	If factor 2 option 2: high 0.4	If factor 3 option 1: low 0.3	0.3*0.4*0.3	= 0.036			

(*continued*)

Table 4. (*continued*)

If factor 1 options	If factor 2 options	If factor 3 options	If factor option combination probabilities		What factor 1 Data Theft $t = 40$ probabilities	What factor 2 Data Theft $t = 80$ probabilities	What factor 2 Data Theft $t = 320$ probabilities
		If factor 3 option 2: high 0.7	0.3*0.4*0.7	= 0.084			
If factor 1 option 3: High 0.3	If factor 2 option 1: low 0.6	If factor 3 option 1: low 0.3	0.3*0.6*0.3	= 0.054		X 0.054	
		If factor 3 option 2: high 0.7	0.3*0.6*0.7	= 0.126		X 0.126	
	If factor 2 option 2: high 0.4	If factor 3 option 1: low 0.3	0.3*0.4*0.3	= 0.036			
		If factor 3 option 2: high 0.7	0.3*0.4*0.7	= 0.084			
Sum of probabilities					0	0.6	0.294
Impact indicator					5	5	5
Risk assessment					0	3	1.47

The analysis of the simulation results reveals a significant insight regarding the relationship between the Threshold of Exploitation (If Factor 2) and the likelihood of Data Theft (see Table 3). Specifically, when the threshold of exploitation is set to a high value, 0.8, the simulation consistently indicates that no data is stolen, as the outcome values do not exceed the critical threshold of 0.3. In contrast, when the If Factor 2 has a low value of 0.2, the probability of data theft increases, particularly when combined with low values for the speed of learning (If Factor 3). However, even in scenarios where the speed of learning is high, above 0.05, the risk of data theft remains below the critical threshold 0.3, provided that the Threshold of Exploitation remains high. This finding underscores the importance of maintaining a high threshold for exploitation within the network's security protocols. This highlights the critical role of robust vulnerability management strategies in cybersecurity, emphasizing that enhancing the threshold of exploitation is a key defense mechanism against potential data breaches. Incorporating this strategy into the overall risk management framework can greatly reduce the likelihood of successful cyberattacks, thereby strengthening the organization's security.

8 Discussion and Further Research

The results from the simulations underscore the importance of adaptive learning in cybersecurity defense strategies. The simulations clearly demonstrated that a higher persistence factor allowed the organization to retain knowledge of effective countermeasures over extended periods. This led to a more rapid and consistent improvement in defensive capabilities, resulting in a higher rate of failed attacks and reduced instances of data breaches.

This work is part of a research programme on computational analysis of safety and security in and through cyberspace. In (Akers, Blomquist, Niedziela, Šebok, Van den Hout, Bouma, Treur, Roelofsma, 2025; Bouma, Hoffmans, Van den Hout, Zwarts, Treur, Roelofsma, 2025; Jeffery, Malsagov, Tijssen, Zeisig, Hoffmans, Treur, Roelofsma, 2025; Keijzer, Lochtenbergh, Marsman, Voorhoeve, Zwarts, Bouma, Treur, Roelofsma, 2025) various other themes within cybersecurity were addressed. Moreover, in (Roelofsma, Jabeen, Taal, Treur, 2025) it was analysed how an AI coach can support safety and security for medical teams in hospitals.

Further research could focus on the exploration of various learning algorithms and techniques, such as reinforcement learning, deep learning, and supervised learning to enhance the model's predictive capabilities and responsiveness to cyber threats.

Additionally, introducing a second level of adaptation to study the behaviors of individuals within the organization can provide deeper insights into how human factors influence learning and outcomes in cybersecurity. This could involve examining how employees' responses to training, adherence to security protocols, and decision-making processes under stress affect the overall effectiveness of the organization's defense mechanisms. Understanding these human factors can lead to the development of more comprehensive and human-centric cybersecurity strategies.

References

Ahmed, Y., Asyhari, A.T., Rahman, M.A.: A cyber kill chain approach for detecting advanced persistent threats. Comput. Mater. Continua **67**(2), 2497–2513 (2021)

Akers, D., et al.: Organizational response to apt attacks: computational analysis by behavioral network modelling. In: Proceedings of the 16th International Conference on Intelligent Human Computer Interaction, IHCI'24. LNCS, Springer Nature (2025)

Aslan, Ö., Samet, R.: Mitigating cyber security attacks by being aware of vulnerabilities and bugs. In: 2017 International Conference on Cyberworlds (cw), pp. 222–225. IEEE (2017)

Bouma, D., Hoffmans, C., Van den Hout, N.J., Zwarts-Wojtowicz, N., Treur, J., Roelofsma, P.H.M.P.: Cyber security in hospitals: a network-oriented model for behavioural learning of employees during phishing simulations. In: Proceedings of the 17th International Conference on Computational Intelligence in Security for Information Systems, CISIS'24. Lecture Notes in Networks and Systems, vol. 957, pp. 98–111. Springer Nature (2024)

Cho, S., et al.: Cyber kill chain based threat taxonomy and its application on cyber common operational picture. In: Proceedings of the 2018 International Conference on Cyber Situational Awareness, Data Analytics and Assessment (Cyber SA), pp. 1–8. IEEE (2018)

Hendrikse, S.C., Treur, J., Koole, S.L.: Modeling emerging interpersonal synchrony and its related adaptive short-term affiliation and long-term bonding: a second-order multi-adaptive neural agent model. Int. J. Neural Syst. **33**(07), 2350038 (2023)

Hoong, L.L., Marthandan, G.: Critical dimensions of disaster recovery planning. Int. J. Bus. Manag. **9**(12), 145 (2014)

Jeffery, E., et al.: Computational analysis of the effectiveness of the intelligence cycle for organizational cyber risk management. In: Proceedings of the 8th International Conference on Computational Methods in Systems and Software, CoMeSySo'24. Lecture Notes in Networks and Systems, Springer Nature (2025)

Jorrigala, V.: Business Continuity and Disaster Recovery Plan for Information Security, Culminating Projects in Information Assurance (2017)

Keijzer, S., et al.: Managing classified information by a third-party contractor: a computational cybersecurity analysis. In: Proceedings of the 8th International Conference on Computer-Human Interaction Research and Applications, CHIRA'24. Communications in Computer and Information Science, vol. 2371, pp. 299–315. Springer Nature (2025)

Mirza, Q.K.A., Brown, M., Halling, O., Shand, L., Alam, A.: Ransomware analysis using cyber kill chain. In: Proceedings of the 8th International Conference on Future Internet of Things and Cloud (FiCloud), pp. 58–65. IEEE (2021)

Refsdal, A., Solhaug, B., Stølen, K.: Cyber-risk management. In: Cyber-Risk Management (pp. 33–47). Springer Briefs in Computer Science. Springer International Publishing, Cham (2015). https://doi.org/10.1007/978-3-319-23570-7_5

Roelofsma, P.H.M.P., Jabeen, F., Taal, H.R., Treur, J. (eds): Using Shared Mental Models and Organisational Learning to Support Safety and Security through Cyberspace: a Computational Analysis Approach. Springer Nature (2025). https://doi.org/10.1007/978-3-031-72075-8

Scala, N.M., Reilly, A.C., Goethals, P.L., Cukierm, M.: Risk and the five hard problems of cybersecurity. Risk Anal. **39**(10), 2119–2126 (2019)

Sebastian, G.: Cyber kill chain analysis of five major US data breaches: lessons learnt and prevention plan. Int. J. Cyber Warfare Terrorism (IJCWT) **12**(1), 1–15 (2022)

Treur, J.: The ins and outs of network-oriented modeling: from biological networks and mental networks to social networks and beyond. Trans. Comput. Collective Intell. **32**, 120–139 (2018a). Springer Berlin Heidelberg

Treur, J.: Network reification as a unified approach to represent network adaptation principles within a network. In: Theory and Practice of Natural Computing. Proceedings of the 7th International Conference TPNC'18. LNCS, vol. 11324, pp. 344–358. Springer International Publishing (2018b)

Treur, J.: Network-Oriented Modeling for Adaptive Networks: Designing Higher-Order Adaptive Biological, Mental and Social Network Models, Springer Nature (2020)

Wexler, J., Pushkarna, M., Bolukbasi, T., Wattenberg, M., Viégas, F., Wilson, J.: The what-if tool: interactive probing of machine learning models. IEEE Trans. Visual Comput. Graphics **26**(1), 56–65 (2019)

Yadav, T., Rao, A.M.: Technical aspects of cyber kill chain. In Security in Computing and Communications: Third International Symposium, SSCC 2015, Kochi, India, August 10–13, 2015. Proceedings 3 (pp. 438–452). Springer International Publishing (2015)

Understanding Insider Threats Behaviour: An Adaptive Network Model of the Evolution of an Insider Threat

Maria Ioana Andreea Ivan[1], Niek Jan van den Hout[2], Jan Treur[1,2(✉)], Sophie C. F. Hendrikse[3], and Peter Roelofsma[2]

[1] Department of Computer Science, Vrije Universiteit Amsterdam, Social AI Group, Amsterdam, Netherlands
j.treur@vu.nl

[2] Centre of Expertise Cybersecurity, Risk Management and Cybersecurity Group, The Hague University of Applied Sciences, The Hague, Netherlands
{N.J.vandenHout,P.H.M.P.Roelofsma}@hhs.nl

[3] Psychological Methods Group, University of Amsterdam, Amsterdam, Netherlands
S.C.F.Hendrikse@tilburguniversity.edu

Abstract. This paper addresses computational analysis of insider threats. Organisational security is significantly at risk from insider threats, which are commonly caused by changes inside the organisation and worsened by unfavourable psychological characteristics and external involvements. In order to provide a comprehensive understanding of the adaptive dynamics that influence insider threats, the presented network model incorporates a variety of factors, including employee satisfaction, interactions, work performance, psychological traits, and technological methods. The simulations demonstrated that organisational changes can have a significant impact on employee behaviour, resulting in altered perceptions of the work environment, reduced satisfaction, reduced communication, and signs of isolation. These factors collectively create an environment where the risk of insider threats is heightened, as disgruntled, isolated, and dissatisfied employees are more likely to engage in malicious or negligent actions. The model's adaptive characteristics improve its capacity to accurately represent the ever-changing interactions between human behaviour and organisational contexts. This study offers significant insights for companies aiming to improve their security frameworks by identifying crucial triggers and action points. Future efforts involve optimising model characteristics and incorporating sophisticated psychological models to improve the accuracy and practicality of predictions in real-life scenarios.

Keywords: Computational analysis · Adaptive · Insider Threats · Cybersecurity · Psychological Traits

1 Introduction

In today's fast-paced, interconnected world, the protection of data is becoming increasingly difficult for governments, enterprises, and organisations. External threats are the primary concern of many people, however, it is becoming increasingly evident that a

D. Singh et al. (Eds.): IHCI 2024, LNCS 15557, pp. 396–410, 2025.
https://doi.org/10.1007/978-3-031-88705-5_33

genuine danger is to be found as well within an organisation. The 2023 Insider Threat Report indicates that insider threats have increased in frequency by 74% compared to the previous years [1]. It conducted a survey of 326 cybersecurity professionals to show that insider threats are a trend that is increasing in severity [1]. An insider threat can be defined as any employee who now or in the past had legitimate access to a company's network, system, or data and who, whether on purpose or by accident, exceeds or misuses that access in a way that compromises the security, privacy, or availability of that company's information or systems [2]. The unfortunate truth is that the quantity of data available on insider threats is limited, however, they have attracted significant attention, as some consider them to be the most severe security issue [3]. The very few high-profile cases of insider threats, such as Edward Snowden, a former computer engineer at the CIA and high-level system administrator at the NSA, who leaked an undisclosed number of digital documents, or Herman Simm, head of the Estonian National Security Authority in the Ministry of Defence (MoD), who disclosed national and international classified information to a foreign government, support this point [4].

In addition to being a significant issue, it also causes the majority of cybersecurity analysts and companies to feel extremely concerned, and a mass majority feels extremely vulnerable in the face of this issue [1]. Detecting an insider threat in a timely manner is a challenging task. Users who are authorised to access sensitive information are legitimate. Insiders can effortlessly abuse their privileges, which enables them to remain undetected until it is too late [5]. In the present day, all organisations implement numerous technical measures to mitigate the risk of a potential breach. However, this may not be sufficient to protect against an insider threat.

Insider threats involve a dynamic and adaptive interplay of many factors that can be described as an adaptive dynamical system. As such a system could be modelled and analysed computationally in a network-oriented format [8], this paper introduces an adaptive network model of the evolution of an insider threat over time. In addition to technical protocols, this model also includes psychological traits such as narcissism, impulsivity, a sense of superiority and decision-making processes, which are triggered by external factors. The created adaptive network model employs first-order and second-order learning associated with the decision-making process and the organisation's capacity to promptly identify a breach. By employing sophisticated modelling approaches and leveraging insights from psychology and behavioural sciences, this research aims to enhance understanding of the behavioural patterns and motivations behind insider threats.

By conducting various simulation experiments based on the implemented model, this paper aims to deepen the understanding of the correlation between psychological and behavioural traits that may lead an employee to become a potential threat. It explores how these processes, when combined with robust technical methods for detecting unusual behaviour, can enhance outcomes and potentially prevent breaches altogether.

2 Background Knowledge and Scenario

We developed an illustrative scenario that played an important role as the basis for deriving valuable conclusions about our adaptive network model. It follows several phases in order to better illustrate the importance of each part. The following is the proposed scenario:

Stage 1: Background Employee X is a software developer for a big technology company. X has worked with the organisation for several years, so X has access to a large amount of sensitive information, such as source code and client personal data. X is quite satisfied with their job, they receive positive feedback, have a good reputation among colleagues and managers, and maintain frequent communication with the team. X enjoys being praised and receiving attention. They actively participate in meetings and collaborations. Employee X exhibits indications of dedication and sociability in their psychological traits, with minimal indications of narcissism currently being observed.

Stage 2: Change in Behaviour As time passed, organisational changes (OC) occurred. The workload has increased, and although Employee X works twice as hard as before, everyone is overwhelmed, and there is an aggravating lack of recognition. This serves as the initial stimulus that affects Employee X's sensory representation (SR) of the work environment, leading to feelings of being unappreciated and neglected. X starts to isolate and detach themselves from their colleagues, socialising with the team becomes increasingly less constant, and they begin missing small deadlines. Psychological traits such as lack of self-control and immaturity become more apparent. Withdrawal symptoms get worse, decreasing communication frequency and work performance greatly.

Stage 3: Communication with External Parties Feeling alone and not receiving the recognition they believe they deserve, Employee X seeks validation elsewhere, looking at choices outside of the organisation. Employee X starts communicating with outside parties, adjusting connection weights to focus more on external stimuli. Psychological traits like narcissism and impulsivity become more obvious and start influencing interactions [10]. External Engagement increases, driven by anti-social behaviour on the job and a desire for affirmation.

Stage 4: The Breach As they become more deeply involved, Employee X's communication with external parties intensifies. Employee X starts accessing and downloading sensitive information from the company's servers, using their insider privileges to go unnoticed. Psychological traits such as paranoia may encourage hiding and evasion of detection methods. Employee X targets sensitive information, such as source code and proprietary algorithms. X attempts to access code segments outside their tasks, triggering alerts of unauthorised access. So far, Employee X has still managed to go unnoticed, but once they started downloading numerous files using external devices, the company started becoming suspicious. Detection mechanisms are put in place, the SIEM system depicts logs and network traffic of unusual patterns, SOC monitors real-time alerts, being aware of anomalies, and colleagues and supervisors start reporting unusual behaviour coming from Employee X.

The above scenario was created based on knowledge of several psychological and social processes, as well as technical cybersecurity knowledge. Although technical controls to prevent undesirable behaviour are becoming more comprehensive, insider threats can still go undetected [6]. For example, Ronald Pelton, a former NSA analyst who was aware of these controls, avoided detection by memorising and writing down information [7]. This model focuses on the events that occur before the breach, aiming to better understand the actions that lead to this threat.

The model was built upon the basic concept of the causal mechanisms addressed in [8]. Employee X's initial response to organisational changes and lack of recognition can be better understood through this concept of decision-making and human mental processes. The organisational changes act as stimuli that affect Employee X's sensory representation, triggering negative emotions and behavioural changes.

Insider threats share common characteristics across different scenarios. For instance, Michael Peri, who pleaded guilty to espionage, admitted that his actions were impulsive. His drastic decision was driven by his emotions, he felt overworked and unappreciated in his Army position, which led him to defect to East Germany. Issues with social skills and decision-making often result in personal needs that can drive harmful actions [9]. This highlights that while strong detection mechanisms are crucial, mental health also plays a significant role in these cases. High levels of stress resulting from outside factors, whether they be personal or work-related, frequently precede insider attacks [9].

Certain psychological features are consistently present in such circumstances. Dark Triad Personality Traits are associated with unpleasant personalities. Narcissism, dishonesty, aggression, and a sense of superiority are all common qualities [10]. Narcissism is a subclinical version of clinical personality disorder that retains the clinical characteristics of grandiosity, entitlement, domination, and superiority [11]. Individuals with narcissism have an exaggerated self-esteem and concentrate mostly on themselves [11]. A recent study has demonstrated that Dark Triad qualities are quite useful in predicting workplace behaviour [13]. Narcissistic qualities and a lack of empathy have been demonstrated to increase the danger of an insider threat [14, 15].

Similarly, it applies in this scenario. This model aims to show that co-workers can play an important role in the detection of insider threats. They are the first to observe these traits. In a normal environment, these traits can go unnoticed in the beginning. Once the individual gets triggered, so do these traits [16, 17]. The scenario highlights the importance of emotional regulation, on the grounds that its absence triggers individuals like Employee X to resort to malicious behaviour. Employee X initially experiences positive emotions due to praise and attention. Those being their main source of satisfaction as an employee, the shift to negative emotions happens easily when organisational changes take place. Emotions play a crucial role in decision-making processes, affecting actions and interactions. The inability to regulate emotions causes Employee X and individuals alike to withdraw within themselves and let the negative perspective take control of them [16]. Faced with dissatisfaction and feeling neglected, Employee X chooses to take action and turn his ideas into reality. The decision-making process involves evaluating the options based on anticipated outcomes and emotional states. The perceived negative work environment determines Employee X's decision to seek validation externally. Validation and attention are some of the most important needs for Employee X, receiving those from external parties makes X more determined to follow this course of action. This represents the psychological part of this scenario.

The technical part is illustrated by the presence of the SIEM analysis software and SOC team that constantly analyse network activity, file access patterns, and usage of external devices over the whole company [18]. The malicious behaviour of Employee X will be detected by these teams. As in the majority of cases, the detection of malicious activities from within the company is detected only after the employee has already put his plan into motion. The SOC team gets numerous warnings daily, from employees who, for example, try to access their old projects but do not have access anymore. Other examples include curious employees who wish to see how certain things were done, but they are outside of their scope, and many other similar cases. The SIEM analysis system has to filter these warnings, pay close attention only to what presents itself as a serious threat, and notify the SOC team [18]. Therefore, an insider threat is harder to catch in time. Employees have the advantage of using insider privileges to go unnoticed in the beginning, giving them the opportunity to access numerous files and even transfer them before even being caught. In most cases, employees will most certainly leave a path behind that will lead to them being held accountable in the end, but the company can face severe consequences.

3 The Network Modelling Approach Used

Network modelling is a powerful approach used to understand and represent complex systems by illustrating them as networks of interconnected states [9]. This section provides a brief description of the network-modelling approach, emphasising its application in modelling certain outcomes of insiders trying to breach their company.

The network-oriented modelling approach represents a dynamical system as a graph, where nodes represent different states of the modelled scenario. The dynamics of state changes of states R and S are influenced by direct links between nodes, enhanced by other network characteristics for connectivity such as connection weights $\boldsymbol{\omega}_{R,S}$, for aggregation such as combination functions $\mathbf{c}_S$, and for timing such as speed factors $\boldsymbol{\eta}_S$, which together determine the model's behaviour over time [8].

At each point in time t, state S will be impacted by its incoming connections from the respective states. If there is just one state R influencing S, at time t the impact $\mathbf{impact}_{R,S}(t)$ will be the activation value $R(t)$ of R times the weight $\boldsymbol{\omega}_{R,S}$ of the connection [8]. The aggregated impact of more than one state is found using a combination function $\mathbf{c}_S$ that can be observed in Eq. (1):

$$\mathbf{aggimpact}_S(t) = \mathbf{c}_S\left(\mathbf{impact}_{R_1,S}(t), \ldots, \mathbf{impact}_{R_k,S}(t)\right) = \mathbf{c}_S\left(\boldsymbol{\omega}_{R_1,S}X_1(t), \ldots, \boldsymbol{\omega}_{R_k,S}X_k(t)\right) \tag{1}$$

where R_i are the states with connections to state S [8]. To model the dynamics of the states, they are simulated using a difference equation framework applied to each state at every time step, as depicted in Eq. (2) [8]:

$$S(t + \Delta t) = S(t) + \boldsymbol{\eta}_S\left[\mathbf{aggimpact}_S(t) - S(t)\right]\Delta t \tag{2}$$

This equation is the basis of the dedicated software environment developed [8]. The software executes these equations in parallel for all states within the network model developed, so a network of n states models a dynamical system that can be described by n coupled difference equations.

To model adaptive dynamical systems, adaptive temporal-causal networks can be used. In that case some of the network characteristics $\boldsymbol{\omega}_{X,Y}$, $\boldsymbol{\eta}_Y$, $\mathbf{c}_Y$ change over time, which can be modeled using the principle of self-modeling or network reification [8].

In this model, first-order adaptive features are introduced through self-model states like $\mathbf{W}_{X,Y}$, representing adaptive connection weights $\boldsymbol{\omega}_{X,Y}$, which can be learned via Hebbian learning functions. Self-modeling can also be applied in an iterative manner. This enables to include second-order adaptive features such as $\mathbf{H}_{\mathbf{W}_{X,Y}}$, representing adaptive speed factors $\boldsymbol{\eta}_{\mathbf{W}_{X,Y}}$ modelling adaptive learning rates. These self-model states interact with the base-level network characteristics, adaptively shaping the model's behaviour [8]. This modelling format covers all adaptive dynamical systems as any (smooth) adaptive dynamical system can be represented in this format, see [19, 20].

4 An Adaptive Network Model of Insider Threats

In this section, the conceptual design of the network model to investigate insider threats is presented. The goal of this model is to simulate how various factors contribute to the likelihood of an insider threat occurring. This includes factors such as employee satisfaction, communication frequency with the team, work performance, external engagement, cybersecurity measures, and decision-making processes. The architecture of the network model is illustrated in Fig. 1, and the states are described in Tables 1 and 2, respectively.

The adaptive network model introduced consists of 33 nodes and closely picked connections in order to illustrate the evolution in time of a possible insider threat enforced by negative psychological traits and external engagement with other parties interested in the company's confidential data. The model includes not just the decision-making process behind employee behaviours, but also psychological traits that might impact the choice for undesirable actions. As previously stated, there are specific psychological characteristics present in these circumstances [13]. In a favourable environment, they may go unnoticed, nevertheless, if the employee develops a bad view of the organisation, these characteristics may be heightened [6]. This might happen for a variety of reasons, but in this case, the most typical one was modelled: organisational changes [6]. Changes must occur in any organisation as time passes, methods change, budget cuts are required, and businesses must stay up with the trend. Once activated, unfavourable psychological features might influence decision-making processes [9]. In this model, the aim is to monitor and understand what happens that leads to a major insider threat, as well as to get useful insights into spotting suspicious conduct before a breach occurs.

The organisational changes are the external factor that started the evolution of the scenario for this model. For this state, a **steponce** function was used to mark the moment that triggers the employee to end up taking such drastic actions. It directly influences the decision-making process of the employee. This process includes several factors that take part in the decision-making process [8]. Sensory representation gives insight into how the employee perceives the environment. In this model, high values represent a positive environment perceived by the employee, while a low value reveals a negative environment perceived. The preparation state shows if the employee prepares to take positive actions if the values are high or negative when the values are low. The predicted effect illustrates how the employee expects his actions to turn out. When taking the expected action, the employee expects to receive positive reinforcement, which will be illustrated by the high values. When the values for the predicted effect start dropping, it shows how the employee does not receive the expected outcome anymore. The same

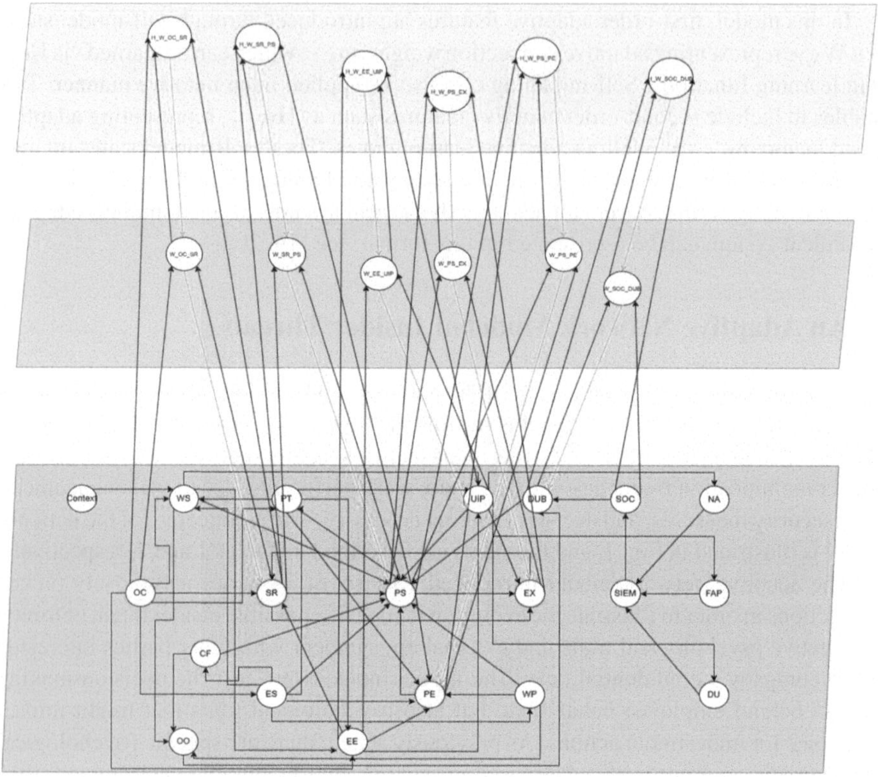

Fig. 1. Conceptual Design of the Adaptive Network Model for Insider Threats

concept was implemented for the execution of actions, states, and psychological traits. High values represent positive actions taken and positive psychological traits, while low values represent negative actions and the presence of negative psychological traits.

As this model is an adaptive network, the most important connections with the highest influence are represented by self-model **W**-states [8]. There are six such states, as shown in Table 2. They represent the first-order self-model level and incorporate Hebbian learning. For each **W**-state, **H**-states were developed to represent the speed factors of the adaptation, and they represent the second-order adaptation level. Such **H**-states can be seen as adaptive learning rates.

Except for the organisational changes state that uses a **steponce** function, and the **W**-states that were incorporated using the function **hebb** for Hebbian learning, all of the other states were assigned the advanced logistic function **alogistic**, see Table 3.

More details of the network model can be found in the Appendix available as Linked Data at https://www.researchgate.net/publication/382975930.

Table 1. Explanation of the base states in the adaptive network model

	State	Explanation	Level
X1	Organizational Changes (OC)	External factors within the organization that initiate the model's evolution	
X2	Sensory Representation (SR)	Provides insight into how the employee perceives the environment. High values represent a positive environment, while low values reveal a negative environment	
X3	Preparation State (PS)	Indicates if the employee prepares to take positive actions (high values) or negative actions (low values)	
X4	Predicted Effect (PE)	Illustrates how the employee expects their actions to turn out. High values indicate positive reinforcement, while low values show the lack of expected outcomes	
X5	Execution of Action (EX)	Represents the actions taken by the employee	
X6	Employee Satisfaction (ES)	Measures the employee's satisfaction level. High values indicate high satisfaction, while low values indicate dissatisfaction	
X7	Communication frequency with the team (CF)	Tracks the frequency of communication with the team	
X8	Work performance (WP)	Assesses the employee's work performance	
X9	Withdrawal signs (WS)	Identifying signs of withdrawal	
X10	External Engagement (EE)	Measures engagement with external parties	Base level
X11	Malicious utilization of internal privileges (UIP)	Tracks the misuse of insider privileges	
X12	Detection of unwanted behaviour (DUB)	Company detects behaviour that deviates from the norm	
X13	Negative Psychological Traits (PT)	Represents negative psychological traits when reaching low values	
X14	Network Activity (NA)	Monitors network activity for anomalies	

(continued)

Table 1. (*continued*)

	State	Explanation	Level
X15	File Access Patterns (FAP)	Analyses patterns in file access	
X16	Device Usage (DU)	Tracks device usage patterns	
X17	SIEM Analysis	Security Information and Event Management analysis for threat detection	
X18	SOC team	Security Operations Center team involvement	
X19	Organizational Observation (OO)	Observations coming from employee's teammates regarding his unusual behaviour	
X20	Context state	Represents contextual information influencing the network model	

Table 2. Explanation of the first- and second-order adaptation states

	State	Explanation	Level
X21	$\mathbf{W}_{OC,SR}$	Represents the adaptive influence of Organizational Changes on Sensory Representation	
X22	$\mathbf{W}_{PS,EX}$	Represents the adaptive influence of Preparation State on Execution of actions	
X23	$\mathbf{W}_{SR,PS}$	Represents the adaptive influence of Sensory Representation on Preparation State	First-order
X24	$\mathbf{W}_{EE,UIP}$	Represents the adaptive influence of External Engagement on the Utilization of Insider Privileges	Reification Level
X25	$\mathbf{W}_{PS,PE}$	Represents the adaptive influence of Psychological State on Predicted Effect	
X26	$\mathbf{W}_{SOC,DUB}$	Represents the adaptive influence of SOC on Detection of Unwanted Behaviour	
X27	$\mathbf{HW}_{OC,SR}$	Represents the adaptive speed factor for $\mathbf{W}_{OC,SR}$	
X28	$\mathbf{HW}_{PS,EX}$	Represents the adaptive speed factor for $\mathbf{W}_{PS,EX}$	Second-order
X29	$\mathbf{HW}_{SR,PS}$	Represents the adaptive speed factor for $\mathbf{W}_{SR,PS}$	Reification
X30	$\mathbf{HW}_{EE,UIP}$	Represents the adaptive speed factor for $\mathbf{W}_{EE,UIP}$	Level
X31	$\mathbf{HW}_{PS,PE}$	Represents the adaptive speed factor for $\mathbf{W}_{PS,PE}$	
X32	$\mathbf{HW}_{SOC,DUB}$	Represents the adaptive speed factor for $\mathbf{W}_{SOC,DUB}$	

Table 3. Description of combination functions used

Name	Notation	Formula	Parameters
Step-once function	$\mathbf{steponce}_{\alpha,\beta}$	1 if $\alpha \leq t \leq \beta$ else 0	p(1) = start time α p(2) = end time β
Advanced logistic	$\mathbf{alogistic}_{\sigma,\tau}(V_1, \ldots, V_k)$	$\left[\frac{1}{1+e^{-\sigma(V_1+\cdots+V_k-\tau)}} - \frac{1}{1+e^{\sigma\tau}}\right](1+e^{-\sigma\tau})$	p(1) = steepness σ p(2) = threshold τ
Hebbian learning	$\mathbf{hebb}_{\mu}(V_1, \ldots, V_k)$	$V_1 V_2(1 - V_3) + \mu V_3$	p(1) = persistence μ

5 Simulation Results

In this section, the base scenario and three variations of it are described. Each simulation evaluates the impact of particular modifications on the organisation's ability to identify and mitigate potential threats and interventions on Employee's X decision making processes. For each simulation, there are two visualisations included. One that displays all states, namely, the base states, first-order reification states, and second-order reification states. The second one includes exclusively the base states. The objective of these visualisations is to provide a thorough comprehension of the manner in which the states interact and develop over time. The simulation scenario presented here demonstrates the impact of organisational changes on Employee X and X's decision-making process. In the beginning, employee satisfaction is high, as is communication frequency with the team and work performance. The sensory representation of the work environment is positive, along with the preparation state, predicted effects of actions, and executed actions. Withdrawal signs are minimal initially, as Employee X is content at work. External engagement is non-existent since Employee X is focused on the company and receives the desired recognition. The malicious utilisation of insider privileges is low, indicating that Employee X is not misusing their access. Psychological traits are positive, as Employee X is initially satisfied with their job. Network activity, file access patterns, and device usage show low values, indicating no abnormal behaviour or suspicious actions from Employee X. Consequently, SIEM analysis and SOC team values are also low in this initial stage, with no red flags from Employee X, and there are no organisational observations from co-workers as Employee X is fulfilling their responsibilities (Figs. 2, 3).

In the initial stage, everything proceeds as expected. Employee X performs the job accordingly, maintains regular communication with the team, and meets all deadlines. The sensory representation of the work environment remains positive, as evidenced by the continuous growth of the sensory representation (SR) state. Furthermore, the constant growth of the preparation state (PS) and predicted effect (PE) shows how Employee X is preparing to take positive actions, such as maintaining frequent communication with the

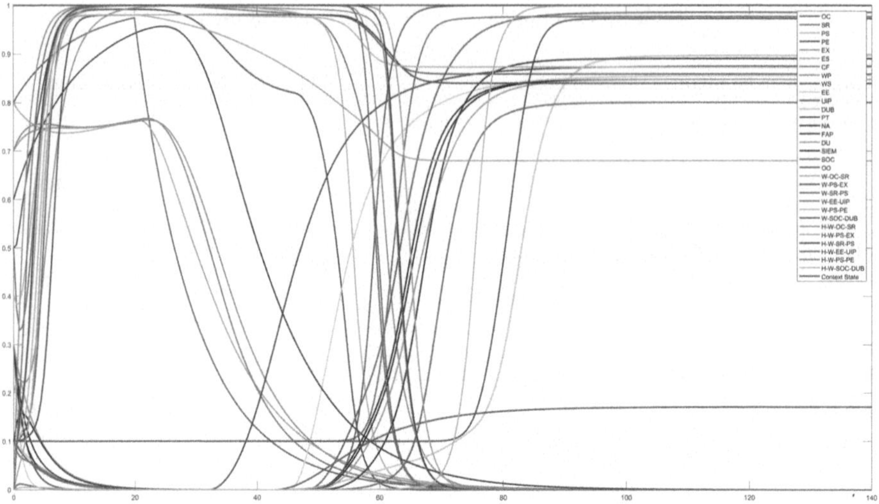

Fig. 2. Graph of the considered simulation scenario including base states, first and second order reification states

team, participating in meetings, and finishing deadlines. The high PE value represents Employee X's expectation of receiving positive feedback and praise for the completed work, which occurs in this stage. This encourages Employee X to continue dedicating to the job, as seen from the constant growth of executed actions.

The second stage is marked by organisational changes that start at $t = 20$. Once these occur, there is a noticeable drop in employee satisfaction (ES), communication frequency (CF), and work performance (WP). Although Employee X no longer receives the usual attention and positive feedback, their perception of the environment remains unchanged at first. They continue to take positive actions, expecting the changes to be temporary. Unfortunately, as the negative environment persists, psychological traits (PT) begin to decline, indicating the emergence of negative traits such as lack of self-control, immaturity, and narcissism. These factors cause Employee X to withdraw, communicate less with the team, and experience a significant drop in work performance, with withdrawal signs (WS) becoming more evident. The predicted effect (PE) of their action changes, leading to confusion about the situation. Isolating themselves from the team drives Employee X to seek external appreciation (Fig. 3).

This marks the third stage, starting at $t = 46$. Employee X starts considering options outside the organisation. As external engagement (EE) with parties interested in the company's secrets increases, Employee X begins contemplating breaching the company. Negative psychological traits become more prominent as PT reaches critical levels. The desire for affirmation and the external validation Employee X receives intensify the external engagement (EE). During this time, Employee X's co-workers began to notice (OO) suspicious behaviour, as indicated by the slight increase in organisational observation (OO). Since the team is unsure of how to respond, they mention it briefly. These organisational observations significantly impact the detection of unusual behaviour (DUB). However, without prompt response from the team, Employee X continues their activities.

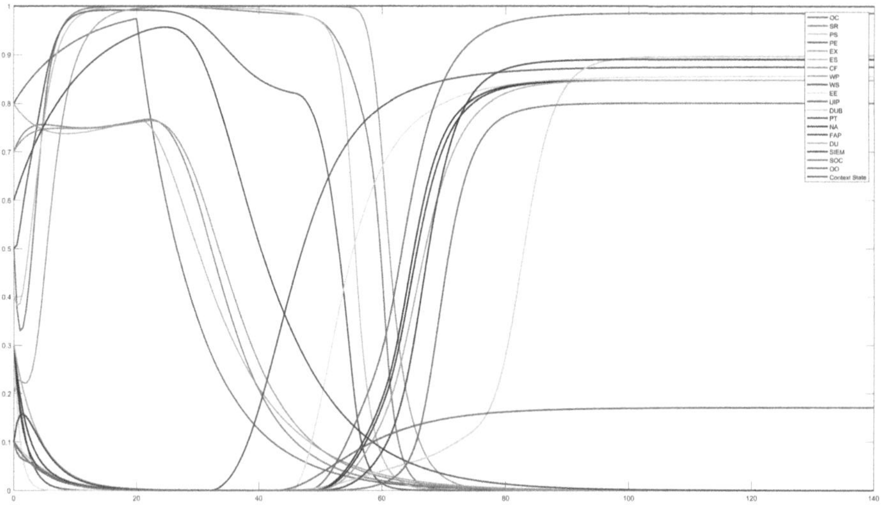

Fig. 3. Graph of the considered simulation scenario only including base states

The final stage is the breach, marked at $t = 50$. Employee X's perception of the company becomes increasingly negative as their external engagement intensifies. The drop in sensory representation reflects this shift. Employee X prepares for negative actions, as seen in the decrease in predicted effect, preparation state, and execution of actions. Employee X starts exploiting their insider privileges (UIP) to access numerous files and uses external devices to download these files. This leads to unusually high network activity (NA) and abnormal file access patterns (FAP). Combined with the use of external devices (DU), these behaviours prompt the SIEM analysis system to pick up on Employee X's activities. Numerous warnings detected by the SIEM system get the attention of the SOC team, leading to the detection of unusual behaviour (DUB) by the company and Employee X's eventual apprehension.

As demonstrated, the detection of unusual behaviour escalates rapidly, but only after Employee X has accessed and downloaded many confidential files, which could already be compromised. Although Employee X is caught and held accountable, the company could face severe consequences.

More simulation scenarios in particular where within the organization earlier detection occurs can be found in the Appendix available as Linked Data at https://www.researchgate.net/publication/382975930.

6 Discussion

The adaptive network model that was developed in this paper offers thorough simulations of the evolution of insider threats within an organisation, with a particular emphasis on the functions of negative psychological traits and external engagement. The model illustrates that organisational changes can serve as initial triggers, resulting in a decrease in employee satisfaction, a decrease in communication frequency, a decrease in work performance, and a negative perception of the environment. These changes have a detrimental impact on the psychological well-being of employees, thereby increasing the probability of malicious behaviour. The simulations have yielded several significant findings, including the significant impact of organisational changes, the significance of negative psychological characteristics such as impulsivity, narcissism, and a lack of self-control, and the impact of increased external engagement on the risk of insider threats. Future work could further explore these psychological characteristics individually to gain deeper insights into their specific contribution to insider threat risks. Furthermore, the model emphasises the significance of the SOC team and coworkers' early identification of strange behaviours in order to mitigate potential threats. The damage caused by insider threats is considerably reduced by improved team responsiveness and direct communication with supervisors, even offering a chance of redemption for employees with behavioural problems. Although direct communication with the SOC team yields promising results in comparison to the base scenario, the team can provide more professional resources, such as notifying the SOC team, when they communicate directly with their supervisors.

One of the main strengths of the model is its comprehensive state representation. It includes a wide range of states covering psychological traits, employee satisfaction, communication patterns, and technical behaviours, providing a holistic view of the factors influencing insider threats. Central to the decision-making process is the Preparation state (PS), which plays a crucial role in shaping employee actions. The Preparation State is influenced by various factors, including sensory representation, predicted effects of actions, employee satisfaction, communication frequency with the team, work performance, withdrawal signs, the negative utilization of insider privileges, detection of unusual behaviour by the company, and psychological traits. This highlights the Preparation States's pivotal role in the model's decision-making framework.

The incorporation of Hebbian learning allows the model to adapt over time, reflecting the dynamic nature of human behaviours and organisational environments. Furthermore, the multiple simulation variants included in the study demonstrate the impact of different interventions, such as improved team responsiveness and opportunities for employee redemption, offering practical insights for organisations.

Future work could involve more dynamic parameter adjustments tailored to each company's case to enhance the model's accuracy and further use. While the model incorporates psychological traits, it simplifies the complexity of human emotions and emotion regulation. Incorporating more nuanced psychological models and comparing different sets of traits could further improve its use. The adaptive network model offers significant potential for improving organisational security by providing a detailed understanding of the factors leading to insider threats. By identifying key triggers and intervention points, organisations can develop more effective strategies to prevent and mitigate these threats. The model highlights the importance of early detection and proactive measures, such as improved team responsiveness and opportunities for employee rehabilitation, in reducing the risk and impact of insider threats.

Therefore, the adaptive network model provides valuable insights into the evolution of insider threats within an organisation. By simulating various scenarios and interventions, the model underscores the importance of understanding psychological traits and external engagements in predicting and preventing malicious activities. Future research and improvements to the model can further enhance its applicability and effectiveness in real-world settings, contributing to more robust organisational security frameworks.

References

1. Schulze, H.: Insider threat report, Cyber-security Insiders, report (2023). https://www.cybersecurity-insiders.com/wp-content/uploads/2023/01/2023_Insider_Threat_Report-16d8d8f7.pdf
2. Cappelli, D.M., Moore, A.P., Trzeciak, R.F.: The CERT Guide to Insider Threats: How to Prevent, Detect, and Respond to Information Technology Crimes (theft, Sabotage, Fraud), Reading, MA. Addison-Wesley, USA (2012)
3. Kim, A., Oh, J., Ryu, J., Lee, K.: A review of insider threat detection approaches with IoT perspective. IEEE Access **8**, 78847–78867 (2020)
4. Kont, M., Pihelgas, M., Wojtkowiak, J., Trinberg, L., Osula, A.-M.: NATO Cooperative Cyber Defence Centre of Excellence: Insider Threat Detection Study. https://afyonluoglu.org/PublicWebFiles/library/ccdcoe/LIB_0023.pdf
5. Homoliak, I., Toffalini, F., Guarnizo, J., Elovici, Y., Ochoa, M.: Insight into insiders and IT. ACM Comput. Surv. **52**(2), 1–40 (2019). https://doi.org/10.1145/3303771
6. Maasberg, M., Warren, J., Beebe, N.L.: The dark side of the insider: detecting the insider threat through examination of dark triad personality traits. In: 48th Hawaii International Conference on System Sciences, Kauai, HI, USA, pp. 3518–3526 (2015). https://doi.org/10.1109/HICSS.2015.423
7. Irvin, J.A., Charney, D.L.: Stopping the Next Snowden. POLITICO Magazine (2014)
8. Treur, J.: Network-Oriented Modeling for Adaptive Networks: Designing Higher-Order Adaptive Biological, Mental and Social Network Models. Springer Nature (2020). https://doi.org/10.1007/978-3-030-31445-3
9. Comparing Insider IT Sabotage and Espionage: A Model-Based Analysis. Retrieved from https://apps.dtic.mil/sti/citations/tr/ADA459911
10. Paulhus, D.L., Williams, K.M.: The dark triad of personality: narcissism, Machiavellianism and psychopathy. J. Res. Pers. **36**(6), 556–563 (2002)
11. Jakobwitz, S., Egan, V.: The dark triad and normal personality traits. Pers. Individ. Differ. **40**(2), 331–339 (2006)
12. Giammarco, E.A., Vernon, P.A.: Vengeance and the dark triad: the role of empathy and perspective taking in trait forgivingness. Pers. Individ. Differ. **67**, 23–29 (2014)
13. Furnham, A., Richards, S., Rangel, L., Jones, D.N.: Measuring malevolence: quantitative issues surrounding the dark triad of personality. Pers. Individ. Differ. **67**, 114–121 (2014)
14. Sarkar, K.R.: Assessing insider threats to information security using technical, behavioral, and organizational measures. Inf. Secur. Tech. Rep. **15**(3), 112–133 (2010)
15. Shaw, E.D., Post, J.M., Ruby, K.G.: Inside the mind of the insider. Secur. Manag. **43**, 34–44 (1999)
16. Posey, C., Bennett, R.J., Roberts, T.L.: Understanding the mindset of the abusive insider: an examination of insiders' causal reasoning following internal security changes. Comput. Secur. **30**(6–7), 486–497 (2011)
17. Nance, K., Marty, R.: Identifying and visualizing the malicious insider threat using bipartite graphs. In: 2011 44th Hawaii International Conference on System Sciences (HICSS), pp. 1–9 (2011)
18. Mughal, A.A.: Building and securing the modern security operations center (SOC). Int. J. Bus. Intell. Big Data Anal. **5**(1), 1–15 (2022)

19. Treur, J.: On the dynamics and adaptivity of mental processes: relating adaptive dynamical systems and self-modeling network models by mathematical analysis. Cogn. Syst. Res. **70**, 93–100 (2021)
20. Hendrikse, S.C.F., Treur, J., Koole, S.L.: Modeling emerging interpersonal synchrony and its related adaptive short-term affiliation and long-term bonding: a second-order multi-adaptive neural agent model. Int. J. Neural Syst. **33**(7), 350038 (2023). https://doi.org/10.1142/S0129065723500387

Organizational Response to APT Attacks: Computational Analysis by Behavioral Network Modelling

Daneel Akers[1], Tanner Blomquist[1], Jan Niedziela[1], Samuel Šebok[1], Niek Jan van den Hout[1], Debby Bouma[1,2], Jan Treur[1,2](✉), and Peter H. M. P. Roelofsma[1]

[1] Centre of Expertise Cybersecurity, Risk Management and Cybersecurity Group, The Hague University of Applied Sciences, The Hague, Netherlands
{N.J.vandenHout,D.Bouma,P.H.M.P.Roelofsma}@hhs.nl, j.treur@vu.nl
[2] Department of Computer Science, Vrije Universiteit Amsterdam, Social AI Group, Amsterdam, Netherlands

Abstract. This paper addresses computational analysis of defence against Advanced Persistent Threats (APTs). The analysis is done by a behavioural network model for simulating the relevant processes, including (1) the APT attack chain itself and adaptivity of it by learning both by the hackers and by the organisation, (2) the understanding of the process by the organisation, and (3) management of the organisational response. Several variations of simulation experiments and a What-If analysis have been applied, leading to risk assessment.

Keywords: APT attack · cyber kill chain · network modelling · adaptive

1 Introduction

The purpose of this paper is to develop a behavioural network model for simulating defence against Advanced Persistent Threats (APTs) on a technical and organisational level. The National Institute of Standards and Technology (Joint Task Force Interagency Working Group, 2020) defines the main characteristics of an APT attacker to be: pursuing its objectives over a long time, adapting to the efforts of the defender, and maintaining the intensity of the attack needed to execute the attack. The goals of an attack of this nature could include stealing confidential data, disturbing operations and modifying critical settings of systems that are in use, all while remaining undetected. These attacks are generally directly carried out by a nation state or sponsored by one and frequently exploit zero-day vulnerabilities. This report will introduce and explain a behavioural network model with the aim of simulating in MATLAB an organizations response to an APT attack as it is detected. Once an attack is detected, the behavioural network model is aimed at correctly understanding the adversarial actions so defender can respond appropriately. This model shows possible APT attack paths in accordance with the Cyber Kill Chain, and the MITRE ATT&CK tactics and techniques as combined by

D. Singh et al. (Eds.): IHCI 2024, LNCS 15557, pp. 411–425, 2025.
https://doi.org/10.1007/978-3-031-88705-5_34

Mat et al. (2024). The Cyber Kill Chain is a model that visualizes the seven steps that a cyber adversary must complete to achieve their goal (Hutchins et al., 2011; Bahrami et al., 2019). MITRE ATT&CK proved useful as it portrays common attack methods and possible exploits per step of the Cyber Kill Chain. Following the design process of the model, multiple methods of analysis were used to showcase its possible application. The first method used was the conceptual what-if analysis to explore multiple variations of the model, followed by a probabilistic what-if analysis. The third and final method of analysis was a What-If analysis and associated risk assessment.

This work is part of a research programme on computational analysis of safety and security in and through cyberspace. In (Abromaityte, Vos, Lucassen, Kruger, Dubero, Van den Hout, Bouma, Treur, Roelofsma, 2025; Börcsök, Bos, Buter, Van Der Hoeven, Velev, Van den Hout, Bouma, Treur, Roelofsma, 2025; Bouma, Den Oudsten, Naffakh, Waliji, Treur, Roelofsma, 2025; Caneva, Oliver, Rietze, Šileris, Barelds, Hoffmans, Treur, Roelofsma, 2025; Capra, Dekker, Hoffmann, Hoffmans, Mueller, Van den Hout, Treur, Roelofsma, 2025; Hoffmans, Van der Does, Van der Kolk, Van Rooyen, Barelds, Treur, Roelofsma, Hoffmans et al., 2025; Jeffery et al, 2025; Keijzer et al., 2025) various other themes within cybersecurity were addressed. Moreover, in (Roelofsma, Jabeen, Taal, Treur, 2025) it was analysed how an AI coach can support safety and security for medical teams in hospitals.

2 Scenario Description and Background Knowledge Used

The scenario explores the progression of attacker and defender decision points throughout the stages of an APT attack. Firstly, the organization is initially exposed to an APT attack, with sensitive data and systems potentially compromised by a malicious actor. Soon after, the APT attack is detected by the organization's internal monitoring system and positively identified as an (APT) attack. After being identified, the organization itself decides on how and in what capacity a response is required. The Cyber Kill Chain, introduced by Hutchins et al. (2011), provides a structured approach to understanding and defending against cyberattacks. This model highlights each stage of a cyberattack, allowing for the identification and mitigation of attacks.

The Cyber Kill Chain is comprised of seven stages, the first of which is reconnaissance, where a threat actor gathers information on a chosen target to discover potential vulnerabilities. Measures to defend against a cyberattack at this stage include the implementation of network monitoring to detect suspicious activities, the deployment well configured firewalls, and the conduction of annual security awareness training to mitigate the risk of information leakage through (spear-) phishing (Hutchins et al., 2011; Vera et al., 2019). Within the weaponization stage, the threat actor combines custom malware with an exploit into a deliverable payload. To defend against this stage, advanced malware tools, regular code reviews and an updated database of threat intelligence to recognize patterns threat actors follow while in the weaponization stage can be deployed to assist in preventing further system intrusions. After the payload has been created, the delivery stage begins. This stage involves the transmission of the payload to the target (Hutchins et al., 2011). This is often accomplished via email attachments, websites or removable media. To counter this stage of the Cyber Kill Chain and limit delivery opportunities, defenders can implement measures in the form of stricter email security protocols, proxy filters, intrusion prevention systems and web content filtering and USB device usage policies (Hutchins et al., 2011).

The exploitation stage, the most critical point within the Cyber Kill Chain, involves the execution of the malware through the exploitation of previously gathered vulnerabilities. To defend against this stage, systems and applications must be maintained with the latest patches and updates and regular malware scanning and vulnerability assessments must be conducted. Once the malware has successfully installed a backdoor or persistent access method within the installation stage, further countermeasures can be deployed. This can be in the form of endpoint detection and response (EDR) techniques which are trained to identify and block unusual activities in conjunction with application whitelisting and endpoint protection measures (Hutchins et al., 2011).

During the command-and-control stage, the threat actor establishes communication channels to remotely control the targeted system. Defence measures that can assist in protecting the targeted network at this point in the Cyber Kill Chain include the utilization of DNS filtering to prevent malicious domains from being accessed. Additionally, continuous monitoring of network traffic for anomalies, the deployment of security information & event management (SIEM) systems and the presence of a dedicated Security Operations Center (SOC) to group, analyse and react to suspicious activity can also assist in defence. Within the final stage of the Cyber Kill Chain, action on objectives, a threat actor achieves their goals. These goals could include the exfiltration of sensitive data or the erasure of critical system functions. Methods of defence that can be employed in this final stage include data loss protection (DLP) tools that safeguard against unauthorised data transfers, file integrity monitoring and the maintenance of timely incident response capabilities to contain and mitigate damage to the system (Hutchins et al., 2011).

To successfully prepare and defend against cyberattacks through the Cyber Kill Chain framework, numerous tactics can be applied. After a breach, reconstructing an intrusion can be very beneficial for a multitude of reasons. It can help the defenders map out the entire attack sequence, including the entry and progression points to determine which cybersecurity measures the organization needs to bolster. It can also assist in uncovering hidden attributes and behaviours of the attack that may be overlooked when focusing on a single phase of the Cyber Kill Chain. Moreover, another tactic involves campaign analysis, through which commonalities and common indicators of compromise (IoCs) are identified. Analysts can use this to compile a system intruder's tactics, technique and procedures (TTP) to gain an idea of how they operate. Once this is understood, an intruder's capabilities, objectives and limitations can then be theorized to enable the correct identification of an intruder's strategy, leading to the correct response by the organization itself (Vera et al., 2019 & Mat, et al., 2024). These tactics are retrospective methods of defence, meaning that prior attempts or successful breach's must have occurred in the past. This could be over the span of multiple years or within a few months.

3 The Network Modelling Approach Used

For the creation of the temporal-causal network model introduced in this paper, the modelling approach used is briefly outlined. Network nodes, also called states, are defined as entities that act as intermediaries that process inputs from other states and generate outputs to other states. A state Y has activation values $Y(t)$ over time t. These states show specific actions and contextual factors in the process of an APT attack that span from the reconnaissance phase to the command-and-control phase as seen in the scenario (Treur, 2020). The interactions between these states are visualized by connections, with the characteristics of these states and their connections being defined by the network designer. These network characteristics define the relationships and interactions between the states within the network and facilitate the dynamics of the model (Treur, 2020):

- *Connectivity characteristics*: connections from state X to state Y to indicate the causal impact $\boldsymbol{\omega}_{X,Y}X(t)$ of X on Y based on their connection weights $\boldsymbol{\omega}_{X,Y}$
- *Aggregation characteristics*: combination functions $\mathbf{c}_Y$ (with some parameters) to aggregate multiple impacts on a given state Y from the incoming connections
- *Timing characteristics*: speed factors $\boldsymbol{\eta}_Y$ specifying how fast a state Y changes upon a given aggregated impact

Initial values represent the starting position or condition of a state at the beginning of the simulation (Treur, 2020). These values remain constant unless explicitly changed, are declared before the system call that executes the system dynamics and are essential for maintaining consistent data throughout the simulation. Based on these network characteristics, the dynamics of the states in the network is defined by the following canonical differential or difference equation for each state Y:

$$dY(t)/dt = \boldsymbol{\eta}_Y\left[\mathbf{c}_Y\left(\boldsymbol{\omega}_{X_1,Y}X_1(t), \ldots, \boldsymbol{\omega}_{X_k,Y}X_k(t)\right) - Y(t)\right]$$

$$Y(t + \Delta t) = Y(t) + \boldsymbol{\eta}_Y\left[\mathbf{c}_Y\left(\boldsymbol{\omega}_{X_1,Y}X_1(t), \ldots, \boldsymbol{\omega}_{X_k,Y}X_k(t)\right) - Y(t)\right]\Delta t$$

In this way a network model with n states defines a dynamical system that can also be described by a system of n coupled differential equations of the above form. In (Treur, 2021) and (Hendrikse, Treur, Koole, 2023) it is shown in detail how any (smooth) dynamical system can be represented as such a temporal-causal network.

An often-used combination function is the function **alogistic**. The key parameters of this function are the excitability threshold $\boldsymbol{\tau}$, showcasing the growth rate change and the steepness $\boldsymbol{\sigma}$, which is how fast the function transfers its value from its lower bound to its upper bound. It is defined as follows:

$$\mathbf{alogistic}_{\sigma,\tau}(V_1, \ldots, V_k)\left[\frac{1}{1 + e^{-\sigma(V_1+\ldots+V_k-\tau)}} - \frac{1}{1 + e^{\sigma\tau}}\right)](1 + e^{-\sigma\tau})$$

Another useful function is the **steponce** function with parameters $\boldsymbol{\alpha}$ for starting time and $\boldsymbol{\beta}$ for end time. This is utilized to model dynamically changing external conditions; it is defined as (where t indicates time).

$$\mathbf{steponce}_{\alpha,\beta}(V_1, \ldots, V_k) = 1\ \textit{if}\ \alpha \le t \le \beta \text{ else } 0$$

The **steponce** function ensures that each state maintains a constant rate during the indicated time interval (which can as well be the entire simulation) and can model fluctuating contextual factors.

To model a network model's adaptivity, some of the network characteristics such as connection weights, speed factor and combination function can be designed to change over time, highlighting the flow of dynamics and adaptivity. To also address such adaptivity, a network model can utilize self-model states for the adaptive network characteristics, thus obtaining a self-modeling network (Treur, 2020). Three often used examples are:

- for connectivity characteristics a self-model state $\mathbf{W}_{X,Y}$ (called a **W**-state) can be added representing an adaptive connection weight $\boldsymbol{\omega}_{X,Y}$
- for aggregation characteristics for the combination function **alogistic** a self-model state $\mathbf{T}_Y$ (called a **T**-state) can be added representing an adaptive excitability threshold $\boldsymbol{\tau}_Y$.
- for timing characteristics a self-model state $\mathbf{H}_Y$ (called an **H**-state) can be added representing an adaptive speed factor $\boldsymbol{\eta}_Y$.

This can also be iterated to obtain multiple orders of adaptivity. In (Treur, 2021) and (Hendrikse, Treur, Koole, 2023) it is also shown in detail how any (smooth) adaptive dynamical system (of any order of adaptivity) can be represented as a self-modeling network. In the adaptive network model introduced here, **T**-states will be used to model adaptation.

4 Design of the Adaptive Network Model

Figures 1 to 3 show the conceptual picture of the designed network model and its parts. It roughly consists of three separate sections that aim to simulate the process of an APT attack and the organizational understanding and response to it. The first section of the model (Fig. 1) aims at simulating the APT attack and its progression through the phases of the Cyber Kill Chain.

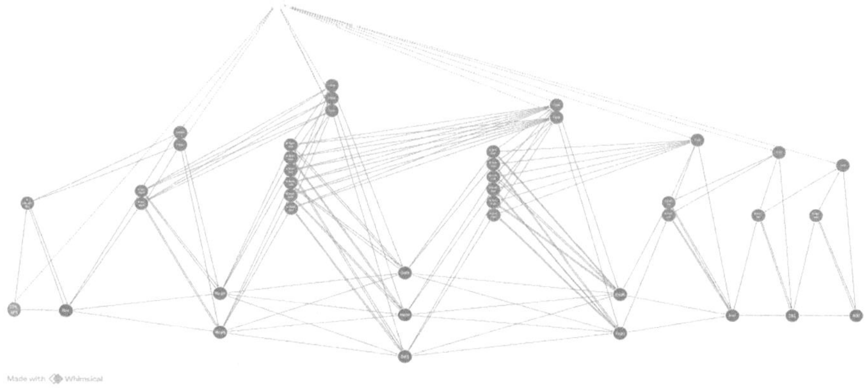

Fig. 1. The APT Attack: Sect. 1 of the network model

The different phases of the Cyber Kill Chain, from reconnaissance (Rec) all the way to action on objectives (AOO), can be seen as base level (pink) nodes. The grey node is the contextual state allowing for the APT attack to begin. The base layer nodes branch out to showcase potential methods of attack to be utilized by the attackers as outlined in (Mat et al., 2024). The adaptation level (blue nodes) consists of two types of nodes. The first one being the **W**-states (representing connection weights) and the second being the **T**-states (representing excitability thresholds) of the base level model (pink states). The thresholds are affected in a downward direction by the attackers as they progress through the Kill Chain due to them learning how to exploit present vulnerabilities (thus decreasing the thresholds) in the process of the attack. In contrast, the organization tries to increase the thresholds, representing patching of a vulnerability, influenced by the organisational understanding of the attack further.

Organisational understanding is represented in the second section of the model showcased in Fig. 2. It simulates the organisational understanding, or awareness, of what methods and tactics the attackers use in the first section of the model (Fig. 1). The base level (pink) nodes represent the prediction that the organisation makes about the potential methods of attack. The first order adaptation level (blue) nodes of this section consists of threshold **T**-states, which represent the degree to which the organisation understands their own vulnerabilities. This understanding influences the organisational prediction of the attack in the base level based on the values of the thresholds. The states in this section have two types of connections to Sect. 1 of the network model that provide inputs for the organisational understanding. The first are the connections from each state of the base level of Sect. 1 to the corresponding state in the base level of Sect. 2.

Fig. 2. Organizational Understanding of the APT Attack**:** Sect. 2 of the network model

This represents correlational system alerts which provide an indication of which method is being used by the attacker. The second group of connections are from the **T**-states representing the adaptive thresholds in the first section to the adaptive thresholds in the second section. These represent the organisation's ability to understand its vulnerabilities. The resulting organisational understanding of the APT threat will then be reflected in the values of the base states in this section. These states are then used to trigger the organisational response in Sect. 3 of the network model.

The third section of the network model (Fig. 3) models the organizational response to the attack. The base level (pink nodes) layout is simple, consisting of a contextual state (grey node) that activates the organisational response to each of the attack methods in the kill chain in Sect. 1. This response is controlled by two types of states in the adaptation level. The first type is the adaptive weight of the organisational response, which is increased or decreased based on the organisational understanding in Sect. 2 of the model, more specifically, the base states. The second type is the adaptive speed of the organisational response, which aims to gradually activate the organisational response when the APT attack reaches the corresponding state of the Kill Chain. This prevents the organisation from suppressing the attack progress where it has not happened yet, which would not be realistic behaviour. Finally, Fig. 4 in the Appendix depicts the overall network model and how the different sections connect to each other.

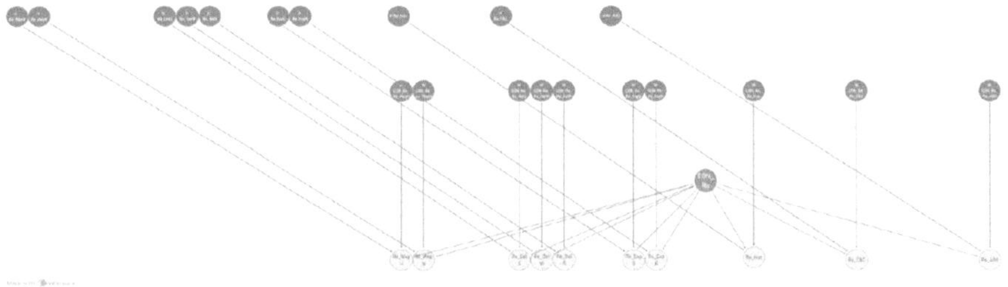

Fig. 3. Organizational Response: Sect. 3 of the network model

Tables 1 and 2 outline the states used in the network model. It includes the states with their corresponding order numbers within the model, as well as the corresponding abbreviation assigned to them. The layer and group of each of the states can be seen on the right side of the table. States X2 through X12 represent the steps and methods utilized by the attackers during the APT attack as outlined by (Mat et al., 2024) in the Cyber Kill Chain. Within these states, states X3 and X4 outline potential methods of attack for weaponization stage, while states X5, X6, X7 showcase the potential methods for the delivery, and X8 and X9 for the exploitation. States X13 to X41 are the adaptive thresholds and weights of the base layer of states X1 through X12 for the first section of the model. For the second section, states X42 through X52 are the base layer with their corresponding adaptive thresholds marked in the table as X53 through X62.

The third section of the model related to management of the organizational response includes states X63 through X83 with states X63 through X73 being the base layer and X74 through X83 being the first order adaptive layer with corresponding adaptive weights. The last grouping of states X84 through X93 are the first order adaptive states regulating the speed of the organizational response. The detailed description of their connections and the reasoning behind the weights of them is described in the Appendix available as Linked Data at

https://www.researchgate.net/publication/382061930.

Table 1. States and their descriptions: Sect. 1 of the adaptive network model

Number	Name	Description	Level	Section
X1	CON_APT	Contextual state to start the APT Attack	Base level	Section 1: APT Attack
X2	Rec	Reconnaissance		
X3	WepN	Weaponization Network-based		
X4	WepH	Weaponization Host-based		
X5	DelS	Delivery Spear Phishing		
X6	DelW	Delivery Watering Hole		
X7	DelR	Delivery Rogue Software		
X8	Exp0	Exploitation 0-Day		
X9	ExpK	Exploitation Known		
X10	Inst	Installation (modify registry key)		
X11	C&C	Command & Control		
X12	AOO	Action on Objectives (data exfil)		
X13-X41	W- and T-states	Representing Adaptive Weights and Thresholds		

5 Simulation Experiments

In this section, initial outcomes of simulation experiments of the model will be examined. Figure 4 portrays the outcome of an example simulation. For detailed analysis, the following figures examine smaller groupings of states to make the graphs more legible. The values of the network characteristics were then tuned to achieve a rough behaviour of the model expected in the conceptual design phase. Several considerations guided the selection of these values, for example:

(1) The organisational prediction must be calculated faster than the time required for the actual APT attack to happen, otherwise it is not a prediction that can guide a response.
(2) The organisational response, if correctly understanding the attack, must be able to suppress it if reaching maximum, or in other words, the power of the attackers and defenders needs to be roughly balanced.
(3) Lastly, the attackers should not be able to decrease a threshold which the organisation is capable of monitoring timely and correctly but should be able to decrease a threshold where the organisational assumption about its value is wrong. This would force the attackers to choose the most optimal attack path by being able to only decrease the thresholds which are not sufficiently monitored and maintained by the organisation.

Table 2. States and their descriptions: Sects. 2 and 3 of the adaptive network model

Number	Name	Description	Level	Section
X42	Rec_S	Reconnaissance Simulation	Base level	Section 2: Organisational Understanding Simulation
X43	WepN_S	Weaponization Network-based Simulation		
X44	WepH_S	Weaponization Host-based Simulation		
X45	DelS_S	Delivery Spear Phishing Simulation		
X46	DelW_S	Delivery Watering Hole Simulation		
X47	DelR_S	Delivery Rogue Software Simulation		
X48	Exp0_S	Exploitation 0-Day Simulation		
X49	ExpK_S	Exploitation Know Simulation		
X50	Inst_S	Installation Simulation		
X51	C&C_S	Command & Control Simulation		
X52	AOO_S	Action on Objectives Simulation		
X53-X62	**W**- and **T**-states	Representing Adaptive Weights and Thresholds		
X63	CON_Re	Start Response	Base level	Section 3: Organisational Response
X64	Re_Wep H	Response to Wep H		
X65	Re_Wep N	Response to Wep N		
X66	Re_Del S	Response to Del S		
X67	Re_Del W	Response to Del W		
X68	Re_Del R	Response to Del R		
X69	Re_Exp 0	Response to Exp 0		
X70	Re_Exp K	Response to Exp K		
X71	Re_Inst	Response to Inst		
X72	Re_C&C	Response to C&C		

(*continued*)

Table 2. (*continued*)

Number	Name	Description	Level	Section
X73	Re_AOO	Response to AOO		
X74-X93	**W**- and **T**-states	Representing Adaptive Weights and Thresholds		

The resulting values in the matrix tables for the simulation are an outcome of the trial and error method aiming to achieve the three considerations described in the paragraph above. Therefore, the current version of the model is not fully optimised. Partial success was achieved when optimising the model for considerations 1 and 2, but not yet for consideration 3. Therefore, other variations are studied in addition. The outcome of this process, the detailed role matrices, can be found in the Appendix mentioned above due to their length.

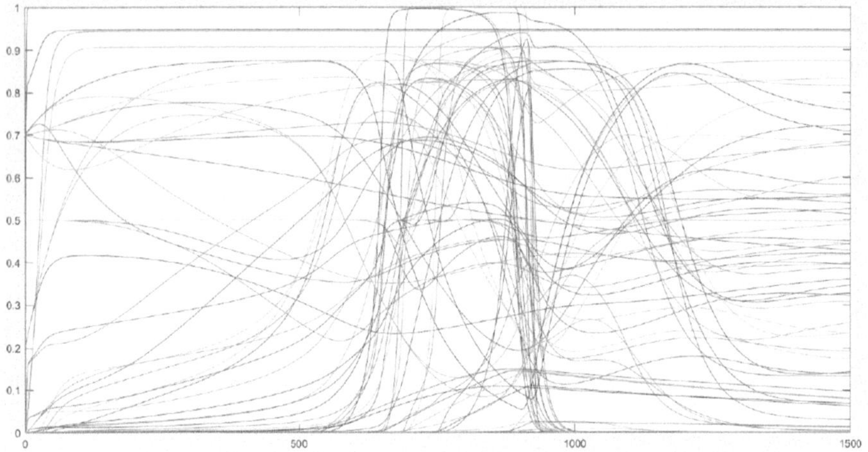

Fig. 4. Outcome of the example network simulation

6 What-If Analysis for Variations of the Scenario

A What-If analysis enables a systematic examination of possible scenarios by altering characteristics like contextual stimuli or model characteristics. This approach allows for observing the impact of changes in characteristics, such as speed factors, on outcomes for certain states. Conducting a thorough evaluation of these variations enhances insights into the model's reliability and sensitivity. These assessments not only deepen the understanding of the APT attacks for organizations, but also offer guidance for organizations on how to respond to such attacks by examining the vulnerabilities and patching them.

Four 'If factors' were selected for the analysis. The first If factor to be changed is the speed factor of the first-order adaptive thresholds of weaponisation network based ($\mathbf{T}_{WepN}$) and weaponisation host based ($\mathbf{T}_{WepH}$). These affect how fast the thresholds representing system vulnerabilities can be affected by both attackers and defenders. The second If factor is the speed factor of the adaptive weight of the connections between the weaponisation stage of the attack to the delivery stage of the APT ($\mathbf{W}_{WepH, DelR}$, $\mathbf{W}_{WepH, DelS}$). These affect the speed by which the attackers progress to the following stage of the APT attack. The first and second If factor affect multiple characteristics to have a symmetrically effect for different attack techniques for the affected step of the kill-chain. The third and fourth If factors are the speed factors of the simulated vulnerability thresholds within the organisational understanding for weaponisation network and host based ($\mathbf{T}_{WepN\_S}$, $\mathbf{T}_{WepH\_S}$). These represent the speed by which the organisation monitors and adjusts their predictions (organisational understanding) to the real-world system vulnerabilities exploited by the attackers. The If factor options are values lower and higher than the original used in the model, together with the probabilities of the If options and their underlying assumptions (See Table 3) factor combinations and the outcomes of this conceptual What-If analysis and the graphs resulting from the simulations can be seen in the Appendix available as Linked Data at

https://www.researchgate.net/publication/382061930.

Table 3. What-If analysis and Risk Assessment

If factor 1 options	If factor 2 options	If factor 3 options	If factor 4 options	If factor option combination probabilities	Outcome 1 (Attackers achieved AOO)	Outcome 2 (Attackers achieved C&C, not AOO)	Outcome 3 (Attackers did not achieve AOO or C&C)
If factor 1 option 1: Low 0.4	If factor 2 option 1: low 0.7	If factor 3 option 1: low 0.6	If factor 4 option 1: low 0.55	0.4 * 0.7 * 0.6 * 0.55 = 0.0924	X 0.0924		
			If factor 4 option 2: high 0.45	0.4 * 0.7 * 0.6 * 0.45 = 0.0756			X 0.0756
		If factor 3 option 2: high 0.4	If factor 4 option 1: low 0.55	0.4 * 0.7 * 0.4 * 0.55 = 0.0616	X 0.0616		
			If factor 4 option 2: high 0.45	0.4 * 0.7 * 0.4 * 0.45 = 0.0504			X 0.0504
	If factor 2 option 2: high 0.3	If factor 3 option 1: low 0.6	If factor 4 option 1: low 0.55	0.4 * 0.3 * 0.6 * 0.55 = 0.0396	X 0.0396		
			If factor 4 option 2: high 0.45	0.4 * 0.3 * 0.6 * 0.45 = 0.0324			X 0.0324

(continued)

Table 3. (*continued*)

If factor 1 options	If factor 2 options	If factor 3 options	If factor 4 options	If factor option combination probabilities	Outcome 1 (Attackers achieved AOO)	Outcome 2 (Attackers achieved C&C, not AOO)	Outcome 3 (Attackers did not achieve AOO or C&C)
		If factor 3 option 2: high 0.4	If factor 4 option 1: low 0.55	0.4 * 0.3 * 0.4 * 0.55 = 0.0264	X 0.0264		
			If factor 4 option 2: high 0.45	0.4 * 0.3 * 0.4 * 0.45 = 0.0216			X 0.0216
If factor 1 option 2: High 0.6	If factor 2 option 1: low 0.7	If factor 3 option 1: low 0.6	If factor 4 option 1: low 0.55	0.6 * 0.7 * 0.6 * 0.55 = 0.1386	X 0.1386		
			If factor 4 option 2: high 0.45	0.6 * 0.7 * 0.6 * 0.45 = 0.1134			X 0.1134
		If factor 3 option 2: high 0.4	If factor 4 option 1: low 0.55	0.6 * 0.7 * 0.4 * 0.55 = 0.0924	X 0.0924		
			If factor 4 option 2: high 0.45	0.6 * 0.7 * 0.4 * 0.45 = 0.0672			X 0.0672
	If factor 2 option 2: high 0.3	If factor 3 option 1: low 0.6	If factor 4 option 1: low 0.55	0.6 * 0.3 * 0.6 * 0.55 = 0.0594	X 0.0594		
			If factor 4 option 2: high 0.45	0.6 * 0.3 * 0.6 * 0.45 = 0.0486		X 0.0486	
		If factor 3 option 2: high 0.4	If factor 4 option 1: low 0.55	0.6 * 0.3 * 0.4 * 0.55 = 0.0396	X 0.0396		
			If factor 4 option 2: high 0.45	0.6 * 0.3 * 0.4 * 0.45 = 0.0324			X 0.0324
				Sum of probabilities:	0.55 (Medium)	0.0486 (Low)	0.4014 (Medium)
				Impact indicator:	9 (High)	5 (Medium)	2 (Low)
				Risk assessment:	4.95 (Medium)	0.243 (Low)	0.8028 (Low)

The remainder of this section focuses on the risk assessment based on the What-If analysis process which is summarized in Table 3. The probabilities used for the If factor options were estimated as shown in the first four columns of Table 3 and the probabilities for their 'and' combinations were determined by the product rule, assuming independence (see the fourth column). The states in our model that are used to estimate the impact (the What factors) are X11 C&C and X12 AOO. These are the steps of the kill-chain which represent attackers taking control of the system and achieving their objectives. If the C&C state achieves values above 0, it would create limited impact on the regular functioning of the attacked system. Therefore, the assumed impact indicator for this is 5 (medium). If the AOO state achieves values above 0, it would mean that the attackers were successful with their attack. The assumed impact indicator for this scenario is 9 (high). In the case neither C&C or AOO were not achieved, the impact on the regular functioning of the attacked system would be minimal, assumed to be 2 (low). Based on this, the last three columns of Table 3 for the What outcomes were created.

7 Discussion and Conclusion

This paper proposed a method of modelling a behavioural network of an APT attack and a response of an organisation based on their understanding of an attack. This computational network model was used to conduct What-If analysis and risk assessment. The risk assessment reveals a medium risk of the attackers reaching the Action on Objectives (AOO). The risks resulting from Command and Control (C&C) and a failed attack is low. When looking for a determinant of whether AOO is present or not present, it can be seen that its presence is fully correlated with If factor 4, Option 1 (the correction speed of the correct organizational understanding of the vulnerability is slower). This suggests that even if an organisation has a correct initial assumption about the level of its vulnerability, it needs to maintain this correct understanding over time. Not only is monitoring of existing vulnerabilities important, monitoring of the system elements not perceived as likely to be threatened can result in a lag of organisational understanding of a developing APT attack. However, these results are limited by the assumptions made for the model.

Further improvement of the model may focus on model behaviour where the attackers are unable to decrease a well-maintained threshold but decrease a threshold which the organisation does not monitor correctly. The resulting model would be capable of modelling an organisational reaction to APT attacks using different methods and tactics. Further optimisation could be done using the Global Optimisation Toolbox from Math Works. The current model and the results of the What-If analysis still reveal important considerations of modelling organisational vulnerabilities to APT attack. It shows that it is not enough to model a vulnerability as a low threshold for the progress of the attackers. Whether an organisation is vulnerable depends not only on a technical vulnerability of a system, but also on the speed of the organisation in detecting it, adjusting to it, and their capacity to respond to it. For example, the outcome of the risk assessment indicated the role of insufficient monitoring of non-vulnerable parts of the system, while untimely monitoring of the vulnerable parts had a lesser effect. Future models may consider this multifaceted nature of vulnerabilities in more detail. This also applies to an organisation exposed to APT attacks. A successful defence of such an attack must involve an interplay between all the above elements, which must be insured by a functioning information security management system, such as ISO/IEC 27001.

References

Abromaityte, M., et al.: Computational analysis of user experience of password-based authentication systems. In: Proceedings of the 16th International Conference on Intelligent Human Computer Interaction, IHCI'24. Lecture Notes in Computer Science, Springer Nature (2025)

Bahrami, P.N., Dehghantanha, A., Dargahi, T., Parizi, R.M., Choo, K.K.R., Javidi, H.H.S.: Cyber kill chain-based taxonomy of advanced persistent threat actors: analogy of tactics, techniques, and procedures. J. Inf. Process. Syst. **15**(4) (2019)

Börcsök, T., et al.: Adaptive defence mechanisms: an adaptive network model to protect HiX systems against cyber attacks. In: Proceedings of the 8th International Conference on Computational Methods in Systems and Software, CoMeSySo'24. Lecture Notes in Networks and Systems, Springer Nature (2025)

Bouma, D., Den Oudsten, T., Naffakh, A., Waliji, M., Treur, J., Roelofsma, P.H.M.P.: Cyber risk management and organisational alignment using Endley's risk management model: computational analysis from an adaptive dynamical system perspective. In: Proceedings of the 8th International Conference on Computational Methods in Systems and Software, CoMeSySo'24. Lecture Notes in Networks and Systems, Springer Nature (2025)

Bouma, D., Hoffmans, C., Van den Hout, N.J., Zwarts-Wojtowicz, N., Treur, J., Roelofsma, P.H.M.P.: Cyber security in hospitals: a network-oriented model for behavioural learning of employees during phishing simulations. In: Proceedings of the 17th International Conference on Computational Intelligence in Security for Information Systems, CISIS'24. Lecture Notes in Networks and Systems, vol. 957, pp. 98–111, Springer Nature (2025)

Caneva, J., et al.: The unforeseen blackout and cyber risk management for business continuity: network-oriented computational analysis. In: Proceedings of the 8th International Conference on Computational Methods in Systems and Software, CoMeSySo'24. Lecture Notes in Networks and Systems, Springer Nature (2025)

Crowdstrike: What is an Advanced Persistent Threat (APT)? [Website]. crowdstrike.com (2024). https://www.crowdstrike.com/cybersecurity-101/advanced-persistent-threat-apt/

Capră, A.M.C., Dekker, E., Hoffmann, L., Hoffmans, C., Hoogstad, M., Mueller, T., van den Hout, N.J., Treur, J., Roelofsma, P.H.M.P.: Computational analysis of human factors in spear-phishing attacks: an adaptive network model. In: Proc. of the 16th International Conference on Intelligent Human Computer Interaction, IHCI'24. Lecture Notes in Computer Science, Springer Nature (2025)

Hendrikse, S.C.F., Treur, J., Koole, S.L.: Modeling emerging interpersonal synchrony and its related adaptive short-term affiliation and long-term bonding: a second-order multi-adaptive neural agent model. Int. J. Neural Syst. **33**(7), 2350038 (2023)

Hoffmans, C.C.A., et al: Analysing the rise of biohacking in a tech-driven city: assessing risks through adaptive network modeling. In: Proceedings of the 16th International Conference on Intelligent Human Computer Interaction, IHCI'24. Lecture Notes in Computer Science, Springer Nature (2025)

Hutchins, E., Cloppert, M., Amin, R.: Intelligence-driven computer network defense informed by analysis of adversary campaigns and intrusion kill chains. Lead. Issues Inf. Warfare Secur. Res. **1**(1), 80 (2011). Lockheed Martin

Jeffery, E., et al.: Computational analysis of the effectiveness of the intelligence cycle for organizational cyber risk management. In: Proceedings of the 8th International Conference on Computational Methods in Systems and Software, CoMeSySo'24. Lecture Notes in Networks and Systems, Springer Nature (2025)

Joint Task Force Interagency Working Group: Security and Privacy Controls for Information Systems and Organizations (Revision 5). National Institute of Standards and Technology (2020). https://doi.org/10.6028/NIST.SP.800-53r5

Keijzer, S., Lochtenbergh, D., Marsman, T., Voorhoeve, S., Zwarts, N., Bouma, D., Treur, J., Roelofsma, P.H.M.P.: Managing classified information by a third-party contractor: a Computational Cybersecurity Analysis. In: Proc. of the 8th International Conference on Computer-Human Interaction Research and Applications, CHIRA'24. Communications in computer and information science, vol 2371, pp 299–315. Springer Nature (2025). https://doi.org/10.1007/978-3-031-83845-3_18

Mat, N.I.C., Jamil, N., Yusoff, Y., Kiah, L.M.: A systematic literature review on advanced persistent threat behaviors and its detection strategy. J. Cybersecurity **10**(1) (2024). https://doi.org/10.1093/cybsec/tyad023

Roelofsma, P.H.M.P., Jabeen, F., Taal, H.R., Treur, J. (eds.).: Using shared mental models and organisational learning to support safety and security through Cyberspace: a computational analysis approach. Springer Nature (2025)

Treur, J.: Network-Oriented modeling for Adaptive networks: Designing Higher-Order Adaptive biological, mental and social network models. Springer Nature (2020). https://doi.org/10.1007/978-3-030-31445-3

Treur, J.: On the dynamics and adaptivity of mental processes: relating adaptive dynamical systems and self-modeling network models by mathematical analysis. Cogn. Syst. Res. **70**, 93–100 (2021)

Vera, R., Shehu, A.F., Dargahi, T., Dehghantanha, A.: Cyber defence triage for multimedia data intelligence: hellsing, desert falcons and lotus blossom APT campaigns as case studies. Int. J. Multimedia Intell. Secu. **3**(3), 221 (2019)

Adversarial Attack Detection in Intrusion Detection System Using Machine Learning

Sanjeev Kumar[2], Saurabh Bharti[1], Rajiv Singh[2], Kapil Kumar[1](✉), and Manju Khari[1]

[1] School of Computer and Systems Sciences, Jawaharlal Nehru University, New Delhi 110067, India
kapil.sharma0942211@gmail.com
[2] Banasthali Vidyapith, Rajasthan, Radha Kishnpura, India

1 Introduction

The Intrusion Detection System acts as a crucial part of guaranteeing the organization's security from shifted interruptions. Intrusion detection can distinguish unapproved frameworks from organized assaults and take convincing defensive measures based on the security required (Dhaliwal et al., 2018). Few of these network intrusion detection systems are watching the information stream to recognize interruptions in any organization. The two primary practices of IDS are Anomaly-based and misuse-based (Sisinni et al., 2018). Intrusion detection could be a basic handle that includes consistent checking of computer systems and frameworks to recognize any potential unauthorized access or malicious action. Cybersecurity is considered a major issue in IoT-based frameworks. Numerous security-based frameworks are being given interior the IoT gadgets. Still, primarily all these security frameworks are based on identifying a particular sort of assault or on the foremost common signature assaults (Sisinni et al., 2018). Numerous assaults are not based upon signature assaults which lead to the disappointment of a security framework. It takes hours and days to distinguish and settle the issue which makes it an unsettling influence in IoT gadgets (Chen et al., 2018). Here a machine learning-based method is proposed to handle the current challenging issue. This requires utilizing progressed program devices and strategies that can rapidly identify security breaches and caution security faculty to require fitting activity. By recognizing and reacting to potential security breaches on time, organizations can minimize the effects of cyber dangers and ensure their important resources. This paper will particularly centre on machine learning-based outline strategies. ML could be a shape of fake in-sights that employ algorithms and models to authorize computers to learn info and make predictions or selections based on it, without the requirement for ex-press programming. These programs empower machines to memorize through involvement, moving forward with their execution of particular assignments. By preparing and testing machine learning models on a specific dataset or issue, they can be utilized in an extent of applications. The whole work is organized as follows. The first section deals with the introduction part consisting of problem statements and research objectives. The second part deals with mainly the literature review. The third section deals with the proposed methodology. The fourth & fifth section deals with Data description and results as well as the analysis part. The last section which is the conclusion and future scope parts.

D. Singh et al. (Eds.): IHCI 2024, LNCS 15557, pp. 426–434, 2025.
https://doi.org/10.1007/978-3-031-88705-5_35

1.1 Problem Statements

The existing IDS is created on a predefined signature and rules-based, which means the detection of available intrusions is unable to detect unknown intrusions as well as it is ineffective against novel & sophisticated attacks. The problem with the current security system is the behaviour of the intrusions, which change from time to time and require retraining of the systems. The system fails to predict intrusion precisely that is the system sometimes recognises normal behaviour as malicious and sometimes recognises malicious behaviour as nor-mal. This is a big issue with the current working IDS. The use of machine learning can solve this issues and also enhance the IDS capabilities by allowing data-driven, adaptive and hands-on detection mechanisms.

1.2 Research Objectives

The centre of this inquiry about consider is basically on cyber security and investigates different network-related assaults that can happen.

To classify the distinctive sorts of assaults, the consider utilizes an extension of machine learning strategies, counting SVC, Naïve Bayes, LR, DT and XGBoost.

Test and train the model on the labelled dataset containing normal and malicious network traffic. Dataset examining strategies are utilized to adjust out the minority classes of assaults.

Evaluate and analyse the performance of each algorithm on the parameters such as overall accuracy, precision and recall. Execution is measured and the results are compared.

2 Literature Review

The literature review aims to give a comprehensive overview of existing research in the field of intrusion detection using machine learning. This section synthesizes key findings, theoretical framework, methodologies and significant trends in the literature, identifying gaps and areas for further investigations.

Xu et al. (2023) a data-driven process to disruption and contradiction finding using robotized ML, dispersed in subtle control, presents a data-centric exercise for stimulating IoT safety. Their consideration not as it were donates to the progress of IoT security events but also highlights the efficiency of robotized ml in trade with compound info designs. Distributed in delicate computation, this inquiry offers a gainful union of data-driven plans, completing the relevance of robotized machine learning for the strong disruption and individuality of location within the forward scene of IoT. ElKashlan et al. (2023) address the burgeoning cybersecurity concerns. Utilizing machine learning methods, the creators propose a strong interruption location framework custom-fitted for this particular setting. Their investigation not as it were contributes to upgrading the security of EVCSs but also underscores the flexibility of machine learning in shielding basic foundations inside the IoT space.

Pandit et al. (2023) the study focuses on utilizing different techniques in executing IDS. The aim of the research work is to improve the effectiveness in detecting assaults by utilising ensemble learning. In this study, a method is used in which multiple algorithms are combined to improve the precision and robustness. The result highlights the important information regarding developing a better intrusion detection system in light of enhancing cyber security threats. Bajpai et al., (2022) the study navigates the complicated challenges inside IoT situations. Through an exhaustive investigation of ML applications, the creators contribute novel experiences to the field. Their work, reported in Neuro Quantology, not as it were underscoring the significance of ML in securing IoT systems but moreover gives viable approaches for executing viable interruption location frameworks. Rani et al. (2022) the proposed work highlights the feasibility of outfit tactics in enhancing IoT security measures. By addressing different arrangements. In this study, various ensemble learning methods is used to enhance the accuracy of the model in intrusion detection practice. To enhance the scene of cybersecurity, creating a good and proficient interruption discovery framework that can easily tackle the challenges faced during the detection of assaults.

In their study, (Kapil & Vishal, 2021) the research investigates algorithmic effectiveness in protecting systems from unauthorized access, The study carefully evaluates and contrasts the output of several machine learning models, highlighting the advantages and disadvantages of each. By providing practitioners with useful insights into the realm of cybersecurity, the work helps in choosing the best intrusion detection. (Trebar et al., 2021) the research analyses the effectiveness of various MLs in classifying abnormalities in edge devices. By performing a performance evaluation of various ML algorithms in resource-constrained edge devices the authors try to highlight the best method for detecting anomalies. This study gives a piece of advice on selecting suitable ML algorithms for detecting anomalies in edge devices that help in the development of edge computing and cybersecurity areas in anomaly detection approaches.

3 Proposed Methodology

In this proposed work, the KDDcup199 dataset is used with various ML practices like SVC, LR, NB, DT and XGBoost as input. To analyse the efficiency of various ML techniques, different evaluation parameters like precision, overall accuracy and recall are applied. The proposed work is presented in Fig. 1 respectively.

The study started with the data collection phase where the kddcup99 data set was used. This attack dataset is utilised for every sorting technique. It is mainly a dedicated dataset that is widely known for analysing intrusion detection that was developed by (Stolfo et al., 2000). The second phase is data preprocessing where first, the traffic of network dataset that is collected is organised and examined and transfer it from multiclass to binary class data for combining entirely intrusion forms into normal. For preprocessing first normalisation is done using standard scaler that is a common processing task in machine learning. It is used to convert numerical data into a standardized scale, The StandardScaler class from sklearn is a calculator that compares features by taking the mean and scaling the variance to unity. After normalisation the next phase comes is Label Encoding. It is very important as dataset some data have object data types that must be encoded into the numeric standards so that the ml model easily identifies and analyses the data. For this, a label encoder is used from sklearn to encode object data

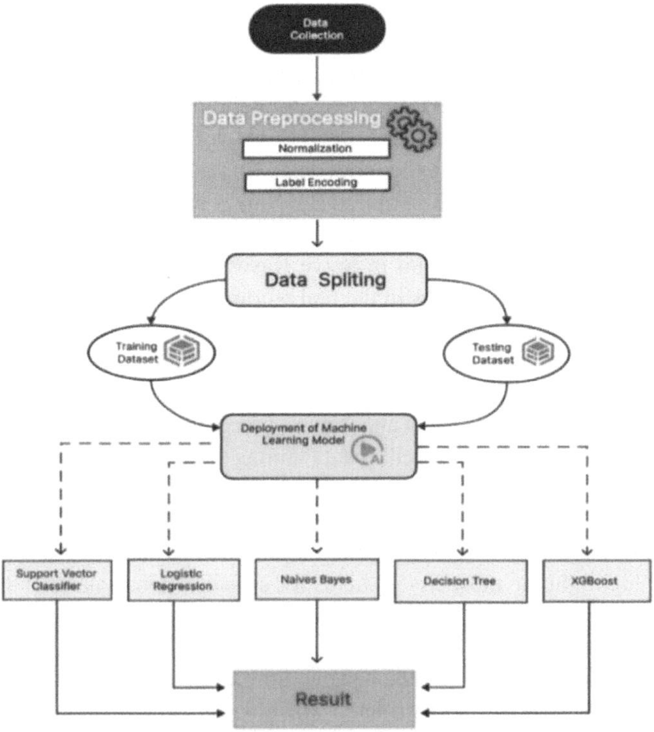

Fig. 1. The proposed work

types in numerical data types. After the label encoding phase, the next phase is data splitting where the dataset is partitioned into 2-parts as test and train data. The training dataset has more data as compared to the test data. After the test and train phase, the next phase is the deployment of ML algorithms. Machine learning methods are generally divided into three types of learning methods: supervised learning provided with labelled data, unsupervised learning provided with unlabeled data, and training provided with labelled and unlabeled data. After train and test phase, the next phase is the deployment of machine learning algorithms and performance is evaluated on evaluation metrics such as overall accuracy, precision and recall.

3.1 Machine Learning Algorithms

A few ML algorithms is utilized to construct an attack prediction model for ex-ample Logistic Regression, Decision-Tree, SVC, Naïve-Bayes, and XGBoost.

Support Vector Classifier (SVC): It is a type of ML algorithm that is built on the statistical theory of learning that gives a higher precision rate as compared to present learning technologies for solving classification-related problems in recognition of patterns. SVC performs better when it comes to solving issues with higher dimensions, non-linearity, and minor samples than other sorting algorithms. For better detection performance, SVC characteristics selection, characteristics weighting, and SVC limit adjustment are therefore essential (Tao et al., 2018).

Logistic Regression (LR): LR was developed by Joseph Berkson, in 1944. The relative entropy damage is applied in diverse classes of circumstances by LR. The balanced mode was used for the parameter class_weight, which balances weights based on the occurrences of different classes in the input data. This tuning is done inversely proportional to the output values (Trebar et al., 2021). Naives Bayes (NB): Naives Bayes was developed by Reverend Thomas Bayes in 1760 (Indika & Harsha, 2021). One popular probabilistic classification algorithm is Naïve Bayes (NB). Simple yet effective, this algorithm finds many practical uses in the real world, from recommending products to diagnosing illnesses to operating self-driving cars. Owing to the inability of actual data to meet NB's assumptions, NB variants are available to accommodate general data.

Decision Tree (DT): It is one of the best classification algorithms are used for the accurate classification of normal and attack class based on selected (Saif et al., 2022). It used by both statistical models as well as data mining for the purpose of prediction with a good precision rate. One of the most famous choices for learning from characteristic-based examples is a DT algorithm. Its layout is often a flowchart, and its benefits include the capacity to handle numerical and categorical attributes, generate rules that are simple to grasp, and highlight the most crucial fields for classification (Salami et al., 2016).

XGBoost: Guestrin & Chen, 2016 developed XGBoost. It is used for both sorting problems and regression problems. It was created for performance and speed with the use of gradient-boosting decision trees. It shows how to apply boosting to machines. This tool is a part of the DMLC, or Distributed Machine Learning Community. XGBoost, also called as Extreme Gradient Boosting, supports making all available hardware as well as memory for tree-based boosting technologies. The important aids it conveys are boosting the algorithms, model fine-tuning, and computer environment applicability (Dhaliwal et al., 2018).

3.2 Evaluation Metrics

These are the evaluation metrics that are used for different ML algorithm's performance analysis.

Accuracy: It presents the ratio of precisely classified examples in a dataset relative to the total count of examples. An equation has demonstrated it.

$Accuracy = \frac{TN\ Normal + TP\ Attack}{FP\ Normal + FN\ Attack + TP\ Attack + TN\ Attack}$

Precision: Prediction of true positive, among all instances that the model predicts as positive. An equation has demonstrated it.

$$Precision = \frac{TP\ Attack}{FP\ Normal + TP\ Attack} \quad (2)$$

Recall: It represents the model's capacity to successfully recognise positive cases accurately while reducing false negatives. An equation has demonstrated it.

$$Recall = \frac{TP\ Attack}{TP\ Attack + FN\ Attack} \quad (3)$$

4 Data Description

The kddcup99 dataset is used that holds nearly 4,900,000 unique correlation vectors with each vector having 41 features and is categorized as normal or at-tack with a specific type of attack. Numerous attacks can infect the network area over some time, the attacks are classified into four categories as debated as follows (Travallaee et al., 2009).

DoS: Type of intrusion where the intruder developed a memory database too engage to serve actual networking queries and hence, rejects actual users' access to a machine. An attacker try to interrupt system resources (Shakir et al., 2022). Examples Ping of death, UDP, Apache, Smurf, mail bomb, Neptune, back etc.

U2R: It is a type of assault in which the attacker starts with gaining access to a normal user system and can use some loophole to acquire root access of the normal user system or targeted system.

R2L: It happens when an intruder uses the skill to send packets to a targeted computer network without having an account on the targeted computer and uses a weakness to gain access to the targeted network as an authentic user on that network.

Probing Assault: An assault to capture the data about a network system for the only purpose of avoiding security controls. Here, the hacker scans the machine or network device for loopholes that can be exploited to compromise the system. This technology is mainly related to data mining: Port sweep, mscan, Satan, Nmap, Saint etc.

4.1 Data Distribution of Kddcup1999

KDD'99 contains a large volume of replicate records with 75% and 78% redundancy for the testing and training data, respectively. Duplicate datasets can compromise assessment results and higher levels of discovery precision. Data share of the KDDcup1999 data set is presented in Fig. 2. This is attached with the details of the replicate log of the KDDcup1999 test and train dataset in Table 1.

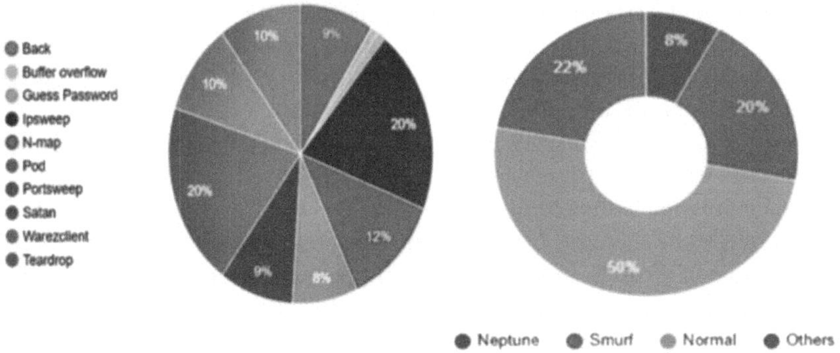

Fig. 2. Data share of the KDDcup1999 data set

The above figure shows the data share percentage of the kddcup1999 dataset. The overall data is divided into four parts such as Neptune, Smurf, Normal and others categories and further it is divided into attack types like back, buffer overflow, guess password, ipsweep, n-map, port sweep etc. with their percentage in the dataset.

Table 1. Duplicate record in kddcup99 test and train dataset

Depiction	Test dataset		Overall	Train Dataset		Overall
	Normal	Malicious		Normal	Malicious	
Unique Records	60591	250436	311027	972781	3925650	4898431
Separate Records	47911	29378	77289	812814	262178	1074992
Duplicate Rate %	20.92	88.26	75.15	16.44	93.32	78.05

The dataset contains duplicate records and redundancy in both test and train data shown in Table 1. Test data is divided into normal and malicious and train data also contains normal and malicious with duplicate record percentages. The test data contain approx. 75 % and train data contain approx. 78% redundancy record.

4.2 Attributes of KDDCup99 Dataset

The whole kddcup99 dataset is divided into 3 attributes separately. The first one consists of basic details of TCP connection, the second is based on domain knowledge connection details and the last one is based on traffic details captured using a 2-second window time.

5 Result and Analysis

This section showcases the outcomes from testing on various ML processes like Support Vector Classifier (SVC), Naïve-Bayes, Logistic Regression, Decision Tree as well as XGBoost. The evaluation of each algorithm is done using metrics like overall accuracy, recall, and precision. Furthermore, we examine the consequences of these findings and evaluate the efficiency of various algorithms in identifying intrusions. Figure 3 represents the result of different ML algorithms with different parameters.

The result shown in Fig. 5 states that XGBoost attained the top accuracy percentage of 98.80%, while SVC and Decision Tree closely followed with 97.00% and 97.20, and logistic regression with 93.00%. While simple, Naive Bayes achieved an accuracy percentage of 87.00%. In terms of recall, XGBoost presented outstanding performance with a recall rate of 98.40%, showcasing its ability to identify intrusion rates. SVC and Decision Tree had high recall rates of 97.22% and 98.10% while logistic regression attained a recall of 92.48%. Even with lower overall accuracy, Naive Bayes presented a recall rate of 83.93%, indicating its capability to detect intrusion instances. Precision is considered by determining the accuracy of appropriately labelled instances within a certain class associated to all instances labelled as that class. XGBoost attained a precision of 98.50%, representing its skill in reducing false positives. The Decision Tree confirmed a precision of 98.00%, while the SVC showed 97.10% precision, and Logistic

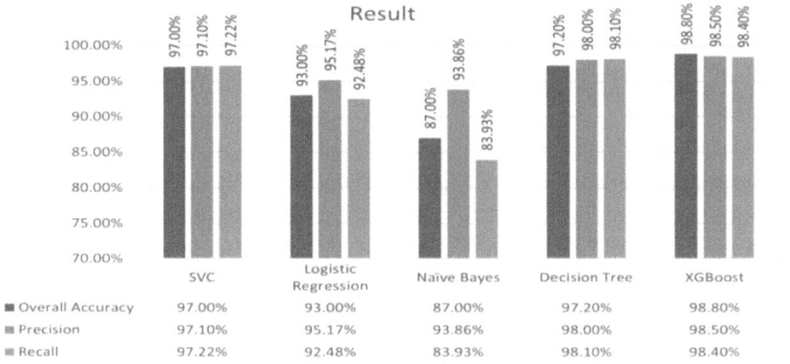

Fig. 3. Represent the result of different ML algorithms with different parameters

Regression had a precision of 95.17%. Despite having inferior overall accuracy, Naive Bayes presented a precision of 93.86%, representing that the intrusions it identified were very likely to be actual positives. The results present that the XGBoost performance is better as compared to other machine-learning algorithms, in terms of precision, recall and overall accuracy. Its ensemble learning method makes it accurately captures complex networks within the dataset and obtains outstanding results

6 Conclusion and Future Scope

The conclusion of this work analysed and showed the effectiveness of various ML algorithms in detecting different types of intrusion with the KDDCUP1999 dataset. For this research work mainly, SVC, NB, LR, DT and lastly XGBoost models are analysed and evaluated. The conclusion shows that the ML algorithm, XGBoost gives better performance in comparison to other algorithms. It shows its reliability for intrusion detection.XGBoost performance is better in comparison to SVC, LR, Naive Bayes, Decision Tree and XGBoost based on parameters such as Precision, recall and overall accuracy for intrusion detection.ML algorithm's performance for intrusion detection depends on the capabilities of handling and processing high-dimensional and complex patterns in the dataset. To effectively differentiate normal from malicious data traffic, preprocessing is a very important step in refining the performance of IDs. The future scope recommended method for detecting intrusions in the network using machine learning has been put into exercise with various ML methods. To enhance the accuracy during implementation different ML as well as deep learning methods can be used. Various domains have shown capable results with Ensemble learning techniques, such as IoT datasets. Compounding forecasts from numerous models can advance the accuracy of IoT datasets through ensemble learning. Through the use of various models and compounding methods like bagging or boosting, ensemble methods can effectively decrease bias and discrepancy, leading to improved overall accuracy. Increasing the assessment to various datasets other than KDDCup1999 may offer a more thorough vision of the adaptableness and flexibility of ML algorithms for detecting intrusions in several network settings.

References

Dhaliwal, S.S., Nahid, A.A., Abbas, R.: Effective intrusion detection system using XGBOOST. Information **9**(7), 149 (2018)

Sisinni, E., Saifullah, A., Han, S., Jennehag, U., Gidlund, M.: Industrial internet of things: challenges, opportunities, and directions. IEEE Trans. Industr. Inf. **14**(11), 4724–4734 (2018)

Chen, Z., Jiang, F., Cheng, Y., Gu, X., Liu, W., Peng, J.: XGBoost classifier for DDoS attack detection and analysis in SDN-based cloud. In: 2018 IEEE International Conference on Big Data and Smart Computing (bigcomp), pp. 251–256 (2018)

Xu, H., Sun, Z., Cao, Y., Bilal, H.: A data-driven approach for intrusion and anomaly detection using automated machine learning for the Internet of Things. Soft. Comput. **27**(19), 14469–14481 (2023)

ElKashlan, M., Elsayed, M.S., Jurcut, A.D., Azer, M.: A machine learning-based intrusion detection system for IoT Electric Vehicle charging Stations (EVCSs). Electronics **12**(4), 1044 (2023)

Pandit, P.V., Bhushan, S., Waje, P.V.: Implementation of intrusion detection system using various machine learning approaches with ensemble learning. In: 2023 International Conference on Advancement in Computation and Computer Technologies (InCACCT), pp. 468–472. IEEE (2023)

Bajpai, S., Sharma, K., Chaurasia, B.K.: Intrusion detection system in IoT network using ML. NeuroQuantology, 3597–3601 (2022)

Rani, D., Gill, N.S., Gulia, P., Chatterjee, J.M.: An Ensemble-based multiclass classifier for intrusion detection using internet of things. Comput. Intell. Neurosci. **2022**, 1–16 (2022)

Kumar, S.K., Bhatnagar, V.: Machine learning algorithms performance evaluation for intrusion detection. J. Inf. Technol. Manag. **13**(1), 42–61 (2021)

Tavallaee, M., Bagheri, E., Lu, W., Ghorbani, A.A.: A detailed analysis of the KDD CUP 99 data set. In: 2009 IEEE Symposium on Computational Intelligence for Security and Défense Applications, Ottawa, ON, Canada, pp. 1–6 (2009)

Stolfo, S.J., Fan, W., Lee, W., Prodromidis, A., Chan, P.K.: Cost-based modeling for fraud and intrusion detection: results from the JAM project. In: Proceedings DARPA Information Survivability Conference and Exposition. DISCEX'00, vol. 2, pp. 130–144. IEEE (2000)

Tao, P., Sun, Z., Sun, Z.: An improved intrusion detection algorithm based on GA and SVM. IEEE Access **6**, 13624–13631 (2018)

Wickramasinghe, I., Kalutarage, H.: Naive Bayes: applications, variations and vulnerabilities: a review of the literature with code snippets for implementation. Soft. Comput. **25**(3), 2277–2293 (2021)

Agarwal, A., Singh, R., Khari, M.: Detection of DDoS attack using ids mechanism: a review. In: 2022 1st International Conference on Informatics (ICI), pp. 36–46. IEEE (2022)

Salami, H.O., Ibrahim, R.S., Yahaya, M.O.: Detecting anomalies in students' results using decision trees. Int. J. Mod. Educ. Comput. Sci. **8**(7), 31 (2016)

Saif, S., Das, P., Biswas, S., Khari, M., Shanmuganathan, V.: HIIDS: hybrid intelligent intrusion detection system empowered with machine learning and metaheuristic algorithms for application in IoT based healthcare. Microprocessors Microsyst. 104622 (2022)

Enhancing Data Privacy with Lattice-Based Encryption Technique

Rohit Meda[1], Intekhab Alam[1(✉)], Priyank Pandey[2], Renu Dalal[3], and Manju Khari[1]

[1] School of Computer and Systems Sciences, Jawaharlal Nehru University, New Delhi 110067, India
intakhabalam575@gmail.com
[2] Graphic Era University Dehradun, Dehradun, Uttarakhand, India
[3] Guru Gobind Singh Indraprastha University, East Delhi Campus Surajmal Vihar, Delhi, India

Abstract. The proposed framework describes a comprehensive approach like Learning with errors, Ring Learning with errors to secure and process sensitive data created in the age of artificial intelligence. Starting with data readiness to ensure data is ready, the process encrypts the data using strong algorithms and transforms it into encrypted string data for secure data transfer to the cloud platforms. Shortest Vector Problem and Closest Vector Problem helps in increasing the complexity of the algorithm to enhance security. Encrypted string data maintains data confidentiality and is only accessible to authorized parties with the proper secret key. Various types of data are accepted, except for visual datasets such as images or video. Encryption converts the data to lattice points. A public key is included to hide the information. Adding noise during encryption protects against eavesdropping. During decryption, noise is separated from the secret key's tiling by the secret key itself. It is important to use a good basis to decryption in order to prevent an intruder from decrypting the wrong ciphertext. The goal of this framework is to ensure data safety and confidentiality through Post-Quantum Cryptography the entire data lifecycle and the results are pretty satisfactory with encryption time of 0.40547 ms and decryption time of 0.71531 ms for public key size of 800 bytes, secret key 1632 bytes and ciphertext 768 bytes. Testing and security measures are implemented to mitigate vulnerabilities.

Keywords: Learning with errors · Ring Learning with errors · Post-Quantum Cryptography · Shortest Vector Problem · Closest Vector Problem

1 Introduction

Data security is of paramount importance in our digitally driven world, serving as the bedrock for preserving the confidentiality, integrity & availability of information. At its core, data security safeguards sensitive data, ranging from personal details to corporate and government information, from falling into unauthorized hands. Confidentiality protection ensures that individual's private data, including financial records and health information, remains shielded from prying eyes. In the face of escalating cyber threats, data security serves as a frontline defense against hacking, malware and phishing attacks.

D. Singh et al. (Eds.): IHCI 2024, LNCS 15557, pp. 435–443, 2025.
https://doi.org/10.1007/978-3-031-88705-5_36

Beyond the corporate realm, national security considerations underscore the importance of safeguarding government data. Moreover, data security empowers individuals, giving them control over their personal information in the digital landscape. Ultimately, data security is not just a technical necessity but a cornerstone of ethical data management, ensuring the benefits of digital age are realized while upholding privacy, trust, and societal well-being.

As we enter the era of AI and ML, the threat landscape shifts to more advanced and automated attacks [3]. Threat actors can use AI to analyze large datasets, identify identity risks, and conduct targeted attacks on a scale never seen before. In addition, the emergence of quantum computing offers both promise and threat as it could disrupt traditional cryptographic algorithms, requiring the development of new, quantum-proof encryption. In short, data threats evolve in tandem with the speed of technological advancement. Advancements bring unprecedented opportunities, but they also bring new challenges that require ongoing innovation in cybersecurity. Understanding these trends is essential for creating proactive and adaptive strategies to protect our data in a rapidly changing digital environment.

1.1 Problem Statements

In the realm of modern cryptosystems, the emergence of Quantum computing threatens the security of traditional cryptographic algorithms. As Post-Quantum cryptosystems continue to advance, traditional encryption methods, reliant on the hardness of complex mathematical problems, risk becoming susceptible for efficient quantum attacks. The urgent need for cryptographic techniques resilient to quantum threats has led to the exploration of lattice-based cryptography [7].

This thesis focuses specifically on the challenges and opportunities presented by lattice cryptography. The fundamental problem addressed is the design, analysis, and implementation of lattice-based encryption which relies on algebraic structure in rings within lattices frameworks. The overarching goal is to develop and assess the viability of ringbased lattice cryptography as a practical and secure alternative in the post-quantum cryptographic landscape.

1.2 Research Objectives

- Investigate and develop efficient algorithms for cryptographic primitives within the context of lattice cryptography, with a particular emphasis on key exchange, Public-Key Encryption.
- Evaluate the computational efficiency and practicality of lattice cryptography in real-world scenarios. Consider factors such as key sizes, computational overhead, and feasibility for implementation in resource-constrained environments.
- Investigate and identify practical use cases for lattice cryptography, particularly in industries and applications where security and post-quantum resilience are paramount. Consider scenarios such as secure communication, data protection, and secure multi-party computation.

By addressing these problem statements, the thesis aims to contribute to the ongoing efforts in developing post-quantum cryptographic solutions. The research will provide insights into the strengths, limitations, and practical implications of lattice cryptography, ultimately advancing our understanding of its potential as a key player in the future of secure communication and data protection.

The rest of this paper is structured as follows. In Sect. 2, literature review is discussed and analysis the existing work and limitations. In Sect. 3, proposed work is described via a diagram that is divided into multiples phases. In Sect. 4, the development of suggested models for the system and the analysis of the outcomes are described. Section 5 providing a conclusion.

2 Literature Review (National and International Review)

(See Table 1)

Table 1. Literature Review

S.No	Title	Author(s)	Contributions	Future Scope
1	"On lattices, learning with errors, random linear codes, and cryptography 2023 [4]"	"O. Regev"	This research focuses on the reduction of the time complexity of lattice hardness as some extend a learning problem. The resulted cryptosystem is found to be quite efficient than the early lattice cryptosystem	Considering some assumptions public key size can be brought down to O(n)
2	"Saber: module-LWR based key exchange, CPA-secure encryption and CCA-secure KEM. 2019 [5]"	"J. P. D'Anvers, A. Karmakar, S. Sinha Roy and F. Vercauteren"	The author's study contributes to the problems related to LWE, LWR and ModLWR. Saber is a fixed time deployed system	Focusing on the optimization of algorithms for better scalability and less computational overhead and complexity makes the it more practically advanced
3	"On Advances of Lattice-Based Cryptographic Schemes and Their Implementations. 2023 [7]"	"Bandara H, Herath Y, Weerasundara T, Alawatugoda J"	Researchers have discussed about the different quantum safe cryptographic schemes. It mainly discussed the LWE problem implementation i.e. CRYSTALS-Kyber	The efficiency of the LWE approaches can be improved using plain lattice structures. Lattice trapdoor can be used to increase efficiency
4	"Lattices in Quantum-ERA Cryptography. 2023 [8]"	"John, M. N., Ozioma, O., Ngozi, O. P., & Egbogho, H. E."	The study is built on the concept of hardness of Ring LWE problem. This can be used in the post quantum cryptography technique for prominent efficiency	Analysis and implementation ca be done to confirm the efficiency and practical view of the schemes

(*continued*)

Table 1. *(continued)*

S.No	Title	Author(s)	Contributions	Future Scope
5	"Evaluation and Comparison of Lattice-Based Cryptosystems for a Secure Quantum Computing Era. 2023 [9]"	"Sabani, M. E., Savvas, I. K., Poulakis, D., Garani, G., & Makris, G. C."	It is a review study which examines the schemes proposed for quantum safe cryptography. Researchers also discussed about lattice-based cryptosystems, LWE, strength and weaknesses	As the time passes there will be multiple innovations in cryptosystems and this can be examined to state which is among those is efficient and effective in real world
6	"The shortest vector in a lattice is hard to approximate to within some constant. 2001 [10]"	"Micciancio, D."	Author have done a detailed study of the Shortest vector problem and stated that approximation for SVP is NP-Hard and it is not computable in polynomial time	Further study can be encourage to make this problem to be solved in satisfactory time complexity

3 Proposed Methodology

The framework here explains how generation of key, key exchange [1, 2], conversion to ciphertext and back to plaintext of the information/data will take place for proposed approach. First the data is encrypted at the start of the process then this encrypted message/information is send to some cloud platform in the form of encoded string this helps in maintaining the confidentiality of information over the cloud database because no cloud user can see what the original information is stored, if they somehow gain access to the encrypted information they only see some random numeric values, this information be then accessed by any party who has the correct secret key and encrypted data and with the help of this two things a person can decode to the original message.

The proposed work can take text, integer, float, string type of data as input, it is not built for visual dataset such as image, video or audio information. The information given is first converted in the form of lattice points and then further these lattice points and the public key together are used to hide the information in the form of encrypted data this encrypted data now can be anywhere on the local computer or cloud storage but it is in the form of numbers that represent lattice points, the cloud user those are working on this encrypted data are unaware of the secret information, this information can only be retrieved using this lattice points and the secret key which only provided to the authorized person and any trespasser tries to change the bits of the information (eavesdropping) then the authorized person would know at the time of decryption that the bits have been changed and someone has attempted to access the information (Fig. 1).

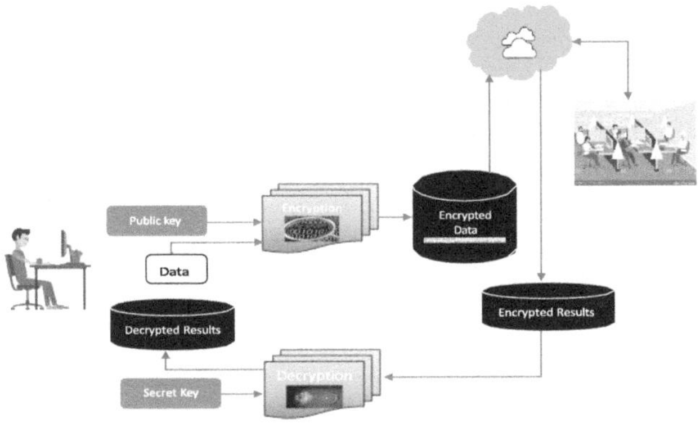

Fig. 1. Architecture of Research Work

In the proposed work the lattices can be represented in 2-dimensions. The lattices that are used in this cryptographic scheme have many more dimensions and with the current knowledge it cannot be represented on a piece of paper because the lattice can have n-dimensions. Lattices can be finitely represented on a computer with the help of their bases.

4 Data Description

RT-IoT is presented as a private dataset derived from an Internet of Things (IoT) real-time infrastructure. This dataset includes a wide variety of IoTs embedded devices and advanced network abuse methods, covering this simple and malicious system behavior to provide a generic overview of real-time situations. Including information from IoT machines (LEDs, Bulb, Temperature) and simulation of net abuse scenarios (Brute-Force, SSH, DDoS using Hping, Slowloris, Nmap patterns) RT-IoTc22 provides an indepth view of the intricate nature of network traffic, bidirectional attributes are carefully extracted with the help of network monitoring tool by Zeek and flowmeter plugin. This dataset can be used by researchers to improve their IDS capabilities, allowing them to develop robust, adaptive IoT network safety solutions in real-time [6]. The data frame has 123117 rows and 85 columns, most of these columns consists of integer values but some of them has text values also. Below is the list of all columns of the dataset (Fig. 2).

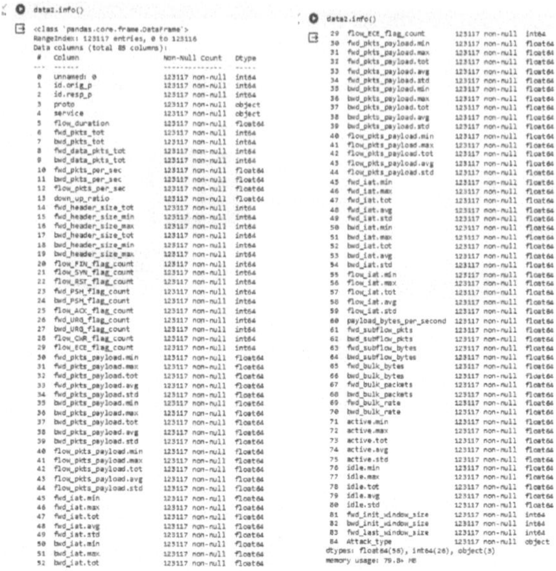

```
data2.info()

<class 'pandas.core.frame.DataFrame'>
RangeIndex: 123117 entries, 0 to 123116
Data columns (total 85 columns):
 #   Column                    Non-Null Count   Dtype
---  ------                    --------------   -----
 0   Unnamed: 0                123117 non-null  int64
 1   id.orig_p                 123117 non-null  int64
 2   id.resp_p                 123117 non-null  int64
 3   proto                     123117 non-null  object
 4   service                   123117 non-null  object
 5   flow_duration             123117 non-null  float64
 6   fwd_pkts_tot              123117 non-null  int64
 7   bwd_pkts_tot              123117 non-null  int64
 8   fwd_data_pkts_tot         123117 non-null  int64
 9   bwd_data_pkts_tot         123117 non-null  int64
 10  fwd_pkts_per_sec          123117 non-null  float64
 11  bwd_pkts_per_sec          123117 non-null  float64
 12  flow_pkts_per_sec         123117 non-null  float64
 13  down_up_ratio             123117 non-null  float64
 14  fwd_header_size_tot       123117 non-null  int64
 15  fwd_header_size_min       123117 non-null  int64
 16  fwd_header_size_max       123117 non-null  int64
 17  bwd_header_size_tot       123117 non-null  int64
 18  bwd_header_size_min       123117 non-null  int64
 19  bwd_header_size_max       123117 non-null  int64
 20  flow_FIN_flag_count       123117 non-null  int64
 21  flow_SYN_flag_count       123117 non-null  int64
 22  flow_RST_flag_count       123117 non-null  int64
 23  fwd_PSH_flag_count        123117 non-null  int64
 24  bwd_PSH_flag_count        123117 non-null  int64
 25  flow_ACK_flag_count       123117 non-null  int64
 26  fwd_URG_flag_count        123117 non-null  int64
 27  bwd_URG_flag_count        123117 non-null  int64
 28  flow_CWR_flag_count       123117 non-null  int64
 29  flow_ECE_flag_count       123117 non-null  int64
 30  fwd_pkts_payload.min      123117 non-null  float64
 31  fwd_pkts_payload.max      123117 non-null  float64
 32  fwd_pkts_payload.tot      123117 non-null  float64
 33  fwd_pkts_payload.avg      123117 non-null  float64
 34  fwd_pkts_payload.std      123117 non-null  float64
 35  bwd_pkts_payload.min      123117 non-null  float64
 36  bwd_pkts_payload.max      123117 non-null  float64
 37  bwd_pkts_payload.tot      123117 non-null  float64
 38  bwd_pkts_payload.avg      123117 non-null  float64
 39  bwd_pkts_payload.std      123117 non-null  float64
 40  flow_pkts_payload.min     123117 non-null  float64
 41  flow_pkts_payload.max     123117 non-null  float64
 42  flow_pkts_payload.tot     123117 non-null  float64
 43  flow_pkts_payload.avg     123117 non-null  float64
 44  flow_pkts_payload.std     123117 non-null  float64
 45  fwd_iat.min               123117 non-null  float64
 46  fwd_iat.max               123117 non-null  float64
 47  fwd_iat.tot               123117 non-null  float64
 48  fwd_iat.avg               123117 non-null  float64
 49  fwd_iat.std               123117 non-null  float64
 50  bwd_iat.min               123117 non-null  float64
 51  bwd_iat.max               123117 non-null  float64
 52  bwd_iat.tot               123117 non-null  float64
```

```
data2.info()

 29  flow_ECE_flag_count       123117 non-null  int64
 30  fwd_pkts_payload.min      123117 non-null  float64
 31  fwd_pkts_payload.max      123117 non-null  float64
 32  fwd_pkts_payload.tot      123117 non-null  float64
 33  fwd_pkts_payload.avg      123117 non-null  float64
 34  fwd_pkts_payload.std      123117 non-null  float64
 35  bwd_pkts_payload.min      123117 non-null  float64
 36  bwd_pkts_payload.max      123117 non-null  float64
 37  bwd_pkts_payload.tot      123117 non-null  float64
 38  bwd_pkts_payload.avg      123117 non-null  float64
 39  bwd_pkts_payload.std      123117 non-null  float64
 40  flow_pkts_payload.min     123117 non-null  float64
 41  flow_pkts_payload.max     123117 non-null  float64
 42  flow_pkts_payload.tot     123117 non-null  float64
 43  flow_pkts_payload.avg     123117 non-null  float64
 44  flow_pkts_payload.std     123117 non-null  float64
 45  fwd_iat.min               123117 non-null  float64
 46  fwd_iat.max               123117 non-null  float64
 47  fwd_iat.tot               123117 non-null  float64
 48  fwd_iat.avg               123117 non-null  float64
 49  fwd_iat.std               123117 non-null  float64
 50  bwd_iat.min               123117 non-null  float64
 51  bwd_iat.max               123117 non-null  float64
 52  bwd_iat.tot               123117 non-null  float64
 53  bwd_iat.avg               123117 non-null  float64
 54  bwd_iat.std               123117 non-null  float64
 55  flow_iat.min              123117 non-null  float64
 56  flow_iat.max              123117 non-null  float64
 57  flow_iat.tot              123117 non-null  float64
 58  flow_iat.avg              123117 non-null  float64
 59  flow_iat.std              123117 non-null  float64
 60  payload_bytes_per_second  123117 non-null  float64
 61  fwd_subflow_pkts          123117 non-null  float64
 62  bwd_subflow_pkts          123117 non-null  float64
 63  fwd_subflow_bytes         123117 non-null  float64
 64  bwd_subflow_bytes         123117 non-null  float64
 65  fwd_bulk_bytes            123117 non-null  float64
 66  bwd_bulk_bytes            123117 non-null  float64
 67  fwd_bulk_packets          123117 non-null  float64
 68  bwd_bulk_packets          123117 non-null  float64
 69  fwd_bulk_rate             123117 non-null  float64
 70  bwd_bulk_rate             123117 non-null  float64
 71  active.min                123117 non-null  float64
 72  active.max                123117 non-null  float64
 73  active.tot                123117 non-null  float64
 74  active.avg                123117 non-null  float64
 75  active.std                123117 non-null  float64
 76  idle.min                  123117 non-null  float64
 77  idle.max                  123117 non-null  float64
 78  idle.tot                  123117 non-null  float64
 79  idle.avg                  123117 non-null  float64
 80  idle.std                  123117 non-null  float64
 81  fwd_init_window_size      123117 non-null  int64
 82  bwd_init_window_size      123117 non-null  int64
 83  fwd_last_window_size      123117 non-null  int64
 84  Attack_type               123117 non-null  object
dtypes: float64(56), int64(26), object(3)
memory usage: 79.8+ MB
```

Fig. 2. Features of dataset

The dataset has entries that consists of following datatypes integer, float and object and Fig. 9 shows the percentage of the entries with respect to their datatypes (Fig. 3).

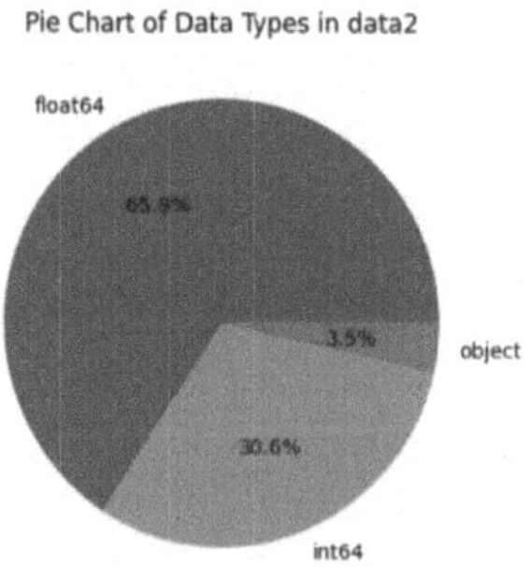

Fig. 3. Data description of dataset

5 Results

Pseudo code for Proposed idea

Key Generation:

i. Parameter Setup:
Select parameters for security n and q and create a lattice $\tilde{A} \subset Z_q^n$ obtained from bases [A]. Impurity distribution d is then selected over Z.

ii. Pub_key generation:
An uncertain Mat., suppose $[S] \overset{\mathbb{R}}{\leftarrow} d^{n \times m}$, generate the Mat. $\mathcal{E} \overset{\mathbb{R}}{\leftarrow} d^{n \times m}$. Thus, the pub_key [A] = $[S] + [\mathcal{E}]$ mod q.

iii. Secret key creation:
Mat. S is the sec_key.

Encryption:

i. Encode and choose message:
Encode the message m to a vector u using an encoding function carefully.

ii. Add Impurity and Encode:
Select an unfixed $\vec{r} \overset{\mathbb{R}}{\leftarrow} Z_q^n$ and impurity $\vec{\mathcal{E}} \overset{\mathbb{R}}{\leftarrow} d^m$. Compute the ciphertext as
$c = (A * \vec{r}) + \vec{\mathcal{E}} + \text{Enc}(u) \bmod q$

Decode:

i. Inner multiplication computation:
Evaluate inner multiplication ‹c, A› mod q to compute an estimation of Enc(u).

ii. Retrieve information:
Validate Similarity:
Check if ‹c, A› mod q is equal to the original information m. Authenticate the pattern when similarity holds; else, deny. Retrieve the original information m by decoding obtained approximation.
Lattice encoding technique relies on assumption that the LWE is very hard to compute. In simple terms, the safety tests show an attacker, if he has access to your pub_key and your encoded information, won't be able to decipher different ciphers of your information.

The safety sketch reduces the issue of decryption to the problem of computing LWE. Consider A is a well-organized adversary that has a non-zero advantage over the encryption scheme. An algorithm B can be built using A to fix LWE logically. As LWE is considered likely difficult, it means that the lattice-based encoding technique is tough to break.

The result of the proposed work is given in the following table, the results obtained are from RT_IoT2022 (Table 2).

Table 2. Results of research work

Size of dataset	Parameters			Encryption time (ms)	Decryption time (ms)	Size (Bytes)		
	n	k	q			sk	pk	ct
52 MB	256	-	3329	0.40547	0.71531	1632	800	768

6 Conclusion and Future Scope

The proposed technique relies upon the theoretical foundation of the LWE problem, known for its presumed hardness and suitability for cryptographic purposes. The scheme leverages this hardness to offer public key encryption in a secure and efficient manner. The security proofs associated with the proposed scheme are crucial as they establish the resilience of the scheme against potential attacks. These proofs typically involve scaling down, where the safety of the cryptographic primitive is compromised to the presumed hardness of underlying mathematical issue (in this case, LWE). Though rigorous mathematical analysis, the scheme demonstrates its resistance to known attacks, providing confidence on its security properties. However, theoretical security alone is not sufficient; practical implementation and efficiency are also essential factors to consider. Real-world cryptographic schemes must not only be secure but also practical to use in terms of computational resources and speed. For this reason, it is encouraged further research and implementation to confirm the effectiveness and practicality of the proposed system. This involves implementing the scheme on various platforms and assessing its performance under different scenarios, such as different key sizes and computational environment.

References

1. Blake-Wilson, S., Menezes, A.: Authenticated Diffie-Hellman key agreement protocols. In: International Workshop on Selected Areas in Cryptography, pp. 339–361. Springer, Heidelberg (1998)
2. Krawczyk, H.: HMQV: a high-performance secure Diffie-Hellman protocol. In: Annual International Cryptology Conference, pp. 546–566. Springer, Heidelberg (2005)
3. Agrawal, A., Singh, R., Khari, M., Vimal, S., Lim, S.: Autoencoder for design of mitigation model for DDOS attacks via M-DBNN. Wirel. Commun. Mob. Comput. (2022)
4. Regev, O.: On-lattices, learning with errors, random linear codes, and cryptography. J. ACM (JACM) **56**(6), 1–40 (2009)
5. D'Anvers, J.P., Karmakar, A., Sinha Roy, S., Vercauteren, F.: Saber: module-LWR based key exchange, CPA-secure encryption and CCA-secure KEM. In: International Conference on Cryptology in Africa, pp. 282–305. Springer, Cham (2018)
6. Singh, P., Khari, M., Vimal, S.: EESSMT: an energy efficient hybrid scheme for securing mobile ad hoc networks using IoT. Wirel. Pers. Commun. **126**(3), 2149–2173 (2022)
7. Bandara, H., Herath, Y., Weerasundara, T., Alawatugoda, J.: On Advances of lattice-based cryptographic schemes and their implementations. Cryptography **6**(4), 56 (2022). https://doi.org/10.3390/cryptography6040056
8. John, M.N., Ozioma, O., Ngozi, O.P., Egbogho, H.E.: Lattices in quantum-ERA cryptography. Journal homepage: www.ijrpr.com, ISSN, 2582, 7421 (2023)

9. Sabani, M.E., Savvas, I.K., Poulakis, D., Garani, G., Makris, G.C.: Evaluation and comparison of lattice-based cryptosystems for a secure quantum computing era. Electronics **12**(12), 2643 (2023)
10. Micciancio, D.: The shortest vector in a lattice is hard to approximate to within some constant. SIAM J. Comput. **30**(6), 2008–2035 (2001)

Improving Threat Recognition in X-ray Security Inspection Using Transformers

Archana Singh[1,2(✉)] and Dhiraj[1,2]

[1] Academy of Scientific and Innovative Research (AcSIR), Ghaziabad 201002, India
archana.ceeri21a@acsir.res.in, dhiraj@ceeri.res.in
[2] CSIR-Central Electronics Engineering Research Institute (CSIR-CEERI), Pilani 333031, India

Abstract. X-ray security screening plays a pivotal role in maintaining safety at airports, railway stations, and other transportation hubs. This technology provides a reliable method to inspect the contents of luggage and other items quickly and non-intrusively. By leveraging intelligent algorithms, modern X-ray imaging technologies significantly enhance the effectiveness of security screening procedures. We experimented with multiple state-of-the-art deep learning object detection architectures based on convolutional neural networks (CNNs) and transformers in order to analyze X-ray baggage images. This research delves further into the use of transformer-based algorithms (DINO and Deformable DETR) for the identification of forbidden objects by varying the number of heads. The comparison includes a few CNN-based object detection algorithms, including YOLO-F and RTMDet. The best performance is achieved by the DINO model with 0.804 mAP for the CLCxray dataset and by Deformable DETR with 0.641 mAP for the PIDray dataset. The results demonstrate that varying the number of heads in Transformer-based algorithms significantly affects their performance in identifying forbidden objects in X-ray security images.

Keywords: X-ray image · Deep learning · Object recognition · Baggage security · Computer vision

1 Introduction

In the realm of security inspection, automatic object detection based on X-ray images is becoming more and more significant. X-ray imaging is an essential tool for secure inspection, particularly in high-security locations such as airports, railway stations, and transportation hubs. The ability of X-rays to penetrate objects and produce perspective images allows for a clear inspection and observation of the contents of the luggage. In X-ray imaging, the process involves the interaction of X-rays with an object and a detector. As the X-rays pass through the object, different materials within it absorb the X-rays to varying degrees. The amount of absorption depends on the density and atomic number

D. Singh et al. (Eds.): IHCI 2024, LNCS 15557, pp. 444–455, 2025.
https://doi.org/10.1007/978-3-031-88705-5_37

of the materials. Usually, green represents inorganic materials, orange represents organic materials, and blue represents metals. This color differentiation helps operators quickly and accurately identify various materials, enhancing the effectiveness of the scanning process. It plays a crucial role in the identification of concealed and contraband items [1]. Traditional methods of baggage screening, which mainly rely on human intervention, indeed raise significant concerns regarding time efficiency and error margins. In order to increase the accuracy of detecting prohibited items in baggage, there has been a noticeable shift in favor of using deep learning techniques.

The application of deep learning techniques, particularly convolutional neural networks (CNNs) and transformers, offers significant advancements in the intelligent analysis of X-ray security inspection images. Image classification, object detection, and image segmentation are three distinct recognition problems that can be used to target this. In image classification, the goal is to assign a label to an image from a predefined set of categories. It involves using algorithms to analyze and categorize visual data into classes. The process of object detection entails not only classifying the objects in an image, but also predicting the location of every object by enclosing it in a bounding box. In order to provide a deeper understanding of an image, image segmentation predicts pixel-by-pixel classifiers to allocate a particular category label to every pixel.

By leveraging the capabilities of deep learning, security agencies can better detect prohibited items, reduce human error, and streamline operations, ultimately ensuring a safer and more efficient security environment [2,3]. Most work in object detection has gone into finding objects-of-interest in standard color imagery through the use of Transformer-based [4,5], multi-stage [6,7], and single-stage [8–10] detectors.

2 Related Work

The detection of prohibited items has been greatly enhanced by recent developments in computer-aided X-ray baggage screening. Initially, researchers used basic image enhancement and manual feature extraction techniques, which often led to subpar outcomes. The advent of deep learning, particularly through the use of CNNs and transfer learning, has significantly enhanced threat detection capabilities. By leveraging the meaningful semantics extracted from CNNs, models can quickly and accurately identify threats, making them invaluable tools in various security domains [11,12]. There are two types of object detection techniques: one-stage and two-stage detectors. In order to predict the category of the proposed region, two-stage detectors first generate a sparse set of proposals using a proposal generator, then extract features from each proposal using region classifiers. Without utilizing the cascaded region classification step, the one-stage detection generates categorical predictions about the objects at each place on the feature maps. Comparatively, one-stage detectors are more effective and two-stage detectors are more accurate for object detection tasks.

Recent developments have concentrated on using advanced techniques such as structured tensors using an encoder-decoder architecture, which improve scan object contours and dramatically increase threat detection in the presence of occlusion [13,14]. The work [15] suggested a module that uses low-level edge and material information to improve the efficacy of state-of-the-art detection techniques. Hu et al. [16] utilized spatial attention and anisotropic fusion in an effort to address the overlapping and scale variation problems in X-ray imagery. To improve outcomes, they made use of multi-scale features and effectively fused per-stage predictions. In order to present an improved method of classifying X-ray images, Ahmed et al. [17] presented balanced affinity loss, which combined max-margin learning with efficient sample utilization. In many domains of computer vision and speech processing, transformer models [18] are well-known for their ability to capture this global context. These methods use the transformer attention mechanism-based feature extraction strategy to display image processing. There are some other contributions for X-ray image testing in the field of baggage screening [19–21].

Threat objects often appear in complex, cluttered environments with occlusions and varying degrees of overlap. This makes distinguishing between benign and harmful objects challenging for models. Training effective models requires large, annotated datasets of X-ray images, which are difficult to collect due to security and privacy concerns. Existing datasets are often too small or lack diversity in object types and configurations. Automatic X-ray testing in baggage screening remains an unsolved issue because of variables like inadequate resilience, adaptiveness, loss of generality, and trade-off between missed detections and false alarms.

3 Methodology

3.1 Dataset

We conducted our experiment using PIDray and CLCXray datasets. For each dataset, we followed the standard split protocols for training and testing, where the model was trained using the training set and assessing its performance on the testing set.

Prohibited Item Detection Dataset (PIDray): In 2021, the PIDray dataset [22], which serves as a publicly accessible benchmark, was introduced. This dataset comprises a total of 47,677 luggage scans, covering a wide array of 12 prohibited item categories, including Bullet, Baton, Gun, Powerbank, Hammer, Wrench, Handcuffs, Pliers, Knife, Lighter, Scissor, and Sprayer. Every image within this dataset is meticulously annotated with both segmentation masks and bounding boxes for precise object identification for X-ray baggage security.

The dataset's test set is further categorized into three subgroups: Hard (consisting of 3,733 scans featuring multiple threat items), Easy (comprising 9,482 scans with single threat items), and Hidden (encompassing 5,005 scans deliberately concealing threat items). The selection of images from this dataset can be visualized in Fig. 1. This dataset are described in detail in Table 1, along with their subsections and classwise distribution.

Fig. 1. PIDray dataset samples (Color figure online)

Table 1. Classwise distribution in PIDray dataset

Categories	Baton	Pliers	Hammer	Scissors	Powerbank	Wrench
Number	2,399	6,814	6,299	7,060	8,116	6,437
Categories	Gun	Bullet	Sprayer	Handcuffs	Knife	Lighter
Number	3,757	2,937	4,227	3,388	5,549	6,157

Cutters and Liquid Containers X-ray Dataset (CLCXray): The CLCXray dataset [23] was developed by Intelligent and Visual Learning Lab in 2022 using X-ray scanner. This dataset consists of 9,565 X-ray images, out of which 5,022 X-ray photographs are scanned from manually created luggage and 4,543 X-ray images (actual data) are taken from a real subway scene. The CLCXray dataset has 12 categories, including 7 different types of liquid containers and 5 different types of cutters.

Table 2. Statistics of the CLCXray dataset

Categories	Blade	Dagger	Knife	Scissors	Tin	Cans
Number	3,539	988	700	2,496	856	789
Categories	Carton Drinks	Glass Bottle	Plastic Bottle	Vacuum Cups	Spray Cans	Swiss Army Knife
Number	1,926	540	5,998	2,166	1,077	1,041

There are five different types of cutters: dagger, knife, scissors, and swiss army knife. Cans, glass bottles, carton drinks, spray cans, plastic bottles, tin, and vacuum cups are seven different types of liquid containers. COCO format is used for the annotations. For evaluation purposes, the dataset is split into train, validation, and test sets at a ratio of 8:1:1. Example images of CLCXray database are given in Fig. 2. The classwise distribution is shown in Table 2.

3.2 Proposed Work

The proposed work introduces a combination of data augmentation, detecting head variation, and hyperparameter tuning to optimize model performance.

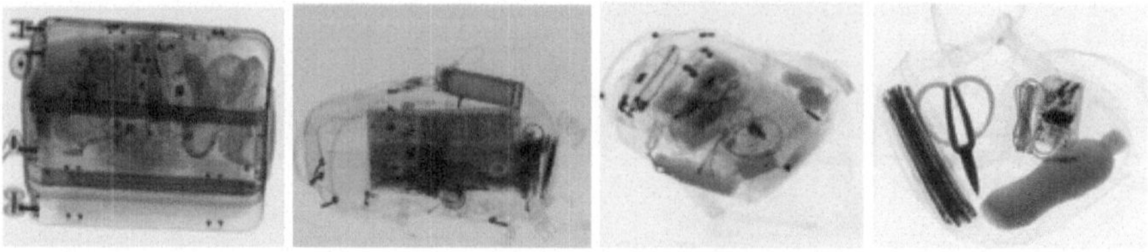

Fig. 2. CLCXray dataset samples (Color figure online)

By varying the number of heads, we can control the granularity and diversity of the attention, potentially improving the model's ability to detect small or overlapping objects. The proposed work also includes the comparison of CNN-based and Transformer-based object detection algorithms.

- **RTMDet:** The real-time model for object detection, named RTMDet [24], is a one-stage object detector that enhances the model's efficiency by investigating large-kernel convolutions in the building blocks of the neck and backbone and adjusting the model's depth, width, and resolution correspondingly.
- **YOLO-F:** YOLOF (You Only Look One-Level Feature) [25] is a deep learning model designed for object detection. It's an evolution of the YOLO series, which are known for their speed and efficiency in detecting objects within images. YOLOF focuses on using a single feature level for both object classification and localization, simplifying the model architecture while maintaining high performance. Employs an encoder that processes features from a single level of the backbone. This encoder typically consists of a series of convolutional layers that further process and refine the features. The decoder is responsible for generating the final predictions. It consists of a detection head that processes the encoded features to predict bounding boxes and class scores for objects in the image.
- **Deformable DETR:** Deformable DETR [26] is an advanced object detection model that builds upon the original DETR framework and replaces traditional convolutional neural networks (CNNs) and region proposal networks (RPNs) with transformer-based architectures. It employs deformable attention module, that allows the model to focus on a sparse set of sampling points around a reference point, and it attends to all points in the feature map. The input feature map $x \in \mathbb{R}^{C\times H\times W}$, let q index a query element with a 2-d reference point p_q and content feature $\mathbb{Z}_q$. The deformable attention feature is given by:

$$\mathrm{DeformAttn}(z_q, p_q, x) = \sum_{m=1}^{M} W_m \sum_{k=1}^{K} A_{mqk} \cdot W_0^m x(p_q + \Delta p_{mqk}) \quad (1)$$

  where K is the total number of sampled keys, m is the attention head, and k is the sampled keys.
- **DINO:** DINO (DETR with Improved DeNoising Anchor Boxes) [27] is an enhanced version of the original DETR architecture, introduces the concept

of denoising anchor boxes to mitigate the issues caused by noisy predictions. These anchor boxes help the model focus on more reliable object proposals, leading to more accurate detections. It consists of a backbone, a multilayer encoder, a multilayer decoder, and multiple prediction heads. The backbone extracts feature maps from the input image and captures hierarchical features. The transformer encoder, processes the feature maps to capture global context and dependencies among features across the entire image. The decoder refines a set of learned object queries based on the encoded features and global context provided by the encoder. It utilizes self-attention and cross-attention mechanisms to enhance object representations and refine predictions.

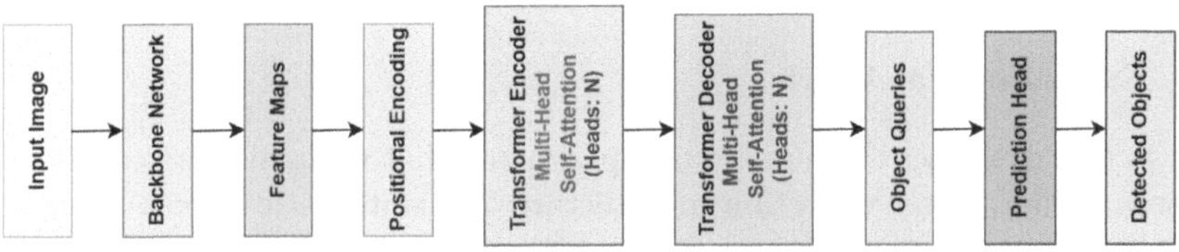

Fig. 3. Proposed Methodology

The block diagram for proposed approach is given in Fig. 3. The dataset is augmented using different data augmentation techniques such as; flipping, scaling, rotation, cropping, and coloring, to balance the different classes of dataset and for better generalization of the model. The backbone network (ResNet101) extracts feature maps from the input image. This step converts raw pixel data into a form that processed by the transformer. Transformers require positional information because they do not have inherent knowledge of spatial locations. Positional encodings are added to the feature maps to provide this information. After that, the self-attention mechanism is applied to process the feature maps to capture relationships between different parts of the image. Multi-head self-attention allows the model to focus on different parts of the input sequence simultaneously by using multiple attention heads. More heads enable the model to attend to multiple aspects of the feature maps concurrently, capturing various spatial relationships and features. A set of learnable object queries that interact with the decoder to generate predictions. Each query represents a potential object in the image. The final output includes bounding boxes, class labels, and scores for detected objects in the image. The contribution of the proposed approach is described below:

- Various data augmentation techniques, including flipping, scaling, rotation, cropping, and coloring, are employed to balance the various dataset classes and to improve the model's generalization. This can be particularly useful for detecting objects with different scales, aspect ratios, or occlusions.
- Transformers use multiheads to allow the model to jointly attend to information from different representation subspaces at different positions. Increasing

the number of heads can enhance the model's ability to focus on various regions and capture more nuanced details, leading to better object localization and classification.

- More heads provide the flexibility to adapt to diverse datasets and detection challenges, enhancing the model's generalization capabilities. This enables parallel processing of different attention patterns, which can improve the model's efficiency and speed in both training and inference stages.
- Hyperparameter tuning is performed with parameters, including learning rates, batch size, number of epochs, and optimizers to improve the detection performance of models. The optimal hyperparameters help to mitigate challenges related to complex imagery and improve feature extraction performance.

3.3 Experimental Work

This section discusses the numerous experiments that we used to assess the effectiveness of the object detection architectures for automatic threat recognition in x-ray images. The average precision (AP) and mean average precision (mAP) at different IoU are utilized to evaluate the performance for object detection methods. IoU is a metric for calculating the overlap between the predicted and the associated ground-truth ROI. AP-S refers to the average precision for small objects (area$< 32^2$), AP-M refers to the average precision for medium-sized objects ($32^2 <$area$< 96^2$), and AP-L refers to the average precision for large objects (area $> 96^2$). These metrics help in understanding how a model performs across multiple scales, and different object sizes. Each experiment is carried out using PyTorch on an Ubuntu system with an NVIDIA GeForce GTX 1080 Ti GPU.

Initially, data augmentation techniques are applied to balance the classes of datasets and to tackle challenges like variations in object scale, aspect ratio, or occlusion. By applying transformations such as flipping, cropping, rotation, or scaling, the model is exposed to diverse versions of the same object, thus improving robustness and generalization across different imaging conditions.

Secondly, to enhance the detection performance of models, hyperparameter tuning is carried out using parameters like learning rates, batch sizes, number of epochs, and optimizers. Fine-tuning hyperparameters is essential for refining model performance, especially when dealing with complex imagery like X-ray scans. The adjustments of learning rates, batch size, epochs, and optimizers can help balance convergence speed and accuracy, ultimately improving feature extraction in challenging scenarios.

Lastly, we experimented with changing the detecting heads to observe how these changes affect the performance of the model across different categories. By increasing the number of heads, the model can allocate more resources and capacity to learn and distinguish between different classes effectively.

Table 3. Experimental findings of threat detection in CLCXray and PIDray datasets

Dataset	Method	AP	AP-50	AP-75	AP-S	AP-M	AP-L
CLCXray dataset	RTMDet	0.782	0.904	0.876	−1.00	0.561	0.787
	YOLO-F	0.733	0.882	0.848	−1.00	0.506	0.747
	DINO	0.804	0.954	0.899	−1.00	−1.00	0.812
	Deformable DETR	0.797	0.956	0.901	−1.00	−1.00	0.804
PIDray dataset	RTMDet	0.618	0.802	0.689	−1.00	0.561	0.787
	YOLO-F	0.627	0.810	0.810	0.715	0.697	−1.00
	DINO	0.638	0.818	0.712	0.756	−1.00	−1.00
	Deformable DETR	0.641	0.759	0.822	0.759	−1.00	−1.00

Table 4. Experimental findings by varying the number of detection heads

Dataset	Nheads	AP	AP-50	AP-75	AP-S	AP-M	AP-L
CLCXray dataset	2	0.735	0.878	0.838	−1.00	0.496	0.737
	4	0.779	0.900	0.873	0.725	0.560	0.787
	8	0.804	0.954	0.899	−1.00	−1.00	0.812
PIDray dataset	2	0.609	0.792	0.679	−1.00	0.561	0.787
	4	0.632	0.839	0.714	0.725	−1.00	−1.00
	8	0.641	0.759	0.822	0.759	−1.00	−1.00

4 Result and Discussion

The simulation results are presented in this section. After comparing RTMDet, YOLO-F, DINO, and Deformable DETR, the overall detection results for threat items in the CLCXray and PIDray datasets are shown in Table 3.

The DINO model achieves best performance for CLCXray dataset, achieving 0.804 mAP. YOLO-F performs the worst in X-ray image analysis. The detection results with class activation maps can be visualized in Fig. 4 for the PIDray and CLCXray datasets. Figure 5a displays the training graphs for mean average precision and loss for the same. With a 0.641 mAP, the Deformable DETR model performs best on the PIDray dataset. RTMDet exhibits the lowest performance in the analysis of X-ray images. The training graphs for mean average precision and loss are shown in Fig. 5b.

In order to investigate the impact of transformer-based models, we experimented with changing the number of detection heads. By default, the transformer's detection head is set to 8. So here, we change the number of detecting heads to see how it affects the trial outcomes. The DINO model heavily relies on the multi-head attention mechanism. With varying numbers of attention heads, Table 4 displays the simulation results for both datasets. These tables show that the performance of the model enhances with an increase in the number of detection heads. Furthermore, we have conducted experiments with several

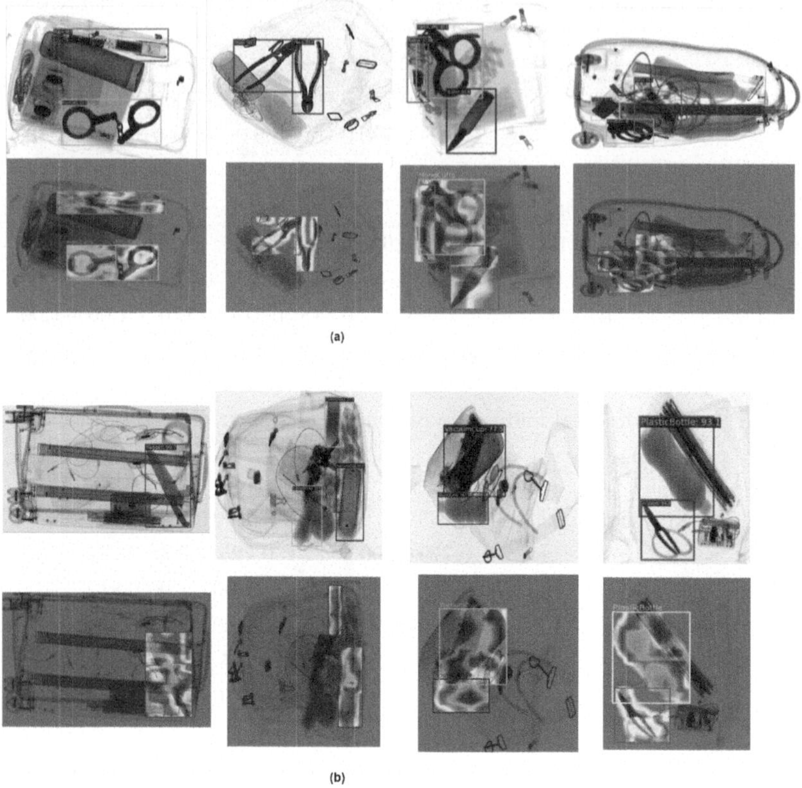

Fig. 4. Class activation maps: (a) PIDray dataset, (b) CLCXray dataset

learning rates (1e−2, 1e−3, and 1e−4), and the model exhibits optimal performance at 1e−3 with the AdamW optimizer. With eight detection heads, the model can identify threat items even in challenging scenarios for both datasets, as presented in Fig. 6a and Fig. 6b.

The improved noising techniques, enhanced loss functions, and refined attention mechanisms in DINO make it more adept at handling these challenging aspects, leading to better performance on the CLCXray dataset. DINO's enhanced positional encoding and optimization strategies contribute to its robustness in diverse and complex environments, typical of security X-ray images. The PIDray dataset might have more regular patterns or larger objects, and it could benefit from the specific design of Deformable DETR. The deformable attention mechanism in Deformable DETR, which focuses on handling varying object scales and aspect ratios, might align better with the characteristics of the PIDray dataset.

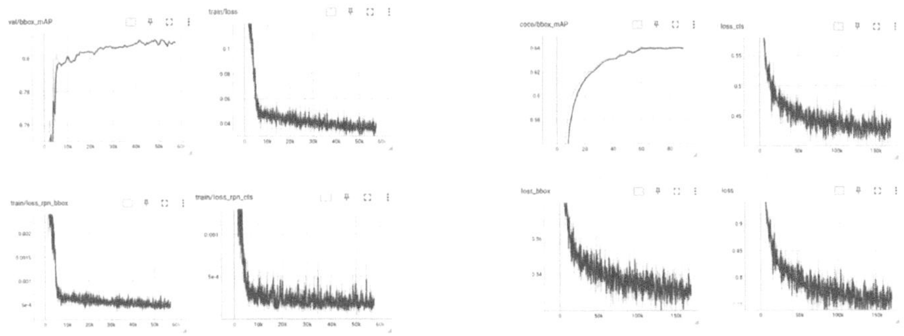

(a) CLCXray dataset (b) PIDray dataset

Fig. 5. Training graphs for top-performing model

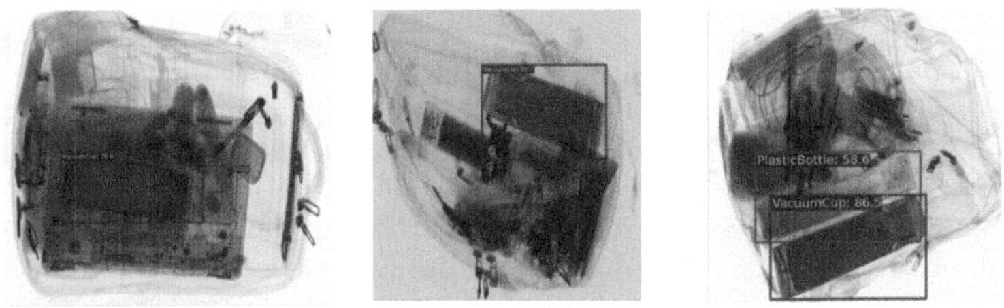

(a) CLCXray dataset

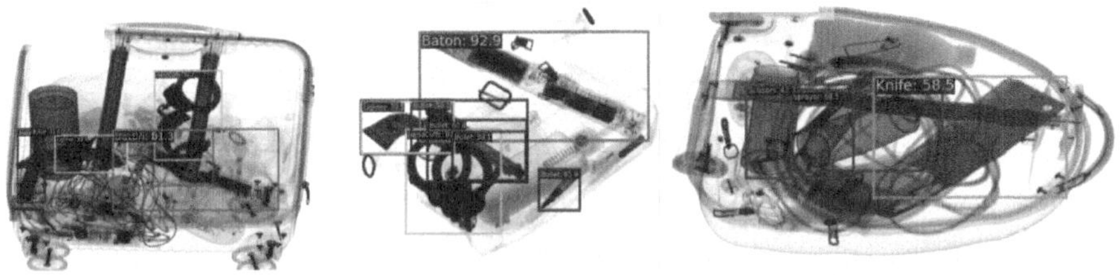

(b) PIDray dataset

Fig. 6. More difficult cases

5 Conclusion

Object detection in X-ray security-checking images involves intelligent analysis to ensure accurate identification of prohibited items. In addition to obtaining the categories of contraband items in X-ray images, the model also needs to locate the prohibited items. In this work, we investigated various CNN and transformer-based deep learning architectures for the purpose of analyzing X-ray images of checked luggage. Deformable DETR performs best with mAP of 0.641 for the PIDray dataset, while the DINO model achieves 0.804 mAP for the CLCxray dataset. As the number of detection heads increases, the model performs better even in difficult situations. Transformer's attention mechanism and ability to handle varying scales and occlusions could be beneficial in enhancing object detection accuracy for X-ray security images. In our future work, we will design customized deep learning architectures with better domain adaptation (difference in data distribution between the source and target domains) to handle multiple variations in threat-based X-ray datasets.

Acknowledgments. The authors thank the Director, CSIR-CEERI, for providing research facilities to carry out the research work. The authors also thank the UGC, New Delhi (F.No. 82-44/2020 (SA-III)), for supporting this research financially.

References

1. Saavedra, D., Banerjee, S., Mery, D.: Detection of threat objects in baggage inspection with x-ray images using deep learning. Neural Comput. Appl. **33**, 7803–7819 (2021)
2. Wu, X., Sahoo, D., Hoi, S.C.H.: Recent advances in deep learning for object detection. Neurocomputing **396**, 39–64 (2020)
3. Akcay, S., Breckon, T.: Towards automatic threat detection: a survey of advances of deep learning within x-ray security imaging. Pattern Recogn. **122**, 108245 (2022)
4. Carion, N., Massa, F., Synnaeve, G., Usunier, N., Kirillov, A., Zagoruyko, S.: End-to-end object detection with transformers. In: European Conference on Computer Vision, pp. 213–229. Springer, Cham (2020)
5. Zhu, X., Su, W., Lu, L., Li, B., Wang, X., Dai, J.: Deformable DeTR: deformable transformers for end-to-end object detection. arXiv preprint arXiv:2010.04159 (2020)
6. Cai, Z., Vasconcelos, N.: Cascade R-CNN: high quality object detection and instance segmentation. IEEE Trans. Pattern Anal. Mach. Intell. **43**(5), 1483–1498 (2019)
7. Ren, S., He, K., Girshick, R., Sun, J.: Faster R-CNN: towards real-time object detection with region proposal networks. Advances in Neural Information Processing Systems, vol. 28 (2015)
8. Redmon, J., Farhadi, A.: YOLOv3: an incremental improvement. arXiv preprint arXiv:1804.02767 (2018)
9. Zhu, C., He, Y., Savvides, M.: Feature selective anchor-free module for single-shot object detection. In: Proceedings of the IEEE/CVF Conference on Computer Vision and Pattern Recognition, pp. 840–849 (2019)
10. Lin, T.-Y., Goyal, P., Girshick, R., He, K., Dollár, P.: Focal loss for dense object detection. In: Proceedings of the IEEE International Conference on Computer Vision, pp. 2980–2988 (2017)
11. Dhiraj, Jain, D.K.: An evaluation of deep learning based object detection strategies for threat object detection in baggage security imagery. Pattern Recogn. Lett. **120**, 112–119 (2019)
12. Kolte, S., Bhowmik, N., Dhiraj: Threat object-based anomaly detection in x-ray images using GAN-based ensembles. Neural Comput. Appl. **35**(31), 23025–23040 (2023)
13. Hassan, T., Akcay, S., Bennamoun, M., Khan, S., Werghi, N.: Tensor pooling-driven instance segmentation framework for baggage threat recognition. Neural Comput. Appl. pp. 1–12 (2022)

14. Hassan, T., Bettayeb, M., Akçay, S., Khan, S., Bennamoun, M., Werghi, N.: Detecting prohibited items in x-ray images: a contour proposal learning approach. In: 2020 IEEE International Conference on Image Processing (ICIP), pp. 2016–2020. IEEE (2020)
15. Wei, Y., Tao, R., Wu, Z., Ma, Y., Zhang, L., Liu, X.: Occluded prohibited items detection: an x-ray security inspection benchmark and de-occlusion attention module. In: Proceedings of the 28th ACM International Conference on Multimedia, pp. 138–146 (2020)
16. Hu, B., Zhang, C., Wang, L., Zhang, Q., Liu, Y.: Multi-label x-ray imagery classification via bottom-up attention and meta fusion. In: Proceedings of the Asian Conference on Computer Vision (2020)
17. Ahmed, A., Velayudhan, D., Hassan, T., Bennamoun, M., Damiani, E., Werghi, N.: Enhancing security in x-ray baggage scans: a contour-driven learning approach for abnormality classification and instance segmentation. Eng. Appl. Artif. Intell. **130**, 107639 (2024)
18. Velayudhan, D., Ahmed, A.H., Hassan, T., Bennamoun, M., Damiani, E., Werghi, N.: Transformers for imbalanced baggage threat recognition. In: 2022 IEEE International Symposium on Robotic and Sensors Environments (ROSE), pp. 1–7. IEEE (2022)
19. Mithal, A., Baser, M., Dhiraj: Automatic threat detection in baggage security imagery using deep learning models. In: 2020 IEEE 15th International Conference on Industrial and Information Systems (ICIIS), pp. 180–185. IEEE (2020)
20. Singh, A., Dhiraj: Improving threat object recognition for x-ray baggage screening using distraction removal techniques. In: 2023 IEEE International Conference on Computer Vision and Machine Intelligence (CVMI), pp. 1–6. IEEE (2023)
21. Tuli, A., Bohra, R., Moghe, T., Chaturvedi, N., Mery, D., Dhiraj: Automatic threat detection in single, stereo (two) and multi view x-ray images. In: 2020 IEEE 17th India Council International Conference (INDICON), pp. 1–7. IEEE (2020)
22. Wang, B., Zhang, L., Wen, L., Liu, X., Wu, Y.: Towards real-world prohibited item detection: a large-scale x-ray benchmark. In: Proceedings of the IEEE/CVF International Conference on Computer Vision, pp. 5412–5421 (2021)
23. Zhao, C., Zhu, L., Dou, S., Deng, W., Wang, L.: Detecting overlapped objects in x-ray security imagery by a label-aware mechanism. IEEE Trans. Inf. Forensics Secur. **17**, 998–1009 (2022)
24. Lyu, C., et al.: An empirical study of designing real-time object detectors. Rtmdet (2022)
25. Chen, Q., Wang, Y., Yang, T., Zhang, X., Cheng, J., Sun, J.: You only look one-level feature. In: IEEE Conference on Computer Vision and Pattern Recognition (2021)
26. Zhu, X., Su, W., Lu, L., Li, B., Wang, X., Dai, J.: Deformable DeTR: deformable transformers for end-to-end object detection. In: International Conference on Learning Representations (2)021
27. Zhang, H., et al.: DeTR with improved denoising anchor boxes for end-to-end object detection. Dino (2022)

Author Index

D. Singh et al. (Eds.): IHCI 2024, LNCS 15557, pp. 457–459, 2025.
https://doi.org/10.1007/978-3-031-88705-5

The manufacturer's authorised representative in the EU is Springer Nature Customer Service Centre GmbH, Europaplatz 3, 69115 Heidelberg, Germany. If you have any concerns regarding our products, please contact ProductSafety@springernature.com

Printed and bound by CPI Group (UK) Ltd, Croydon, CR0 4YY

30/09/2026

02225585-0013